INSTRUCTOR'S GUIDE

FOR

Precalculus: Functions and Graphs
A Graphing Approach

AND

Precalculus With Limits:
A Graphing Approach

Larson/Hostetler/Edwards

Bruce H. Edwards

University of Florida
Gainesville, Florida

Dianna L. Zook

Indiana University
Purdue University at Fort Wayne, Indiana

D. C. Heath and Company

Lexington, Massachusetts   Toronto

Address editorial correspondence to:
D. C. Heath and Company
125 Spring Street
Lexington, MA 02173

Published simultaneously in Canada.

Printed in the United States of America.

International Standard Book Number: 0–669–35207–1

10 9 8 7 6 5 4

PREFACE

This *Instructor's Guide* is a supplement to the textbook *Precalculus: Functions and Graphs A Graphing Approach* by Roland E. Larson, Robert P. Hostetler, and Bruce H. Edwards.

The first part of this guide consists of chapter summaries. Each summary lists the topics covered in that section, along with chapter comments based on our years of teaching experience.

The second portion of the guide consists of the solutions to the even-numbered exercises in the book, as well as solutions to the cumulative tests. Finally, we have included answers to the discussion problems.

This *Instructor's Guide* is the result of the efforts of Dan Bruce, Dorene Phenneger, Jennifer O'Neil, Laurie Sontheimer, Lena Tanner and Michele Bliss. Thanks to Lisa Edwards for her help in preparing this project, and to Edward Schlindwein.

If you have any corrections or suggestions for improving this *Instructor's Guide*, we would appreciate hearing from you.

Bruce H. Edwards
University of Florida
Gainesville, Florida 32611
(be@ math.ufl.edu)

Dianna L. Zook
Indiana University,
Purdue University
at Fort Wayne, Indiana 46805

CONTENTS

Part 1　CHAPTER SUMMARIES

Pedagogical Suggestions

Encourage students to work together, both in and out of class, to learn how to use a graphics calculator or computer graphing utility as well as to communicate about the mathematics. In class, consider having students work some problems in teams or pairs as you circulate around the room (if feasible), providing guidance as needed. Discussion problems can be one source of group activity. Anticipate a higher noise level than usual.

In presenting the material, you may want to include the following approaches to problem solving:
- Solving a problem algebraically and then verifying the work with a graphical approach.
- Solving a problem graphically, then verifying the result algebraically or numerically.
- Include cases during the instruction where a graphical or numerical approach is advantageous and cases where an algebraic approach is more efficient.
- Convey to students that each approach—graphical, numerical, or algebraic—can be equally valid by discussing the advantages and disadvantages of the methods used.
- In class, encourage active participation. Ask open-ended questions and give adequate time for students to respond. Once a student responds, ask another student if they agree with the comment and why. If graphics calculators are used, have your students work through a problem with you.
- Encourage students to try the Discovery features before class and before attempting the exercises. This will provide additional experience investigating the behaviors of functions and give an intuitive basis for learning the concepts.
- If your students will be using computer graphing utilities, they may prefer a program that can easily print results. Review with your class the basics of using the computer and its print capabilities. (BestGrapher is available through D.C. Heath as an option for both site license and individual purchase for either IBM-PC or Macintosh computers.)
- Be aware of the resources available to students. In the text, there is an introduction to calculators and problem solving in the preface and a summary section on graphing utilities in the appendix. In addition, a *Graphing Technology Guide* and a *Study and Solutions Guide* are available as supplements.
- You may want to consider using more than in-class tests for assessment. Take-home tests and graded homework assignments can be useful as they allow for plenty of time for completion.

CHAPTER P
Prerequisites: Review of Basic Algebra

Section Topics

P.1 **The Real Number System**—the real number system; the real number line; ordering the real numbers; the absolute value of a real number; the distance between two real numbers

P.2 **Properties of Real Numbers and the Basic Rules of Algebra**—algebraic expressions; basic rules of algebra; equations; exponents; scientific notation

P.3 **Radicals and Rational Exponents**—radicals and and properties of radicals; simplifying radicals; rationalizing denominators and numerators; rational exponents; radicals and calculators

P.4 **Polynomials and Special Products**—polynomials; operations with polynomials; special products

P.5 **Factoring**—introduction; factoring special polynomial forms; trinomials with binomial factors; factoring by grouping

P.6 **Fractional Expressions**—domain of algebraic expression; simplifying rational expressions; operations with rational expressions; compound fractions

P.7 **The Cartesian Plane**—the Cartesian plane; the distance between two points in the plane; the midpoint formula; the equation of a circle

P.8 **Exploring Data**—line plots; stem and leaf plots; histograms and frequency distributions; line graphs

Chapter Comments

This prerequisites chapter is designed as a review chapter. For well-prepared students, a quick review of each section is all that should be needed. For students not as well-prepared, caution should still be taken not to get bogged down. Expect difficulty with radicals, rational expressions, and fractional expressions. You may want to quiz students on those sections before going on. The special factorization formulas can be reinforced by referencing their use in later chapters.

Section P.7, The Cartesian Plane, provides insight into some of the ways the rectangular coordinate system helps students to visualize relationships between variables x and y, and reviews some geometric concepts.

Since this is a review chapter, move through it as quickly as possible so that you have adequate time to cover new topics. Alternatively, the content could be covered as needed later in the course.

Section P.8 discusses several ways to organize data. These ideas will be especially useful if you choose to cover Sections 2.6, 4.6 and 9.7–9.9 later in the course.

CHAPTER ONE
Functions and Graphs

Section Topics

1.1 **Graphs and Graphing Utilities**—the graph of an equation; using a graphing utility; determining a viewing rectangle; applications

1.2 **Lines in the Plane**—the slope of a line; the point-slope form of the equation of a line; sketching graphs of lines; changing the viewing rectangle; parallel and perpendicular lines

1.3 **Functions**—introduction to functions; function notation; the domain of a function; function keys on a graphing utility; applications

1.4 **Graphs of Functions**—the graph of a function; increasing and decreasing functions; relative minimum and maximum values; step functions; even and odd functions

1.5 **Shifting, Reflecting, and Stretching Graphs**—summary of graphs of common functions; vertical and horizontal shifts; reflections; nonrigid transformations

1.6 **Combinations of Functions**—arithmetic combinations of functions; compositions of functions; applications

1.7 **Inverse Functions**—the inverse of a function; the graph of the inverse of a function; the existence of an inverse function; finding the inverse of a function

Chapter Comments

Chapter 1 deals with the basics of functions and an introduction to graphing functions. This is an excellent place for students to establish good graphing habits with and without a graphing utility. Students should learn the basic skills that are transferable to both graphs of algebraic and transcendental functions. An overhead projector or similar teaching aid is helpful in this chapter.

In Section 1.2 be sure to carefully explain each form of the equation of a line and to identify which form is more efficient for a given style of problem. For instance, the point-slope equation is more efficient in problems in which two points are given. Knowing and recognizing the equations for vertical and horizontal lines sets the stage for finding vertical and horizontal asymptotes in Chapter 3. A good way to remember slopes of perpendicular lines is to verbally note that "the slopes are negative reciprocals of each other." Don't let students get so tangled up with the mechanics of $m_1 = -1/m_2$ that they miss the intuitive relationship of the slopes. Have students explore and investigate this concept with a graphing utility.

In Sections 1.4, 1.5, and 1.6, give special attention to concepts and terms such as function, domain, range, minimum, maximum, increasing or decreasing functions, and inverse functions. Take time to relate the concepts of domain and range to a viewing rectangle. It is useful to discuss an example where the domain is restricted due to the context of an application. These suggestions and concepts, along with the decomposition of composite functions, are all important to students planning to take calculus.

Be ready to answer questions about functions defined by two equations like Example 4 in Section 1.3. They tend initially to be confusing to students.

Sections 1.3, 1.5, and 1.7 contain programs. These can be treated as optional and will not affect the continuity of the course.

C H A P T E R   T W O
Intercepts, Zeros, and Solutions

Section Topics

2.1 **Linear Equations and Modeling**—equations and solutions of equations; linear equations; using mathematical models to solve problems; common formulas

2.2 **Solving Equations Graphically**—intercepts, zeros, and solutions; finding solutions graphically; viewing rectangles, scale, and accuracy; points of intersection of two graphs

2.3 **Complex Numbers**—the imaginary unit i; operations with complex numbers; complex conjugates and division; applications

2.4 **Solving Equations Algebraically**—quadratic equations; polynomial equations of higher degree; equations involving radicals; equations involving fractions or absolute values; applications

2.5 **Solving Inequalities Algebraically and Graphically**—properties of inequalities; solving a linear inequality; inequalities involving absolute values; polynomial inequalities; rational inequalities; applications

2.6 **Exploring Data: Linear Models and Scatter Plots**—scatter plots; fitting a line to data

Chapter Comments

Chapter 2 covers methods for solving equations and inequalities. An understanding and knowledge of how to generate equivalent equations is important (add the same quantity to both sides, etc.). This should then be carried over in solving inequalities.

Students should be urged to organize the solutions of equations in a vertical manner. This also applies to solving inequalities. Discuss the Discovery feature on page 136, which encourages verifying solutions with a graphical approach.

Be sure your students understand the differences between the techniques used to *combine fractional expressions* (Prerequisites Section 6), *solve fractional equations* (Section 2.1), and *solve fractional inequalities* (Section 2.5). By the time you reach Section 2.5 students often have combined the methods into a "home-made" technique that doesn't work. Remind students to distinguish between *simplifying* an expression and *solving* an equation.

In Section 2.2, discuss the close relationship between the concepts of x-intercepts, zeros of functions, and solutions. Demonstrate how to approximate solutions and find x-intercepts by algebraic techniques and graphical approximations. Point out that the accuracy of a graphic approximation will be no more than the distance between tic marks in the viewing rectangle. Note that the text approximates real solutions with an error of at most 0.01, unless stated otherwise.

Completing the square in Section 2.4 is important because it sets the background for many situations later on. These include solving quadratic equations, writing standard equations for conics (Chapter 10), and rewriting algebraic expressions for calculus. If time is a factor, omit the latter use.

In Section 2.4, it is helpful if each student commits the quadratic formula to memory by knowing the verbal statement, "Negative b plus or minus the square root of b squared minus $4ac$, all divided by $2a$." A good motivation or challenge to the better student is to be able to derive this formula. A program is available in a Technology note.

Section 2.3 contains a discussion of the Mandelbrot Set and fractals as a recent application of complex numbers. Graphing calculator programs for some such figures are provided in the Appendix.

C H A P T E R   T H R E E
Polynomial and Rational Functions

Section Topics

3.1   **Quadratic Functions**—the graph of a quadratic function; the standard form of a quadratic function; applications

3.2   **Polynomial Functions of Higher Degree**—graphs of polynomial functions; the leading coefficient test; zeros of polynomial functions; the intermediate value theorem

3.3   **Polynomial Division and Synthetic Division**—long division of polynomials; synthetic division; the remainder and factor theorems; Descartes's Rule of Signs; the rational zero test; bounds for real zeros of polynomial functions; bisection method

3.4   **Complex Zeros and the Fundamental Theorem of Algebra**—the fundamental theorem of algebra; conjugate pairs; factoring a polynomial

3.5   **Rational Functions and Asymptotes**—introduction to rational functions; horizontal and vertical asymptotes; applications

3.6   **Graphs of Rational Functions**—the graph of a rational function; slant asymptotes; application

Chapter Comments

Since polynomial functions are the most often used functions in algebra, students need to become very familiar with their general characteristics. In Section 3.1, you should review completing the square when explaining how to rewrite a quadratic function into the standard form $y = a(x-h)^2 + k$. Pay special attention to problems in which $a \neq 1$ or a is a negative number.

In Section 3.2, review the shapes of the graphs of polynomial functions of degree 0, 1, 2, 3, 4, and 5 and then use these to illustrate the Leading Coefficient Test. This test, combined with information learned from Section 1.4, should add efficiency to recognizing characteristics of a graph of a polynomial.

Section 3.3 contains long division and synthetic division algorithms. Division for the purpose of finding linear factors and zeros of polynomial functions is illustrated in the chapter. Division to rewrite a rational function as a sum of terms is used in Section 3.6 to find slant asymptotes. This skill is also needed frequently in calculus. When division yields a remainder, insist that your students write the remainder term correctly.

Finding the zeros of polynomial functions is a very important concept in algebra. The rational zero test, synthetic division, and the fact that complex zeros occur in conjugate pairs are important aids in this process.

CHAPTER FOUR
Exponential and Logarithmic Functions

Section Topics

4.1 Exponential Functions and Their Graphs—exponential functions; graphs of exponential functions; the natural base e; compound interest; other applications

4.2 Logarithmic Functions and Their Graphs—logarithmic functions; graphs of logarithmic functions; the natural logarithmic function; applications

4.3 Properties of Logarithms—change of base; properties of logarithms; rewriting logarithmic expressions

4.4 Solving Exponential and Logarithmic Equations—introduction; solving exponential equations; solving logarithmic equations; approximating solutions; applications

4.5 Applications of Exponential and Logarithmic Functions—compound interest; growth and decay; logistics growth models; logarithmic models

4.6 Exploring Data: Nonlinear Models—classifying scatter plots; fitting nonlinear models to data; applications

Chapter Comments

Chapter 4 starts the study of transcendental functions—these are often unfamiliar to students. The calculator is used extensively to assist students in their work with these functions.

In Section 4.1, when analyzing the graphs of exponential functions, be sure to draw from previously learned material in Section 1.4 and Section 2 of the Prerequisites chapter. A clear discussion of the irrational number e is valuable. This may be the first time your students have encountered it and they will have many questions. Provide experiences in evaluating using the natural base e and opportunities to numerically and graphically approximate e.

In Section 4.2, capitalize both graphically and algebraically on the fact that the logarithmic function is the inverse of the exponential function. (The logarithmic function is one of the most difficult for students to deal with.) When explaining the logarithmic function, try to resolve the mystery that is associated with the function by continually reminding your students that a *logarithm is an exponent*. Converting back and forth from logarithmic form to exponential form supports this concept. Also, your students should understand the three basic properties of logarithms and how they follow from the definition. To verify a graph of a logarithmic function, have your students set up a table of values and remind them to use the graphing techniques learned in Section 1.4. Be sure your students are very familiar with the graphs of both $f(x) = e^x$ and $f(x) = \ln x$.

The properties of logarithms listed in Section 4.3 are extremely important. A good way to teach them is to explain their similarity to the properties of exponents.

Some common errors to watch for are rewriting $\log x - \log y$ as $\log x / \log y$ instead of $\log(x/y)$, or writing $\log ax^n$ as $n \log ax$. Identify these errors and resolve them before proceeding to Section 4.4.

When solving logarithmic or exponential equations (Section 4.4), encourage your students to first graph the function using a graphing utility, then solve for the unknown algebraically and use their calculators to find the numerical form of the answer. To accomplish this, you can require a graph and an exact algebraic solution, as well as an approximate numerical one.

Point out that equations that involve combinations of different types of functions can be difficult to solve algebraically. In such cases, a graphing utility can be used to obtain an approximate solution. Encourage students to double check the settings on their graphing utility for the selected equation to be sure they are graphing the appropriate function.

C H A P T E R   F I V E
Trigonometric Functions

Section Topics

5.1   **Radian and Degree Measure**—angles; radian measure; degree measure; applications

5.2   **The Trigonometric Functions and the Unit Circle**—the unit circle, the trigonometric functions; domain and period of sine and cosine; evaluating trig functions with a calculator

5.3   **Trigonometric Functions and Right Triangles**—trigonometric functions of an acute angle; trigonometric identities; applications involving right triangles

5.4   **Trigonometric Functions of Any Angle**—trigonometric functions of any angle; reference angles

5.5   **Graphs of Sine and Cosine Functions**—basic sine and cosine curves; key points on basic sine and cosine curves; amplitude and period of sine and cosine curves; translations of sine and cosine curves

5.6   **Other Trigonometric Graphs**—graph of tangent and cotangent functions; graphs of the reciprocal functions; graphs of combinations of trigonometric functions; combinations of algebraic and trigonometric functions; damped trigonometric graphs

5.7   **Inverse Trigonometric Functions**—inverse sine function; other inverse trigonometric functions; compositions of trigonometric and inverse trigonometric functions

5.8   **Applications of Trigonometry**—applications involving right triangles; trigonometry and bearings; harmonic motion

Chapter Comments

Chapter 5 starts the study of trigonometry. To be successful in this chapter, understanding all of the basic definitions is essential. One way for students to do this is to put the identities on a 3 × 5 card. They can then add to this list as new definitions and identities arise. Trigonometric functions are introduced from a unit circle approach. (You may need to review the Pythagorean Theorem; see Prerequisites, Section 7.) Your students need to become as familiar with radian measure of angles as they are with degree measure. Conversions from one form of measure to another can be helpful here. As an aid to learning the right triangle definitions of the trigonometric functions, point out the reciprocal relationships of the functions.

In Section 5.4, the definitions of the trigonometric functions of any angle are given. Be sure your students see the importance of reference angles in evaluating trigonometric functions of angles greater than $90°$. Be ready to explain how to evaluate $\sec 1.2$, or to find the angle that satisfies $\cot \theta = 3.1$, using reciprocal functions. Students often find this to be confusing. Have your students rewrite these expressions in terms of sine, cosine, or tangent, whichever is applicable, before evaluating. Evaluate with a calculator as well, and compare results.

A spool and thread is a good visual aid when discussing trigonometric functions such as wrapping functions of real numbers.

When discussing the graph of the sine and cosine functions in Section 5.5, emphasize how to locate the key points (intercepts, maximum, and minimum). Remind students that the standard viewing rectangle is generally not appropriate for graphing most sine and cosine functions. This section is an opportunity to pull together ideas from several previous sections and to take advantage of a graphing utility. Reinforce the fact that interpretation of a graph is essential. Encourage experimentation with a wide variety of complicated graphs.

Be sure to emphasize the domain/range restrictions used in defining the inverse trigonometric functions. Graphic illustrations and reflections about the line $y = x$ can be useful in this explanation. Be sure your students understand that $y = \arcsin x$ means "y is the angle whose sine is x." This becomes clearer when written in the equivalent form $\sin y = x$.

CHAPTER SIX
Additional Topics in Trigonometry

Section Topics

6.1 Applications of Fundamental Identities—introduction; some uses of the fundamental identities

6.2 Verifying Trigonometric Identities—introduction; verifying trigonometric identities

6.3 Solving Trigonometric Equations—introduction; equations of quadratic type; functions involving multiple angles; using inverse functions and a calculator

6.4 Sum and Difference Formulas—introduction; using sum and difference formulas

6.5 Multiple-Angle and Product-to-Sum Formulas—multiple-angle formulas; power-reducing formulas; half-angle formulas; product-to-sum formulas

6.6 Law of Sines—law of sines; the ambiguous case (SSA); application; the area of an oblique triangle

6.7 Law of Cosines—law of cosines; Heron's formula

6.8 DeMoivre's Theorem and Nth Roots—trigonometric form of a complex number; multiplication and division of complex numbers; powers of complex numbers; roots of complex numbers

Chapter Comments

In Chapter 6, you need to clearly explain to your students that they are now going to study trigonometric functions more from an algebraic than a geometric perspective. However, many problems in the chapter can be solved graphically, analytically, or numerically. Encourage students to verify solutions by solving problems in another way. Students will learn to simplify trigonometric expressions, verify trigonometric identities, and use trigonometric identities to evaluate trigonometric functions.

In Section 6.2, be sure the student understands what it means to *verify an identity* and doesn't try to *solve an identity* as an equation. Here again the student should assemble a list of techniques for rewriting trigonometric expressions, such as those demonstrated in Examples 1 through 7. A graphing utility can be used to confirm trigonometric identities.

To solve the trigonometric equations in Section 6.3, remind the students that the techniques of factoring polynomials and solving quadratic equations are useful here. Much of the success in solving these equations analytically lies in the proper use of algebra. Expect some difficulties with solutions to trigonometric equations involving multiple angles such as $\sin 2\theta = -1/2$. Encourage graphing of the equations as a first step in the solution or to approximate a solution.

Be sure that your students correctly interpret sum and difference formulas as well as the double angle formulas. For instance, a common error is to write $\sin(x+y) = \sin x + \sin y$, or $\sin 2x = 2\sin x$.

The Law of Sines and the Law of Cosines in Sections 6.6 and 6.7 are fairly straightforward. Expect some difficulty with the ambiguous case. Explain in detail how to match up the given parts with the law to be used. Also, for cases in which the Law of Cosines must be used, the largest angle should be solved for first. Then to finish the problem, point out that either the Law of Sines or the Law of Cosines can be used. Insist on a detailed analysis of each problem, including a sketch of the triangle, the method of solution, and the number of solutions.

Give reasons for the trigonometric form of complex numbers early in the discussion; namely to efficiently multiply, divide, raise to powers, and find roots of complex numbers. The latter two are discussed in Section 6.8.

CHAPTER SEVEN
Linear Models and Systems of Equations

Section Topics

7.1 Solving Systems of Equations Algebraically and Graphically—the method of substitution; graphical approach to finding solutions; applications

7.2 Systems of Linear Equations in Two Variables—the method of elimination; graphical interpretation of solutions; applications

7.3 Systems of Linear Equations in More Than Two Variables—row-echelon form and back-substitution; Gaussian elimination; graphical interpretation of a linear system in three variables; nonsquare systems; applications

7.4 Partial Fractions—introduction to partial fractions; partial fraction decomposition

7.5 Systems of Inequalities—the graph of an inequality; systems of inequalities; applications

7.6 Linear Programming—the graphical approach to linear programming; applications

Chapter Comments

When solving systems of equations algebraically, be sure to reinforce the solution graphically (if applicable). To be efficient, students need to know and understand both the substitution and the elimination methods. Point out the difference between the two methods, and the advantage of one over the other. Using a graphical approach first is sometimes useful to identify the number of solutions.

As demonstrated in Section 7.3, encourage students to use back-substitution for efficiency in solving systems of equations with more than two variables. Emphasize the usefulness of row-echelon form in determining the number (one, many, or none) of solutions.

It is important when teaching linear programming that students immediately become familiar with the terms. Emphasize the fact that this technique is powerful in that it quickly identifies those few points, out of many, that can be tested to find the point that gives the maximum (or minimum) value.

CHAPTER EIGHT
Matrices and Determinants

Section Topics

8.1   **Matrices and Systems of Equations**—matrices, elementary row operations; gaussian elimination with back-substitution; Gauss–Jordan elimination

8.2   **Operations with Matrices**—equality of matrices; matrix addition and scalar multiplication; matrix multiplication; applications

8.3   **Inverse Matrices and Systems of Linear Equations**—the inverse of a matrix; the inverse of a 2×2 matrix (Quick Method); systems of linear equations

8.4   **The Determinant of a Square Matrix**—the determinant of a 2×2 matrix; minors and cofactors of a square matrix; the determinant of a square matrix; triangular matrices

8.5   **Applications of Matrices and Determinants**—Cramer's rule; area of a triangle; lines in the plane, cryptography

Chapter Comments

In Section 8.1 matrices are introduced as an aid for solving systems of linear equations. Make clear to your students the match up of a matrix with its system of linear equations. Also, point out that the elementary row operations on matrices in Section 8.1 and Gaussian elimination with back-substitution in Section 7.3 are essentially the same. Encourage your students to write down the row operation performed in each step. This is an effective way to check their work.

Students may have difficulty with matrix notation. The application in Example 7 of Section 8.2 can be used as a good motivation for matrix multiplication. Emphasize the scheme (following Example 5) that identifies the condition for which multiplication of two matrices is defined. Review with students how to enter a matrix in their calculator or computer utility and encourage them to use technology to find inverses of matrices 3×3 or larger, and to check matrix operations.

CHAPTER NINE
Sequences, Probability, and Statistics

Section Topics

9.1   **Sequences and Summation Notation**—sequences; factorial notation; summation notation; the sum of an infinite sequence; applications

9.2   Arithmetic Sequences—arithmetic sequences; the sum of an arithmetic sequence; arithmetic mean; applications

9.3   Geometric Sequences—geometric sequences; the sum of a finite geometric sequence; application

9.4   Mathematical Induction—mathematical induction; sums of powers of integers; pattern recognition; finite differences

9.5   The Binomial Theorem—binomial coefficients; Pascal's triangle; binomial expansions

9.6   Counting Principles, Permutations, Combinations—simple counting problems; counting principles; permutations; combinations

9.7   Probability—sample spaces; the probability of an event; mutually exclusive events; independent events; the complement of an event

9.8   Exploring Data: Measures of Central Tendency—mean, median, and mode; working with organized data; choosing a measure of central tendency

9.9   Exploring Data: Measures of Dispersion—variance and standard deviation; computing standard deviation; box-and-whisker plots

Chapter Comments

When studying sequences, series, and probability, be sure your students understand the notation and know the terminology well. Carefully go over sequences in which $d < 0$ or $r < 0$. As an aid to understanding the formula for finding the nth term of an arithmetic sequence, have your students intuitively derive the formula. Application programs will help the students see the need for studying sequences.

Students need many demonstrations of proof by mathematical induction. You may want to add some proofs of your own.

Be sure your students see the importance of the Fundamental Counting Principle. It is the one from which the other counting procedures arise.

Careful detail and numerous examples are needed for students to understand when to count different orders and when not to. This is a necessary distinction to make between permutations and combinations.

CHAPTER TEN
Conics, Parametric Equations, and Polar Coordinates

Section Topics

10.1   Introduction to Conics: Parabolas—conics; parabolas; applications

10.2   Ellipses—ellipses; application; eccentricity

10.3   Hyperbolas—hyperbolas; asymptotes of a hyperbola; applications

10.4   Rotation and Systems of Quadratic Equations—rotation; invariants under rotation; systems of quadratic equations

10.5   Plane Curves and Parametric Equations—plane curves; sketching a plane curve; eliminating the parameter; finding parametric equations for a graph

10.6   Polar Coordinates—introduction; coordinate conversion; equation conversion

10.7   **Graphs of Polar Equations**—introduction; using a graphing utility; symmetry; maximum *r*-values; special polar graphs

10.8   **Polar Equations of Conics**—alternative definition of conics; polar equations of conics; applications

Chapter Comments

Conics are a very important topic for calculus. Students should use their calculators to sketch the graphs of parabolas, ellipses, and hyperbolas.

Students should notice how a curve is traced out, clockwise or counterclockwise, and learn that if the parameter represents time, then two different sets of parametric equations may have the same graph. This is readily demonstrated on a graphing utility. Encourage students to solve problems analytically and graphically.

Students should learn to analyze graphs of polar equations with regard to symmetry, tangents at the pole, and maximum *r*-values. Have them investigate and compare the graphs of $r = 3\cos 2\theta$, $r = 3\cos 2.1\theta$, $r = 3\cos 2.2\theta$, or $r = 3\cos 2\theta$, $r = 3.1\cos 2\theta$, and $r = 3.2\cos 2\theta$.

Graphing polar equations by hand is especially difficult. This is one of the advantages of a graphing utility. Polar equations of conics are useful in calculus when driving Kepler's Laws of planetary motion.

CHAPTER ELEVEN
Vectors in the Plane

Section Topics

11.1   **Vectors in the Plane**—vectors in the plane; component form of a vector; vector operations; unit vectors; direction angles; applications of vectors

11.2   **The Dot Product of Two Vectors**—the dot product of two vectors; angle between two vectors; finding vector components; work

11.3   **The Three-Dimensional Coordinate System**—the three-dimensional coordinate system; the distance and midpoint formulas; the equation of a sphere

11.4   **Vectors in Space**—vectors in space; parallel vectors; application

11.5   **The Cross Product of Two Vectors**—the cross product; geometric properties of the cross product; the triple scalar product

11.6   **Equations of Lines and Planes in Space**—lines in space; planes in space; sketching planes in space; distance between a point and a plane

Chapter Comments

Students often have trouble distinguishing between a point in the plane and a vector. Emphasize that two points P and Q determine a vector from P to Q. Vectors have both magnitude (length) and direction. Show the geometric interpretations of addition and scalar multiplication.

The dot product leads to the measurement of angle between two vectors. The dot product is used frequently in calculus and engineering applications, such as work.

Take time to practice graphing points in the three-dimensional coordinate system. Most operations in $\mathbf{R}^3$ are direct generalizations of operations in $\mathbf{R}^2$.

The cross product will remind students how to calculate determinants. The properties of the cross product are tedious to prove, but it is important to illustrate them with examples. For instance, verify with an example that the cross product of two vectors is orthogonal to each vector. The cross product will be used to derive the equation of a plane in the next section.

Students often have difficulty with lines in space. Give a specific example and have the students plot various points on the line for different t values. Example 2 could be used in this way; for instance, the students could find the t values that correspond to the original two points that determine the line.

CHAPTER TWELVE
Limits and an Introduction to Calculus

Section Topics

12.1 **Introduction to Limits**—the limit concept; definition of limit; limits that fail to exist; properties of limits

12.2 **Techniques for Evaluating Limits**—limits of polynomial and rational functions; cancellation technique/rationalizing technique; using technology; one-sided limits; a limit from calculus

12.3 **The Tangent Line Problem**—tangent line to a graph; slope of a graph; slope and the limit process; the derivative of a function

12.4 **Limits at Infinity and Limits of Sequences**—limits at infinity and horizontal asymptotes; limits of sequences

12.5 **The Area Problem**—limits of summations; the area problem

Chapter Comments

This chapter presents an elementary introduction to the fundamental ideas of calculus. The emphasis is on understanding the underlying concepts of limit, tangent line, and area, rather than the development of computational formulas.

The definition of limit is informal. Emphasize how limits can fail to exist. For example, a graphing utility can be quite impressive showing how the function $y = \sin(1/x)$ behaves near $x = 0$.

The chapter emphasizes that you can estimate a limit three ways: numerically, graphically and algebraically. You might have students work in teams on a particular example, showing how to calculate a limit using the three methods. The example $\lim\limits_{x \to 0} \dfrac{\sin x}{x}$ is particularly important in calculus.

The tangent line problem motivates the definition of derivative. Again, the emphasis is on understanding, not developing formulas. Have the students use the limit definition to calculate slopes, and graph the function together with its tangent line.

The limit of a sequence leads directly to the area problem, the central theme of integral calculus. Again, there is little emphasis in this section on formulas for integrals, but rather on understanding how the area problem can be handled using limits. The summation formula proofs provide a nice exercise in mathematical induction.

Part 2   SOLUTIONS TO EVEN-NUMBERED EXERCISES

CHAPTER P
Prerequisites: Review of Basic Algebra

P.1 | The Real Number System

2. (a) Natural numbers: None

(b) Integers: $\{-7, 0\}$

(c) Rational numbers: $\left\{-7, -\frac{7}{3}, 0, 3.12, \frac{5}{4}\right\}$

(d) Irrational numbers: $\{\sqrt{5}\}$

4. (a) Natural numbers: $\left\{\frac{8}{2}, 9\right\}$

(b) Integers: $\left\{\frac{8}{2}, -4, 9\right\}$

(c) Rational numbers: $\left\{\frac{8}{2}, -\frac{8}{3}, -4, 9, 14.2\right\}$

(d) Irrational Numbers: $\{\sqrt{10}\}$

6. (a) Natural numbers: $\{25, \sqrt{9}\}$

(b) Integers: $\{25, -17, \sqrt{9}\}$

(c) Rational numbers: $\left\{25, -17, \sqrt{9}, -\frac{12}{5}, 3.12\right\}$

(d) Irrational numbers: $\left\{\frac{1}{2}\pi\right\}$

8. $-3.5 < 1$

10. $1 < \frac{16}{3}$

12. $-\frac{8}{7} < -\frac{3}{7}$

14. The interval $(4, 10]$ denotes all real numbers greater than 4 and less than or equal to 10. That is, $4 < x \le 10$, which is bounded.

16. The interval $(-6, \infty)$ denotes all real numbers greater than -6. That is, $x > -6$, which is unbounded.

18. The inequality $x \ge -2$ denotes all real numbers greater than or equal to -2. That is, $[-2, \infty)$.

20. The inequality $x > 3$ denotes all real numbers greater than 3. That is, $(3, \infty)$.

22. The inequality $x < 2$ denotes all real numbers less than 2. That is, $(-\infty, 2)$.

24. The inequality $0 \le x \le 5$ denotes all real numbers between 0 and 5, including 0 and 5. That is, $[0, 5]$.

26. The inequality $0 < x \le 6$ denotes all real numbers between 0 and 6, not including 0 but including 6. That is, $(0, 6]$.

28. $z \ge 10$, $[10, \infty)$

30. $5 < y \leq 12$, $(5, 12]$ **32.** $Y \leq 45$, $(-\infty, 45]$ **34.** $0 < p \leq 1.35$, $(0, 1.35]$

36. $|4 - \pi| = 4 - \pi \approx 0.8584$

38. $|-1| - |-2| = -(-1) + (-2) = 1 - 2 = -1$

40. $2|33| = 66$

42. Since $|-4| = 4$, we have $|-4|$ $\boxed{=}$ $|4|$.

44. Since $-|-6| = -6$ and $|-6| = 6$, we have $-|-6|$ $\boxed{<}$ $|-6|$.

46. Since $-(-2) = 2$, we have $-(-2)$ $\boxed{>}$ -2.

48. $d\left(\frac{1}{4}, \frac{11}{4}\right) = \left|\frac{11}{4} - \frac{1}{4}\right| = \frac{10}{4} = \frac{5}{2}$

50. $d(-126, -75) = |-75 - (-126)| = 51$

52. $d\left(\frac{16}{5}, \frac{112}{75}\right) = \left|\frac{112}{75} - \frac{16}{5}\right| = \frac{128}{75}$

54. Since $d(x, -10) = |x - (-10)|$ and $d(x, -10) \geq 6$, we have $|x + 10| \geq 6$.

56. Since $d(z, 0) = |z - 0| = |z|$ and $d(z, 0) < 8$, we have $|z| < 8$.

58. Since $d(y, a) = |y - a|$ and $d(y, a) \leq 2$, we have $|y - a| \leq 2$.

60. $\frac{26}{15} \approx 1.733333333$

$\sqrt{3} \approx 1.732050808$

1.7320

$\frac{381}{220} \approx 1.731818182$

$\sqrt{10} - \sqrt{2} \approx 1.748064098$

$\frac{381}{220} < 1.7320 < \sqrt{3} < \frac{26}{15} < \sqrt{10} - \sqrt{2}$

62. $\frac{1}{3} = 0.333333\ldots$

64. $\frac{6}{11} = 0.545454\ldots$

66. True. Since each integer p can be written as p/q, where $q = 1$, then p is also rational.

68. False. $|0| = 0$.

P.2   Properties of Real Numbers and the Basic Rules of Algebra

2. Terms: -5, $3x$

4. $3x^2 - 8x - 11 = 3x^2 + (-8x) + (-11)$
Terms: $3x^2$, $-8x$, -11

6. Terms: $3x^4$, $3x^3$

8. (a) $9 - 7(-3) = 9 + 21 = 30$
(b) $9 - 7(3) = 9 - 21 = -12$

10. (a) $-(-1)^2 + 5(-1) - 4 = -1 - 5 - 4 = -10$
(b) $-(1)^2 + 5(1) - 4 = -1 + 5 - 4 = 0$

12. (a) $\dfrac{2}{2+2} = \dfrac{2}{4} = \dfrac{1}{2}$
(b) $\dfrac{-2}{-2+2} = \dfrac{-2}{0}$ which is undefined.
You cannot divide by zero.

14. $(x+3) - (x+3) = 0$,
Inverse Property of Addition

16. $2\left(\frac{1}{2}\right) = 1$,
Inverse Property of Multiplication

18. $(z-2) + 0 = z - 2$,
Identity Property of Addition

20. $x + (y + 10) = (x + y) + 10$,
Associative Property of Addition

22. $\frac{1}{7}(7 \cdot 12) = \left(\frac{1}{7} \cdot 7\right) 12$ by the Associative Property of Multiplication
$\qquad = 1 \cdot 12 \qquad$ Inverse Property of Multiplication
$\qquad = 12 \qquad$ Identity Property of Multiplication

24. $10(23 - 30 + 7) = 10 \cdot 0 = 0$

26. $15 - \dfrac{3-3}{5} = 15 - \dfrac{0}{5} = 15 - 0 = 15$

28. $-3(5-2) = -3(3) = -9$

30. $\dfrac{27 - 35}{4} = \dfrac{-8}{4} = -2$

32. $\dfrac{6}{7} - \dfrac{4}{7} = \dfrac{6-4}{7} = \dfrac{2}{7}$

34. $\dfrac{10}{11} + \dfrac{6}{33} - \dfrac{13}{66} = \dfrac{10 \cdot 6}{11 \cdot 6} + \dfrac{6 \cdot 2}{33 \cdot 2} - \dfrac{13}{66}$
$\qquad = \dfrac{60 + 12 - 13}{66} = \dfrac{59}{66}$

36. $\dfrac{2}{3} \cdot \dfrac{5}{8} \cdot \dfrac{5}{16} \cdot \dfrac{3}{4} = \dfrac{5 \cdot 5}{4 \cdot 16 \cdot 4} = \dfrac{25}{256}$

38. $\dfrac{11}{16} \div \dfrac{3}{4} = \dfrac{11}{16} \times \dfrac{4}{3} = \dfrac{11 \cdot \cancel{4}}{4 \cdot \cancel{4} \cdot 3} = \dfrac{11}{12}$

40. $\left(\dfrac{3}{5} \div 3\right) - \left(6 \times \dfrac{4}{8}\right) = \left(\dfrac{3}{5} \times \dfrac{1}{3}\right) - \left(\dfrac{24}{8}\right)$
$\qquad = \dfrac{1}{5} - 3 = \dfrac{1 - 15}{5} = -\dfrac{14}{5}$

42. $3\left(\dfrac{-5}{12} + \dfrac{3}{8}\right) = 3\left(\dfrac{-1}{24}\right) = \dfrac{-3}{24} \approx -0.13$

44. $\dfrac{\left(\frac{1}{5}\right)(-8-9)}{-\frac{1}{3}} = \dfrac{\frac{1}{5}(-17)}{-\frac{1}{3}} = \dfrac{\frac{17}{5}}{\frac{1}{3}} = \dfrac{51}{5} = 10.20$

46. $\left(\dfrac{3}{2}\right)^{-2} - \left(\dfrac{3}{4}\right)^3 = \left(\dfrac{2}{3}\right)^2 - \left(\dfrac{3}{4}\right)^3$
$\qquad = \dfrac{4}{9} - \dfrac{27}{64}$
$\qquad = \dfrac{4 \cdot 64 - 27 \cdot 9}{9 \cdot 64}$
$\qquad = \dfrac{13}{576} \approx 0.02$

48. $\dfrac{1-3^{-5}}{0.5+(0.3)^4} = \dfrac{1-(1/3)^5}{0.5+(0.3)^4} \approx \dfrac{0.9959}{0.5081} \approx 1.96$ **50.** $3 \cdot 3^3 = 3^{1+3} = 3^4 = 81$

52. $-3^2 = -9$

54. $\left(-\dfrac{3}{5}\right)^3 \left(\dfrac{5}{3}\right)^2 = (-1)^3 \left(\dfrac{3^3}{5^3}\right) \left(\dfrac{5^2}{3^2}\right)$

$$= -1\left(\dfrac{3}{5}\right) = -\dfrac{3}{5}$$

56. $(-2)^0 = 1$

58. $7(4)^{-2} = \dfrac{7}{4^2} = \dfrac{7}{16}$

60. $5(-3)^3 = 5(-1)^3(3)^3$

$$= 5(-1)(27)$$

$$= -135$$

62. $(3x)^2 = (3)^2 x^2 = 9x^2$

64. $(-z)^3(3z^4) = (-1)^3 z^3 (3z^4)$

$$= -3z^7$$

66. $\dfrac{25y^8}{10y^4} = \dfrac{5}{2}y^4$

68. $\dfrac{r^4}{r^6} = \dfrac{1}{r^2}$

70. $\left(\dfrac{4}{y}\right)^3 \left(\dfrac{3}{y}\right)^4 = \left(\dfrac{4^3}{y^3}\right) \left(\dfrac{3^4}{y^4}\right)$

$$= \dfrac{4^3 \cdot 3^4}{y^{3+4}}$$

$$= \dfrac{64 \cdot 81}{y^7} = \dfrac{5184}{y^7}$$

72. $(2x^5)^0 = 1, \; x \neq 0$

74. $(4y^{-2})(8y^4) = 32y^{-2+4} = 32y^2$

76. $\left(\dfrac{x^{-3}y^4}{5}\right)^{-3} = \left(\dfrac{5}{x^{-3}y^4}\right)^3 = \left(\dfrac{5x^3}{y^4}\right)^3 = \dfrac{125x^9}{y^{12}}$ **78.** $\dfrac{x^2 \cdot x^n}{x^3 \cdot x^n} = \dfrac{1}{x}$

80. $9{,}461{,}000{,}000{,}000{,}000 = 9.461 \times 10^{15}$

82. $0.00003937 = 3.937 \times 10^{-5}$

84. $1.3 \times 10^7 = 13{,}000{,}000$

86. $9.0 \times 10^{-4} = 0.0009$

88. (a) $(9.3 \times 10^6)^3(6.1 \times 10^{-4}) \approx 4.907 \times 10^{17}$

    $\boxed{(}$ 9.3 $\boxed{\text{EXP}}$ 6 $\boxed{)}$ $\boxed{y^x}$ 3 $\boxed{\times}$ $\boxed{(}$ 6.1 $\boxed{\text{EXP}}$ 4 $\boxed{+/-}$ $\boxed{)}$ $\boxed{=}$

   (b) $\dfrac{(2.414 \times 10^4)^6}{(1.68 \times 10^5)^5} \approx 1.479$

    $\boxed{(}$ 2.414 $\boxed{\text{EXP}}$ 4 $\boxed{)}$ $\boxed{y^x}$ 6 $\boxed{\div}$ $\boxed{(}$ 1.68 $\boxed{\text{EXP}}$5 $\boxed{)}$ $\boxed{y^x}$ 5 $\boxed{=}$

P.3 Radicals and Rational Exponents

2. Radical form: $\sqrt[3]{64} = 4$
Rational exponent form: $64^{1/3} = 4$

4. Radical form: $-\sqrt{144} = -12$
Rational exponent form: $-(144^{1/2}) = -12$

6. Radical form: $\sqrt[3]{614.125} = 8.5$
Rational exponent form: $(614.125)^{1/3} = 8.5$

8. Radical form: $\sqrt[5]{-243} = -3$
Rational exponent form: $(-243)^{1/5} = -3$

10. Radical form: $(\sqrt[4]{81})^3 = 27$
Rational exponent form: $(81^{1/4})^3 = 81^{3/4} = 27$

12. $\sqrt{49} = 7$

14. $\sqrt[3]{64} = \sqrt[3]{4^3} = 4$

16. $\sqrt[3]{\dfrac{27}{8}} = \dfrac{\sqrt[3]{27}}{\sqrt[3]{8}} = \dfrac{3}{2}$

18. $\sqrt[3]{0} = 0$

20. $\dfrac{\sqrt[4]{81}}{3} = \dfrac{\sqrt[4]{3^4}}{3} = \dfrac{3}{3} = 1$

22. $\sqrt[4]{562^4} = 562$

24. $(27)^{1/3} = \sqrt[3]{27} = \sqrt[3]{3^3} = 3$

26. $100^{-3/2} = (\sqrt{100})^{-3}$
$= 10^{-3} = \dfrac{1}{10^3} = \dfrac{1}{1000}$

28. $\left(\dfrac{9}{4}\right)^{-1/2} = \left(\dfrac{4}{9}\right)^{1/2}$
$= \dfrac{\sqrt{4}}{\sqrt{9}} = \dfrac{2}{3}$

30. $\left(-\dfrac{125}{27}\right)^{-1/3} = \left(-\dfrac{27}{125}\right)^{1/3}$
$= \dfrac{\sqrt[3]{-1}\,\sqrt[3]{27}}{\sqrt[3]{125}}$
$= \dfrac{-\sqrt[3]{3^3}}{\sqrt[3]{5^3}} = -\dfrac{3}{5}$

32. $\dfrac{\sqrt[3]{16}}{\sqrt[3]{27}} = \dfrac{\sqrt[3]{8}\,\sqrt[3]{2}}{3}$
$= \dfrac{2\sqrt[3]{2}}{3}$

34. $\sqrt{4.5 \times 10^9} = \sqrt{4.5}\sqrt{(10^4)^2}\sqrt{10}$
$= \sqrt{45} \cdot 10^4$
$= \sqrt{9}\sqrt{5} \times 10^4$
$= 3\sqrt{5} \times 10^4$

36. $\sqrt{54xy^4} = \sqrt{9 \cdot 6 \cdot x \cdot y^4}$
$= 3y^2\sqrt{6x}$

38. $\sqrt[4]{(3x^2)^4} = 3x^2$

40. $\sqrt[5]{96x^5} = \sqrt[5]{32 \cdot 3 \cdot x^5}$

$\qquad = 2x\sqrt[5]{3}$

42. $\dfrac{5}{\sqrt{10}} = \dfrac{5}{\sqrt{10}} \cdot \dfrac{\sqrt{10}}{\sqrt{10}}$

$\qquad = \dfrac{5\sqrt{10}}{10}$

$\qquad = \dfrac{\sqrt{10}}{2}$

44. $\dfrac{5}{\sqrt[3]{(5x)^2}} = \dfrac{5}{\sqrt[3]{(5x)^2}} \cdot \dfrac{\sqrt[3]{(5x)}}{\sqrt[3]{(5x)}}$

$\qquad = \dfrac{5\sqrt[3]{(5x)}}{5x}$

$\qquad = \dfrac{\sqrt[3]{5x}}{x}$

46. $\dfrac{5}{\sqrt{14} - 2} = \dfrac{5}{\sqrt{14} - 2} \cdot \dfrac{\sqrt{14} + 2}{\sqrt{14} + 2}$

$\qquad = \dfrac{5(\sqrt{14} + 2)}{14 - 4}$

$\qquad = \dfrac{5(\sqrt{14} + 2)}{10}$

$\qquad = \dfrac{\sqrt{14} + 2}{2}$

48. $\dfrac{5}{2\sqrt{10} - 5} = \dfrac{5}{2\sqrt{10} - 5} \cdot \dfrac{2\sqrt{10} + 5}{2\sqrt{10} + 5}$

$\qquad = \dfrac{5(2\sqrt{10} + 5)}{4 \cdot 10 - 25}$

$\qquad = \dfrac{5(2\sqrt{10} + 5)}{15}$

$\qquad = \dfrac{2\sqrt{10} + 5}{3}$

50. $\dfrac{\sqrt{2}}{3} = \dfrac{\sqrt{2}}{3} \cdot \dfrac{\sqrt{2}}{\sqrt{2}}$

$\qquad = \dfrac{2}{3\sqrt{2}}$

52. $\dfrac{\sqrt{3} - \sqrt{2}}{2} = \dfrac{\sqrt{3} - \sqrt{2}}{2} \cdot \dfrac{\sqrt{3} + \sqrt{2}}{\sqrt{3} + \sqrt{2}}$

$\qquad = \dfrac{3 - 2}{2(\sqrt{3} + \sqrt{2})}$

$\qquad = \dfrac{1}{2(\sqrt{3} + \sqrt{2})}$

$\qquad = \dfrac{-1}{2(\sqrt{7} + 3)}$

54. $\dfrac{2\sqrt{3} + \sqrt{3}}{3} = \dfrac{2\sqrt{3} + \sqrt{3}}{3} \cdot \dfrac{\sqrt{3}}{\sqrt{3}}$

$\qquad = \dfrac{2 \cdot 3 + 3}{3\sqrt{3}}$

$\qquad = \dfrac{9}{3\sqrt{3}}$

$\qquad = \dfrac{3}{\sqrt{3}}$

56. $\sqrt[6]{x^3} = x^{3/6}$

$\qquad = x^{1/2}$

$\qquad = \sqrt{x}$

58. $\sqrt[4]{(3x^2)^4} = (3x^2)^{4/4}$

$\qquad = (3x^2)^1$

$\qquad = 3x^2$

60. $\sqrt{\sqrt{243(x+1)}} = \sqrt{\sqrt{81 \cdot 3(x+1)}}$

$$= \sqrt{9\sqrt{3(x+1)}} = \sqrt{9 \cdot 3^{1/2}(x+1)^{1/2}}$$

$$= 3 \cdot (3^{1/2})^{1/2}[(x+1)^{1/2}]^{1/2} = 3 \cdot 3^{1/4}(x+1)^{1/4} = 3\sqrt[4]{3(x+1)}$$

62. $\sqrt{\sqrt[3]{10a^7b}} = [(10a^7b)^{1/3}]^{1/2}$

$$= (10a^7b)^{1/6} = \sqrt[6]{10a^7b} = a\sqrt[6]{10ab}$$

64. $3\sqrt{x+1} + 10\sqrt{x+1} = (3+10)\sqrt{x+1}$

$$= 13\sqrt{x+1}$$

66. $4\sqrt{27} - \sqrt{75} = 4 \cdot 3\sqrt{3} - 5\sqrt{3}$

$$= (12-5)\sqrt{3}$$

$$= 7\sqrt{3}$$

68. $7\sqrt{80x} - 2\sqrt{125x} = 7\sqrt{16 \cdot 5x} - 2\sqrt{25 \cdot 5x}$

$$= 28\sqrt{5x} - 10\sqrt{5x}$$

$$= 18\sqrt{5x}$$

70. $\dfrac{5^{-1/2} \cdot 5x^{5/2}}{(5x)^{3/2}} = \dfrac{5x^{5/2}}{5^{1/2}5^{3/2}x^{3/2}}$

$$= 5^{1-1/2-3/2}x^{5/2-3/2}$$

$$= 5^{-1}x = \frac{x}{5}$$

72. $(-2u^{3/5}v^{-1/5})^3 = \left(\dfrac{-2u^{3/5}}{v^{1/5}}\right)^3 = \dfrac{-8u^{9/5}}{v^{3/5}}$

74. $\dfrac{a^{3/4} \cdot a^{1/2}}{a^{5/2}} = \dfrac{a^{5/4}}{a^{10/4}} = \dfrac{1}{a^{5/4}}$

76. $\left(\dfrac{3m^{1/6}n^{1/3}}{4n^{-2/3}}\right)^2 = \left(\dfrac{3m^{1/6}n}{4}\right)^2 = \dfrac{9m^{1/3}n^2}{16}$

78. $(k^{-1/3})^{3/2} = \left(\dfrac{1}{k^{1/3}}\right)^{3/2} = \dfrac{1}{k^{1/2}} = \dfrac{1}{\sqrt{k}}$

80. $\sqrt[3]{45^2} \approx 12.651$

45 $\boxed{x^2}$ $\boxed{y^x}$ $\boxed{(}$ 1 $\boxed{\div}$ 3 $\boxed{)}$ $\boxed{=}$

82. $(15.25)^{-1.4} \approx 0.022$

15.25 $\boxed{y^x}$ 1.4 $\boxed{+/-}$ $\boxed{=}$

84. $(2.65 \times 10^{-4})^{1/3} \approx 0.064$

$\boxed{(}$ 2.65 $\boxed{\text{EXP}}$ 4 $\boxed{+/-}$ $\boxed{)}$ $\boxed{y^x}$ $\boxed{(}$ 1 $\boxed{\div}$ 3 $\boxed{)}$ $\boxed{=}$

86. $\sqrt{3} - \sqrt{2} \approx 0.318$

$\sqrt{3-2} = 1$

Therefore, $\sqrt{3} - \sqrt{2} < \sqrt{3-2}$.

88. $\sqrt{3^2 + 4^2} = \sqrt{9+16} = \sqrt{25} = 5$

Therefore, $5 = \sqrt{3^2 + 4^2}$.

90. $\sqrt{\dfrac{3}{11}} = \dfrac{\sqrt{3}}{\sqrt{11}}$

92. $\left(\dfrac{2}{\sqrt{5}}\right)^2 = \dfrac{2^2}{(\sqrt{5})^2} = \dfrac{4}{5}$

No, rationalizing the denominator produces a number that is equal to the original fraction; squaring does not.

P.4 Polynomials and Special Products

2. Standard form: $-3x^4 + 2x^2 - 5$
Degree: 4
Leading coefficient: -3

4. Standard form: 3
Degree: 0
Leading coefficient: 3

6. Standard form: $2x$
Degree: 1
Leading coefficient: 2

8. $2x^3 + x - 3x^{-1}$
This is not a polynomial because of the **negative exponent.**

10. $\dfrac{x^2 + 2x - 3}{2} = \dfrac{x^2}{2} + \dfrac{2x}{2} - \dfrac{3}{2}$

$\qquad = \dfrac{1}{2}x^2 + x - \dfrac{3}{2}$

This is a polynomial.
Standard form: $\frac{1}{2}x^2 + x - \frac{3}{2}$

12. $\sqrt{y^2 - y^4}$
This is not a polynomial because of the square root.

14. $(2x^2 + 1) - (x^2 - 2x + 1) = 2x^2 + 1 - x^2 + 2x - 1$

$\qquad\qquad\qquad\qquad = (2x^2 - x^2) + 2x + (1 - 1)$

$\qquad\qquad\qquad\qquad = x^2 + 2x$

16. $-(5x^2 - 1) - (-3x^2 + 5) = -5x^2 + 1 + 3x^2 - 5$

$\qquad\qquad\qquad\qquad = (-5x^2 + 3x^2) + (1 - 5)$

$\qquad\qquad\qquad\qquad = -2x^2 - 4$

18. $(15x^4 - 18x - 19) - (13x^4 - 5x + 15) = 15x^4 - 18x - 19 - 13x^4 + 5x - 15$

$\qquad\qquad\qquad\qquad\qquad\qquad = (15x^4 - 13x^4) + (-18x + 5x) + (-19 - 15)$

$\qquad\qquad\qquad\qquad\qquad\qquad = 2x^4 - 13x - 34$

20. $(y^3 + 1) - [(y^2 + 1) + (3y - 7)] = y^3 + 1 - (y^2 + 1 + 3y - 7) = y^3 + 1 - y^2 - 1 - 3y + 7 = y^3 - y^2 - 3y + 7$

22. $y^2(4y^2 + 2y - 3) = y^2(4y^2) + y^2(2y) + y^2(-3)$

$\qquad\qquad\qquad = 4y^4 + 2y^3 - 3y^2$

24. $-4x(3 - x^3) = -4x(3) + (-4x)(-x^3)$

$\qquad\qquad\qquad = -12x + 4x^4$

26. $(1 - x^3)(4x) = 1(4x) - x^3(4x)$

$\qquad = 4x - 4x^4$

28. $(x - 5)(x + 10) = x^2 + 10x - 5x - 50$

$\qquad = x^2 + 5x - 50$

30. $(7x - 2)(4x - 3) = 28x^2 - 21x - 8x + 6$

$\qquad = 28x^2 - 29x + 6$

32. $(3x - 2)^2 = (3x)^2 + 2(3x)(-2) + (-2)^2$

$\qquad = 9x^2 - 12x + 4$

34. $(5 - 8x)^2 = 5^2 - 2(5)(8x) + (8x)^2$

$\qquad = 25 - 80x + 64x^2$

36. $[(x + 1) - y]^2 = (x + 1)^2 - 2(x + 1)y + y^2$

$\qquad = x^2 + 2x + 1 - 2xy - 2y + y^2$

$\qquad = x^2 - 2xy + y^2 + 2x - 2y + 1$

38. $(2x + 3)(2x - 3) = 4x^2 - 9$

40. $(2x + 3y)(2x - 3y) = 4x^2 - 9y^2$

42. $(x + y + 1)(x + y - 1) = [(x + y) + 1][(x + y) - 1]$

$\qquad = (x + y)^2 - 1$

$\qquad = x^2 + 2xy + y^2 - 1$

44. $(3a^3 - 4b^2)(3a^3 + 4b^2) = 9a^6 - 16b^4$

46. $(x - 2)^3 = x^3 - 3x^2(2) + 3x(2)^2 - (2)^3$

$\qquad = x^3 - 6x^2 + 12x - 8$

48. $(3x + 2y)^3 = (3x)^3 + 3(3x)^2(2y) + 3(3x)(2y)^2 + (2y)^3 = 27x^3 + 54x^2y + 36xy^2 + 8y^3$

50. $(5 + \sqrt{x})(5 - \sqrt{x}) = 25 - (\sqrt{x})^2 = 25 - x$

52. $(8x + 3)^2 = (8x)^2 + 2(8x)3 + 3^2$

$\qquad = 64x^2 + 48x + 9$

54. $(x - 2)(x^2 + 2x + 4) = (x - 2)x^2 + (x - 2)2x + (x - 2)4 = x^3 - 2x^2 + 2x^2 - 4x + 4x - 8 = x^3 - 8$

56. By the Vertical Method we have:

$$
\begin{array}{r}
x^2 + 3x - 2 \\
x^2 - 3x - 2 \\
\hline
x^4 + 3x^3 - 2x^2 \\
- 3x^3 - 9x^2 + 6x \\
- 2x^2 - 6x + 4 \\
\hline
x^4 + 0x^3 - 13x^2 + 0x + 4
\end{array}
$$

which equals $x^4 - 13x^2 + 4$.

58. $(2x - 1)(x + 3) + 3(x + 3) = 2x^2 + 6x - x - 3 + 3x + 9$
$$= 2x^2 + 8x + 6$$

60. $(x + y)(x - y)(x^2 + y^2) = (x^2 - y^2)(x^2 + y^2)$
$$= x^4 - y^4$$

62. $(a_m x^m + a_{m-1} x^{m-1} + \cdots + a_0) + (b_n x^n + b_{n-1} x^{n-1} + \cdots + b_0)$

Since $n > m$, then the term with the highest degree is $b_n x^n$.

Therefore, the degree of the sum is n.

P.5 Factoring

2. $5y - 30 = 5(y - 6)$

4. $4x^3 - 6x^2 + 12x = 2x(2x^2 - 3x + 6)$

6. $3x(x + 2) - 4(x + 2) = (x + 2)(3x - 4)$

8. $x^2 - \frac{1}{4} = x^2 - \left(\frac{1}{2}\right)^2 = \left(x + \frac{1}{2}\right)\left(x - \frac{1}{2}\right)$

10. $49 - 9y^2 = (7)^2 - (3y)^2 = (7 + 3y)(7 - 3y)$

12. $25 - (z + 5)^2 = 5^2 - (z + 5)^2$
$$= [5 + (z + 5)][5 - (z + 5)]$$
$$= (z + 10)(-z)$$
$$= -z(z + 10)$$

14. $x^2 + 10x + 25 = x^2 + 2(5)(x) + 5^2 = (x + 5)^2$

16. $9x^2 - 12x + 4 = (3x)^2 - 2(3x)(2) + 2^2 = (3x - 2)^2$

18. $z^2 + z + \frac{1}{4} = z^2 + 2(z)\left(\frac{1}{2}\right) + \left(\frac{1}{2}\right)^2 = \left(z + \frac{1}{2}\right)^2$

20. $x^2 + 5x + 6 = (x + 3)(x + 2)$ since $(3)(2) = 6$ and $(3) + (2) = 5$.

22. $t^2 - t - 6 = (t - 3)(t + 2)$ since $(-3)(2) = -6$ and $(-3) + (2) = -1$.

24. $z^2 - 5z - 24 = (z - 8)(z + 3)$ since $(-8)(3) = -24$ and $(-8) + (3) = -5$.

26. $x^2 - 13x + 42 = (x - 7)(x - 6)$ since $(-7)(-6) = 42$ and $(-7) + (-6) = -13$.

28. $2x^2 - x - 1 = (2x + 1)(x - 1)$ since $(1)(-1) = -1$ and $(-2) + (1) = -1$.

30. $12x^2 + 7x + 1 = (3x + 1)(4x + 1)$ since $(1)(1) = 1$ and $(3) + (4) = 7$.

32. $5u^2 + 13u - 6 = (5u - 2)(u + 3)$ since $(-2)(3) = -6$ and $(15) + (-2) = 13$.

34. $x^3 - 27 = x^3 - 3^3 = (x - 3)(x^2 + 3x + 9)$

36. $z^3 + 125 = z^3 + 5^3 = (z + 5)(z^2 - 5z + 25)$

38. $27x^3 + 8 = (3x)^3 + 2^3 = (3x + 2)(9x^2 - 6x + 4)$

40. $x^3 + 5x^2 - 5x - 25 = (x^3 + 5x^2) + (-5x - 25)$
$$= x^2(x + 5) - 5(x + 5)$$
$$= (x + 5)(x^2 - 5)$$

42. $5x^3 - 10x^2 + 3x - 6 = (5x^3 - 10x^2) + (3x - 6)$
$$= 5x^2(x - 2) + 3(x - 2)$$
$$= (x - 2)(5x^2 + 3)$$

44. $x^5 + 2x^3 + x^2 + 2 = (x^5 + 2x^3) + (x^2 + 2)$
$$= x^3(x^2 + 2) + (x^2 + 2)$$
$$= (x^2 + 2)(x^3 + 1)$$

46. $2x^2 + 9x + 9 = 2x^2 + 6x + 3x + 9$
$$= (2x^2 + 6x) + (3x + 9)$$
$$= 2x(x + 3) + 3(x + 3)$$
$$= (x + 3)(2x + 3)$$

48. $6x^2 - x - 15 = 6x^2 + 9x - 10x - 15$
$$= (6x^2 + 9x) + (-10x - 15)$$
$$= 3x(2x + 3) - 5(2x + 3)$$
$$= (2x + 3)(3x - 5)$$

50. $12x^2 - 13x + 1 = 12x^2 - 12x - x + 1$
$$= (12x^2 - 12x) + (-x + 1) = 12x(x - 1) - (x - 1) = (x - 1)(12x - 1)$$

52. $x^2 + 4x + 3 = (x + 3)(x + 1)$

54. $x^2 + 3x + 2 = (x + 2)(x + 1)$

56. $12x^2 - 48 = 12(x^2 - 4) = 12(x - 2)(x + 2)$

58. $6x^2 - 54 = 6(x^2 - 9) = 6(x - 3)(x + 3)$

60. $16 + 6x - x^2 = -(x^2 - 6x - 16) = -(x - 8)(x + 2) = (8 - x)(x + 2)$

62. $9x^2 - 6x + 1 = (3x - 1)^2$

64. $2y^3 - 7y^2 - 15y = y(2y^2 - 7y - 15)$
$$= y(2y + 3)(y - 5)$$

66. $13x + 6 + 5x^2 = 5x^2 + 13x + 6$
$$= (5x + 3)(x + 2)$$

68. $5 - x + 5x^2 - x^3 = (5 - x) + x^2(5 - x)$
$$= (5 - x)(1 + x^2)$$

70. $3u - 2u^2 + 6 - u^3 = 3u + 6 - u^3 - 2u^2$
$$= 3(u + 2) - u^2(u + 2)$$
$$= (u + 2)(3 - u^2)$$

72. $(t - 1)^2 - 49 = [(t - 1) + 7][(t - 1) - 7]$
$$= (t + 6)(t - 8)$$

74. $(x^2 + 8)^2 - 36x^2 = [(x^2 + 8) + 6x][(x^2 + 8) - 6x]$
$$= (x^2 + 6x + 8)(x^2 - 6x + 8)$$
$$= (x + 2)(x + 4)(x - 2)(x - 4)$$

76. $5x^3 + 40 = 5(x^3 + 8) = 5(x + 2)(x^2 - 2x + 4)$

78. $5(3 - 4x)^2 - 8(3 - 4x)(5x - 1) = (3 - 4x)[5(3 - 4x) - 8(5x - 1)]$
$$= (3 - 4x)(15 - 20x - 40x + 8)$$
$$= (3 - 4x)(23 - 60x)$$

80. $7(3x + 2)^2(1 - x)^2 + (3x + 2)(1 - x)^3 = (3x + 2)(1 - x)^2[7(3x + 2) + (1 - x)]$
$$= (1 - x)^2(3x + 2)(21x + 14 + 1 - x)$$
$$= (1 - x)^2(3x + 2)(20x + 15)$$
$$= 5(1 - x)^2(3x + 2)(4x + 3)$$

P.6 Fractional Expressions

2. The domain of the polynomial $2x^2 + 5x - 2$ is the set of all real numbers.

4. The domain of the polynomial $6x^2 + 7x - 9$, $x > 0$ is the set of positive real numbers, since the polynomial is restricted to that set.

6. The domain of $\dfrac{x + 1}{2x + 1}$ is the set of all real numbers except $x = -\dfrac{1}{2}$, which would produce an undefined division by zero.

8. The domain of $\dfrac{2x + 1}{(x + 3)(x - 3)}$ is the set of all real numbers except $x = 3$ and $x = -3$, which would produce an undefined division by zero.

10. The domain of $\dfrac{1}{\sqrt{x + 1}}$ is the set of all real numbers greater than -1 since $x + 1 > 0$ when $x > -1$.

12. $\dfrac{3}{4} = \dfrac{3(x+1)}{4(x+1)},\ x \neq -1$

14. $\dfrac{3y-4}{y+1} = \dfrac{(3y-4)(y-1)}{(y+1)(y-1)}$
$= \dfrac{(3y-4)(y-1)}{y^2-1},\ y \neq 1$

16. $\dfrac{1-z}{z^2} = \dfrac{(1-z)(1+z)}{z^2(1+z)}$
$= \dfrac{(1-z)(1+z)}{z^2+z^3},\ z \neq -1$

18. $\dfrac{18y^2}{60y^5} = \dfrac{6y^2(3)}{6y^2(10y^3)} = \dfrac{3}{10y^3}$

20. $\dfrac{9x^2+9x}{2x+2} = \dfrac{9x(x+1)}{2(x+1)} = \dfrac{9x}{2},\ x \neq -1$

22. $\dfrac{x^2-25}{5-x} = \dfrac{(x+5)(x-5)}{-(x-5)} = -(x+5),\ x \neq 5$

24. $\dfrac{x^2+8x-20}{x^2+11x+10} = \dfrac{(x+10)(x-2)}{(x+10)(x+1)}$
$= \dfrac{x-2}{x+1},\ x \neq -10$

26. $\dfrac{10+x}{x^2+11x+10} = \dfrac{10+x}{(x+10)(x+1)}$
$= \dfrac{1}{x+1},\ x \neq -10,-1$

28. $\dfrac{x^2-9}{x^3+x^2-9x-9} = \dfrac{x^2-9}{(x^3-9x)+(x^2-9)}$
$= \dfrac{x^2-9}{x(x^2-9)+(x^2-9)}$
$= \dfrac{x^2-9}{(x^2-9)(x+1)}$
$= \dfrac{1}{x+1},\ x \neq \pm 3$

30. $\dfrac{y^3-2y^2-3y}{y^3+1} = \dfrac{y(y^2-2y-3)}{y^3+1}$
$= \dfrac{y(y-3)(y+1)}{(y+1)(y^2-y+1)}$
$= \dfrac{y(y-3)}{y^2-y+1},\ y \neq -1$

32. $\dfrac{(x+5)(x-3)}{x+2} \cdot \dfrac{1}{(x+5)(x+2)} = \dfrac{x-3}{(x+2)^2},\ x \neq -5$

34. $\dfrac{4y-16}{5y+15} \cdot \dfrac{2y+6}{4-y} = \dfrac{4(y-4)}{5(y+3)} \cdot \dfrac{-2(y+3)}{y-4} = -\dfrac{8}{5},\ y \neq -3,4$

36. $\dfrac{x^3-1}{x+1} \cdot \dfrac{x^2+1}{x^2-1} = \dfrac{(x-1)(x^2+x+1)}{x+1} \cdot \dfrac{x^2+1}{(x+1)(x-1)} = \dfrac{(x^2+1)(x^2+x+1)}{(x+1)^2},\ x \neq 1$

38. $\dfrac{x+2}{5(x-3)} \div \dfrac{x-2}{5(x-3)} = \dfrac{x+2}{5(x-3)} \cdot \dfrac{5(x-3)}{x-2} = \dfrac{x+2}{x-2},\ x \neq 3$

40. $\dfrac{\left(\dfrac{x^2-1}{x}\right)}{\left(\dfrac{(x-1)^2}{x}\right)} = \dfrac{x^2-1}{x} \div \dfrac{(x-1)^2}{x}$

$\qquad = \dfrac{x^2-1}{x} \cdot \dfrac{x}{(x-1)^2}$

$\qquad = \dfrac{(x+1)(x-1)}{x} \cdot \dfrac{x}{(x-1)^2}$

$\qquad = \dfrac{x+1}{x-1}, \; x \neq 0$

42. $\dfrac{2x-1}{x+3} + \dfrac{1-x}{x+3} = \dfrac{(2x-1)+(1-x)}{x+3} = \dfrac{x}{x+3}$

44. $\dfrac{3}{x-1} - 5 = \dfrac{3}{x-1} - \dfrac{5(x-1)}{(x-1)}$

$\qquad = \dfrac{3-5(x-1)}{x-1} = \dfrac{8-5x}{x-1}$

46. $\dfrac{2x}{x-5} - \dfrac{5}{5-x} = \dfrac{2x}{x-5} + \dfrac{5}{x-5}$

$\qquad = \dfrac{2x+5}{x-5}$

48. $\dfrac{x}{x^2+x-2} - \dfrac{1}{x+2} = \dfrac{x}{(x+2)(x-1)} - \dfrac{1}{x+2}$

$\qquad = \dfrac{x}{(x+2)(x-1)} - \dfrac{x-1}{(x+2)(x-1)}$

$\qquad = \dfrac{x-(x-1)}{(x+2)(x-1)} = \dfrac{1}{(x+2)(x-1)}$

50. $\dfrac{2}{x+1} + \dfrac{2}{x-1} + \dfrac{1}{x^2-1} = \dfrac{2}{x+1} + \dfrac{2}{x-1} + \dfrac{1}{(x+1)(x-1)}$

$\qquad = \dfrac{2(x-1)}{(x+1)(x-1)} + \dfrac{2(x+1)}{(x+1)(x-1)} + \dfrac{1}{(x+1)(x-1)}$

$\qquad = \dfrac{2x-2+2x+2+1}{(x+1)(x-1)} = \dfrac{4x+1}{(x+1)(x-1)}$

52. $(1-x^3)^{1/3} - x^3(1-x^3)^{-2/3} = (1-x^3)^{-2/3}\left((1-x^3)-x^3\right) = \dfrac{1-2x^3}{(1-x^3)^{2/3}}$

54. $2(2+x)^{-1/2} + (1-x)(2+x)^{-3/2} = (2+x)^{-3/2}\left(2(2+x)+(1-x)\right) = \dfrac{x+5}{(2+x)^{3/2}}$

56. $\dfrac{\dfrac{(x-4)}{x}}{\dfrac{x}{4} - \dfrac{4}{x}} = \dfrac{(x-4)}{\dfrac{x}{4} - \dfrac{4}{x}} \cdot \dfrac{4x}{4x}$

$\qquad = \dfrac{4x(x-4)}{x^2-16} = \dfrac{4x(x-4)}{(x-4)(x+4)}$

$\qquad = \dfrac{4x}{x+4}, \; x \neq 4$

58. $\dfrac{\left(\dfrac{5}{y} - \dfrac{6}{2y+1}\right)}{\left(\dfrac{5}{y} + 4\right)} = \dfrac{\dfrac{5(2y+1)-6y}{y(2y+1)}}{\dfrac{5+4y}{y}}$

$\qquad = \dfrac{4y+5}{y(2y+1)} \cdot \dfrac{y}{5+4y}$

$\qquad = \dfrac{1}{2y+1}, \; y \neq 0, \; -\dfrac{5}{4}$

60. $\dfrac{\dfrac{x+h}{x+h+1} - \dfrac{x}{x+1}}{h} = \dfrac{\dfrac{(x+h)(x+1) - x(x+h+1)}{(x+h+1)(x+1)}}{h}$

$\qquad = \dfrac{\dfrac{x^2 + x + hx + h - x^2 - hx - x}{(x+h+1)(x+1)}}{h}$

$\qquad = \dfrac{h}{(x+h+1)(x+1)} \cdot \dfrac{1}{h}$

$\qquad = \dfrac{1}{(x+h+1)(x+1)}, \ h \neq 0$

62. $\dfrac{z^{-1} - 2(z+2)^{-1}}{z^{-2}} = \left(z^{-1} - 2(z+2)^{-1}\right) z^2 = z - 2z^2(z+2)^{-1} = z - \dfrac{2z^2}{(z+2)}$

64. $\dfrac{3x^{1/3} - x^{-2/3}}{3x^{-2/3}} = \dfrac{3x^{1/3} - x^{-2/3}}{3x^{-2/3}} \cdot \dfrac{x^{2/3}}{x^{2/3}} = \dfrac{3x - 1}{3}$

66. $\dfrac{-x^3(1-x^2)^{-1/2} - 2x(1-x^2)^{1/2}}{x^4} = \dfrac{-x^3(1-x^2)^{-1/2} - 2x(1-x^2)^{1/2}}{x^4} \cdot \dfrac{(1-x^2)^{1/2}}{(1-x^2)^{1/2}}$

$\qquad = \dfrac{-x^3 - 2x(1-x^2)}{x^4(1-x^2)^{1/2}}$

$\qquad = \dfrac{x^3 - 2x}{x^4(1-x^2)^{1/2}}$

$\qquad = \dfrac{x^2 - 2}{x^3(1-x^2)^{1/2}}$

68. $\dfrac{\sqrt{z-3} - \sqrt{z}}{3} = \dfrac{\sqrt{z-3} - \sqrt{z}}{3} \cdot \dfrac{\sqrt{z-3} + \sqrt{z}}{\sqrt{z-3} + \sqrt{z}}$

$\qquad = \dfrac{(z-3) - z}{3(\sqrt{z-3} + \sqrt{z})}$

$\qquad = -\dfrac{3}{3(\sqrt{z-3} + \sqrt{z})}$

$\qquad = -\dfrac{1}{\sqrt{z-3} + \sqrt{z}}$

P.7 The Cartesian Plane

2.

4.
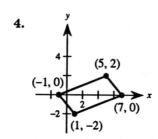

6. Since the points $(1, 4)$ and $(8, 4)$ lie on the same vertical line, the distance between the points is given by the absolute value of the difference of their x-coordinates.

$$d = |8 - 1| = |7| = 7$$

8. Since the points $(-3, -4)$ and $(-3, 6)$ lie on the same horizontal line, the distance between the points is given by the absolute value of the difference of their y-coordinates.

$$d = |6 - (-4)| = |6 + 4| = |10| = 10$$

10. (a) $a = |13 - 1| = 12$

$\qquad b = |6 - 1| = 5$

$\qquad c = \sqrt{12^2 + 5^2} = \sqrt{144 + 25} = \sqrt{169} = 13$

(b) $d = \sqrt{(13 - 1)^2 + (6 - 1)^2}$

$\qquad = \sqrt{12^2 + 5^2} = \sqrt{169} = 13$

12. (a) $a = |6 - 2| = 4$

$\qquad b = |5 - (-2)| = 7$

$\qquad c = \sqrt{4^2 + 7^2} = \sqrt{16 + 49} = \sqrt{65}$

(b) $d = \sqrt{(6 - 2)^2 + (-2 - 5)^2}$

$\qquad = \sqrt{4^2 + (-7)^2} = \sqrt{16 + 49} = \sqrt{65}$

14. (a)

(b) $d = \sqrt{(6 - 1)^2 + (0 - 12)^2}$

$\qquad = \sqrt{5^2 + (-12)^2}$

$\qquad = \sqrt{169} = 13$

(c) $m = \left(\dfrac{1 + 6}{2}, \dfrac{12 + 0}{2} \right) = \left(\dfrac{7}{2}, 6 \right)$

16. (a)
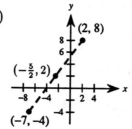

(b) $d = \sqrt{(-7 - 2)^2 + (-4 - 8)^2}$

$\qquad = \sqrt{81 + 144} = 15$

(c) $m = \left(\dfrac{-7 + 2}{2}, \dfrac{-4 + 8}{2} \right) = \left(-\dfrac{5}{2}, 2 \right)$

18. (a)

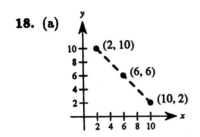

(b) $d = \sqrt{(2-10)^2 + (10-2)^2}$

$= \sqrt{64 + 64} = 8\sqrt{2}$

(c) $m = \left(\dfrac{2+10}{2}, \dfrac{10+2}{2}\right) = (6, 6)$

20. (a)

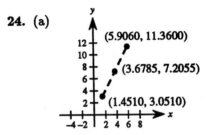

(b) $d = \sqrt{\left(-\dfrac{1}{3} + \dfrac{1}{6}\right)^2 + \left(-\dfrac{1}{3} + \dfrac{1}{2}\right)^2}$

$= \sqrt{\dfrac{1}{36} + \dfrac{1}{36}} = \dfrac{\sqrt{2}}{6}$

(c) $m = \left(\dfrac{-\frac{1}{3} - \frac{1}{6}}{2}, \dfrac{-\frac{1}{3} - \frac{1}{2}}{2}\right) = \left(-\dfrac{1}{4}, -\dfrac{5}{12}\right)$

22. (a)

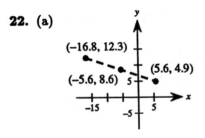

(b) $d = \sqrt{(5.6 + 16.8)^2 + (4.9 - 12.3)^2}$

$= \sqrt{501.76 + 54.76}$

$= \sqrt{556.52}$

(c) $m = \left(\dfrac{-16.8 + 5.6}{2}, \dfrac{12.3 + 4.9}{2}\right)$

$= (-5.6, \ 8.6)$

24. (a)

(b) $d = \sqrt{(5.906 - 1.451)^2 + (11.360 - 3.051)^2}$

$= \sqrt{88.886506}$

(c) $m = \left(\dfrac{1.451 + 5.906}{2}, \dfrac{3.051 + 11.360}{2}\right)$

$= (3.6785, \ 7.2055)$

26. $\dfrac{4,200,000 + 5,650,000}{2} = 4,925,000$

The estimated sales for 1991 is $4,925,000.

28. $d_1 = \sqrt{(1-3)^2 + (-3-2)^2} = \sqrt{29}$

$d_2 = \sqrt{(1+2)^2 + (-3-4)^2} = \sqrt{58}$

$d_3 = \sqrt{(3+2)^2 + (2-4)^2} = \sqrt{29}$

Since $d_1 = d_3$, we can conclude that the triangle is an isosceles triangle.

30. $d_1 = \sqrt{(0-3)^2 + (1-7)^2} = \sqrt{45}$

$d_2 = \sqrt{(0-1)^2 + (1+2)^2} = \sqrt{10}$

$d_3 = \sqrt{(4-3)^2 + (4-7)^2} = \sqrt{10}$

$d_4 = \sqrt{(4-1)^2 + (4+2)^2} = \sqrt{45}$

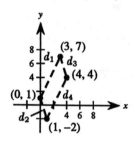

Since $d_1 = d_4$ and $d_2 = d_3$, then we can conclude that the points form the vertices of a parallelogram.

32. $\sqrt{(x+8)^2 + (5-0)^2} = 13$

$\sqrt{x^2 + 16x + 64 + 25} = 13$

$x^2 + 16x + 89 = 169$

$x^2 + 16x - 80 = 0$

$(x+20)(x-4) = 0$

$x = -20 \quad \text{or} \quad x = 4$

34. $\sqrt{(7+8)^2 + (y-4)^2} = 17$

$\sqrt{225 + (y-4)^2} = 17$

$225 + (y-4)^2 = 289$

$(y-4)^2 = 64$

$y - 4 = \pm 8$

$y = 4 \pm 8$

$y = -4 \quad \text{or} \quad y = 12$

36. The distance between $(3, \frac{5}{2})$ and (x, y) is equal to the distance between $(-7, 1)$ and (x, y).

$$\sqrt{(x-3)^2 + (y-\tfrac{5}{2})^2} = \sqrt{(x+7)^2 + (y-1)^2}$$

$$(x-3)^2 + (y-\tfrac{5}{2})^2 = (x+7)^2 + (y-1)^2$$

$$x^2 - 6x + 9 + y^2 - 5y + \tfrac{25}{4} = x^2 + 14x + 49 + y^2 - 2y + 1$$

$$-20x - 3y - \tfrac{139}{4} = 0$$

$$80x + 12y + 139 = 0$$

38. $x < 0 \Rightarrow x$ lies in Quadrant II or in Quadrant III.

$y < 0 \Rightarrow y$ lies in Quadrant III or in Quadrant IV.

$x < 0$ and $y < 0 \Rightarrow (x, y)$ lies in Quadrant III.

40. $x < 0 \Rightarrow x$ lies in Quadrant II or in Quadrant III.

$y > 0 \Rightarrow y$ lies in Quadrant I or in Quadrant II.

$x < 0$ and $y > 0 \Rightarrow (x, y)$ lies in Quadrant II.

42. $x > 2 \Rightarrow x$ is positive $\Rightarrow x$ lies in Quadrant I or Quadrant IV.

$y = 3 \Rightarrow y$ is positive $\Rightarrow y$ lies in Quadrant I or Quadrant II.

$x > 2$ and $y = 3 \Rightarrow (x, y)$ lies in Quadrant I.

44. $x > 4 \Rightarrow x$ is positive $\Rightarrow x$ lies in either Quadrant I or Quadrant IV.

46. If $xy < 0$, then x and y have opposite signs. Hence, (x, y) lies in either Quadrant II or Quadrant IV.

48. Since $(-x, y)$ is in Quadrant IV, we know that $-x > 0$ and $y < 0$. If $-x > 0$, then $x < 0$.

$x < 0 \Rightarrow x$ lies in Quadrant II or in Quadrant III.

$y < 0 \Rightarrow y$ lies in Quadrant III or in Quadrant IV.

$x < 0$ and $y < 0 \Rightarrow (x, y)$ lies in Quadrant III.

50.

The points are reflected through the x-axis.

52. $(x - 0)^2 + (y - 0)^2 = 5^2$

$x^2 + y^2 = 25$

54. $(x - 0)^2 + \left(y - \frac{1}{3}\right)^2 = \left(\frac{1}{3}\right)^2$

$x^2 + \left(y - \frac{1}{3}\right)^2 = \frac{1}{9}$

56. $(x - 3)^2 + (y + 2)^2 = r^2$

$(-1 - 3)^2 + (1 + 2)^2 = r^2 \Rightarrow r^2 = 25$

$(x - 3)^2 + (y + 2)^2 = 25$

58. $r = \frac{1}{2}\sqrt{(4 + 4)^2 + (1 + 1)^2} = \sqrt{17}$

Center $= \left(\frac{-4 + 4}{2}, \frac{-1 + 1}{2}\right) = (0, 0)$

$x^2 + y^2 = 17$

P.8 Exploring Data: Representing Data Graphically

2. By scanning the data, we see that the largest number is 25 and the smallest number is 14. We construct the line plot on the interval [14, 25] as follows.

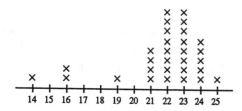

The scores of 22 and 23 occurred with the greatest frequency (8).

4. By scanning the data, we see that the largest number is 90 and the smallest number is 59. We construct the line plot on the interval [59, 90] as follows.

The score of 68 occurred with the greatest frequency (4).

6. Since the scores on Exams 1 and 2 range from 59 to 100, the stems are 5, 6, 7, 8, 9, and 10.

Leaves (Exam 2)	Stems	Leaves (Exam 1)
9	5	
8 8 8 8 7 7 6 6 4 3	6	
8 8 8 7 7 6 4 3 3 1 1 0 0	7	0 5 5 5 7 7 8 8 8
7 4 1 0 0	8	1 1 1 1 2 3 4 5 5 5 5 9 8 9 9 9
0	9	0 2 8
	10	0 0

By comparing the leaves you see that the scores on Exam 1 were higher.

8. Since the snowfall ranges from 41.2 to 142.8, you could use a stem-and-leaf plot as follows.

Stems	Leaves
4	1.2 2.5
5	3.7 5.2
6	6.0 6.7 8.2 8.6 8.6 9.2
7	1.3 5.9 6.5 6.5 9.7
8	4.8 5.6 9.4
9	2.3 2.9
10	3.5 6.3 7.9
11	0.0 1.5 4.9 5.9
12	0.0 4.9
13	
14	2.8

10.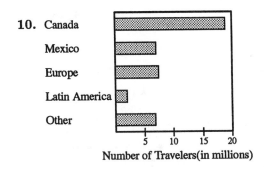

12. Let the vertical axis represent the number of students and the horizontal axis the year. Plot the 8 points and connect them with line segments as follows.

14. Let the vertical axis represent the receipts (in billions of dollars) and the horizontal axis the year. Plot the 21 points and connect them with line segments as follows.

CHAPTER ONE
Functions and Graphs

1.1 Graphs and Graphing Utilities

2. (a) $2^2 - 3(2) + 2 = 0$, yes
(b) $(-2)^2 - 3(-2) + 2 = 12 \neq 8$, no

4. (a) $3^2 + (-2)^2 = 13 \neq 20$, no
(b) $(-4)^2 + 2^2 = 20$, yes

6. (a) $\dfrac{1}{0^2 + 1} = 1 \neq 0$, no
(b) $\dfrac{1}{3^2 + 1} = 0.1$, yes

8. $y = Cx^3$
$8 = C(-4)^3$
$C = -\dfrac{8}{64}$
$C = -\dfrac{1}{8}$

10. $x + C(y + 2) = 0$
$4 + C(3 + 2) = 0$
$4 + 5C = 0$
$5C = -4$
$C = -\dfrac{4}{5}$

12. $y = 4 - x^2$

x	-2	-1	0	2	3
y	0	3	4	0	-5
$(x,\ y)$	$(-2,\ 0)$	$(-1,\ 3)$	$(0,\ 4)$	$(2,\ 0)$	$(3,\ -5)$

14. $y = x^2 + 2x$
x-intercepts: $(0,\ 0)$, $(-2,\ 0)$
y-intercept: $(0,\ 0)$
Matches graph (f)

```
RANGE
Xmin=-4
Xmax=2
Xscl=1
Ymin=-2
Ymax=4
Yscl=1
```

16. $y = \sqrt{x}$
x-intercept: $(0,\ 0)$
y-intercept: $(0,\ 0)$
Matches graph (a)

```
RANGE
Xmin=-1
Xmax=8
Xscl=1
Ymin=-1
Ymax=4
Yscl=1
```

18. $y = |x| - 2$
x-intercepts: $(2,\ 0)$, $(-2,\ 0)$
y-intercept: $(0,\ -2)$
Matches graph (b)

```
RANGE
Xmin=-4
Xmax=4
Xscl=1
Ymin=-3
Ymax=3
Yscl=1
```

20. $y = 2x - 3$

Intercepts: $(0, -3)$, $\left(\frac{3}{2}, 0\right)$

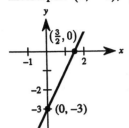

22. $y = x^2 - 1$

Intercepts:

$(-1, 0)$, $(0, -1)$, $(1, 0)$

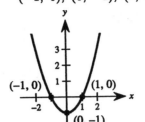

24. $y = x^3 - 1$

Intercepts: $(0, -1)$, $(1, 0)$

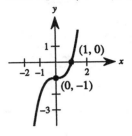

26. $y = x(x - 5)$

Intercepts: $(0, 0)$, $(5, 0)$

28. $y = \sqrt{1 - x}$

Intercepts: $(0, 1)$, $(1, 0)$

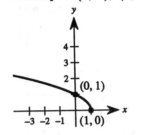

30. $y = 4 - |x|$

Intercepts:

$(-4, 0)$, $(0, 4)$, $(4, 0)$

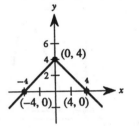

32. $y = 4 - 4x - x^2$

The graph intersects the x-axis twice and the y-axis once.

34. $y = \frac{1}{27}\left(x^4 + 4x^3\right)$

The graph intersects the x-axis twice and the y-axis once.

36. $y = x\sqrt{4 - x^2}$

The graph intersects the x-axis 3 times and the y-axis once.

38. $y = \dfrac{10}{x^2 + 1}$

The graph intersects the y-axis once.

40. $y = 4x^3 - x^4$

The graph intersects the x-axis twice and the y-axis once.

42. $y = 100x\sqrt{25 - x^2}$

The graph intersects the x-axis 3 times and the y-axis once.

44. $2x^3 - 100x - 15{,}625 + 250y = 0 \;\Rightarrow\; 250y = -2x^3 + 100x + 15{,}625$

$$\Rightarrow\; y = \frac{-1}{125}x^3 + \frac{2}{5}x + \frac{125}{2}$$

The graph intersects the x-axis once and the y-axis once.

46. $x^2 + y^2 = 49 \Rightarrow y^2 = 49 - x^2$

$\Rightarrow y = \pm\sqrt{49 - x^2}$

$y_1 = \sqrt{49 - x^2},\ y_2 = -\sqrt{49 - x^2}$

48. $x^2 + 9y^2 = 81 \Rightarrow 9y^2 = 81 - x^2$

$\Rightarrow y = \pm\frac{1}{3}\sqrt{81 - x^2}$

$y_1 = \frac{1}{3}\sqrt{81 - x^2},\ y_2 = -\frac{1}{3}\sqrt{81 - x^2}$

50. $0.75x - 3y + 200 = 0 \Rightarrow 3y = \frac{3}{4}x + 200$

$\Rightarrow y = \frac{1}{4}x + \frac{200}{3}$

Since the y-intercept is $\frac{200}{3} \approx 66.67$, we can use the following range.

```
RANGE
Xmin=-1
Xmax=100
Xscl=5
Ymin=50
Ymax=100
Yscl=5
```

(Answers not unique.)

52. $y = \sqrt{x^3 + 8}$

Since the equation is defined for $x \geq -2$, we can use the following range.

```
RANGE
Xmin=-3
Xmax=10
Xscl=1
Ymin=-2
Ymax=20
Yscl=2
```

54. $y = 2.44x - \dfrac{x^2}{20,000} - 5000$

The second viewing rectangle is preferred because it shows a greater increase.

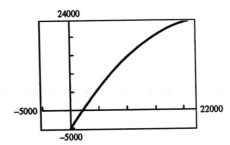

56. $y = x^3(3x + 4)$

(a)

x	-1	0	1
y	-1	0	7

(b)

x	-1	$-\frac{3}{4}$	$-\frac{1}{2}$	$-\frac{1}{4}$	0	$\frac{1}{4}$	$\frac{1}{2}$	$\frac{3}{4}$	1
y	-1	-0.74	-0.31	-0.05	0	0.07	0.69	2.64	7

(c)

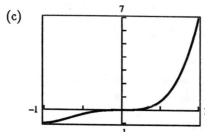

58. $y = x^3(x - 3)$

(a) $(2.25, y)$, $y = -8.54$

(b) $(x, 20)$, $x = 3.48$ and $x = -1.63$

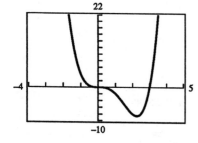

60. $y = |x^2 - 6x + 5|$

(a) $(2, y)$, $y = 3$

(b) $(x, 1.5)$, $x = 0.65, 1.42, 4.58, 5.35$

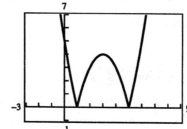

62. (a) Perimeter $= 12 = 2x + 2w$, which implies that $w = 6 - x$. Hence, the area is $y = xw = x(6 - x)$.

(b) $0 \leq x \leq 6$

(c)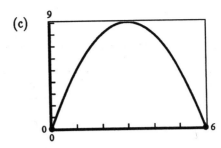

(d) Dimensions for maximum area are $x = w = 3$ (a square).

64. $y = \dfrac{x + 66.94}{0.01x + 1}$

(a)

66. $y = \dfrac{10{,}770}{x^2} - 0.37$, $5 \leq x \leq 100$

When $x = 50$, $y = 3.938$.

(b) Using the zoom and trace features, we have $y = 77.04$ when $x = 44$ (year 1994).

1.2 Lines in the Plane

2. slope $= \dfrac{\text{rise}}{\text{run}} = \dfrac{2}{1} = 2$

4. slope $= \dfrac{\text{rise}}{\text{run}} = \dfrac{-1}{1} = -1$

6. slope $= \dfrac{\text{rise}}{\text{run}} = \dfrac{3}{2}$

8.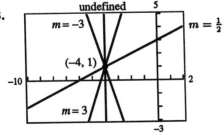

10. slope $= \dfrac{-4 - 4}{4 - 2} = -4$

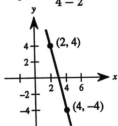

12. slope $= \dfrac{-10-0}{0+4} = -\dfrac{5}{2}$

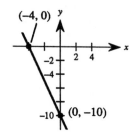

14. slope $= \dfrac{-\frac{1}{4}-\frac{3}{4}}{\frac{5}{4}-\frac{7}{8}} = \dfrac{-1}{\frac{3}{8}} = -\dfrac{8}{3}$

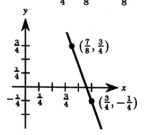

16. Since m is undefined, the line is vertical, and since the line passes through $(-4, 1)$, all other points on the line will be of the form $(-4, y)$. Three additional points are: $(-4, 0)$, $(-4, 3)$, $(-4, 5)$.

18. Since $m = -1$, y decreases by 1 for every one unit increase in x. Three points are: $(0, 4)$, $(9, -5)$, $(11, -7)$.

20. The slope of L_1 is $m_1 = \dfrac{5-(-1)}{1-(-2)} = \dfrac{6}{3} = 2$.

The slope of L_2 is $m_2 = \dfrac{3-(-5)}{1-5} = \dfrac{8}{-4} = -2$.

Since m_1 and m_2 are neither equal to each other nor negative reciprocals of each other, the lines are neither parallel nor perpendicular.

22. The slope of L_1 is $m_1 = \dfrac{8-2}{4-(-4)} = \dfrac{6}{8} = \dfrac{3}{4}$.

The slope of L_2 is $m_2 = \dfrac{\frac{1}{3}-(-5)}{-1-3} = \dfrac{\frac{16}{3}}{-4} = -\dfrac{4}{3}$.

Since m_1 and m_2 are negative reciprocals of each other, the lines are perpendicular.

24. The points are not collinear, since the slope between the first two points is $\dfrac{0-1}{-1-(-2)} = -1$, while the slope between the second and third points is $\dfrac{-2-0}{2-(-1)} = -\dfrac{2}{3}$.

26. Since the slope of the roof is 3/4, the height y of the attic satisfies $3/4 = y/15$, which gives $y = 45/4 = 11.25$ feet.

28. $2x + 3y - 9 = 0$

$$3y = -2x + 9$$

$$y = -\frac{2}{3}x + 3$$

Slope: $m = -\frac{2}{3}$

y-intercept: $(0, 3)$

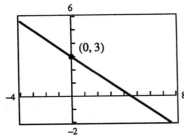

30. $3y + 5 = 0$

$$3y = -5$$

$$y = -\frac{5}{3}$$

Slope: $m = 0$

y-intercept: $\left(0, -\frac{5}{3}\right)$

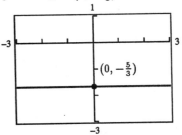

32. $m = \dfrac{3 - (-4)}{4 - (-4)} = \dfrac{7}{8}$

$$y - 3 = \frac{7}{8}(x - 4)$$

$$8y - 24 = 7(x - 4)$$

$$8y - 24 = 7x - 28$$

$$7x - 8y - 4 = 0$$

34. $m = \dfrac{4 - 4}{6 - (-1)} = \dfrac{0}{7} = 0$

$$y - 4 = 0(x + 1)$$

$$y - 4 = 0$$

36. $m = \dfrac{0.6 - (-2.4)}{-8 - 2}$

$$= \frac{3}{-10} = -\frac{3}{10}$$

$$y - 0.6 = -\frac{3}{10}(x + 8)$$

$$10y - 6 = -3(x + 8)$$

$$10y - 6 = -3x - 24$$

$$3x + 10y + 18 = 0$$

38. $y - 0 = 4(x - 0)$

$$y = 4x$$

$$4x - y = 0$$

40. $y + 5 = \frac{3}{4}(x + 2)$

$$4y + 20 = 3(x + 2)$$

$$4y + 20 = 3x + 6$$

$$3x - 4y - 14 = 0$$

42. $y - 4 = 0(x + 10)$

$$y - 4 = 0$$

44. $y = -8x + 5$

 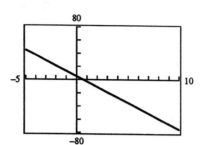

46. $y = \dfrac{2}{3}x$

$y = -\dfrac{3}{2}x$

$y = \dfrac{2}{3}x + 2$

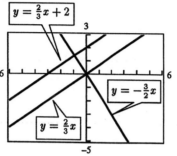

48. $y = x - 8$

$y = x + 1$

$y = -x + 3$

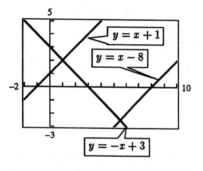

50. $\dfrac{x}{-6} + \dfrac{y}{2} = 1 \Rightarrow \dfrac{y}{2} = \dfrac{x}{6} + 1$

$\Rightarrow y = \dfrac{1}{3}x + 2$

$a = -6$ is the x-intercept and $b = 2$ is the y-intercept.

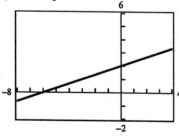

52. $\dfrac{x}{-1/6} + \dfrac{y}{-2/3} = 1$

$-6x - \dfrac{3}{2}y = 1$

$-12x - 3y = 2$

$12x + 3y + 2 = 0$

54. $5x + 3y = 0$

$$3y = -5x$$

$$y = -\frac{5}{3}x$$

The slope of the given line is $m_1 = -\frac{5}{3}$.

(a) The slope of the parallel line is $m_2 = m_1 = -\frac{5}{3}$.

$$y - \frac{3}{4} = -\frac{5}{3}(x - \frac{7}{8})$$

$$y - \frac{3}{4} = -\frac{5}{3}x + \frac{35}{24}$$

$$24y - 18 = -40x + 35$$

$$40x + 24y - 53 = 0$$

(b) The slope of the perpendicular line is $m_2 = -1/m_1 = 3/5$.

$$y - \frac{3}{4} = \frac{3}{5}\left(x - \frac{7}{8}\right)$$

$$y - \frac{3}{4} = \frac{3}{5}x - \frac{21}{40}$$

$$40y - 30 = 24x - 21$$

$$24x - 40y + 9 = 0$$

56. $x = 4$

The slope of the given line is undefined.

(a) The slope of the parallel line is undefined.

$$x = 2$$

$$x - 2 = 0$$

(b) The slope of the perpendicular line is 0.

$$y - 5 = 0(x - 2)$$

$$y - 5 = 0$$

58. Value = \$156 + (\$4.50)(the number of years t after 1990)

$$V = 4.50t + 156, \quad 0 \le t \le 5$$

60. Value = \$245,000 + (\$5600)(the number of years t after 1990)

$$V = 5600t + 245,000, \ 0 \le t \le 5$$

62. An employee is paid \$12.50 per hour plus \$1.50 for each unit produced per hour. Matches graph (c). The slope is $m = \$1.50$ and the y-intercept is \$12.50.

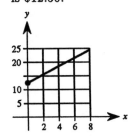

64. A typewriter purchased for \$600 depreciates \$100 per year. Matches graph (d). The slope is -100 and the y-intercept is 600.

66. $F = \frac{9}{5}C + 32$

C	$-17.8°$	$-10°$	$10°$	$20°$	$32.2°$	$177°$
F	$0°$	$14°$	$50°$	$68°$	$90°$	$350.6°$

68. Let $t = 0$ represent 1990.

$(0, 2546), (2, 2702)$

$$E - 2546 = \frac{2702 - 2546}{2 - 0}(t - 0)$$

$$E = 78t + 2546$$

When $t = 7$, $E = 78(7) + 2546 = 3092$ students.

70. (a) $y = 0.75x + 11.50$

(b)

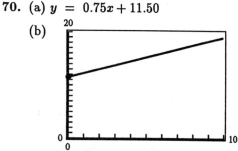

(c) When $x = 6$, $y = 16.00$.

(d) When $y = 18.25$, $x = 9.00$.

72. Daily cost = Cost for lodging and meals
$+ 0.26$ (number of miles driven)

$$C = 120 + 0.26x$$

74. (a) $x - 50 = \dfrac{47 - 50}{425 - 380}(p - 380)$

$$= \frac{-1}{15}(p - 380)$$

$$x = \frac{-1}{15}p + \frac{226}{3}$$

(b)

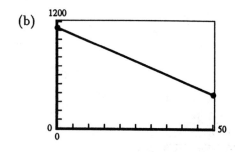

(c) If $p = 455$, $x = 45$ units.

(d) If $p = 395$, $x = 49$ units.

76. (Many answers possible.) The weight of a block of melting ice is 40 lbs at noon, and 25 lbs two hours later. When will it melt completely?

1.3 Functions

2. (a) The element c in A is matched with two elements, 2 and 3 of B, so it does not represent a function.

(b) Each element of A is matched with exactly one element of B, so it does represent a function.

(c) Since the implied input values are elements of B, $\{(1, a), (0, a), (2, c), (3, b)\}$ does not represent a function.

(d) Each element of A is matched with exactly one element of B, so it does represent a function.

4. y is not a function of x since some values of x give two values for y. For example, if $x = 9$, then $y = \pm 3$.

6. y is not a function of x since some values of x give two values for y. For example, if $x = -12$, then $y = \pm 4$.

8.
$$x^2 + y^2 - 2x - 4y + 1 = 0$$
$$(x^2 - 2x + 1) + (y^2 - 4y + 4) = -1 + 1 + 4$$
$$(x - 1)^2 + (y - 2)^2 = 4$$
$$(y - 2)^2 = 4 - (x - 1)^2$$
$$y - 2 = \pm\sqrt{4 - (x - 1)^2}$$
$$y = 2 \pm \sqrt{4 - (x - 1)^2}$$

y is not a function of x since some values of x give two values for y. For example, if $x = 1$, then $y = 2 \pm 2 = 0$ and 4.

10. y is a function of x. No value of x yields more than one value of y.

12. (a) $g(2) = (2)^2 - 2(2) = 0$

(b) $g(-3) = (-3)^2 - 2(-3) = 15$

(c) $g(t + 1) = (t + 1)^2 - 2(t + 1) = t^2 - 1$

(d) $g(x + h) = (x + h)^2 - 2(x + h) = (x + h)(x + h - 2)$

14. (a) $f(3) = \sqrt{25 - 3^2} = \sqrt{25 - 9} = \sqrt{16} = 4$

(b) $f(5) = \sqrt{25 - 5^2} = \sqrt{25 - 25} = 0$

(c) $f(x + 5) = \sqrt{25 - (x + 5)^2} = \sqrt{25 - x^2 - 10x - 25} = \sqrt{-x^2 - 10x}$

(d) $f(2 + h) = \sqrt{25 - (2 + h)^2} = \sqrt{25 - 4 - 4h - h^2} = \sqrt{21 - 4h - h^2}$

16. (a) $V(3) = \dfrac{4}{3}\pi(3)^3 = 36\pi$

(b) $V(0) = \dfrac{4}{3}\pi(0)^3 = 0$

(c) $V\left(\dfrac{3}{2}\right) = \dfrac{4}{3}\pi\left(\dfrac{3}{2}\right)^3 = \dfrac{9\pi}{2}$

(d) $V(2r) = \dfrac{4}{3}\pi(2r)^3 = \dfrac{32\pi r^3}{3}$

18. (a) $f(-8) = \sqrt{-8 + 8} + 2 = 2$

(b) $f(1) = \sqrt{1 + 8} + 2 = 5$

(c) $f(x - 8) = \sqrt{x - 8 + 8} + 2 = \sqrt{x} + 2$

(d) $f(h + 8) = \sqrt{h + 8 + 8} + 2 = \sqrt{h + 16} + 2$

20. (a) $q(2) = \dfrac{2(2)^2 + 3}{(2)^2} = \dfrac{11}{4}$

(b) $q(0) = \dfrac{2(0)^2 + 3}{(0)^2}$ is undefined.

(c) $q(x) = \dfrac{2x^2 + 3}{(x^2)}$

(d) $q(-x) = \dfrac{2(-x)^2 + 3}{(-x)^2} = \dfrac{2x^2 + 3}{x^2}$

22. (a) $f(-2) = (-2)^2 + 2 = 6$

(b) $f(0) = (0)^2 + 2 = 2$

(c) $f(1) = (1)^2 + 2 = 3$

(d) $f(2) = 2(2)^2 + 2 = 10$

24. All real numbers t

26. $1 - x^2 \geq 0$

$-x^2 \geq -1$

$x^2 \leq 1$

$-1 \leq x \leq 1$

Therefore, the domain is $-1 \leq x \leq 1$.

28. Since $x^2 - 2x \neq 0$, $x \neq 0$, 2. Therefore, the domain is all real numbers except $x = 0$, 2.

30. Since $s - 4 \neq 0$, $s \neq 4$.
Since $s - 1 \geq 0$, $s \geq 1$.
Therefore, the domain is $s \geq 1$ except $s = 4$.

32. $\{(-2, f(-2)), (-1, f(-1)), (0, f(0)), (1, f(1)), (2, f(2))\}$

$\left\{\left(-2, -\tfrac{4}{5}\right), (-1, -1), (0, 0), (1, 1), \left(2, \tfrac{4}{5}\right)\right\}$

34. $\{(-2, f(-2)), (-1, f(-1)), (0, f(0)), (1, f(1)), (2, f(2))\}$

$\{(-2, 1), (-1, 0), (0, 1), (1, 2), (2, 3)\}$

36. (a) $\sqrt{232} = 15.232$

(b) $\sqrt[3]{2500} = 13.572$

38. (a) $|-326.8| = 326.800$

(b) $10^{3.8} = 6309.573$

40. The data fits the linear model $y = cx$, with $c = \tfrac{1}{4}$.

42. The data fits the model $y = c\sqrt{|x|}$, with $c = 3$.

44.
$$f(x) = 5x - x^2$$
$$f(5 + h) = 5(5 + h) - (5 + h)^2$$
$$= 25 + 5h - (25 + 10h + h^2)$$
$$= -h^2 - 5h$$
$$f(5) = 5(5) - (5)^2 = 0$$
$$f(5 + h) - f(5) = -h^2 - 5h$$
$$\frac{f(5 + h) - f(5)}{h} = -h - 5, \; h \neq 0$$

46.
$$f(t) = \frac{1}{t}$$
$$f(1) = 1$$
$$f(t) - f(1) = \frac{1}{t} - 1 = \frac{1 - t}{t}$$
$$\frac{f(t) - f(1)}{t - 1} = \frac{(1 - t)/t}{t - 1} = \frac{1 - t}{t} \cdot \frac{1}{t - 1}$$
$$= \frac{-(t - 1)}{t} \cdot \frac{1}{t - 1} = -\frac{1}{t}, \; t \neq 1$$

48. $A = \frac{1}{2}sh,\ \ h^2 + \left(\frac{s}{2}\right)^2 = s^2$

$h^2 = \frac{3s^2}{4}$

$h = \frac{\sqrt{3}s}{2}$

$A = \frac{1}{2}s\left(\frac{\sqrt{3}}{2}s\right) = \frac{\sqrt{3}}{4}s^2$

50. $A = l \cdot w = 2xy$

Since $y = \sqrt{25 - x^2}$, $A = 2x\sqrt{25 - x^2}$. Since $A > 0$, $x > 0$ and $25 - x^2 > 0$. Therefore, the domain is $0 < x < 5$.

52. $V = l \cdot w \cdot h = (12 - 2x)(12 - 2x)(x)$

$= 4x(6 - x)^2$

$V > 0$	$l > 0$
$4x(6-x)^2 > 0$	$12 - 2x > 0$
$x > 0$	$-2x > -12$
	$x < 6$

Therefore, the domain is $0 < x < 6$.

54. $1978:\ p(-2) = 19{,}503.6 + 1753.6(-2)$

$= 15{,}996.4$

Therefore, the average price of a mobile home in 1978 was \$15,996.

$1988:\ p(8) = 19{,}838.8 + 81.11(8)^2$

$= 25{,}029.84$

Therefore, the average price of a mobile home in 1988 was \$25,030.

56. (a) Revenue = (Number of people)(Rate per person)

$R = n(\text{Rate})$

$R = n[8 - 0.05(n - 80)],\ n \geq 80$

$R = 8n - 0.05n^2 + 4n$

$R = 12n - 0.05n^2,\ n \geq 80$

(b)

n	90	100	110	120	130	140	150
$R(n)$	675	700	715	720	715	700	675

(c)

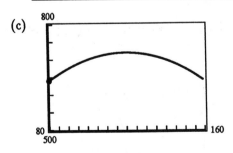

A maximum revenue of $R = 720$ is produced when $n = 120$.

1.4 | Graphs of Functions

2. From the graph we see that the x-values are all real numbers. Therefore, the domain is $(-\infty, \infty)$. Similarly, the y-values are less than or equal to 4. Therefore, the range is $(-\infty, 4]$.

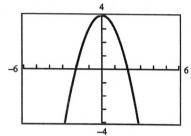

4. From the graph we see that the x-values are all real numbers. Therefore, the domain is $(-\infty, \infty)$. Similarly, the y-values are greater than or equal to 0. Therefore, the range is $[0, \infty)$.

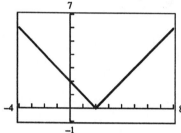

6. From the graph we see that the x-values are less than 2 and greater than 2. Therefore, the domain is $(-\infty, 2)$, $(2, \infty)$. Similarly, the y-values are -1 and 1. Therefore, the range is -1 and 1.

8. Since no vertical line would ever cross the graph more than one time, y is a function of x.

```
RANGE
Xmin=-3
Xmax=3
Xscl=1
Ymin=-3
Ymax=1
Yscl=1
```

10. Some vertical lines cross the graph more than once. For example, the vertical line $x = 0$ crosses the graph at $(0, \ 3)$ and $(0, \ -3)$. Therefore, y is not a function of x.

```
RANGE
Xmin=-6
Xmax=6
Xscl=1
Ymin=-4
Ymax=4
Yscl=1
```

12. Some vertical lines cross the graph more than once. For example, the vertical line $x = 1$ crosses the graph at $(1, \ 1)$ and $(1, \ -1)$. Therefore, y is not a function of x.

```
RANGE
Xmin=-1
Xmax=4
Xscl=1
Ymin=-2
Ymax=2
Yscl=1
```

14. $f(x) = 6[x - (0.1x)^5]$

(c) shows the most complete graph.

16. $f(x) = 10x\sqrt{400 - x^2}$

(c) shows the most complete graph.

18. By its graph we see that f is decreasing on $(-\infty, 1)$ and is increasing on $(1, \infty)$.

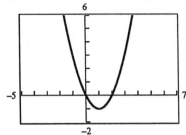

20. By its graph we see that f is decreasing on $(-\infty, -2)$ and is increasing on $(2, \infty)$.

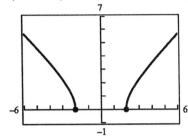

22. By its graph we see that f is decreasing on $(-\infty, 0)$ and is increasing on $(0, \infty)$.

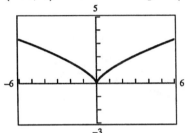

24. By its graph we see that f is decreasing on $(-\infty, -1)$, is constant on $(-1, 1)$, and increasing on $(1, \infty)$.

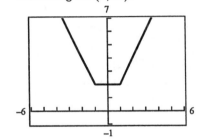

26. Using the zoom and trace features, we find that $(-1.00, 4.00)$ is a relative maximum, and $(1.00, 0.00)$ is a relative minimum.

28. Using the zoom and trace features, we find that $(0.00, 15.00)$ is a relative maximum, and $(4.00, -17.00)$ is a relative minimum.

30. Using the zoom and trace features, we find that $(2.67, 3.08)$ is a relative maximum. (The point $(4, 0)$ is not really a relative minimum.)

32. (a) The cost underwater is \$8 per foot times 5280 feet per mile, times $\sqrt{x^2 + \frac{1}{4}}$ miles. The cost overland is \$6 per foot times 5280 feet per mile, times $(6 - x)$ miles. Hence, the total cost is the sum
$$T = 8(5280)\sqrt{x^2 + \tfrac{1}{4}} + 6(5280)(6 - x).$$

(b)

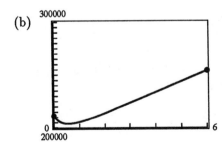

(c) Using the zoom and trace features, the value of x which minimizes the cost is $x = 0.57$.

34. Assume that $p = ax + b$. Then

$$25 = 800a + b$$
$$30 = 775a + b.$$

Hence, $5 = -25a$, or $a = -\frac{1}{5}$, and $b = 185$. Thus, $p = -\frac{1}{5}x + 185$, and $R = xp = x(-\frac{1}{5}x + 185) = -\frac{1}{5}x^2 + 185x$.

For $x = 462.5$ and $p = 92.5$, the revenue is maximized.

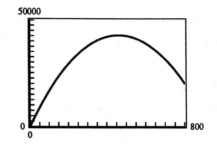

36. $f(x) = \begin{cases} \sqrt{4+x}, & x < 0 \\ \sqrt{4-x}, & x \geq 0 \end{cases}$

38. $f(x) = \begin{cases} 1 - (x-1)^3, & x \leq 2 \\ \sqrt{x-2}, & x > 2 \end{cases}$

40. $g(x) = 6 - [\![x]\!]$

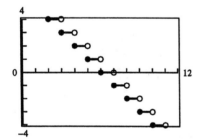

(Hint: use dot mode, not connected)
Domain: all x

Range: $\{\ldots, -2, -1, 0, 1, 2, \ldots\}$

42. (a) $C = 9.80 + 2.50[\![x]\!]$

(b)

(c) \$37.30 for a package weighing 12 pounds

44. (a) $(3, 7)$
   (b) $(3, -7)$

46. (a) $\left(-\frac{3}{4}, -\frac{7}{8}\right)$
   (b) $\left(-\frac{3}{4}, \frac{7}{8}\right)$

48. $g(x) = x$

$g(-x) = -x = -g(x)$

g is odd.

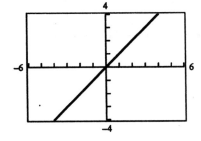

50. $h(x) = x^2 - 4$

$h(-x) = (-x)^2 - 4$

$\qquad = x^2 - 4 = h(x)$

h is even.

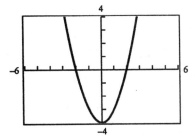

52. $f(t) = -t^4$

$f(-t) = -(-t)^4$

$\qquad = -t^4 = f(t)$

f is even.

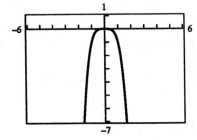

54. $f(x) = x^2 \sqrt[3]{x}$

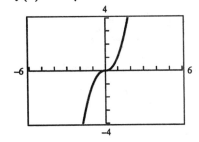

$$f(x) = x^2 \sqrt[3]{x}$$
$$f(-x) = (-x)^2 \sqrt[3]{-x}$$
$$\qquad = x^2 (-\sqrt[3]{x})$$
$$\qquad = -f(x)$$

f is odd.

56. $f(x) = |x + 2|$

$f(-x) = |-x + 2|$

$\neq f(x) \neq -f(x)$

f is neither odd nor even.

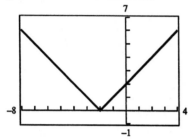

58. $h(x) = x^3 - 5$

$h(-x) = (-x)^3 - 5$

$= -x^3 - 5$

$\neq -h(x) \neq h(x)$

h is neither even nor odd.

60. $f(x) = x\sqrt{1 - x^2}$

$f(-x) = -x\sqrt{1 - (-x)^2}$

$= -x\sqrt{1 - x^2}$

$= -f(x)$

f is odd.

62. $g(s) = 4s^{2/3}$

$g(-s) = 4(-s)^{2/3}$

$= 4s^{2/3}$

$= g(s)$

g is even.

64. $f(x) \geq 0$

$x^2 - 4x \geq 0$

$x(x - 4) \geq 0$

$x \leq 0$ or $x \geq 4$

$f(x) \geq 0$ on the intervals
$(-\infty, 0]$, $[4, \infty)$.

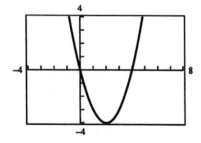

66. $f(x) \geq 0$

$\sqrt{x + 2} \geq 0$

$x + 2 \geq 0$

$x \geq -2$

$f(x) \geq 0$ on the interval
$[-2, \infty)$.

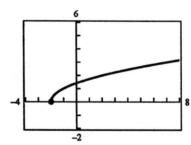

68.
$$f(x) \geq 0$$
$$-(1 + |x|) \geq 0$$
$$1 + |x| \leq 0$$
$$|x| \leq -1$$

$f(x) < 0$ for all x.

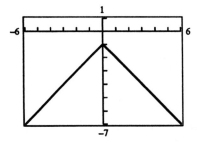

70. $h = $ top $-$ bottom
$$h = 4 - (4x - x^2)$$

72. $h = $ top $-$ bottom
$$h = 4 - \sqrt{2x}$$

74. $L = $ right $-$ left
$$L = 2 - \sqrt{y}$$

76. $f(x) = a_{2n+1}x^{2n+1} + a_{2n-1}x^{2n-1} + \cdots + a_3x^3 + a_1x$

$f(-x) = a_{2n+1}(-x)^{2n+1} + a_{2n-1}(-x)^{2n-1} + \cdots + a_3(-x)^3 + a_1(-x)$

$\qquad = -a_{2n+1}x^{2n+1} - a_{2n-1}x^{2n-1} - \cdots - a_3x^3 - a_1x$

$\qquad = -f(x)$

Therefore, $f(x)$ is odd.

1.5 | Shifting, Reflecting, and Stretching Graphs

2.

4.

6.

8.

10. Since the vertex of g is at $(-1, 0)$, and g opens downward, $g(x) = -(x+1)^2$.
Since the vertex of h is at $(-2, -3)$, $h(x) = (x+2)^2 - 3$.

12. Since the vertex of g is $(100, 100)$, and g opens downward, $g(x) = -(x-100)^2 + 100$.
Since the vertex of h is at $(25, 50)$, $h(x) = (x-25)^2 + 50$.

14. $f(x) = x^3 - 3x^2 + 2$
$g(x) = f(x-1) = (x-1)^3 - 3(x-1)^2 + 2$
(horizontal shift 1 unit to right)
$h(x) = f(2x) = (2x)^3 - 3(2x)^2 + 2$
(shrink horizontally by factor of 2)

16. $f(x) = x^3 - 3x^2 + 2$
$g(x) = -f(x) = -(x^3 - 3x^2 + 2)$
(reflection in the x-axis)
$h(x) = f(-x) = (-x)^3 - 3(-x)^2 + 2$
(reflection in the y-axis)

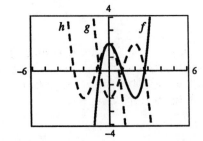

18. The graph of g is obtained from that of f by first shifting horizontally two units to the right, and then vertically upward one unit. Hence, $g(x) = (x-2)^3 - 3(x-2)^2 + 1$.

20. $y = -\sqrt{x}$ is $f(x)$ reflected in the y-axis.

22. $y = \sqrt{x+3}$ is $f(x)$ shifted left three units.

24. $y = \sqrt{-x}$ is $f(x)$ reflected in the y-axis.

26. $y = \sqrt[3]{x+1}$ is $f(x)$ shifted left one unit.

28. $y = \sqrt[3]{x-2}$ is $f(x)$ shifted right one unit and reflected in the x-axis.

30. $y = \frac{1}{2}\sqrt[3]{x}$ is a vertical shrink of f by $\frac{1}{2}$.

32. $f(x+2)$

34. $f(x) - 1$

36. $\frac{1}{2}f(x)$

38. $g(x) = (x-4)^3$ is obtained from $f(x)$ by a horizontal shift of 4 units to the right.

40. $h(x) = -2(x-1)^3 + 3$ is obtained from $f(x)$ by a right shift of one unit, a vertical stretch by a factor of two, a reflection in the x-axis, and a vertical shift three units upward.

42. $p(x) = [3(x-2)]^3$ is obtained from $f(x)$ by a right shift of two units, followed by a horizontal stretch.

44. (a) $H(x) = 0.002x^2 + 0.005x$
$- 0.029, \ 10 \leq x \leq 100$

(b) $H(x) = 0.002(1.6x)^2 + 0.005(1.6x) - 0.029$

$= 0.00512x^2 + 0.008x - 0.029$

This is a horizontal shrink.

46. These three functions are odd. As the exponents increase, the graphs become flatter in the interval $(-1, 1)$, while they grow more rapidly as x tends to infinity.

48. $f(x) = (x-3)^3$

$g(x) = (x+1)^2$

$h(x) = (x-4)^5$

50. $f(x) = x^3(x-6)^2$

52. $f(x) = x^3(x-6)^3$

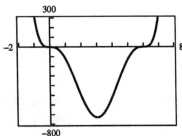

1.6 Combinations of Functions

2. (a) $(f+g)(x) = f(x) + g(x) = (2x-5) + 5 = 2x$

(b) $(f-g)(x) = f(x) - g(x) = (2x-5) - 5 = 2x - 10$

(c) $(fg)(x) = f(x) \cdot g(x) = (2x-5) \cdot 5 = 10x - 25$

(d) $\left(\dfrac{f}{g}\right)(x) = \dfrac{f(x)}{g(x)} = \dfrac{2x-5}{5} = \dfrac{2}{5}x - 1, \quad -\infty < x < \infty$

The domain of f/g is $(-\infty, \infty)$.

4. (a) $(f+g)(x) = f(x) + g(x) = \sqrt{x^2 - 4} + \dfrac{x^2}{x^2 + 1}$

(b) $(f-g)(x) = f(x) - g(x) = \sqrt{x^2 - 4} - \dfrac{x^2}{x^2 + 1}$

(c) $(fg)(x) = f(x) \cdot g(x) = \dfrac{x^2\sqrt{x^2 - 4}}{x^2 + 1}$

(d) $\left(\dfrac{f}{g}\right)(x) = \dfrac{f(x)}{g(x)} = \dfrac{(x^2 + 1)\sqrt{x^2 - 4}}{x^2},$

$|x| \geq 2$

The domain of f/g is $(-\infty, -2), (2, \infty)$.

6. (a) $(f+g)(x) = f(x) + g(x)$

$= \dfrac{x}{x+1} + x^3 = \dfrac{x^4 + x^3 + x}{x + 1}$

(b) $(f-g)(x) = f(x) - g(x)$

$= \dfrac{x}{x+1} - x^3 = \dfrac{-x^4 - x^3 + x}{x + 1}$

(c) $(fg)(x) = f(x) \cdot g(x) = \dfrac{x^4}{x + 1}$

(d) $\left(\dfrac{f}{g}\right)(x) = \dfrac{f(x)}{g(x)}$

$= \dfrac{x}{x+1}\left(\dfrac{1}{x^3}\right)$

$= \dfrac{1}{x^2(x+1)}, \quad x \neq 0, -1$

The domain of f/g is
$(-\infty, -1), (-1, 0), (0, \infty)$.

8. $(f-g)(-2) = f(-2) - g(-2)$

$= [(-2)^2 + 1] - (-2 - 4)$

$= 5 + 6$

$= 11$

10. $(f+g)(t-1) = f(t-1) + g(t-1)$

$= [(t-1)^2 + 1] + [(t-1) - 4]$

$= t^2 - 2t + 1 + 1 + t - 1 - 4$

$= t^2 - t - 3$

12. $(fg)(-6) = f(-6)g(-6)$

$= [(-6)^2 + 1][-6 - 4]$

$= (36 + 1)(-10)$

$= -370$

14. $\left(\dfrac{f}{g}\right)(0) = \dfrac{f(0)}{g(0)}$

$= \dfrac{0^2 + 1}{0 - 4}$

$= \dfrac{1}{-4}$

$= -\dfrac{1}{4}$

16. $(2f)(5) = 2f(5)$

$= 2(5^2 + 1)$

$= 2(26)$

$= 52$

18.

20.

22.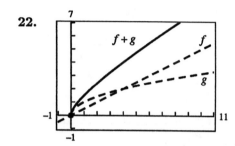

24. $R_1 = 500 - 0.8t^2$, $R_2 = 250 + 0.78t$
$R = R_1 + R_2 = 750 + 0.78t - 0.8t^2$

The total sales have been decreasing.

When $0 \le x \le 2$, g contributes the most.
When $x > 5$, f contributes the most.

26. (a) $(f \circ g)(x) = f(g(x)) = f(x^3 + 1) = \sqrt[3]{(x^3 + 1) - 1} = \sqrt[3]{x^3} = x$

(b) $(g \circ f)(x) = g(f(x)) = g\left(\sqrt[3]{x - 1}\right) = \left(\sqrt[3]{x - 1}\right)^3 + 1 = x - 1 + 1 = x$

(c) $(f \circ f)(x) = f(f(x)) = f\left(\sqrt[3]{x - 1}\right) = \sqrt[3]{\sqrt[3]{x - 1} - 1}$

28. (a) $(f \circ g)(x) = f(g(x)) = f\left(\dfrac{1}{x}\right) = \left(\dfrac{1}{x}\right)^3 = \dfrac{1}{x^3}$

(b) $(g \circ f)(x) = g(f(x)) = g(x^3) = \dfrac{1}{x^3}$

(c) $(f \circ f)(x) = f(f(x)) = f(x^3) = (x^3)^3 = x^9$

30. (a) $(f \circ g)(x) = f(x^5 - 2) = \sqrt[5]{x^5 - 2 + 1} = \sqrt[5]{x^5 - 1}$

(b) $(g \circ f)(x) = g(\sqrt[5]{x + 1}) = [\sqrt[5]{x + 1}]^5 - 2 = x + 1 - 2 = x - 1$

32. (a) $f \circ g = f(g(x)) = f(\sqrt{x}) = \sqrt{\sqrt{x}} = \sqrt[4]{x}$

(b) $g \circ f = g(f(x)) = g(\sqrt{x}) = \sqrt{\sqrt{x}} = \sqrt[4]{x}$

34. (a) $f \circ g = f(g(x)) = f(x^6) = (x^6)^{2/3} = x^4$

(b) $g \circ f = g(f(x)) = g(x^{2/3}) = (x^{2/3})^6 = x^4$

36. (a) $(f - g)(1) = f(1) - g(1) = 2 - 3 = -1$

(b) $(fg)(4) = f(4)g(4) = 4(0) = 0$

38. (a) $(f \circ g)(1) = f(g(1)) = f(3) = 2$

(b) $(g \circ f)(3) = g(f(3)) = g(2) = 2$

40. Let $f(x) = x^3$ and $g(x) = 1 - x$,
then $(f \circ g)(x) = h(x)$.

42. Let $f(x) = \sqrt{x}$ and $g(x) = 9 - x$,
then $(f \circ g)(x) = h(x)$.

44. Let $f(x) = \dfrac{4}{x^2}$ and $g(x) = 5x + 2$,
then $(f \circ g)(x) = h(x)$.

46. Let $f(x) = x^{3/2}$ and $g(x) = x + 3$, then $(f \circ g)(x) = h(x)$.

48. (a) The domain of $f(x) = 1/x$ is all real numbers except $x = 0$.
(b) The domain of $g(x) = x + 3$ is all real numbers.
(c) $f \circ g = f(g(x)) = f(x + 3) = 1/(x + 3)$; The domain of $f \circ g = 1/(x + 3)$ is all real numbers except $x = -3$.

50. (a) The domain of $f(x) = 2x + 3$ is all real numbers.
(b) The domain of $g(x) = x/2$ is all real numbers.
(c) $(f \circ g)(x) = f(g(x)) = f(x/2) = 2(x/2) + 3 = x + 3$

The domain of $f \circ g = x + 3$ is all real numbers.

52. (a) $r(x) = \dfrac{x}{2}$

(b) $A(r) = \pi r^2$

(c) $(A \circ r)(x) = A(r(x))$

$$= A\left(\frac{x}{2}\right) = \pi\left(\frac{x}{2}\right)^2 = \frac{1}{4}\pi x^2$$

$A \circ r$ represents the area of the circular base of the tank with radius $x/2$.

54. $x = 150 \text{ miles} - (450 \text{ mph})(t \text{ hours})$

$y = 200 \text{ miles} - (450 \text{ mph})(t \text{ hours})$

$s = \sqrt{x^2 + y^2}$

$$= \sqrt{(150 - 450t)^2 + (200 - 450t)^2}$$

56. Let $f(x)$ and $g(x)$ be two odd functions and define $h(x) = f(x)g(x)$. Then

$$h(-x) = f(-x)g(-x)$$
$$= [-f(x)][-g(x)]$$
$$= f(x)g(x)$$
$$= h(x).$$

Thus, h is even.

Let $f(x)$ and $g(x)$ be two even functions and define $h(x) = f(x)g(x)$. Then
$$h(-x) = f(-x)g(-x)$$
$$= f(x)g(x)$$
$$= h(x).$$

Thus, h is even.

58. $g(x) = \frac{1}{2}[f(x) + f(-x)]$

$g(-x) = \frac{1}{2}[f(-x) + f(-(-x))]$
$$= \frac{1}{2}[f(-x) + f(x)]$$
$$= g(x), \text{ even}$$

$h(x) = \frac{1}{2}[f(x) - f(-x)]$

$h(-x) = \frac{1}{2}[f(-x) - f(-(-x))]$
$$= \frac{1}{2}[f(-x) - f(x)]$$
$$= -h(x), \text{ odd}$$

1.7   Inverse Functions

60. (a) $f(x) = x^2 - 2x + 1$

$$= \frac{1}{2}[(x^2 - 2x + 1) + (x^2 + 2x + 1)] + \frac{1}{2}[(x^2 - 2x + 1) - (x^2 + 2x + 1)]$$

$$= \frac{1}{2}[2x^2 + 2] + \frac{1}{2}[-4x]$$

$$= (x^2 + 1) + (-2x)$$

(b) $f(x) = \dfrac{1}{x+1}$

$$= \frac{1}{2}\left[\frac{1}{x+1} + \frac{1}{-x+1}\right] + \frac{1}{2}\left[\frac{1}{x+1} - \frac{1}{-x+1}\right]$$

$$= \frac{1}{2}\left[\frac{2}{(x+1)(-x+1)}\right] + \frac{1}{2}\left[\frac{-2x}{(x+1)(-x+1)}\right]$$

$$= \frac{1}{1-x^2} + \frac{-x}{1-x^2}$$

2. To "undo" $\frac{1}{5}$ times x, let f^{-1} multiply x times 5.

$$f^{-1}(x) = 5x$$

Now, we have

$$f(f^{-1}(x)) = f(5x) \quad \text{and} \quad f^{-1}(f(x)) = f^{-1}\left(\tfrac{1}{5}x\right)$$
$$= \tfrac{1}{5}(5x) \qquad\qquad\qquad = 5\left(\tfrac{1}{5}x\right)$$
$$= x \qquad\qquad\qquad\qquad\quad = x.$$

4. To "undo" x minus 5, let f^{-1} equal x plus 5.

$$f^{-1}(x) = x + 5$$

Now, we have

$$f(f^{-1}(x)) = f(x+5)$$
$$= x + 5 - 5$$
$$= x$$

and

$$f^{-1}(f(x)) = f^{-1}(x-5)$$
$$= x - 5 + 5$$
$$= x.$$

6. To "undo" x to the fifth power, we let f^{-1} equal the fifth root of x.

$$f^{-1}(x) = \sqrt[5]{x}$$

Now, we have

$$f(f^{-1}(x)) = f\left(\sqrt[5]{x}\right) \quad \text{and} \quad f^{-1}(f(x)) = f^{-1}(x^5)$$
$$= \left(\sqrt[5]{x}\right)^5 \qquad\qquad\qquad\qquad = \sqrt[5]{x^5}$$
$$= x \qquad\qquad\qquad\qquad\qquad\quad = x.$$

8. (a) $f(g(x)) = f(x+5) = (x+5) - 5 = x$

$$g(f(x)) = g(x-5) = (x-5) + 5 = x$$

(b)

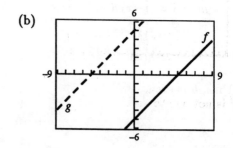

10. (a) $f(g(x)) = f\left(\dfrac{3-x}{4}\right) = 3 - 4\left(\dfrac{3-x}{4}\right) = x$

$g(f(x)) = g(3 - 4x) = \dfrac{3 - (3 - 4x)}{4} = x$

(b)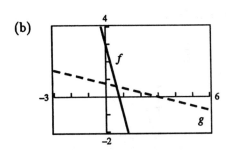

12. (a) $f(g(x)) = f\left(\dfrac{1}{x}\right) = \dfrac{1}{1/x} = x$

$g(f(x)) = g\left(\dfrac{1}{x}\right) = \dfrac{1}{1/x} = x$

(b)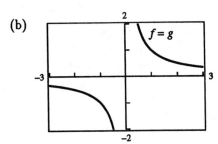

14. (a) $f(g(x)) = f\left(\sqrt[3]{1-x}\right) = 1 - \left(\sqrt[3]{1-x}\right)^3 = x$

$g(f(x)) = g(1 - x^3) = \sqrt[3]{1 - (1 - x^3)} = x$

(b)

16. (a) $f(g(x)) = f\left(\dfrac{1-x}{x}\right),\ 0 < x \le 1$

$= \dfrac{1}{1 + (1-x)/x} = \dfrac{x}{x + 1 - x} = x$

$g(f(x)) = g\left(\dfrac{1}{1+x}\right),\ x \ge 0$

$= \dfrac{1 - 1/(1+x)}{1/(1+x)} = \dfrac{1 + x - 1}{1} = x$

(b)

18. $f(x) = 10$

f is not one-to-one.

20. $g(x) = (x+5)^3$

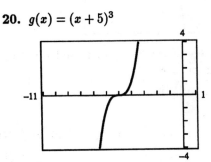

g is one-to-one.

22. $f(x) = \frac{1}{8}(x+2)^2 - 1$

f is not one-to-one.

24.
$$f(x) = 3x$$
$$y = 3x$$
$$x = 3y$$
$$y = \frac{x}{3}$$
$$f^{-1}(x) = \frac{x}{3}$$

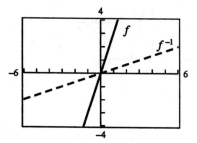

26.
$$f(x) = x^3 + 1$$
$$y = x^3 + 1$$
$$x = y^3 + 1$$
$$y^3 = x - 1$$
$$y = \sqrt[3]{x - 1}$$
$$f^{-1}(x) = \sqrt[3]{x - 1}$$

28.
$$f(x) = x^2, \quad x \geq 0$$
$$y = x^2$$
$$x = y^2$$
$$y = \sqrt{x}$$
$$f^{-1}(x) = \sqrt{x}$$

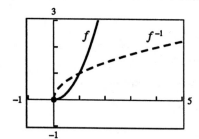

30. $f(x) = \dfrac{4}{x}$

$y = \dfrac{4}{x}$

$x = \dfrac{4}{y}$

$xy = 4$

$y = \dfrac{4}{x}$

$f^{-1}(x) = \dfrac{4}{x}$

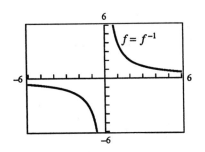

32. $f(x) = x^{3/5}$

$y = x^{3/5}$

$x = y^{3/5}$

$y = x^{5/3}$

$f^{-1}(x) = x^{5/3}$

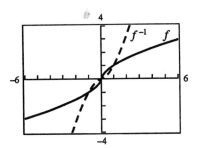

34. Since $f(2) = \frac{1}{4}$ and $f(-2) = \frac{1}{4}$, f is not one-to-one and does not have an inverse.

36. Since $f(0) = -3$ and $f(2) = -3$, f is not one-to-one and does not have an inverse.

38. Since $q(9) = 16$ and $q(1) = 16, q$ is not one-to-one and does not have an inverse.

40. $\quad f(a) = f(b)$

$|a - 2| = |b - 2|$

$a - 2 = b - 2$, since $a, b \leq 2$

$a = b$

Therefore, f is one-to-one.

$f(x) = |x - 2|, \ x \leq 2$

$y = -(x - 2)$

$x = -(y - 2)$

$y - 2 = -x$

$y = 2 - x$

$f^{-1}(x) = 2 - x, \ x \geq 0$

42. $\quad f(a) = f(b)$

$\sqrt{a - 2} = \sqrt{b - 2}$

$a - 2 = b - 2$

$a = b$

Therefore, f is one-to-one.

$f(x) = \sqrt{x - 2}$

$y = \sqrt{x - 2}$

$x = \sqrt{y - 2}$

$y - 2 = x^2$

$y = x^2 + 2$

$f^{-1}(x) = x^2 + 2, \ x \geq 0$

44. Since $f(1) = \frac{1}{2}$ and $f(-1) = \frac{1}{2}$, f is not one-to-one and does not have an inverse.

46.
$$f(c) = f(d)$$
$$ac + b = ad + b$$
$$ac = ad$$
$$c = d$$

Therefore, f is one-to-one.
$$f(x) = ax + b, \ a \neq 0$$
$$y = ax + b$$
$$x = ay + b$$
$$ay = x - b$$
$$y = \frac{x - b}{a}$$
$$f^{-1}(x) = \frac{x - b}{a}$$

$$f(x) = 16 - x^4, \ x \geq 0$$
$$y = 16 - x^4$$
$$x = 16 - y^4$$
$$y^4 = 16 - x$$
$$y = \sqrt[4]{16 - x}$$
$$f^{-1}(x) = \sqrt[4]{16 - x}, \ x \leq 16$$

48. Delete the part corresponding to $x < 0$.

$$f(x) = |x - 3|, \ x \geq 3$$
$$f(x) = x - 3$$
$$y = x - 3$$
$$x = y - 3$$
$$y = x + 3$$
$$f^{-1}(x) = x + 3, \ x \geq 0$$

50. Delete the part corresponding to $x < 3$.

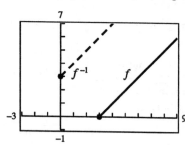

52.

x	0	2	4	6
$f^{-1}(x)$	6	2	0	−2

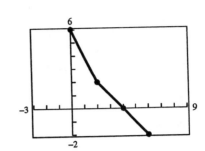

In Exercises 54 and 56, $f(x) = \frac{1}{8}x - 3$, $f^{-1}(x) = 8(x+3)$, $g(x) = x^3$, $g^{-1}(x) = \sqrt[3]{x}$.

54. $(g^{-1} \circ f^{-1})(-3) = g^{-1}(f^{-1}(-3))$
$$= g^{-1}\left[8\left(-3 + 3\right)\right]$$
$$= \sqrt[3]{8(-3+3)}$$
$$= 0$$

56. $(g^{-1} \circ g^{-1})(-4) = g^{-1}(g^{-1}(-4))$
$$= g^{-1}\left(\sqrt[3]{-4}\right)$$
$$= \left[(-4)^{1/3}\right]^{1/3}$$
$$= -\sqrt[9]{4}$$

In Exercises 58 and 60, $f(x) = x + 4$, $f^{-1}(x) = x - 4$, $g(x) = 2x - 5$, $g^{-1}(x) = \frac{1}{2}(x+5)$.

58. $(f^{-1} \circ g^{-1})(x) = f^{-1}\left(\frac{1}{2}(x+5)\right)$
$$= \frac{1}{2}(x+5) - 4$$
$$= \frac{1}{2}x - \frac{3}{2}$$

60. $(g \circ f)(x) = g(x+4) = 2(x+4) - 5 = 2x + 3$

Hence,
$$(g \circ f)^{-1}(x) = \frac{x-3}{2} = \frac{1}{2}x - \frac{3}{2}.$$

Alternatively, observe that
$$(g \circ f)^{-1}(x) = (f^{-1} \circ g^{-1})(x) = \frac{1}{2}x - \frac{3}{2}.$$
(Exercise 58)

62. When $y = 500$, we have
$$500 = 0.03x^2 + 254.50$$
$$0.03x^2 = 245.5$$
$$x^2 = \frac{24,550}{3} = 8183.33$$
$$x \approx 90.5\%.$$

64. If the inverse of f exists, then we know that f is one-to-one. Since f is one-to-one, if the point $(0, a)$ lies on the graph of f, the point $(a, 0)$ lies on the graph of f^{-1}. Therefore, the y-intercept of f is the x-intercept of f^{-1}. The statement is true.

66. The statement is false. For example, $f(x) = x$ and $g(x) = x$ are inverses of each other and $f(x) = g(x)$.

Review Exercises for Chapter 1

2. $3x + 2y + 6 = 0$

4. $y = 8 - |x|$

6. $y = \sqrt{x + 2}$

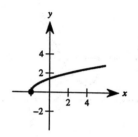

8. $y = x^2 - 4x$

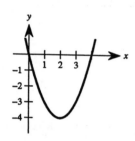

10. $x^2 + y^2 = 10$

12. $y = 4 - (x - 4)^2$

Intersects the x-axis twice and the y-axis once

14. $y = \frac{1}{4}x^3 - 3x$

Intersects the x-axis three times and the y-axis once

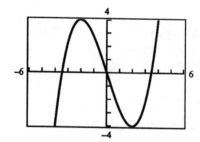

16. $y = x\sqrt{x+3}$

Intersects the x-axis twice and the y-axis once

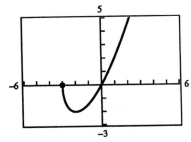

18. $y = |x+2| + |3-x|$

Does not intersect the x-axis; intersects the y-axis once

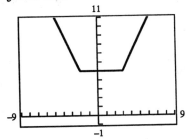

20. $(x+2)^2 + y^2 = 16$

$$y = \pm\sqrt{16 - (x+2)^2}$$

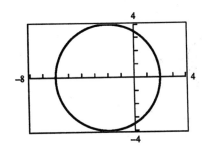

22. The line through $(-6,\ 1)$ and $(10,\ 5)$ is:

$$y - 1 = \frac{5-1}{10+6}(x+6)$$

$$y - 1 = \frac{1}{4}(x+6)$$

$$4y - 4 = x + 6$$

$$x - 4y = -10$$

For $(1,\ t)$ to be on this line also, it must satisfy the equation $x - 4y = -10$.

$$(1) - 4(t) = -10$$

$$-4t = -11$$

Thus, $t = \frac{11}{4}$.

24. The line through $(-3,\ 3)$ and $(8,\ 6)$ is:

$$y - 3 = \frac{6-3}{8+3}(x+3)$$

$$y - 3 = \frac{3}{11}(x+3)$$

$$11y - 33 = 3(x+3)$$

$$11y - 33 = 3x + 9$$

$$3x - 11y = -42.$$

For $(t,\ -1)$ to be on this line also, it must satisfy the equation $3x - 11y = -42$.

$$3(t) - 11(-1) = -42$$

$$3t = -53$$

Thus, $t = -\frac{53}{3}$.

26. $y - 0 = \dfrac{4-0}{-1-2}(x-2) = \dfrac{4}{-3}(x-2) \Rightarrow$

$$y = -\frac{4}{3}x + \frac{8}{3}$$

28. $y - 2 = \dfrac{-10 - 2}{3 + 2}(x + 2) = -\dfrac{12}{5}(x + 2) \Rightarrow$

$\qquad y = -\dfrac{12}{5}x - \dfrac{14}{5}$

30. $y - 6 = \dfrac{2 - 6}{4 - 1}(x - 1) = -\dfrac{4}{3}(x - 1) \Rightarrow$

$\qquad y = -\dfrac{4}{3}x + \dfrac{22}{3}$

32. Since the slope is zero, the line is horizontal. Since the line passes through $(-2, 6)$, the equation is $y = 6$ or $y - 6 = 0$.

34. Since the slope is undefined, the line is **vertical**. Since the line passes through $(5, 4)$, the equation is $x = 5$ or $x - 5 = 0$.

36. $2x + 3y = 5$

$\qquad 3y = -2x + 5$

$\qquad y = -\frac{2}{3}x + \frac{5}{3} \;\Rightarrow\; m_1 = -\frac{2}{3}$

(a) Parallel slope $m_2 = -\frac{2}{3}$

$\qquad y - 3 = -\frac{2}{3}(x - (-8))$

$\qquad y - 3 = -\frac{2}{3}x - \frac{16}{3}$

$\qquad 3y - 9 = -2x - 16$

$\qquad 2x + 3y + 7 = 0$

(b) Perpendicular slope $m_2 = \frac{3}{2}$

$\qquad y - 3 = \frac{3}{2}(x - (-8))$

$\qquad y - 3 = \frac{3}{2}x + 12$

$\qquad 2y - 6 = 3x + 24$

$\qquad 3x - 2y + 30 = 0$

38. (a) $V = 3.75t + 85$, $t \geq 0$

(b)

(c) For 1995 we let $t = 5$, and then we have

$$V = 3.75(5) + 85$$
$$= \$103.75.$$

40. $g(x) = x^{4/3}$

(a) $g(8) = 8^{4/3} = 16$

(b) $g(t + 1) = (t + 1)^{4/3}$

(c) $\dfrac{g(8) - g(1)}{8 - 1} = \dfrac{16 - 1}{8 - 1} = \dfrac{15}{7}$

(d) $g(-x) = (-x)^{4/3} = x^{4/3}$

42. $f(t) = \sqrt[4]{t}$

(a) $f(16) = \sqrt[4]{16} = 2$

(b) $f(t + 5) = \sqrt[4]{t + 5}$

(c) $\dfrac{f(16) - f(0)}{16} = \dfrac{2 - 0}{16} = \dfrac{1}{8}$

(d) $f(t + h) = \sqrt[4]{t + h}$

44. (a) $f(x) = x^3 + c$

$c = -2: \quad f(x) = x^3 - 2$

$c = 0: \quad f(x) = x^3$

$c = 2: \quad f(x) = x^3 + 2$

(b) $f(x) = (x - c)^3$

$c = -2: \quad f(x) = (x + 2)^3$

$c = 0: \quad f(x) = x^3$

$c = 2: \quad f(x) = (x - 2)^3$

(c) $f(x) = (x - 2)^3 + c$

$c = -2: \quad (x - 2)^3 - 2$

$c = 0: \quad (x - 2)^3$

$c = 2: \quad (x - 2)^3 + 2$

46. The domain is all real numbers.

50. The domain is all real numbers.

48. $x^2 + 8x \geq 0$

$x(x + 8) \geq 0$

Test intervals: $(-\infty, -8) \Rightarrow x(x + 8) > 0$

$(-8, 0) \Rightarrow x(x + 8) < 0$

$(0, \infty) \Rightarrow x(x + 8) > 0$

Solution interval:

$(-\infty, -8] \cup [0, \infty)$

The domain is

$(-\infty, -8] \cup [0, \infty).$

52. (a) Total cost = variable costs + fixed costs

$$C(x) = 5.35x + 16{,}000$$

(b) Profit = Revenue − cost

$$P(x) = 8.20x - (5.35x + 16{,}000)$$

$$P(x) = 2.85x - 16{,}000$$

54. $C = 0.70 + 0.38[[x]]$

56. (a) Increasing on $(-2,\ 0)$ and $(2,\ \infty)$
Decreasing on $(-\infty,\ -2)$ and $(0,\ 2)$

(b) Relative maximum at $(0,\ 16)$
Relative minima at $(\pm 2,\ 0)$

(c) Since $f(-x) = ((-x)^2 - 4)^2 = (x^2 - 4)^2 = f(x)$, f is even.

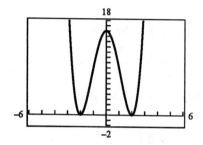

58. (a) Increasing on $(-\infty,\ -3)$ and $(-1,\ \infty)$
Decreasing on $(-3,\ -1)$

(b) Relative maximum at $(-3,\ 0)$
Relative minima at $(-1,\ -1.59)$

(c) g is neither even nor odd.

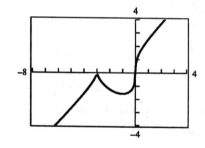

60. (a) $R = n\,[8 - 0.05(n - 80)]$, $n \geq 80$

(b) The number of persons to maximize the
revenue is $n = 120$, for a revenue of $720.00.

62. $f(x) = 5x - 7$

(a) $f^{-1}(x) = \dfrac{x + 7}{5}$

(b)
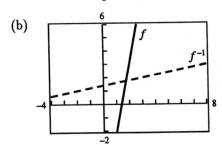

(c) $f^{-1}(f(x)) = f^{-1}(5x - 7)$

$$= \frac{5x - 7 + 7}{5} = x$$

$$f(f^{-1}(x)) = f\left(\frac{x + 7}{5}\right)$$

$$= 5\left(\frac{x + 7}{5}\right) - 7 = x$$

64. $f(x) = x^3 + 2$

(a) $f^{-1}(x) = \sqrt[3]{x - 2}$

(b)

(c) $f^{-1}(f(x)) = f^{-1}(x^3 + 2)$

$$= \sqrt[3]{x^3 + 2 - 2} = x$$

$$f(f^{-1}(x)) = f(\sqrt[3]{x - 2})$$

$$= (x - 2) + 2 = x$$

66. $f(x) = \sqrt[3]{x + 1}$

(a) $f^{-1}(x) = x^3 - 1$

(b)

(c) $f^{-1}(f(x)) = f^{-1}(\sqrt[3]{x + 1})$

$$= (x + 1) - 1$$

$$= x$$

$$f(f^{-1}(x)) = f(x^3 - 1)$$

$$= \sqrt[3]{x^3 - 1 + 1}$$

$$= x$$

68. $f(x) = |x - 2|$ is increasing on the interval $[2, \infty)$.

$$f(x) = (x - 2)$$

$$y = x - 2$$

$$x = y - 2$$

$$y = x + 2$$

$$f^{-1}(x) = x + 2, \quad x \geq 0$$

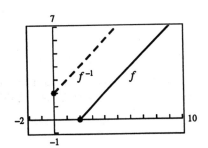

70. $f(x) = x^{4/3}$ is increasing on the interval $[0, \infty)$.

$$f(x) = x^{4/3}, \quad x \geq 0$$
$$y = x^{4/3}, \quad x \geq 0$$
$$x = y^{4/3}$$
$$y = x^{3/4}$$
$$f^{-1}(x) = x^{3/4}, \quad x \geq 0$$

72. $(f + h)(5) = f(5) + h(5)$
$$= (3 - 2 \cdot 5) + (3 \cdot 5^2 + 2)$$
$$= 70$$

74. $\left(\dfrac{g}{h}\right)(1) = \dfrac{g(1)}{h(1)}$
$$= \dfrac{\sqrt{1}}{3 \cdot 1^2 + 2}$$
$$= \dfrac{1}{5}$$

76. $(g \circ f)(-2) = g(f(-2))$
$$= g(3 - 2[-2])$$
$$= \sqrt{7}$$

78. $f^{-1}(x) = \dfrac{3 - x}{2}$
$$(h \circ f^{-1})(1) = h(f^{-1}(1))$$
$$= h\left(\dfrac{3 - 1}{2}\right)$$
$$= 3 \cdot 1^2 + 2$$
$$= 5$$

CHAPTER TWO
Intercepts, Zeros and Solutions

2.1 | Linear Equations and Modeling

2. $3(x+2) = 3x + 4$

$3x + 6 = 3x + 4$

$6 \neq 4$

No solution; conditional

4. $3(x+2) - 5 = 3x + 1$

$3x + 6 - 5 = 3x + 1$

$3x + 1 = 3x + 1$

Identity

6. $-7(x-3) + 4x = 3(7-x)$

$-7x + 21 + 4x = 21 - 3x$

$21 - 3x = 21 - 3x$

Identity

8. $x^2 + 2(3x - 2) = x^2 + 6x - 4$

$x^2 + 6x - 4 = x^2 + 6x - 4$

Identity

10. $\dfrac{5}{x} + \dfrac{3}{x} = 24$

$\dfrac{5}{x} + \dfrac{3}{x} = \dfrac{24x}{x}$

$5 + 3 = 24x$

$8 = 24x$

$x = \dfrac{1}{3}$

Conditional

12. $7 - 3x = 5x - 17$

(a) $7 - 3(-3) \overset{?}{=} 5(-3) - 17$

$16 \neq -32$

$x = -3$ *is not* a solution.

(c) $7 - 3(8) \overset{?}{=} 5(8) - 17$

$17 \neq 23$

$x = 8$ *is not* a solution.

(b) $7 - 3(0) \overset{?}{=} 5(0) - 17$

$7 \neq -17$

$x = 0$ *is not* a solution.

(d) $7 - 3(3) \overset{?}{=} 5(3) - 17$

$-2 = -2$

$x = 3$ *is a* solution.

14. $\sqrt[3]{x - 8} = 3$

(a) $\sqrt[3]{2 - 8} \overset{?}{=} 3$

$-\sqrt[3]{6} \neq 3$

$x = 2$ *is not* a solution.

(b) $\sqrt[3]{-2 - 8} \overset{?}{=} 3$

$-\sqrt[3]{10} \neq 3$

$x = -2$ *is not* a solution.

(c) $\sqrt[3]{35-8} \overset{?}{=} 3$

$\qquad 3 = 3$

$x = 35$ *is* a solution.

(d) $\sqrt[3]{8-8} \overset{?}{=} 3$

$\qquad 0 \ne 3$

$x = 8$ *is not* a solution.

16. $7 - x = 18$

$\qquad -x = 11$

$\qquad x = -11$

18. $7x + 2 = 16$

$\qquad 7x = 14$

$\qquad x = 2$

20. $7x + 3 = 3x - 13$

$\qquad 4x = -16$

$\qquad x = -4$

22. $2(13t - 15) + 3(t - 19) = 0$

$\qquad 26t - 30 + 3t - 57 = 0$

$\qquad\qquad\qquad 29t = 87$

$\qquad\qquad\qquad\quad t = 3$

24. $8(x + 2) - 3(2x + 1) = 2(x + 5)$

$\qquad 8x + 16 - 6x - 3 = 2x + 10$

$\qquad\qquad 2x + 13 = 2x + 10$

$\qquad\qquad\qquad 13 = 10$

Not possible

26. $\dfrac{x}{5} - \dfrac{x}{2} = 3$

$10\left(\dfrac{x}{5}\right) - 10\left(\dfrac{x}{2}\right) = 10(3)$

$\qquad\qquad 2x - 5x = 30$

$\qquad\qquad\quad -3x = 30$

$\qquad\qquad\qquad x = -10$

28. $0.25x + 0.75(10 - x) = 3$

$\qquad 0.25x + 7.5 - 0.75x = 3$

$\qquad\qquad\qquad -0.5x = -4.5$

$\qquad\qquad\qquad\qquad x = 9$

30. $3(x + 3) = 5(1 - x) - 1$

$\qquad 3x + 9 = 5 - 5x - 1$

$\qquad 3x + 9 = 4 - 5x$

$\qquad\quad 8x = -5$

$\qquad\quad\ x = -\dfrac{5}{8}$

32. $\dfrac{17 + y}{y} + \dfrac{32 + y}{y} = 100$

$(y)\left(\dfrac{17 + y}{y}\right) + (y)\left(\dfrac{32 + y}{y}\right) = (y)100$

$\qquad\quad 17 + y + 32 + y = 100y$

$\qquad\qquad\qquad\quad 49 = 98y$

$\qquad\qquad\qquad\quad\ y = \dfrac{1}{2}$

34. $\dfrac{10x+3}{5x+6} = \dfrac{1}{2}$

$$2(10x+3) = 5x+6$$

$$20x+6 = 5x+6$$

$$15x = 0$$

$$x = 0$$

36. $\dfrac{15}{x} - 4 = \dfrac{6}{x} + 3$

$$(x)\dfrac{15}{x} - (x)4 = (x)\dfrac{6}{x} + (x)3$$

$$15 - 4x = 6 + 3x$$

$$9 = 7x$$

$$x = \dfrac{9}{7}$$

38. $\dfrac{1}{x-2} + \dfrac{3}{x+3} = \dfrac{4}{x^2+x-6}$

$$\dfrac{1}{x-2}(x-2)(x+3) + \dfrac{3}{x+3}(x-2)(x+3) = \dfrac{4}{x^2+x-6}(x-2)(x+3)$$

$$x+3+3(x-2) = 4$$

$$x+3+3x-6 = 4$$

$$4x = 7$$

$$x = \dfrac{7}{4}$$

40. $\dfrac{2}{(x-4)(x-2)} = \dfrac{1}{(x-4)} + \dfrac{2}{(x-2)}$

$$\dfrac{2}{(x-4)(x-2)}(x-4)(x-2) = \dfrac{1}{(x-4)}(x-4)(x-2) + \dfrac{2}{(x-2)}(x-4)(x-2)$$

$$2 = x-2+2(x-4)$$

$$2 = x-2+2x-8$$

$$12 = 3x$$

$$x = 4$$

A check reveals that $x = 4$ is an extraneous solution, so there is no solution.

42. $\dfrac{4}{u-1} + \dfrac{6}{3u+1} = \dfrac{15}{3u+1}$

$$\dfrac{4}{u-1}(u-1)(3u+1) + \dfrac{6}{3u+1}(u-1)(3u+1) = \dfrac{15}{3u+1}(u-1)(3u+1)$$

$$4(3u+1) + 6(u-1) = 15(u-1)$$

$$12u+4+6u-6 = 15u-15$$

$$3u = -13$$

$$u = -\dfrac{13}{3}$$

44. $\dfrac{6}{x} - \dfrac{2}{x+3} = \dfrac{3(x+5)}{x(x+3)}$

$$\dfrac{6}{x}(x)(x+3) - \dfrac{2}{x+3}(x)(x+3) = \dfrac{3(x+5)}{x(x+3)}(x)(x+3)$$

$$6(x+3) - 2x = 3(x+5)$$

$$6x + 18 - 2x = 3x + 15$$

$$x = -3$$

A check reveals that $x = -3$ is an extraneous solution, so there is no solution.

46. $3 = 2 + \dfrac{2}{z+2}$

$$3(z+2) = 2(z+2) + \dfrac{2}{z+2}(z+2)$$

$$3z + 6 = 2z + 4 + 2$$

$$z = 0$$

48. $(2x+1)^2 = 4(x^2 + x + 1)$

$$4x^2 + 4x + 1 = 4x^2 + 4x + 4$$

$$1 = 4$$

Not possible; no solution

50. $6x + ax = 2x + 5$

$$4x + ax = 5$$

$$x(4 + a) = 5$$

$$x = \dfrac{5}{4+a}, \ a \neq -4$$

52. $5 + ax = 12 - bx$

$$ax + bx = 7$$

$$x(a + b) = 7$$

$$x = \dfrac{7}{a+b}, \ a \neq -b$$

54. $2.763 - 4.5(2.1x - 5.1432) = 6.32x + 5$

$$2.763 - (4.5)(2.1)x + (4.5)(5.1432) = 6.32x + 5$$

$$2.763 + (4.5)(5.1432) - 5 = 6.32x + (4.5)(2.1)x$$

$$2.763 + (4.5)(5.1432) - 5 = (6.32 + (4.5)(2.1))x$$

$$x = \dfrac{2.763 + (4.5)(5.1432) - 5}{6.32 + (4.5)(2.1)}$$

$$x \approx 1.326$$

56. $\dfrac{2}{7.398} - \dfrac{4.405}{x} = \dfrac{1}{x}$

$$\dfrac{2}{7.398}(7.398)x - \dfrac{4.405}{x}(7.398)x = \dfrac{1}{x}(7.398)x$$

$$2x - 4.405(7.398) = 7.398$$

$$2x = 7.398 + 4.405(7.398)$$

$$x = \dfrac{7.398 + 4.405(7.398)}{2}$$

$$x \approx 19.993$$

58. $ax + b = cx$

$$ax - cx = -b$$

$$x(a - c) = -b$$

$$x = \frac{-b}{a - c}$$

Since $x = 1/3$, we have $1/3 = -b/(a - c)$. Choose any combination of a, b, and c that satisfies this equation. One possibility is $a = 2$, $b = 5$, and $c = 17$. The equation is $2x + 5 = 17x$.

60. *Verbal Model*   Perimeter $= 2$(Width) $+ 2$(Height)

Labels   Perimeter $= 3$ feet
Width (in feet) $= 0.62$ (height) $= 0.62x$
Height (in feet) $= x$

Algebraic Equation $3 = 2(0.62x) + 2x$

$$x = \frac{3}{3.24} \approx 0.93$$

Width $= 0.62x \approx 0.57$ feet, height $= x \approx 0.93$ feet

62. Distance $=$ (First rate)(Time at first rate) $+$ (Second rate)(Time at second rate)

$$317 = 58 \cdot t + 52(5.75 - t)$$

$$317 = 58t + 299 - 52t$$

$$18 = 6t$$

$$t = 3$$

Time at 58 miles per hour $= 3$ hours
Time at 52 miles per hour $= 5.75 - 3 = 2.75 = 2$ hours and 45 minutes

64. (Distance) = Rate × Time

$$d_1 = 45 \text{ miles per hour} \times t$$

$$d_2 = 55 \text{ miles per hour} \times \left(t - \frac{1}{2}\right)$$

(Distance traveled by first car) = (Distance traveled by second car)

$$d_1 = d_2$$

$$45t = 55\left(t - \frac{1}{2}\right)$$

$$-10t = -27.5$$

$$t = 2.75 \text{ hours after first car left}$$

$$\text{Time} = \frac{\text{Distance}}{\text{Rate}}$$

$$t = \frac{135}{45} = 3 \text{ hours}$$

Yes. It takes the first car 3 hours to reach the game and the second car catches up in 2.75 hours.

66. $\text{Time} = \dfrac{\text{Distance}}{\text{Speed}}$

$$t_1 = \frac{200 \text{ miles}}{55 \text{ mph}} \quad \text{(trip out)}$$

$$t_2 = \frac{200 \text{ miles}}{40 \text{ mph}} \quad \text{(return trip)}$$

$$\text{Speed} = \frac{\text{Round trip distance}}{\text{Round trip time}}$$

$$r = \frac{2 \times 200 \text{ miles}}{t_1 + t_2}$$

$$r = \frac{400}{(200/55) + (200/40)}$$

$$r = \frac{400(40)(55)}{200(40 + 55)} = \frac{4400}{95} = \frac{800}{19} \approx 46.3 \text{ mph}$$

68. $d = rt$

$$1.5 \times 10^{11} = 3.0 \times 10^8 \cdot t$$

$$t = 500 \text{ seconds}$$

$$= \tfrac{25}{3} \text{ minutes} \approx 8.33 \text{ minutes}$$

70. $$\frac{\text{Height of tree}}{\text{Tree's shadow}} = \frac{\text{Height of lamp post}}{\text{Lamp post's shadow}}$$

$$\frac{x}{25} = \frac{5}{2}$$

$$x = 25\left(\frac{5}{2}\right) = 62.5 \text{ feet}$$

72. By similar triangles,

$$\frac{\text{Height of silo}}{\text{Silo's shadow}} = \frac{\text{Height of person}}{\text{Person's shadow}}$$

$$\frac{50}{32 + x} = \frac{6}{x}$$

$$50x = 6(32 + x)$$

$$50x = 192 + 6x$$

$$44x = 192$$

$$x \approx 4.36 \text{ feet}$$

74. *Verbal Model* $\qquad \dfrac{\text{Sum of scores}}{\text{Sum of possible points}} = \text{Average}$

Labels $\qquad$ Sum of scores $= 87 + 92 + 84 + x$
Sum of possible points $= 500$
Average $= 0.90$

Algebraic Equation $\qquad \dfrac{87 + 92 + 94 + x}{500} = 0.90$

$$\frac{263 + x}{500} = 0.90$$

$$263 + x = 450$$

$$x = 187 \text{ (or greater)}$$

76. Interest $=$ Interest rate $\times$ Principal

$$i_1 = 11\% \times \$x$$

$$i_2 = 12.5\% \times \$(25{,}000 - x)$$

Total interest $=$ (Interest in first account) $+$ (Interest in second account)

$$2975 = i_1 + i_2$$

$$2975 = 0.11x + 0.125(25{,}000 - x)$$

$$2975 = -0.015x + 3125$$

$$x = \frac{150}{0.015} = 10{,}000$$

You have \$10,000 in the 11% fund and \$15,000 in the $12\frac{1}{2}$% fund.

78. Interest = Interest rate × Principal × Portion of year

$i_1 = x \times \$10{,}000 \times \frac{3}{4}$ year

$i_2 = (x - 1.5\%) \times \$10{,}000 \times \frac{1}{4}$ year

Total interest = (First three quarter's interest) + (Last quarter's interest)

$\$1112.50 = i_1 + i_2$

$1112.50 = 7500x + 2500(x - 0.015)$

$x = \frac{1150}{10{,}000} = 0.115 = 11.5\%$

The first three quarter's rate was 11.5% and the last quarter's rate was 10%.

80. (Final concentration)(Amount)

=(Solution 1 concentration)(Amount) + (Solution 2 concentration)(Amount)

$\left(\frac{40}{41}\right)(2 + x \text{ gal}) = \left(\frac{32}{33}\right)(2 \text{ gal}) + (100\%)x$

$\frac{80}{41} + \frac{40x}{41} = \frac{64}{33} + x$

$\frac{16}{(41)(33)} = \frac{x}{41}$

$x = \frac{16}{33} \approx 0.48$ gallon

82.   Cost = (Fixed costs) + (Variable cost)(Number of units)

$\$85{,}000 = \$10{,}000 + \$9.30x$

$x = \frac{75{,}000}{9.3} \approx 8064.52$ units

At most, the company can manufacture 8064 units.

84. $V = \pi r^2 h$

$\frac{V}{\pi r^2} = h$

86. $A = P\left(1 + \frac{R}{n}\right)^{nT}$

$A\left(1 + \frac{R}{n}\right)^{-nT} = P$

88. $A = \frac{\pi r^2 \theta}{360}$

$\frac{360A}{\pi r^2} = \theta$

90. $h = v_0 t + \frac{1}{2}at^2$

$h - v_0 t = \frac{1}{2}at^2$

$2(h - v_0 t) = at^2$

$\frac{2(h - v_0 t)}{t^2} = a$

92.
$$C = \frac{1}{(1/C_1) = (1/C_2)}$$
$$C = \frac{C_1 C_2}{C_2 + C_1}$$
$$C_1 C_2 = CC_2 + CC_1$$
$$-CC_2 = CC_1 - C_1 C_2$$
$$C_1 = \frac{CC_2}{C_2 - C}$$

94.
$$V = \frac{1}{6}H(S_0 + 4S_1 + S_2)$$
$$\frac{6V}{H} = S_0 + 4S_1 + S_2$$
$$\frac{6V}{H} - S_0 - S_2 = 4S_1$$
$$\frac{6V - S_0 H - S_2 H}{4H} = S_1$$

96. (Person's wt) $\cdot$ (Person's distance from fulcrum) $=$ (Wt of rock) $\cdot$ (Distance from rock to fulcrum)
$$200 \cdot x = 550(5 - x)$$
$$200x = 2750 - 550x$$
$$750x = 2750$$
$$x = 3\tfrac{2}{3} \text{ feet}$$

▮ 2.2 ▮ Solving Equations Graphically

2. $y = (x - 1)(x - 3)$

Let $y = 0$: $(x - 1)(x - 3) = 0 \Rightarrow x = 1, 3 \Rightarrow (1, 0), (3, 0)$ x-intercepts
Let $x = 0$: $y = (0 - 1)(0 - 3) = 3 \Rightarrow (0, 3)$ y-intercept

4. $y = 4 - x^2$

Let $y = 0$: $0 = 4 - x^2 \Rightarrow x = 2, -2 \Rightarrow (2, 0), (-2, 0)$ x-intercepts
Let $x = 0$: $y = 4 - 0^2 = 4 \Rightarrow (0, 4)$ y-intercept

6. $xy = 4 \Rightarrow y = 4/x$

Neither y nor x can be zero. Hence, there are no intercepts.

8. $x^2 y - x^2 + 4y = 0$

Let $y = 0$: $-x^2 = 0 \Rightarrow x = 0$
$\Rightarrow (0, 0)$ x-intercept

Let $x = 0$: $4y = 0 \Rightarrow y = 0$
$\Rightarrow (0, 0)$ y-intercept

10. $f(x) = 3(x - 5) + 9$

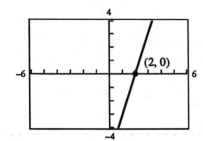

12. $f(x) = x^3 - 9x^2 + 18x$

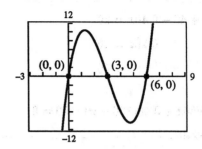

14. $f(x) = x - 3 - \dfrac{10}{x}$

16.
$$1200 = 300 + 2(x - 500)$$
$$900 = 2x - 1000$$
$$1900 = 2x$$
$$2x - 1900 = 0$$

18.
$$\frac{x - 3}{25} = \frac{x - 5}{12}$$
$$\frac{x - 3}{25} - \frac{x - 5}{12} = 0$$

20.
$$\frac{6}{x} + \frac{8}{x + 5} = 10$$
$$\frac{6}{x} - \frac{8}{x + 5} - 10 = 0$$

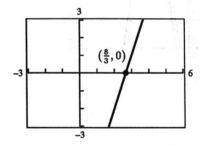

22. $3.5x - 8 = 0.5x$
$$3x = 8$$
$$x = \frac{8}{3}$$

$$f(x) = 3.5x - 8 - 0.05x = 0$$
$$x = 2.667$$

24. $0.60x + 0.40(100 - x) = 50$

$0.60x + 40 - 0.40x = 50$

$0.20x = 10$

$x = 50$

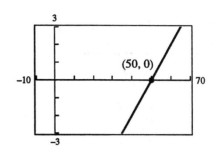

$f(x) = 0.60x + 0.40(100 - x) - 50 = 0$

$x = 50.0$

26. $(x + 1)^2 + 2(x - 2) = (x + 1)(x - 2)$

$x^2 + 2x + 1 + 2x - 4 = x^2 - x - 2$

$5x = 1$

$x = \frac{1}{5}$

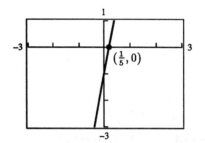

$f(x) = (x + 1)^2 + 2(x - 2) - (x + 1)(x - 2) = 0$

$x = 0.20$

28. $-2(x^2 - 6x + 6) = 0$

$x = 1.268,\ 4.732$

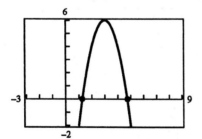

30. $\frac{1}{9}x^3 + x + 4 = 0$

$x = -2.422$

32. $4x^3 + 12x^2 - 26x - 24 = 0$

$x = -4.206, -0.735, 1.941$

34. $x^5 = 3 + 2x^3$

$x^5 - 3 - 2x^3 = 0$

$x = 1.638$

36. $\dfrac{5}{x} = 1 + \dfrac{3}{x+2}$

$\dfrac{5}{x} - 1 - \dfrac{3}{x+2} = 0$

$x = -3.162, 3.162$

38. $(x, y) = (-1.222, 8.222)$

40. $(x, y) = (1.670, 1.660)$

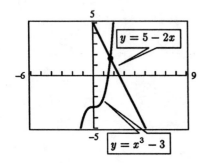

42. $(x, y) = (2.050, 32)$

44. $(x, y) = (0, 2), (3, 5)$

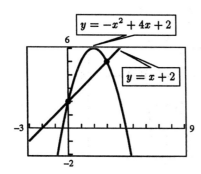

46. $(x, y) = (0, 0), (3, -3)$

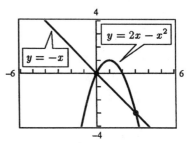

48. $\dfrac{1 + 0.86603}{1 - 0.86603}$

(a) 13.93

(b) $\dfrac{1.87}{0.13} \approx 14.38$

The second method introduced an additional round-off error.

50. $\dfrac{1.73205 - 1.19195}{3 - (1.73205)(1.19195)}$

(a) 0.58

(b) $\dfrac{0.54}{0.94} \approx 0.57$

The second method did introduce a slight additional round-off error.

52. (a) $C = 18.65x + 25{,}000$

(b) If $C \le 200{,}000$, then

$$200{,}000 = 18.65x + 25{,}000$$
$$175{,}000 = 18.65x$$
$$x = 9383 \text{ units.}$$

This problem is easier to solve algebraically.

54. (a) Since $2x + 2y = 230$, $y = 115 - x$ and area $= A = x(115 - x)$.

(b) Domain $0 \le x \le 115$

(c) If $A = 2000$, then $x = 93.6$ or 21.4.
The larger answer is appropriate, since x is the length.

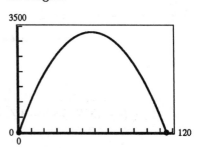

56. If $G = 4600$, then

$$4600 = 8000 - \tfrac{1}{2}x$$
$$x = 6800$$

58. If $S = 10{,}500$, then

$$10{,}500 = \tfrac{1}{2}x + 8000$$
$$x = 5000$$

and $G = 8000 - \tfrac{1}{2}x$

$$= 8000 - \tfrac{1}{2}(5000)$$
$$= 5500.$$

2.3 Complex Numbers

2. (a) $i^{40} = (i^4)^{10} = (1)^{10} = 1$

(b) $i^{25} = i^{24} \cdot i = (i^4)^6 \cdot i = (1)^6 \cdot i = i$

(c) $i^{50} = i^{48} \cdot i^2 = (i^4)^{12} \cdot i^2 = (1)^{12}(-1) = -1$

(d) $i^{67} = i^{64} \cdot i^3 = (i^4)^{16} \cdot i^3 = (1)^{16} \cdot (-i) = -i$

4. $a + bi = 13 + 4i$

$a = 13$

$b = 4$

6. $(a+6) + 2bi = 6 - 5i$

$a + 6 = 6 \Rightarrow a = 0$

$2b = -5 \Rightarrow b = -\frac{5}{2}$

8. $3 + \sqrt{-16} = 3 + \sqrt{16}\, i$

$\qquad = 3 + 4i$

10. $1 + \sqrt{-8} = 1 + \sqrt{8}\, i$

$\qquad = 1 + 2\sqrt{2}\, i$

12. 45 is in standard form. The imaginary part is 0.

14. $-4i^2 + 2i = -4(-1) + 2i$

$\qquad = 4 + 2i$

16. $(-i)^3 = (-1)^3(i)^3$

$\qquad = -1(-i)$

$\qquad = i$

18. $(\sqrt{-4})^2 - 5 = (2i)^2 - 5$

$\qquad = 4i^2 - 5$

$\qquad = 4(-1) - 5$

$\qquad = -9$

20. $(13 - 2i) + (-5 + 6i) = 13 - 2i - 5 + 6i$

$\qquad = (13 - 5) + (-2 + 6)i$

$\qquad = 8 + 4i$

22. $(3 + 2i) - (6 + 13i) = 3 + 2i - 6 - 13i$

$\qquad = (3 - 6) + (2 - 13)i$

$\qquad = -3 - 11i$

24. $(8 + \sqrt{-18}) - (4 + 3\sqrt{2}\, i) = 8 + 3\sqrt{2}\, i - 4 - 3\sqrt{2}\, i$

$\qquad = (8 - 4) + (3\sqrt{2} - 3\sqrt{2})i$

$\qquad = 4$

26. $(1.6 + 3.2i) + (-5.8 + 4.3i) = 1.6 + 3.2i - 5.8 + 4.3i$

$\qquad = (1.6 - 5.8) + (3.2 + 4.3)i$

$\qquad = -4.2 + 7.5i$

28. The complex conjugate of $9 - 12i$ is $9 + 12i$.

$$(9 - 12i)(9 + 12i) = 9^2 - (12i)^2$$
$$= 81 - 144i^2$$
$$= 81 + 144$$
$$= 225$$

30. The complex conjugate of $20i$ is $-20i$.

$$(20i)(-20i) = -400i^2 = 400$$

32. $\sqrt{-5} \cdot \sqrt{-10} = (\sqrt{5}\,i)(\sqrt{10}\,i)$
$$= \sqrt{50}\,i^2$$
$$= -5\sqrt{2}$$

34. $(\sqrt{-75})^2 = (\sqrt{75}\,i)^2$
$$= (\sqrt{75})^2(i)^2$$
$$= (75 - 1)$$
$$= -75$$

36. $(6 - 2i)(2 - 3i) = 12 - 18i - 4i + 6i^2$
$$= 12 - 22i - 6$$
$$= 6 - 22i$$

38. $-8i(9 + 4i) = -72i - 32i^2 = 32 - 72i$

40. $(3 + \sqrt{-5})(7 - \sqrt{-10}) = (3 + \sqrt{5}\,i)(7 - \sqrt{10}\,i)$
$$= 21 - 3\sqrt{10}\,i + 7\sqrt{5}\,i - \sqrt{50}\,i^2$$
$$= 21 - 3\sqrt{10}\,i + 7\sqrt{5}\,i + 5\sqrt{2} = (21 + 5\sqrt{2}) + (7\sqrt{5} - 3\sqrt{10})i$$

42. $(2 - 3i)^2 = (2 - 3i)(2 - 3i)$
$$= 4 - 12i - 9$$
$$= -5 - 12i$$

44. $(1 - 2i)^2 - (1 + 2i)^2 = [(1 - 2i) + (1 + 2i)][(1 - 2i) - (1 + 2i)]$
$$= (2)(-4i)$$
$$= -8i$$

46. $\dfrac{3}{1 - i} = \dfrac{3}{1 - i} \cdot \dfrac{1 + i}{1 + i}$
$$= \dfrac{3 + 3i}{1 + 1}$$
$$= \dfrac{3}{2} + \dfrac{3}{2}i$$

48. $\dfrac{8 - 7i}{1 - 2i} = \dfrac{8 - 7i}{1 - 2i} \cdot \dfrac{1 + 2i}{1 + 2i}$
$$= \dfrac{8 + 16i - 7i - 14i^2}{1 + 4}$$
$$= \dfrac{22 + 9i}{5}$$
$$= \dfrac{22}{5} + \dfrac{9}{5}i$$

50. $\dfrac{8+20i}{2i} = \dfrac{8+20i}{2i} \cdot \dfrac{-2i}{-2i}$

$\qquad = \dfrac{-16i - 40i^2}{4}$

$\qquad = \dfrac{40 - 16i}{4}$

$\qquad = 10 - 4i$

52. $\dfrac{1}{i^3} = \dfrac{1}{-i}$

$\qquad = \dfrac{1}{-i} \cdot \dfrac{i}{i}$

$\qquad = \dfrac{i}{-i^2}$

$\qquad = \dfrac{i}{1}$

$\qquad = i$

54. $\dfrac{(2-3i)(5i)}{2+3i} \cdot \dfrac{(2-3i)}{(2-3i)} = \dfrac{(2-3i)^2(5i)}{4+9}$

$\qquad\qquad = \dfrac{(20i - 60i^2 + 45i^3)}{13}$

$\qquad\qquad = \dfrac{20i + 60 - 45i}{13}$

$\qquad\qquad = \dfrac{60 - 25i}{13}$

$\qquad\qquad = \dfrac{60}{13} - \dfrac{25}{13}\,i$

56.

58.

60.

62. The complex number 2 is not in the Mandelbrot Set since for $c = 2$, the corresponding Mandelbrot sequence is 2, 6, 38, 1446, 2090918, $\approx 4.37 \times 10^{12}$, ... which is unbounded.

64. The complex number $-i$ is in the Mandelbrot Set since for $c = -i$, the corresponding Mandelbrot sequence is $-i$, $-1-i$, i, $-1-i$, i, $-1-i$, which is bounded.

66. The complex number -1 is in the Mandelbrot Set since for $c = -1$, the corresponding Mandelbrot sequence is -1, 0, -1, 0, -1, 0, which is bounded.

68.
$$2^4 = 16$$
$$(-2)^4 = 16$$
$$(2i)^4 = 2^4 i^4 = 16(1) = 16$$
$$(-2i)^4 = (-2)^4 i^4 = 16(1) = 16$$

All four numbers are fourth roots of 16.

70. $(a + bi) - (a - bi) = a + bi - a + bi$
$$= 2bi \text{ which is an imaginary number.}$$
$(a - bi) - (a + bi) = a - bi - a - bi$
$$= -2bi \text{ which is an imaginary number.}$$

72. $(a_1 + b_1 i)(a_2 + b_2 i) = a_1 a_2 + a_1 b_2 i + a_2 b_1 i + b_1 b_2 i^2$
$$= (a_1 a_2 - b_1 b_2) + (a_1 b_2 + a_2 b_1)i$$

The conjugate of the product is $(a_1 a_2 - b_1 b_2) - (a_1 b_2 + a_2 b_1)i$, and the product of the conjugates is
$$(a_1 - b_1 i)(a_2 - b_2 i) = a_1 a_2 - a_1 b_2 i - a_2 b_1 i + b_1 b_2 i^2$$
$$= (a_1 a_2 - b_1 b_2) - (a_1 b_2 + a_2 b_1)i.$$

Thus, the conjugate of the product is the product of the conjugates.

2.4 Solving Equations Algebraically

2.
$$9x^2 - 1 = 0$$
$$(3x + 1)(3x - 1) = 0$$

$$3x + 1 = 0 \Rightarrow x = -\tfrac{1}{3}$$
$$3x - 1 = 0 \Rightarrow x = \tfrac{1}{3}$$

4. $x^2 - 10x + 9 = 0$
$$(x - 9)(x - 1) = 0$$

$$x - 9 = 0 \Rightarrow x = 9$$
$$x - 1 = 0 \Rightarrow x = 1$$

6. $16x^2 + 56x + 49 = 0$
$$4x + 7 = 0 \Rightarrow x = -\tfrac{7}{4}$$

8.
$$2x^2 = 19x + 33$$
$$2x^2 - 19x - 33 = 0$$
$$(2x + 3)(x - 11) = 0$$

$$2x + 3 = 0 \Rightarrow x = -\tfrac{3}{2}$$
$$x - 11 = 0 \Rightarrow x = 11$$

10.
$$x(8 - x) = 12$$
$$-x^2 + 8x = 12$$
$$-x^2 + 8x - 12 = 0$$
$$x^2 - 8x + 12 = 0$$
$$(x - 6)(x - 2) = 0$$

$$x - 6 = 0 \Rightarrow x = 6$$
$$x - 2 = 0 \Rightarrow x = 2$$

12.
$$(x + a)^2 - b^2 = 0$$
$$(x + a + b)(x + a - b) = 0$$

$$x + a + b = 0 \Rightarrow x = -a - b$$
$$x + a - b = 0 \Rightarrow x = -a + b$$

14. $9x^2 = 25$
$$x^2 = \tfrac{25}{9}$$
$$x = \pm\sqrt{\tfrac{25}{9}}$$
$$x = \pm\tfrac{5}{3} \approx \pm 1.67$$

16. $(x + 13)^2 = 21$
$$x + 13 = \pm\sqrt{21}$$
$$x = -13 \pm \sqrt{21}$$
$$x \approx -8.42 \text{ or}$$
$$x \approx -17.58$$

18. $(x - 5)^2 = 20$
$$x - 5 = \pm\sqrt{20}$$
$$x = 5 \pm 2\sqrt{5}$$
$$x \approx 9.47 \text{ or } x \approx 0.53$$

20. $(x + 5)^2 = (x + 4)^2$
$$x + 5 = \pm(x + 4)$$
For $x + 5 = +(x + 4)$
$$5 = 4, \quad \text{No solution}$$
For $x + 5 = -(x + 4)$
$$x + 5 = -x - 4$$
$$2x = -9$$
$$x = -\tfrac{9}{2} = -4.50$$

22. $x^2 + 8x + 14 = 0$

$x^2 + 8x = -14$

$x^2 + 8x + 4^2 = -14 + 4^2$

$(x + 4)^2 = 2$

$x + 4 = \pm\sqrt{2}$

$x = -4 \pm \sqrt{2}$

24. $9x^2 - 12x - 14 = 0$

$x^2 - \frac{4}{3}x - \frac{14}{9} = 0$

$x^2 - \frac{4}{3}x = \frac{14}{9}$

$x^2 - \frac{4}{3}x + \left(\frac{2}{3}\right)^2 = \frac{14}{9} + \left(\frac{2}{3}\right)^2$

$\left(x - \frac{2}{3}\right)^2 = 2$

$x - \frac{2}{3} = \pm\sqrt{2}$

$x = \frac{2}{3} \pm \sqrt{2}$

26. $4x^2 - 4x - 99 = 0$

$x^2 - x - \frac{99}{4} = 0$

$x^2 - x = \frac{99}{4}$

$x^2 - x + \left(\frac{1}{2}\right)^2 = \frac{99}{4} + \left(\frac{1}{2}\right)^2$

$\left(x - \frac{1}{2}\right)^2 = 25$

$x - \frac{1}{2} = \pm 5$

$x = \frac{1}{2} \pm 5$

$x = \frac{11}{2}$ or

$x = -\frac{9}{2}$

28. $2x^2 - x - 1 = 0; \ a = 2, \ b = -1, \ c = -1$

$x = \dfrac{-(-1) \pm \sqrt{(-1)^2 - 4(2)(-1)}}{2(2)}$

$= \dfrac{1 \pm \sqrt{9}}{4} = \dfrac{1 \pm 3}{4}$

$x = \dfrac{1 + 3}{4} = 1$ or $x = \dfrac{1 - 3}{4} = -\dfrac{1}{2}$

30. $25x^2 - 20x + 3 = 0; \ a = 25, \ b = -20, \ c = 3$

$x = \dfrac{-(-20) \pm \sqrt{(-20)^2 - 4(25)(3)}}{2(25)}$

$= \dfrac{20 \pm \sqrt{100}}{50} = \dfrac{20 \pm 10}{50}$

$x = \dfrac{20 + 10}{50} = \dfrac{3}{5}$ or $x = \dfrac{20 - 10}{50} = \dfrac{1}{5}$

32. $x^2 - 10x + 22 = 0; \ a = 1, \ b = -10, \ c = 22$

$x = \dfrac{-(-10) \pm \sqrt{(-10)^2 - 4(1)(22)}}{2(1)}$

$= \dfrac{10 \pm \sqrt{12}}{2} = \dfrac{10 \pm 2\sqrt{3}}{2}$

$x = 5 \pm \sqrt{3}$

34. $6x = 4 - x^2$

$x^2 + 6x - 4 = 0; \ a = 1, \ b = 6, \ c = -4$

$x = \dfrac{-6 \pm \sqrt{6^2 - 4(1)(-4)}}{2(1)}$

$= \dfrac{-6 \pm \sqrt{52}}{2} = \dfrac{-6 \pm 2\sqrt{13}}{2}$

$x = -3 \pm \sqrt{13}$

36. $4x^2 - 4x - 4 = 0; \ a = 4, \ b = -4, \ c = -4$

$x = \dfrac{-(-4) \pm \sqrt{(-4)^2 - 4(4)(-4)}}{2(4)}$

$= \dfrac{4 \pm \sqrt{80}}{8} = \dfrac{4 \pm 4\sqrt{5}}{8}$

$x = \dfrac{1}{2} \pm \dfrac{\sqrt{5}}{2}$

38. $16x^2 + 22 = 40x$

$16x^2 - 40x + 22 = 0$

$8x^2 - 20x + 11 = 0; \ a = 8, \ b = -20, \ c = 11$

$$x = \frac{-(-20) \pm \sqrt{(-20)^2 - 4(8)(11)}}{2(8)}$$

$$= \frac{20 \pm \sqrt{48}}{16} = \frac{20 \pm 4\sqrt{3}}{16}$$

$$x = \frac{5}{4} \pm \frac{\sqrt{3}}{4}$$

40. $3x + x^2 - 1 = 0$

$x^2 + 3x - 1 = 0; \ a = 1, \ b = 3, \ c = -1$

$$x = \frac{-3 \pm \sqrt{3^2 - 4(1)(-1)}}{2(1)} = \frac{-3 \pm \sqrt{13}}{2}$$

$$x = -\frac{3}{2} \pm \frac{\sqrt{13}}{2}$$

42. $16x^2 - 40x + 5 = 0; \ a = 16, \ b = -40, \ c = 5$

$$x = \frac{-(-40) \pm \sqrt{(-40)^2 - 4(16)(5)}}{2(16)}$$

$$= \frac{40 \pm \sqrt{1280}}{32} = \frac{40 \pm 16\sqrt{5}}{32}$$

$$x = \frac{5}{4} \pm \frac{\sqrt{5}}{2}$$

44. $9x^2 + 24x + 16 = 0; \ a = 9, \ b = 24, \ c = 16$

$$x = \frac{-24 \pm \sqrt{(24)^2 - 4(9)(16)}}{2(9)}$$

$$= \frac{-24 \pm \sqrt{0}}{18} = -\frac{4}{3}$$

46. $8t = 5 + 2t^2$

$2t^2 - 8t + 5 = 0; \ a = 2, \ b = -8, \ c = 5$

$$x = \frac{-(-8) \pm \sqrt{(-8)^2 - 4(2)(5)}}{2(2)}$$

$$= \frac{8 \pm \sqrt{24}}{4} = \frac{8 \pm 2\sqrt{6}}{4}$$

$$x = 2 \pm \frac{\sqrt{6}}{2}$$

48. $(z+6)^2 = -2z$

$z^2 + 12z + 36 = -2z$

$z^2 + 14z + 36 = 0; \quad a = 1, \; b = 14, \; c = 36$

$x = \dfrac{-14 \pm \sqrt{14^2 - 4(1)(36)}}{2(1)}$

$= \dfrac{-14 \pm \sqrt{52}}{2} = \dfrac{-14 \pm 2\sqrt{13}}{2}$

$x = -7 \pm \sqrt{13}$

50. $2x^4 - 15x^3 + 18x^2 = 0$

$x^2(2x^2 - 15x + 18) = 0$

$x^2(2x - 3)(x - 6) = 0$

$x^2 = 0 \Rightarrow x = 0$

$2x - 3 = 0 \Rightarrow x = \frac{3}{2}$

$x - 6 = 0 \Rightarrow x = 6$

52. $20y^3 - 125y = 0$

$5y(4y^2 - 25) = 0$

$5y(2y + 5)(2y - 5) = 0$

$5y = 0 \Rightarrow y = 0$

$2y + 5 = 0 \Rightarrow y = -\frac{5}{2}$

$2y - 5 = 0 \Rightarrow y = \frac{5}{2}$

54. $x^3 + 2x^2 + 3x + 6 = 0$

$x^2(x + 2) + 3(x + 2) = 0$

$(x + 2)(x^2 + 3) = 0$

$x + 2 = 0 \Rightarrow x = -2$

$x^2 + 3 = 0 \Rightarrow x = \pm\sqrt{3}\,i$

56. $36t^4 + 29t^2 - 7 = 0$

$(36t^2 - 7)(t^2 + 1) = 0$

$(6t + \sqrt{7})(6t - \sqrt{7})(t^2 + 1) = 0$

$6t + \sqrt{7} = 0 \Rightarrow t = -\dfrac{\sqrt{7}}{6}$

$6t - \sqrt{7} = 0 \Rightarrow t = \dfrac{\sqrt{7}}{6}$

$t^2 + 1 = 0 \Rightarrow t = \pm i$

58. $6\left(\dfrac{s}{s+1}\right)^2 + 5\left(\dfrac{s}{s+1}\right) - 6 = 0$

Let $u = s/(s+1)$.

$u = \left(\dfrac{s}{s+1}\right)$

$6u^2 + 5u - 6 = 0$

$(3u - 2)(2u + 3) = 0$

$u = \dfrac{2}{3} \Rightarrow s = 2$

$u = -\dfrac{3}{2} \Rightarrow s = -\dfrac{3}{5}$

60. $9t^{2/3} + 24t^{1/3} + 16 = 0$

$(3t^{1/3} + 4)^2 = 0$

$t^{1/3} = -\frac{4}{3} \Rightarrow t = -\frac{64}{27}$

62. $\sqrt{5 - x} - 3 = 0$

$\sqrt{5 - x} = 3$

$5 - x = 9$

$x = -4$

64. $\sqrt[3]{3x + 1} - 5 = 0$

$\sqrt[3]{3x + 1} = 5$

$3x + 1 = 125$

$3x = 124$

$x = \frac{124}{3}$

66. $\sqrt{x+5} = \sqrt{x-5}$

$\ x + 5 = x - 5$

$\ 5 = -5$

No solution

68. $\sqrt{x} - \sqrt{x-5} = 1$

$\sqrt{x} - 1 = \sqrt{x-5}$

$(\sqrt{x} - 1)^2 = (\sqrt{x-5})^2$

$x - 2\sqrt{x} + 1 = x - 5$

$-2\sqrt{x} = -6$

$\sqrt{x} = 3$

$x = 9$

70. $2\sqrt{x+1} - \sqrt{2x+3} = 1$

$2\sqrt{x+1} = 1 + \sqrt{2x+3}$

$(2\sqrt{x+1})^2 = (1 + \sqrt{2x+3})^2$

$4(x+1) = 1 + 2\sqrt{2x+3} + 2x + 3$

$x = \sqrt{2x+3}$

$x^2 = 2x + 3$

$x^2 - 2x - 3 = 0$

$(x-3)(x+1) = 0$

$x - 3 = 0 \;\Rightarrow\; x = 3$

$x + 1 = 0 \;\Rightarrow\; x = -1, \text{ extraneous}$

72. $\dfrac{4}{x} - \dfrac{5}{3} = \dfrac{x}{6}$

$(6x)\dfrac{4}{x} - (6x)\dfrac{5}{3} = 6x\left(\dfrac{x}{6}\right)$

$24 - 10x = x^2$

$x^2 + 10x - 24 = 0$

$(x+12)(x-2) = 0$

$x + 12 = 0 \;\Rightarrow\; x = -12$

$x - 2 = 0 \;\Rightarrow\; x = 2$

74. $\dfrac{x}{x^2-4} + \dfrac{1}{x+2} = 3$

$(x+2)(x-2)\dfrac{x}{x^2-4} + (x+2)(x-2)\dfrac{1}{x+2} = (x+2)(x-2)3$

$x + x - 2 = 3x^2 - 12$

$3x^2 - 2x - 10 = 0$

$a = 3,\ b = -2,\ c = -10$

$x = \dfrac{-(-2) \pm \sqrt{(-2)^2 - 4(3)(-10)}}{2(3)} = \dfrac{2 \pm \sqrt{124}}{6} = \dfrac{2 \pm 2\sqrt{31}}{6} = \dfrac{1}{3} \pm \dfrac{\sqrt{31}}{3}$

76.
$$\frac{x+1}{3} - \frac{x+1}{x+2} = 0$$
$$3(x+2)\frac{x+1}{3} - 3(x+2)\frac{x+1}{x+2} = 0$$
$$(x+1)(x+2) - 3(x+1) = 0$$
$$x^2 + 3x + 2 - 3x - 3 = 0$$
$$x^2 - 1 = 0$$
$$(x+1)(x-1) = 0$$
$$x + 1 = 0 \;\Rightarrow\; x = -1$$
$$x + 1 = 0 \;\Rightarrow\; x = 1$$

78. $|3x + 2| = 7$
$$3x + 2 = 7 \;\Rightarrow\; x = \tfrac{5}{3}$$
$$-(3x + 2) = 7 \;\Rightarrow\; x = -3$$

80. $|x^2 + 6x| = 3x + 18$

$$x^2 + 6x = 3x + 18$$
$$x^2 + 3x - 18 = 0$$
$$(x+6)(x-3) = 0$$
$$x + 6 = 0 \;\Rightarrow\; x = -6$$
$$x - 3 = 0 \;\Rightarrow\; x = 3$$

$$-(x^2 + 6x) = 3x + 18$$
$$-x^2 - 6x = 3x + 18$$
$$x^2 + 9x + 18 = 0$$
$$(x+3)(x+6) = 0$$
$$x + 3 = 0 \;\Rightarrow\; x = -3$$
$$x + 6 = 0 \;\Rightarrow\; x = -6$$

82. $x^4 + 2x^3 - 8x - 16 = 0$
Using the zoom and trace features, we find that $x = 2, -2$.

84. $x^4 - 4x^2 + 3 = 0$
Using the zoom and trace features, we find that $x = 1, -1, \; 1.73, \; -1.73$.

86. $6x - 7\sqrt{x} - 3 = 0$
Using the zoom and trace features, we find that $x = 2.13$.

88. $x + \sqrt{31 - 9x} = 5$
Using the zoom and trace features, we find that $x = 3, -2$.

90. $3\sqrt{x} - \dfrac{4}{\sqrt{x}} = 4$
Using the zoom and trace features, we find that $x = 4$.

92. $4x + 1 = \dfrac{3}{x}$
Using the zoom and trace features, we find that $x = 0.75, -1$.

94. $x + \dfrac{9}{x+1} = 5$
Using the zoom and trace features, we find that $x = 2$.

96. $|x + 1| = x^2 - 5$
Using the zoom and trace features, we find that $x = 3, -2.56$.

98. $7.08x^6 + 4.15x^3 - 9.6 = 0$
Using the zoom and trace features, we find that $x = -1.14, \; 0.97$.

100. $4x^{2/3} + 8x^{1/3} + 3.6 = 0$
Using the zoom and trace features, we find that $x = -2.28, \; -0.32$

102. (a) $0 = -16t^2 + 32{,}000$

$16t^2 = 32{,}000$

$t^2 = 2000$

$t = \pm\sqrt{2000}$

$t = \pm 20\sqrt{5} \approx \pm 44.72$

Since $t > 0$, it will take $20\sqrt{5}$ or ≈ 44.72 seconds for the bomb to hit the ground.

(b) Distance $=$ rate $\times$ time

$$d = 600\frac{\text{miles}}{\text{hour}} \times \left(20\sqrt{5}\ \text{seconds} \times \frac{1\ \text{hour}}{3600\ \text{seconds}}\right)$$

$$d = \frac{600(20\sqrt{5})}{3600}$$

$$d \approx 7.5$$

The bomb traveled a horizontal distance of ≈ 7.5 miles.

104. $2x^2 = 200$

$x^2 = 100$

$x = \pm 10$

Choosing a positive value we have $x = 10$. The original piece was $x + 2 + 2 = 14$ inches square.

106. Amount of monthly rent $= R$

$$\frac{R}{3} - \frac{R}{4} = 75$$

$$12\left(\frac{R}{3}\right) - 12\left(\frac{R}{4}\right) = 12(75)$$

$$4R - 3R = 900$$

$$R = 900$$

The monthly rent is $900.

108. Distance out $= 1080 = rt$

Distance back $= 1080 = (r-6)\left(t + \frac{5}{2}\right)$

$$t = \frac{1080}{r}$$

$$(r-6)\left(\frac{1080}{r} + \frac{5}{2}\right) = 1080$$

$$(r-6)(2160 + 5r) = 2160r$$

$$5r^2 - 30r - 12{,}960 = 0$$

$$r = \frac{30 \pm \sqrt{(-30)^2 - 4(5)(-12{,}960)}}{2(5)}$$

$$= \frac{30 \pm 510}{10} = 3 \pm 51$$

The speed on the way to the lodge was $r = 54$ mph.

110. (a) $C = \sqrt{0.2x + 1}$

(b) If $C = 2.5$, then $x = 26.25$, or 26,250 passengers.

112.
$$13.95 = 30 - \sqrt{0.0001x + 1}$$
$$\sqrt{0.0001x + 1} = 16.05$$
$$0.0001x + 1 = 257.6025$$
$$0.0001x = 256.6025$$
$$x = 2{,}566{,}025$$

The number of units demanded each day is 2,566,025.

2.5 Solving Inequalities Algebraically and Graphically

2. $x \geq 6$ indicates all points to the right of and including $x = 6$; graph (h).

4. $0 \leq x \leq \frac{7}{2}$ indicates all points between $x = 0$ and $x = \frac{7}{2}$, including the endpoints; graph (e).

6. $|x| > 3$ indicates all points more than 3 units from $x = 0$; graph (a).

8. $|x + 6| < 3$

$-3 < x + 6 < 3$

$-9 < x < -3$

Graph (d)

10. $2x > 3$

$x > \frac{3}{2}$

12. $-6x > 15$

$x < -\frac{5}{2}$

14. $x + 7 \leq 12$

$x \leq 5$

16. $3x + 1 \geq 2$

$3x \geq 1$

$x \geq \frac{1}{3}$

18. $-8 \leq 1 - 3(x - 2) < 13$

$-9 \leq \quad -3(x - 2) \quad < 12$

$3 \geq \quad\quad x - 2 \quad > -4$

$5 \geq \quad\quad\quad x \quad\quad > -2$

20. $0 \leq \dfrac{x + 3}{2} < 5$

$0 \leq x + 3 < 10$

$-3 \leq \quad x \quad < 7$

22. $-1 < -\dfrac{x}{3} < 1$

$3 > \quad x \quad > -3$

24. $|2x| < 6$

$-6 < 2x < 6$

$-3 < \quad x < 3$

26. $|5x| > 10$

$5x < -10 \quad \text{or} \quad 5x > 10$

$x < -2 \quad\quad\quad\quad x > 2$

28. $|x - 7| < 6$

$-6 < x - 7 < 6$

$1 < \quad x \quad < 13$

30. $|x + 14| + 3 > 17$

$|x + 14| > 14$

$x + 14 < -14 \quad \text{or} \quad x + 14 > 14$

$x < -28 \quad\quad\quad\quad x > 0$

32. $\left| 1 - \dfrac{2x}{3} \right| < 1$

$-1 < 1 - \dfrac{2x}{3} < 1$

$-2 < \quad -\dfrac{2x}{3} \quad < 0$

$3 > \quad\quad x \quad\quad > 0$

34. $|x - 5| \geq 0$

All real numbers x

36. Graphing $y_1 = 2x + 7 - 3 = 2x + 4$ on the standard viewing rectangle shows that the solution to $2x + 7 < 3$ is $x < -2$.

38. Graphing

$$2x - 9 - \frac{-4x + 3}{3}$$

on the standard viewing rectangle shows that the solution is $x < 3$.

40. Graphing

$$y_1 = -1 - 13 + 2x = -14 + 2x$$

and

$$y_2 = 13 - 2x - 9 = 4 - 2x$$

on the same viewing rectangle shows that the solution to $-1 \le 13 - 2x \le 9$ is $2 \le x \le 7$.

42. Graphing $y_1 = |1 - 2x| - 5$ on the standard viewing rectangle shows that the solution is $-2 < x < 3$.

44. Graphing $y_1 = 3|4 - 5x| - 9$ on $[-5, 5] \times [-10, 10]$ shows that the solution is $0.2 \le x \le 1.4$.

46. $|x| > 2$

48. $|x + 2| \le 4$

50. $|x - 8| \ge 5$

52. $|x - (-6)| \le 7$

$|x + 6| \le 7$

54.
$$(x - 3)^2 \ge 1$$
$$x^2 - 6x + 8 \ge 0$$
$$(x - 2)(x - 4) \ge 0$$

Critical numbers: $x = 2$, $x = 4$
Test intervals:
$$(-\infty,\ 2) \Rightarrow (x - 2)(x - 4) > 0$$
$$(2,\ 4) \Rightarrow (x - 2)(x - 4) < 0$$
$$(4,\ \infty) \Rightarrow (x - 2)(x - 4) > 0$$
Solution intervals: $(-\infty,\ 2] \cup [4,\ \infty)$

56.
$$x^2 - 6x + 9 < 16$$
$$x^2 - 6x - 7 < 0$$
$$(x + 1)(x - 7) < 0$$

Critical numbers: $x = -1$, $x = 7$
Test intervals:
$$(-\infty,\ -1) \Rightarrow (x + 1)(x - 7) > 0$$
$$(-1,\ 7) \Rightarrow (x + 1)(x - 7) < 0$$
$$(7,\ \infty) \Rightarrow (x + 1)(x - 7) > 0$$
Solution interval: $(-1,\ 7)$

58.
$$x^2 + 2x > 3$$
$$x^2 + 2x - 3 > 0$$
$$(x + 3)(x - 1) > 0$$

Critical numbers: $x = -3$, $x = 1$
Test intervals:
$$(-\infty,\ -3) \Rightarrow (x + 3)(x - 1) > 0$$
$$(-3,\ 1) \Rightarrow (x + 3)(x - 1) < 0$$
$$(1,\ \infty) \Rightarrow (x + 3)(x - 1) > 0$$
Solution intervals: $(-\infty,\ -3) \cup (1,\ \infty)$

60. $6(x + 2)(x - 1) < 0$
Critical numbers: $x = -2$, $x = 1$
Test intervals:
$$(-\infty,\ -2) \Rightarrow 6(x + 2)(x - 1) > 0$$
$$(-2,\ 1) \Rightarrow 6(x + 2)(x - 1) < 0$$
$$(1,\ \infty) \Rightarrow 6(x + 2)(x - 1) > 0$$
Solution interval: $(-2,\ 1)$

62. $x^2 - 4x - 1 > 0$

$$x = \frac{4 \pm \sqrt{16+4}}{2} = 2 \pm \sqrt{5}$$

Critical numbers: $x = 2 - \sqrt{5}$, $x = 2 + \sqrt{5}$
Test intervals:

$$(-\infty,\ 2 - \sqrt{5}) \Rightarrow x^2 - 4x - 1 > 0$$
$$(2 - \sqrt{5},\ 2 + \sqrt{5}) \Rightarrow x^2 - 4x - 1 < 0$$
$$(2 + \sqrt{5},\ \infty) \Rightarrow x^2 - 4x - 1 > 0$$

Solution intervals: $(-\infty,\ 2 - \sqrt{5}) \cup (2 + \sqrt{5},\ \infty)$

64. $4x^3 - 12x^2 > 0$

$$4x^2(x - 3) > 0$$

Critical numbers: $x = 0$, $x = 3$
Test intervals:

$$(-\infty,\ 0) \Rightarrow 4x^2(x - 3) < 0$$
$$(0,\ 3) \Rightarrow 4x^2(x - 3) < 0$$
$$(3,\ \infty) \Rightarrow 4x^2(x - 3) > 0$$

Solution interval: $(3,\ \infty)$

66. $2x^3 - x^4 \leq 0$

$$x^3(2 - x) \leq 0$$

Critical numbers: $x = 0$, $x = 2$
Test intervals:

$$(-\infty,\ 0) \Rightarrow x^3(2 - x) < 0$$
$$(0,\ 2) \Rightarrow x^3(2 - x) > 0$$
$$(2,\ \infty) \Rightarrow x^3(2 - x) < 0$$

Solution intervals: $(-\infty,\ 0] \cup [2,\ \infty)$

68. $\dfrac{1}{x} < 4$

$$\frac{1}{x} - 4 < 0$$
$$\frac{1 - 4x}{x} < 0$$

Critical numbers: $x = 0$, $x = \frac{1}{4}$
Test intervals:

$$(-\infty,\ 0) \Rightarrow \frac{1 - 4x}{x} < 0$$
$$\left(0,\ \frac{1}{4}\right) \Rightarrow \frac{1 - 4x}{x} > 0$$
$$\left(\frac{1}{4},\ \infty\right) \Rightarrow \frac{1 - 4x}{x} < 0$$

Solution interval: $(-\infty,\ 0) \cup \left(\frac{1}{4},\ \infty\right)$

70. $\dfrac{x+12}{x+2} \geq 3$

$\dfrac{x+12}{x+2} - 3 \geq 0$

$\dfrac{x+12-3(x+2)}{x+2} \geq 0$

$\dfrac{6-2x}{x+2} \geq 0$

Critical numbers: $x = -2$, $x = 3$
Test intervals:

$(-\infty, -2) \Rightarrow \dfrac{6-2x}{x+2} < 0$

$(-2, 3) \Rightarrow \dfrac{6-2x}{x+2} > 0$

$(3, \infty) \Rightarrow \dfrac{6-2x}{x+2} < 0$

Solution interval: $(-2, 3]$

74. $\dfrac{1}{x-3} \leq \dfrac{9}{4x+3}$

$\dfrac{4x+3-9(x-3)}{(x-3)(4x+3)} \leq 0$

$\dfrac{30-5x}{(x-3)(4x+3)} \leq 0$

Critical numbers: $x = -\frac{3}{4}$, $x = 3$, $x = 6$
Test intervals:

$\left(-\infty, -\dfrac{3}{4}\right) \Rightarrow \dfrac{30-5x}{(x-3)(4x+3)} > 0$

$\left(-\dfrac{3}{4}, 3\right) \Rightarrow \dfrac{30-5x}{(x-3)(4x+3)} < 0$

$(3, 6) \Rightarrow \dfrac{30-5x}{(x-3)(4x+3)} > 0$

$(6, \infty) \Rightarrow \dfrac{30-5x}{(x-3)(4x+3)} < 0$

Solution intervals: $\left(-\frac{3}{4}, 3\right) \cup [6, \infty)$

72. $\dfrac{5+7x}{1+2x} < 4$

$\dfrac{5+7x-4(1+2x)}{1+2x} < 0$

$\dfrac{1-x}{1+2x} < 0$

Critical numbers: $x = -\frac{1}{2}$, $x = 1$
Test intervals:

$\left(-\infty, -\dfrac{1}{2}\right) \Rightarrow \dfrac{1-x}{1+2x} < 0$

$\left(-\dfrac{1}{2}, 1\right) \Rightarrow \dfrac{1-x}{1+2x} > 0$

$(1, \infty) \Rightarrow \dfrac{1-x}{1+2x} < 0$

Solution intervals: $\left(-\infty, -\frac{1}{2}\right) \cup (1, \infty)$

76. $|x^3 + 6x - 1| \geq 2x + 4$

Graphing $y_1 = |x^3 + 6x - 1|$ and $y_2 = 2x + 4$ on the same viewing rectangle, we obtain $x \geq 1$ or $x \leq -0.369$.

78. $x^4(x-3) \leq 0$

Using the zoom and trace features, we find that $x \leq 3$.

80. $1.2x^2 + 4.8x + 3.1 < 5.3$

Using the zoom and trace features, we find that $-4.42 < x < 0.42$.

82. $\dfrac{1}{x} \geq \dfrac{1}{x+3}$

Using the zoom and trace features, we find that $x < -3$ or $x > 0$.

84. $\dfrac{x^2 + x - 6}{x} \geq 0$

Using the zoom and trace features, we find that $-3 \leq x < 0$ or $x \geq 2$.

86. The radicand of $\sqrt{x-10}$ is $x - 10$.

$$x - 10 \geq 0$$

$$x \geq 10$$

Therefore, the interval is $[10, \infty)$.

88. The radicand of $\sqrt[4]{6x+15}$ is $6x + 15$.

$$6x + 15 \geq 0$$

$$6x \geq -15$$

$$x \geq -\tfrac{5}{2}$$

Therefore, the interval is $\left[-\tfrac{5}{2}, \infty\right)$.

90.
$$x^2 - 4 \geq 0$$

$$(x + 2)(x - 2) \geq 0$$

Critical numbers: $x = -2$, $x = 2$
Test intervals:

$$(-\infty, -2) \Rightarrow (x+2)(x-2) > 0$$

$$(-2, 2) \Rightarrow (x+2)(x-2) < 0$$

$$(2, \infty) \Rightarrow (x+2)(x-2) > 0$$

Domain: $(-\infty, -2] \cup [2, \infty)$

92.
$$144 - 9x^2 \geq 0$$

$$9(4 - x)(4 + x) \geq 0$$

Critical numbers: $x = -4$, $x = 4$
Test intervals:

$$(-\infty, -4) \Rightarrow 9(4-x)(4+x) < 0$$

$$(-4, 4) \Rightarrow 9(4-x)(4+x) > 0$$

$$(4, \infty) \Rightarrow 9(4-x)(4+x) < 0$$

Domain: $[-4, 4]$

94. $\sqrt{x^2 + 4}$
$x^2 + 4 \geq 0$ is true for all x. Hence, the domain is all real numbers.

96. $\dfrac{x}{x+4} - \dfrac{3x}{x-1} \geq 0$
Using the zoom and trace features, the domain is $(-6.5, -4) \cup [0, 1)$.

98. $C = 15.40x + 150{,}000$
$R = 24.55x$

Using the zoom and trace features, we find that $R > C$ when $x > 16{,}393.4$, or $x \geq 16{,}394$ units.

100. $10,000 > 0.32m + 2300$

$7700 > 0.32m$

$24,062.5 > m$

102. $10,000 > 0.32m + 2300$

$S > 35,000$ if $t \geq 12.623$ years.

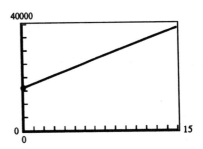

104. $|h - 50| \leq 30$

$-30 \leq h - 50 \leq 30$

$20 \leq \quad h \quad \leq 80$

106. $2L + 2W = 100$

$$W = \frac{100 - 2L}{2} = 50 - L$$

$LW \geq 500$

$L(50 - L) \geq 500$

$50L - L^2 \geq 500$

$0 \geq L^2 - 50L + 500$

Using a graphing utility to graph $y_1 = x^2 - 50x + 500$ on the viewing rectangle $[0, 50] \times [-150, 150]$, we see that $y_1 \leq 0$ on $[13.82, 36.18]$. Algebraically, we obtain:

Critical numbers:

$$L = \frac{50 \pm \sqrt{(50)^2 - 4(500)}}{2} = \frac{50 \pm \sqrt{500}}{2} = \frac{50 \pm 10\sqrt{5}}{2} = 25 \pm 5\sqrt{5}$$

Test intervals:

$$(0, \ 25 - 5\sqrt{5}) \Rightarrow L^2 - 50L + 500 > 0$$

$$(25 - 5\sqrt{5}, \ 25 + 5\sqrt{5}) \Rightarrow L^2 - 50L + 500 < 0$$

$$(25 + 5\sqrt{5}, \ 50) \Rightarrow L^2 - 50L + 500 > 0$$

Solution interval: $[25 - 5\sqrt{5}, \ 25 + 5\sqrt{5}]$ or 13.8197 meters $\leq L \leq 36.1803$ meters

108. (a) $R = x(50 - 0.0002x)$

$C = 12x + 150{,}000$

$P = R - C$

$= x(50 - 0.0002x) - (12x + 150{,}000)$

(b)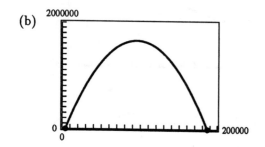

(c) If $P \geq 1{,}650{,}000$, then $90{,}000 \leq x \leq 100{,}000$.

2.6 | Exploring Data: Linear Models and Scatter Plots

2. Since y tends to increase as x increases, the points are positively correlated.

4. No, the model ∪ not accurate for large values of x.

6. Since v tends to decrease as h increases, the points are negatively correlated.

8. Since the model yields $v = 834$ ft/sec when $h = 70$, we see that the model is inaccurate for large values of h.

10. (a)

(b) $y = -x + 4$ fits the data well, and the sum of the squared differences is

$0^2 + 0^2 + 0^2 + 1^2 = 1$

(c)

x	y	xy	x^2
0	4	0	0
1	3	3	1
2	2	4	4
4	1	4	16
$\sum x_i = 7$	$\sum y_i = 10$	$\sum x_i y_i = 11$	$\sum x_i^2 = 21$

Using the formula for the least squares regression line with $n = 4$ produces

$$a = \frac{4(11) - 7(10)}{4(21) - (7)^2} = \frac{-26}{35}$$

$$b = \frac{1}{4}\left(10 - \frac{-26}{35}(7)\right) = \frac{1}{4}\left(10 + \frac{26}{5}\right) = \frac{76}{20} = \frac{19}{5}$$

Thus, the least squares regression line is $y = \dfrac{-26}{35}x + \dfrac{19}{5}$.

The sum of the squared differences is

$$(4 - 3.8)^2 + \left(3 - \frac{107}{35}\right)^2 + \left(2 - \frac{81}{35}\right)^2 + \left(1 - \frac{29}{35}\right)^2 \approx 0.17143$$

12. (a)

(b)

x	y	xy	x^2
-3	1	-3	9
-1	2	-2	1
0	2	0	0
1	3	3	1
3	5	15	9
$\sum x_i = 0$	$\sum y_i = 13$	$\sum x_i y_i = 13$	$\sum x_i^2 = 20$

$$a = \frac{5(13) - (0)(13)}{5(20) - (0)^2} = \frac{65}{100} = .65$$

$$b = \frac{1}{5}(13 - .65(0)) = \frac{13}{5} = 2.6$$

$$y = 0.65x + 2.6$$

14. (a)

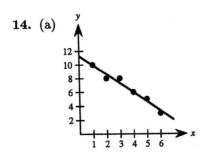

(b)

x	y	xy	x^2
1	10	10	1
2	8	16	4
3	8	24	9
4	6	24	16
5	5	25	25
6	3	18	36
$\sum x_i = 21$	$\sum y_i = 40$	$\sum x_i y_i = 117$	$\sum x_i^2 = 91$

$$a = \frac{6(117) - 21(40)}{6(91) - (21)^2} \approx -1.3143$$

$$b = \frac{1}{6}(40 - (-1.3143)(21)) \approx 11.2667$$

$$y = -1.3143x + 11.2667$$

16. $y = 2.2286x + 43.0857$

18. $y = -6.0512x + 148.0349$

20. (a) $y = 0.0989t + 10.2214$

When $t = 12$ (1992),
$y \approx 11.41$.

(b)

(c) Correlation coefficient
≈ 0.9752

Review Exercises for Chapter 2

2. $3(x-2)+2x = 2(x+3)$

$3x - 6 + 2x = 2x + 6$

$5x - 6 = 2x + 6$

$3x = 12$

$x = 4$ Conditional

4. (a) $6 + \dfrac{3}{4-4} \overset{?}{=} 5$

$6 + \dfrac{3}{0} \neq 5$

$x = 4$ *is not* a solution.

(c) $6 + \dfrac{3}{-2-4} \overset{?}{=} 5$

$5\frac{1}{2} \neq 5$

$x = -2$ *is not* a solution.

(b) $6 + \dfrac{3}{0-4} \overset{?}{=} 5$

$5\frac{1}{4} \neq 5$

$x = 0$ *is not* a solution.

(d) $6 + \dfrac{3}{1-4} \overset{?}{=} 5$

$5 = 5$

$x = 1$ *is* a solution.

6. $4(x+3) - 3 = 2(4 - 3x) - 4$

$4x + 12 - 3 = 8 - 6x - 4$

$10x = -5$

$x = -\frac{1}{2}$

8. $\dfrac{1}{x-2} = 3$

$1 = 3(x-2)$

$1 = 3x - 6$

$7 = 3x$

$\dfrac{7}{3} = x$

10. $15 + x - 2x^2 = 0$

$2x^2 - x - 15 = 0$

$(2x+5)(x-3) = 0$

$2x + 5 = 0 \ \Rightarrow \ x = -\frac{5}{2}$

$x - 3 = 0 \ \Rightarrow \ x = \ \ 3$

12. $16x^2 = 25$

$x^2 = \frac{25}{16}$

$x = \pm\frac{5}{4}$

14. $5x^4 - 12x^3 = 0$

$x^3(5x - 12) = 0$

$x^3 = 0 \ \Rightarrow \ x = \ \ 0$

$5x - 12 = 0 \ \Rightarrow \ x = \frac{12}{5}$

16. $2 - x^{-2} = 0$

$2 = \dfrac{1}{x^2}$

$2x^2 = 1$

$x^2 = \dfrac{1}{2}$

$x = \pm\dfrac{\sqrt{2}}{2}$

18.
$$\frac{4}{x-3} - \frac{4}{x} = 1$$

$$x(x-3)\frac{4}{x-3} - x(x-3)\frac{4}{x} = x(x-3)(1)$$

$$4x - 4x + 12 = x^2 - 3x$$

$$x^2 - 3x - 12 = 0$$

$$x = \frac{-(-3) \pm \sqrt{(-3)^2 - 4(1)(-12)}}{2(1)}$$

$$x = \frac{3 \pm \sqrt{57}}{2}$$

20. $\sqrt{x-2} - 8 = 0$

$$(\sqrt{x-2})^2 = 8^2$$

$$x - 2 = 64$$

$$x = 66$$

22. $\sqrt{2x+3} + \sqrt{x-2} = 2$

$$(\sqrt{2x+3})^2 = (2 - \sqrt{x-2})^2$$

$$2x + 3 = 4 - 4\sqrt{x-2} + x - 2$$

$$x + 1 = -4\sqrt{x-2}$$

$$(x+1)^2 = (-4\sqrt{x-2})^2$$

$$x^2 + 2x + 1 = 16(x-2)$$

$$x^2 - 14x + 33 = 0$$

$$(x-3)(x-11) = 0$$

$x - 3 = 0 \Rightarrow x = 3$, extraneous

$x - 11 = 0 \Rightarrow x = 11$, extraneous

No solution

24. $(x+2)^{3/4} = 27$

$$x + 2 = 27^{4/3}$$

$$x = -2 + 81$$

$$= 79$$

26.
$$8x^2(x^2-4)^{1/3} + (x^2-4)^{4/3} = 0$$

$$(x^2-4)^{1/3}[8x^2 + (x^2-4)] = 0$$

$$(x^2-4)^{1/3}(9x^2-4) = 0$$

$$(x+2)^{1/3}(x-2)^{1/3}(3x+2)(3x-2) = 0$$

$x + 2 = 0 \Rightarrow x = -2$

$x - 2 = 0 \Rightarrow x = 2$

$3x + 2 = 0 \Rightarrow x = -\frac{2}{3}$

$3x - 2 = 0 \Rightarrow x = \frac{2}{3}$

28. $|x^2 - 6| = x$

$$x^2 - 6 = x \qquad\qquad \text{or} \qquad\qquad -(x^2 - 6) = x$$

$$x^2 - x - 6 = 0 \qquad\qquad\qquad\qquad x^2 + x - 6 = 0$$

$$(x - 3)(x + 2) = 0 \qquad\qquad\qquad (x + 3)(x - 2) = 0$$

$$x - 3 = 0 \;\Rightarrow\; x = 3 \qquad\qquad x + 3 = 0 \;\Rightarrow\; x = -3, \;\text{extraneous}$$

$$x + 2 = 0 \;\Rightarrow\; x = -2, \;\text{extraneous} \qquad x - 2 = 0 \;\Rightarrow\; x = 2$$

30. $12t^3 - 84t^2 + 120t = 0$
Using the zoom and trace features, we find that $t = 0, 2, 5$.

32. $\sqrt{3x - 2} = 4 - x$
Using the zoom and trace features, we find that $x = 2.0$.

34. $\dfrac{1}{(t + 1)^2} = 1$
Using the zoom and trace features, we find that $t = 0, -2$.

36. $|2x + 3| = 7$
Using the zoom and trace features, we find that $x = 2, -5$.

38. $Z = \sqrt{R^2 - X^2}$

$$Z^2 = R^2 - X^2$$

$$X^2 = R^2 - Z^2$$

$$X = \pm\sqrt{R^2 - Z^2}$$

40.
$$E = 2kw\left(\frac{v}{2}\right)^2$$

$$E = \frac{kw}{2}(v)^2$$

$$\frac{2E}{kw} = v^2$$

$$\pm\sqrt{\frac{2E}{kw}} = v$$

42. $\left(\dfrac{\sqrt{2}}{2} - \dfrac{\sqrt{2}}{2}i\right) - \left(\dfrac{\sqrt{2}}{2} + \dfrac{\sqrt{2}}{2}i\right) = -\sqrt{2}\,i$

44. $i(6 + i)(3 - 2i) = (6i - 1)(3 - 2i)$

$$= 18i + 12 - 3 + 2i$$

$$= 9 + 20i$$

46. $\dfrac{3 + 2i}{5 + i} = \dfrac{3 + 2i}{5 + i} \cdot \dfrac{5 - i}{5 - i}$

$$= \frac{15 + 2 + 10i - 3i}{25 + 1}$$

$$= \frac{17}{26} + \frac{7}{26}i$$

48. $f(x) = (x - 5)[x - (1 - \sqrt{2})][x - (1 + \sqrt{2})]$

$$= (x - 5)[(x - 1)^2 - 2]$$

$$= (x - 5)(x^2 - 2x - 1)$$

$$= x^3 - 7x^2 + 9x + 5$$

50. $f(x) = (x - 2)(x + 3)[x - (1 - 2i)][x - (1 + 2i)]$

$$= (x^2 + x - 6)[(x - 1)^2 + 4]$$

$$= (x^2 + x - 6)(x^2 - 2x + 5)$$

$$= x^4 - x^3 - 3x^2 + 17x - 30$$

52.
$$x^2 - 2x \geq 3$$
$$x^2 - 2x - 3 \geq 0$$
$$(x+1)(x-3) \geq 0$$

Test intervals:
$$(-\infty, -1) \Rightarrow x^2 - 2x > 3$$
$$(-1, 3) \Rightarrow x^2 - 2x < 3$$
$$(3, \infty) \Rightarrow x^2 - 2x > 3$$

Solution intervals: $(-\infty, 1] \cup [3, \infty)$

54.
$$\frac{2}{x+1} \leq \frac{3}{x-1}$$
$$\frac{2(x-1) - 3(x+1)}{(x+1)(x-1)} \leq 0$$
$$\frac{-(x+5)}{(x+1)(x-1)} \leq 0$$

Test intervals:
$$(-\infty, -5) \Rightarrow \frac{-(x+5)}{(x+1)(x-1)} > 0$$
$$(-5, -1) \Rightarrow \frac{-(x+5)}{(x+1)(x-1)} < 0$$
$$(-1, 1) \Rightarrow \frac{-(x+5)}{(x+1)(x-1)} > 0$$
$$(1, \infty) \Rightarrow \frac{-(x+5)}{(x+1)(x-1)} < 0$$

Solution intervals: $[-5, -1) \cup (1, \infty)$

56. $|x| \leq 4$
$$-4 \leq x \leq 4 \text{ or } [-4, 4]$$

58. $|x+3| > 4$
$$x + 3 < -4 \text{ or } x + 3 > 4 \text{ or } (-\infty, -7) \cup$$
$$x < -7 \qquad x > 1$$

$$(1, \infty)$$

60. $2x^2 + x \geq 15$
Using the zoom and trace features, we find that $(-\infty, -3] \cup [2.5, \infty)$.

62. $|x(x-6)| < 5$
Using the zoom and trace features, we find that $(5, 6.742) \cup (-0.742, 1)$.

64. $x(x-4) \geq 0$
Test intervals:
$$(-\infty, 0) \Rightarrow x(x-4) > 0$$
$$(0, 4) \Rightarrow x(x-4) < 0$$
$$(4, \infty) \Rightarrow x(x-4) > 0$$

Solution intervals: $(-\infty, 0] \cup [4, \infty)$

66. Discount $= p$
$$(340 + 85)p = 85$$
$$p = \frac{85}{425}$$
$$p = 0.2$$

Discount $= p = 0.2 = 20\%$

68. Let $l = $ the distance of the straight portion of the track. Let $r = $ the radius of the semicircle of the inner track and $r + 3 = $ the radius of the semicircle of the outer track. Let $d_1 = $ the distance of the inner track.

$$d_1 = 2l + 2\pi r$$

Let $d_2 = $ the distance of the outer track.

$$d_2 = 2l + 2pi(r - 3)$$

Let $d = $ the distance between starting positions.

$$d = d_1 - d_2$$
$$= 2l + 2\pi r - [2l + 2\pi(r - 3)]$$
$$= 2l + 2\pi r - (2l + 2\pi r - 6\pi)$$
$$= 6\pi \text{ feet}$$

70. Let $r = $ speed in mph on the return trip. Let $t = $ time in hours for the return trip.

$$d = rt$$
$$56 = rt$$
$$t = \frac{56}{r}$$

Let $(r - 8) = $ speed in mph going to the service call. Let $\left(t + \frac{1}{6}\right) = $ time in hours going to the service call.

$$56 = (r - 8)\left(t + \frac{1}{6}\right)$$
$$\frac{56}{r - 8} = t + \frac{1}{6}$$
$$\frac{56}{r - 8} - \frac{1}{6} = t$$
$$\frac{56}{r - 8} - \frac{1}{6} = \frac{56}{r}$$
$$(6r)(r - 8)\frac{56}{r - 8} - (6r)(r - 8)\frac{1}{6} = (6r)(r - 8)\frac{56}{r}$$
$$336r - (r^2 - 8r) = 336r - 2688$$
$$r^2 - 8r - 2688 = 0$$
$$(r + 48)(r - 56) = 0$$

$$r + 48 = 0 \implies r = -48$$
$$r - 56 = 0 \implies r = 56$$

Since $r > 0$, the average speed of the return trip was 56 mph.

72. Let x = the number of current investors.

Cost per investor = $90,000/x$

If three more investors join, the cost per investor will be $90,000/(x+3)$.

$$\frac{90,000}{x} - \frac{90,000}{x+3} = 2500$$

$$90,000(x+3) - 90,000(x) = 2500(x)(x+3)$$

$$36(x+3) - 36x = x(x+3)$$

$$36x + 108 - 36x = x^2 + 3x$$

$$x^2 + 3x - 108 = 0$$

$$(x+12)(x-9) = 0$$

$x + 12 = 0 \Rightarrow x = -12,$ extraneous

$x - 9 = 0 \Rightarrow x = \quad 9$

There are currently 9 investors.

74. $1000(1+r)^5 > 1400$

$(1+r)^5 > 1.4$

$1 + r > 1.0696$

$r > 0.0696 = 6.96\%$

76. (a) $M = 500x(20 - x)$

Domain: $0 \le x \le 20$

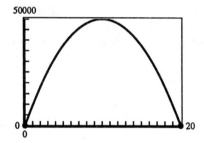

(b) $x = 0,\ x = 20$

(c) Greatest when $x = 10,\ M = 50,000$

(d) $M < 40,000$ for $x < 5.53$ or $x > 14.47$

78. (a) $y = 0.07974x - 7.4541$

correlation coefficient = 0.7823

(c) If $x = 128$, then $y \approx 2.75$ grade point average.

(d) IQ and grade point average are not linearly related.

(b)

Cumulative Test, Chapters 1—2

1. $\left(\dfrac{12}{5} \div 9\right) - \left(10 - \dfrac{1}{45}\right) = \left(\dfrac{12}{5} \cdot \dfrac{1}{9}\right) - \left(\dfrac{450 - 1}{45}\right)$

$\qquad\qquad\qquad\qquad\qquad = \dfrac{12}{45} - \dfrac{449}{45}$

$\qquad\qquad\qquad\qquad\qquad = -\dfrac{437}{45}$

2. $\dfrac{8x^2 y^{-3}}{30x^{-1}y^2} = \dfrac{8x^3}{30y^5} = \dfrac{4x^3}{15y^5}$

3. $\sqrt{24x^4 y^3} = \sqrt{2 \cdot 2 \cdot 2 \cdot 3 x^4 y^3} = 2x^2 |y| \sqrt{6y}$

4. $(x - 2)(x^2 + x - 3) = x^3 + x^2 - 3x - 2x^2 - 2x + 6$

$\qquad\qquad\qquad\qquad = x^3 - x^2 - 5x + 6$

5. $x - 5x^2 - 6x^3 = -x(6x^2 + 5x - 1)$

$\qquad\qquad\qquad = -x(x + 1)(6x - 1)$

6. $\dfrac{2}{s+3} - \dfrac{1}{s+1} = \dfrac{2(s+1) - 1(s+3)}{(s+3)(s+1)}$

$\qquad\qquad\qquad = \dfrac{s - 1}{(s+3)(s+1)}$

7. $\dfrac{\dfrac{2x+1}{2\sqrt{x}} - \sqrt{x}}{2x + 1} = \dfrac{2x + 1 - 2x}{(2x+1)2\sqrt{x}} = \dfrac{1}{2(2x+1)\sqrt{x}}$

8. (a) $x - 3y + 12 = 0$

$\qquad\quad -3y = -x - 12$

$\qquad\qquad y = \dfrac{x}{3} + 4$

(b) $y = x^2 - 4x + 1$

$\qquad = x^2 - 4x + 4 - 3$

$\qquad = (x - 2)^2 - 3$

–CONTINUED ON NEXT PAGE–

8. –CONTINUED–

(c) $y = \sqrt{4-x}$

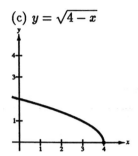

(d) $y = 3 - |x-2|$

9.
$$2x - C(y+1) = 10$$
$$2(2) - C(-3+1) = 10$$
$$4 + 2C = 10$$
$$2C = 6$$
$$C = 3$$

10. Since $m = -\frac{2}{3}$, the line falls 2 units for every 3 units to the right. Given the point $(4,\ 9)$ on the line, 3 additional points are $(7,\ 7)$, $(10,\ 5)$ and $(1,\ 11)$.

11. The line passing through $(-\frac{1}{2},\ 1)$ and $(3,\ 8)$ has slope
$$m = \frac{8-1}{3-(-1/2)} = \frac{7}{7/2} = 2.$$
Hence,
$$y - 8 = 2(x-3)$$
$$y = 2x + 2 \quad \text{or} \quad 2x - y + 2 = 0.$$

12. $f(x) = \dfrac{x}{x-2}$

(a) $f(6) = \dfrac{6}{6-2} = \dfrac{6}{4} = \dfrac{3}{2}$

(b) $f(2)$ is undefined (division by zero).

(c) $f(2t) = \dfrac{2t}{2t-2} = \dfrac{t}{t-1}$

(d) $f(s+2) = \dfrac{s+2}{(s+2)-2} = \dfrac{s+2}{s}$

13. The domain consists of all x such that $3 - x > 0$, or $x < 3$. The range is all $y > 0$.

14. $A = \dfrac{1}{2}bh$, where $b = s$ and $h^2 + \left(\dfrac{s}{2}\right)^2 = s^2$ or $h = \dfrac{\sqrt{3}}{2}s$. Hence, $A = \dfrac{1}{2}s \cdot \dfrac{\sqrt{3}}{2}s = \dfrac{\sqrt{3}}{4}s^2$.

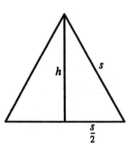

15. For some values of x there corresponds more than one value of y.

16. $y = \sqrt[3]{x}$

    (a) f is a horizontal stretch by a factor of 2.

    (b) r is a vertical shrink by a factor of $1/2$.

    (c) h is a vertical translation of 2 units upward.

    (d) g is a horizontal translation of 2 units to the left.

17. $f(x) = \sqrt{x}$, $g(x) = x^2 + 3$

    $f \circ g(x) = f(g(x))$

    $= f(x^2 + 3)$

    $= \sqrt{x^2 + 3}$

18. $h(x) = 5x - 2$ is one-to-one, since if $5a - 2 = 5b - 2$, then $5a = 5b$ and $a = b$. To find the inverse of h, let

$$y = 5x - 2$$

$$x = 5y - 2 \qquad \text{Interchange } x \text{ and } y.$$

$$x + 2 = 5y$$

$$y = \tfrac{1}{5}(x + 2) \qquad \text{Solve for } y.$$

$$h^{-1}(x) = \tfrac{1}{5}(x + 2)$$

19. (a) $2x - 3(x - 4) = 5$

    $2x - 3x + 12 = 5$

    $-x = -7$

    $x = 7$

(b) $\dfrac{2}{t - 3} + \dfrac{2}{t - 2} = \dfrac{10}{t^2 - 5t + 6}$

    $2(t - 2) + 2(t - 3) = 10$

    $4t - 10 = 10$

    $4t = 20$

    $t = 5$

(c) $3y^2 + 6y + 2 = 0$

$$y = \frac{-6 \pm \sqrt{36 - 4(3)(2)}}{2(3)}$$

$$= \frac{-6 \pm \sqrt{12}}{6}$$

$$= \frac{-3 \pm \sqrt{3}}{3}$$

(d) $\sqrt{x + 10} = x - 2$

    $x + 10 = (x - 2)^2$

    $= x^2 - 4x + 4$

    $0 = x^2 - 5x - 6$

    $= (x - 6)(x + 1)$

    $x = 6$

    ($x = -1$ is not a solution.)

20. $\sqrt{x+4} + \sqrt{x} = 6$

$f(x) = \sqrt{x+4} + \sqrt{x} - 6 = 0$

Using the zoom and trace features, there is one solution, $x = 7.1$.

21. $\left|\dfrac{x-2}{3}\right| < 1 \Leftrightarrow |x-2| < 3$

$\Leftrightarrow -3 < x - 2 < 3$

$\Leftrightarrow -1 < x < 5$

22. $x^2 - x < 6 \Leftrightarrow x^2 - x - 6 < 0$

$\Leftrightarrow (x-3)(x+2) < 0$

$\Leftrightarrow -2 < x < 3$

23. $\qquad x + y = 12{,}000 \Rightarrow y = 12{,}000 - x$

$0.075x + 0.09y = 960 \Rightarrow 7.5x + 9y = 96{,}000$

Hence,

$7.5x + 9(12{,}000 - x) = 96{,}000$

$-1.5x + 108{,}000 = 96{,}000$

$12{,}000 = 1.5x$

$8000 = x$

$\Rightarrow y = 4000.$

$8000 at 7.5%, $4000 at 9%

24. Let x be the amount of water to be added.

(Final concentration) (final amount) = (original concentration) (amount) + water

$(0.75)(3 + x) = (0.60)(3) + (1.0)x$

$2.25 + 0.75x = 1.8 + x$

$0.45 = 0.25x$

$x = 1.8 \text{ gallons}$

25. Number of people $= n$; $f = $ cost per person

$$fn = 36,000 \Rightarrow f = \frac{36,000}{n}$$

$$(f - 1000)(n + 3) = 36,000$$

$$\left(\frac{36,000}{n} - 1000\right)(n + 3) = 36,000$$

$$(36,000 - 1000n)(n + 3) = 36,000n$$

$$36,000n + 3(36,000) - 1000n^2 - 3000n = 36,000n$$

$$-1000n^2 - 3000n + 108,000 = 0$$

$$n^2 + 3n - 108 = 0$$

$$(n - 9)(n + 12) = 0 \quad \Rightarrow \quad n = 9 \text{ or } -12.$$

Since $n > 0$, $n = 9$ people.

26.

x	y	xy	x^2
0	2	0	0
2	4	8	4
4	5	20	16
$\sum x_1 = 6$	$\sum y_1 = 11$	$\sum x_1 y_1 = 28$	$\sum x_1^2 = 20$

$$a = \frac{3(28) - 5(11)}{3(20 - 6^2} = 0.75$$

$$b = \frac{1}{3}(11 - 0.75(6)) = 2.1\overline{6}$$

$$y = 0.75x + 2.1\overline{6} = \frac{3}{4}x + \frac{13}{6}$$

CHAPTER THREE
Polynomial and Rational Functions

3.1 │ Quadratic Functions

2. $f(x) = (x+5)^2$ opens upward and has vertex $(-5, 0)$; graph (d)

```
RANGE
Xmin=-8
Xmax=1
Xscl=1
Ymin=-1
Ymax=6
Yscl=1
```

4. $f(x) = 5 - x^2$ opens downward and has vertex $(0, 5)$; graph (e)

```
RANGE
Xmin=-4
Xmax=4
Xscl=1
Ymin=-1
Ymax=6
Yscl=1
```

6. $f(x) = (x+2)^2 - 2$ opens upward and has vertex $(-2, -2)$; graph (a)

```
RANGE
Xmin=-5
Xmax=1
Xscl=1
Ymin=-3
Ymax=3
Yscl=1
```

8. (a) f opens downward and the vertex is shifted upward 4 units. The graph is broader than $y = x^2$.
  (b) g opens upward and the vertex is shifted downward 5 units. The graph is narrower than $y = x^2$.

10. (a) f is shifted 4.5 units to the right and 5.2 units upward. It opens upward and is broader than $y = x^2$.
  (b) g is shifted 1.4 units to the left and 3.8 units upward. It opens downward and is broader than $y = x^2$.

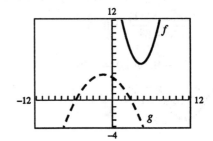

12. Vertex: $(0, 4)$

$$y = a(x-0)^2 + 4$$
$$0 = a(2)^2 + 4$$
$$a = -1$$
$$y = -x^2 + 4$$

14. Vertex: $(-2, -2)$

$$y = a(x+2)^2 - 2$$
$$2 = a(0+2)^2 - 2$$
$$a = 1$$
$$y = (x+2)^2 - 2$$

16. Vertex: $(3, 0)$

$$y = a(x-3)^2 + 0$$
$$2 = a(4-3)^2$$
$$a = 2$$
$$y = 2(x-3)^2$$

18. $h(x) = 25 - x^2$

Vertex: $(0, 25)$

Intercepts: $(-5, 0), (0, 25), (5, 0)$

20. $f(x) = (x - 6)^2 + 3$

Vertex: $(6, 3)$

Intercept: $(0, 39)$

22. $f(x) = x^2 + 3x + \frac{1}{4}$

$\qquad = \left(x^2 + 3x + \frac{9}{4}\right) + \frac{1}{4} - \frac{9}{4}$

$\qquad = \left(x + \frac{3}{2}\right)^2 - 2$

Vertex: $\left(-\frac{3}{2}, -2\right)$

Intercepts: $\left(-\frac{3}{2} - \sqrt{2}, 0\right), \left(-\frac{3}{2} + \sqrt{2}, 0\right), \left(0, \frac{1}{4}\right)$

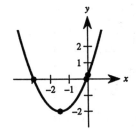

24. $f(x) = -x^2 - 4x + 1$

$\qquad = -(x^2 + 4x + 4) + 1 + 4$

$\qquad = -(x + 2)^2 + 5$

Vertex: $(-2, 5)$

Intercepts: $(-2 - \sqrt{5}, 0), (0, 1), (-2 + \sqrt{5}, 0)$

26. $f(x) = 2x^2 - x + 1$

$\quad = 2\left(x^2 - \frac{1}{2}x + \frac{1}{16}\right) + 1 - \frac{1}{8}$

$\quad = 2\left(x - \frac{1}{4}\right)^2 + \frac{7}{8}$

Vertex: $\left(\frac{1}{4}, \frac{7}{8}\right)$

Intercept: $(0, 1)$

28. $g(x) = x^2 + 8x + 11$

$\quad = x^2 + 8x + 16 + 11 - 16$

$\quad = (x + 4)^2 - 5$

Vertex: $(-4, -5)$

Intercepts: $(-1.76, 0)$, $(-6.24, 0)$, $(0, 11)$

30. $g(x) = \frac{1}{2}(x^2 + 4x - 2)$

$\quad = \frac{1}{2}(x^2 + 4x + 4) - 3$

$\quad = \frac{1}{2}(x + 2)^2 - 3$

Vertex: $(-2, -3)$

Intercepts: $(0.45, 0)$, $(-4.45, 0)$, $(0, -1)$

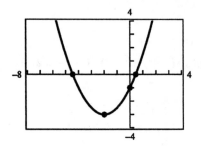

32. $y = a(x - 2)^2 + 3$

$\quad 2 = a(0 - 2)^2 + 3$

$\quad 4a = -1$

$\quad a = -\frac{1}{4}$

$\quad y = -\frac{1}{4}(x - 2)^2 + 3$

34. $y = a(x + 2)^2 - 2$

$\quad 0 = a(-1 + 2)^2 - 2$

$\quad a = 2$

$\quad y = 2(x + 2)^2 - 2$

36. $y = (2x + 5)(x - 2) = 2x^2 + x - 10$

$\quad y = -(2x + 5)(x - 2) = -2x^2 - x + 10$

38. $y = (x - 4)(x - 8) = x^2 - 12x + 32$

$\quad y = -(x - 4)(x - 8) = -x^2 + 12x - 32$

40. $y = (x + 5)(x - 5) = x^2 - 25$

$y = -(x + 5)(x - 5) = -x^2 + 25$

42. Let x and Sx be the two numbers. Then the product is given by

$$f(x) = x(S - x)$$

$$= -x^2 + Sx$$

$$= -\left(x^2 - Sx + \frac{S^2}{4}\right) + \frac{S^2}{4}$$

$$= -\left(x - \frac{S}{2}\right)^2 + \frac{S^2}{4}$$

Vertex: $\left(\dfrac{S}{2}, \dfrac{S^2}{4}\right)$

The two numbers are the same: $x = S/2$ and $S - x = S/2$.

44. (a) $2x + 2w = 36$

$$w = 18 - x$$

$$A(x) = x(18 - x)$$

Domain: $0 < x < 18$

(b)

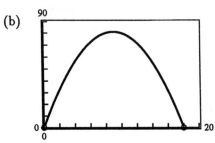

(c) $x = w = 9$ for maximum area (square).

46. $P = 200 = 2x + \pi y$

$$\pi y = 200 - 2x$$

$$y = \frac{200 - 2x}{\pi}$$

$$A = xy = x\left(\frac{200 - 2x}{\pi}\right)$$

$$= -\frac{2}{\pi}x^2 + \frac{200}{\pi}x$$

$$= -\frac{2}{\pi}(x^2 - 100x)$$

$$= -\frac{2}{\pi}(x^2 - 100x + 2500) + \frac{5000}{\pi}$$

$$= -\frac{2}{\pi}(x - 50)^2 + \frac{5000}{\pi}$$

The maximum area occurs when the length is $x = 50$ meters and the width is

$$y = \frac{200 - 2(50)}{\pi} = \frac{100}{\pi} \text{ meters.}$$

48. $R = 100x - 0.0002x^2$

$\quad = -0.0002(x^2 - 500{,}000 + (250{,}500)^2) + 12{,}500{,}000$

$\quad = -0.0002(x - 250{,}000)^2 + 12{,}500{,}000$

Vertex: $(250{,}000,\ 12{,}500{,}000)$

Maximum when $x = 250{,}000$ units.

50. $P = 230 + 20x - 0.5x^2$

$\quad = -0.5(x^2 - 40x + 400) + 230 + 200$

$\quad = -0.5(x - 20)^2 + 430$

Vertex: $(20,\ 430)$

Maximum when $x = \$2000$.

52. $y = -\frac{4}{9}x^2 + \frac{24}{9}x + 10$

$\quad = -\frac{4}{9}(x^2 - 6x + 9) + 10 + 4$

$\quad = -\frac{4}{9}(x - 3)^2 + 14$

Vertex: $(3,\ 14)$

Maximum when $y = 14$ feet.

54. $y = 0.002x^2 + 0.005x - 0.0029$,

$\quad 0 \le x \le 100$

(a)

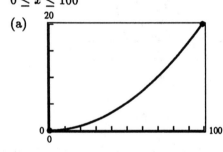

(b) If $y = 10$, $x = 69.57$.

56. Exercise 21:

$\quad a = 1,\ b = -8,\ c = 16$

$\quad \text{Vertex} = \left(\dfrac{-(-8)}{2},\ -\dfrac{(-8)^2 - 4(16)}{4} \right) = (4,\ 0)$

Exercise 22:

$\quad a = 1,\ b = 3,\ c = 1/4$

$\quad \text{Vertex} = \left(-\dfrac{3}{2},\ -\dfrac{(3)^2 - 4(1/4)}{4} \right) = \left(-\dfrac{3}{2},\ -2 \right)$

58. (a) $f(x) = \dfrac{1}{2}(x - 3)^2 - 2 = \dfrac{1}{2}x^2 - 3x + \dfrac{5}{2}$

$\quad$ Zeros: $(1, 0),\ (5, 0)$

$\quad$ Vertex: $(3,\ -2)$

$\quad$ Average of zeros $= \dfrac{1}{2}[1 + 5] = \dfrac{6}{2} = 3$

(b) $f(x) = 6 - \dfrac{2}{3}(x + 1)^2$

$\quad$ Zeros: $(2, 0),\ (-4, 0)$

$\quad$ Vertex: $(-1,\ 6)$

$\quad$ Average of zeros $= \dfrac{1}{2}[2 + (-4)] = \dfrac{-2}{2} = -1$

3.2 | Polynomial Functions of Higher Degree

2. (a) $f(x) = (x + 3)^4$ is a left shift of 3 units of $y = x^4$.

(b) $f(x) = x^4 - 3$ is a downward shift of 3 units of $y = x^4$.

(c) $f(x) = 4 - x^4$ is a reflection in the x-axis, and an upward shift of 4 units, of $y = x^4$.

(d) $f(x) = \frac{1}{2}(x - 1)^4$ is a right shift of 1 unit, and wider, compared to $y = x^4$.

4. $f(x) = x^2 - 2x$ is a parabola with x-intercepts $(0, 0)$ and $(2, 0)$; graph (c).

```
RANGE
Xmin=-2
Xmax=4
Xscl=1
Ymin=-2
Ymax=3
Yscl=1
```

6. $f(x) = 3x^3 - 9x + 1$ is a cubic with y-intercept $(0, 1)$; graph (f).

```
RANGE
Xmin=-3
Xmax=3
Xscl=1
Ymin=-6
Ymax=8
Yscl=2
```

8. $f(x) = -\frac{1}{4}x^4 + 2x^2$ has intercepts $(0, 0)$ and $(\pm 2\sqrt{2}, 0)$; graph (g).

```
RANGE
Xmin=-4
Xmax=4
Xscl=1
Ymin=-2
Ymax=5
Yscl=1
```

10. $f(x) = x^5 - 5x^3 + 4x = x(x^2 - 1)(x^2 - 4)$ has intercepts $(0, 0)$, $(\pm 1, 0)$ and $(\pm 2, 0)$; graph (h).

```
RANGE
Xmin=-3
Xmax=3
Xscl=1
Ymin=-4
Ymax=4
Yscl=1
```

12.

14.

16. $g(x) = 5 - \frac{7}{2}x - 3x^2$

The degree is even and the leading coefficient is negative. The graph falls to the left and right.

18. $f(s) = -\frac{7}{8}(s^3 + 5s^2 - 7s + 1)$

The degree is odd and the leading coefficient is negative. The graph rises to the left and falls to the right.

20. $f(x) = \frac{3x^4 - 2x + 5}{4}$

The degree is even and the leading coefficient is positive. The graph rises to the left and right.

22. $f(x) = 49 - x^2$

$= (7 + x)(7 - x)$

Thus, the real zeros are -7 and 7.

24. $f(x) = x^2 + 10x + 25$

$= (x + 5)^2$

Thus, the real zero is -5.

26. $f(x) = \frac{1}{2}x^2 + \frac{5}{2}x - \frac{3}{2}$

$= \frac{1}{2}(x^2 + 5x - 3)$

$= \frac{-5 \pm \sqrt{25 + 12}}{2} = \frac{-5 \pm \sqrt{37}}{2}$

Thus, the real zeros are

$$\frac{-5 \pm \sqrt{37}}{2} \approx 0.54, \ -5.54.$$

28. $g(x) = 5(x^2 - 2x - 1)$

$x = \frac{2 \pm \sqrt{4 + 4}}{2} = 1 \pm \sqrt{2}$

Thus, the real zeros are $1 \pm \sqrt{2}$.

30. $f(x) = x^4 - x^3 - 20x^2$

$= x^2(x + 4)(x - 5)$

Thus, the real zeros are -4, 0, and 5.

32. $f(x) = x^5 + x^3 - 6x$

$\qquad = x(x^2 + 3)(x^2 - 2)$

$\qquad = x(x^2 + 3)(x + \sqrt{2})(x - \sqrt{2})$

Thus, the real zeros are $-\sqrt{2}$, 0, and $\sqrt{2}$.

34. $g(t) = t^5 - 6t^3 + 9t$

$\qquad = t(t^2 - 3)^2$

$\qquad = t(t + \sqrt{3})^2(t - \sqrt{3})^2$

Thus, the real zeros are $-\sqrt{3}$, 0, and $\sqrt{3}$.

36. $f(x) = x^3 - 4x^2 - 25x + 100$

$\qquad = x^2(x - 4) - 25(x - 4)$

$\qquad = (x + 5)(x - 5)(x - 4)$

Thus, the real zeros are -5, 4, and 5.

38. $f(x) = (x - 0)(x + 3)$

$\qquad = x^2 + 3x$

40. $f(x) = (x + 4)(x - 5)$

$\qquad = x^2 - x - 20$

42. $f(x) = (x - 0)(x - 2)(x - 5)$

$\qquad = x(x^2 - 7x + 10)$

$\qquad = x^3 - 7x^2 + 10x$

44. $f(x) = (x + 2)(x + 1)(x - 0)(x - 1)(x - 2)$

$\qquad = x(x^2 - 4)(x^2 - 1)$

$\qquad = x(x^4 - 5x^2 + 4)$

$\qquad = x^5 - 5x^3 + 4x$

46. $f(x) = (x - 2)(x - (4 + \sqrt{5}))(x - (4 - \sqrt{5}))$

$\qquad = (x - 2)[(x - 4) - \sqrt{5}][(x - 4) + \sqrt{5}]$

$\qquad = (x - 2)[(x - 4)^2 - 5]$

$\qquad = (x - 2)(x^2 - 8x + 16 - 5)$

$\qquad = x^2 - 2x - 2$

48.

50.

52.

54.

56.

58.

60.

62.

64. $[0, 1]$, $[6, 7]$, $[11, 12]$

66. $[-4, -3]$, $[-1, 0]$, $[0, 1]$, $[3, 4]$

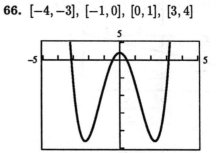

68. (a) Volume $= (12 - 2x)(12 - 4x)x$

$$= 8x(3 - x)(6 - x)$$

(b) Domain $0 < x < 3$

(c) $V(x)$ is maximum for $x = 1.27$.

70. $G = -0.003t^3 + 0.137t^2 + 0.458t - 0.839$

$$2 \le t \le 34$$

G is increasing most rapidly when x is approximately 15.

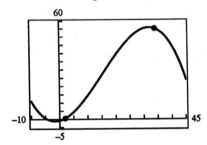

3.3 Real Zeros of Polynomial Functions

2.
$$
\begin{array}{r}
5x + 3 \\
x - 4 \overline{)\ 5x^2 - 17x - 12} \\
\underline{5x^2 - 20x} \\
3x - 12 \\
\underline{3x - 12} \\
0
\end{array}
$$

4.
$$
\begin{array}{r}
2x^2 - 4x + 3 \\
3x - 2 \overline{)\ 6x^3 - 16x^2 + 17x - 6} \\
\underline{6x^3 - 4x^2} \\
-12x^2 + 17x \\
\underline{-12x^2 + 8x} \\
9x - 6 \\
\underline{9x - 6} \\
0
\end{array}
$$

6.

$$
\begin{array}{r}
x + 4 \\
x^2 - 3 \overline{\smash{\big)}\ x^3 + 4x^2 - 3x - 12} \\
\underline{x^3\ \ \ \ \ \ \ - 3x} \\
4x^2\ \ \ \ \ \ - 12 \\
\underline{4x^2\ \ \ \ \ \ - 12} \\
0
\end{array}
$$

8.

$$
\begin{array}{r}
4 \\
2x + 1 \overline{\smash{\big)}\ 8x - 5} \\
\underline{8x + 4} \\
-9
\end{array}
$$

$$\frac{8x - 5}{2x + 1} = 4 - \frac{9}{2x + 1}$$

10.

$$
\begin{array}{r}
x \\
x^2 + 1 \overline{\smash{\big)}\ x^3\ \ \ \ \ - 9} \\
\underline{x^3 + x} \\
-x - 9
\end{array}
$$

$$\frac{x^3 - 9}{x^2 + 1} = x - \frac{x + 9}{x^2 + 1}$$

12.

$$
\begin{array}{r}
x^2 \\
x^3 - 1 \overline{\smash{\big)}\ x^5\ \ \ \ \ + 7} \\
\underline{x^5 - x^2} \\
x^2 + 7
\end{array}
$$

$$\Rightarrow \quad \frac{x^5 + 7}{x^3 - 1} = x^2 + \frac{x^2 + 7}{x^3 - 1}$$

14. Observe that $(x - 1)^3 = x^3 - 3x^2 + 3x - 1$.

$$
\begin{array}{r}
x + 3 \\
x^3 - 3x^2 + 3x - 1 \overline{\smash{\big)}\ x^4} \\
\underline{x^4 - 3x^3 + 3x^2 - x} \\
3x^3 - 3x^2 + x \\
\underline{3x^3 - 9x^2 + 9x - 3} \\
6x^2 - 8x + 3
\end{array}
\quad \Rightarrow \quad \frac{x^4}{(x - 1)^3} = x + 3 + \frac{6x^2 - 8x + 3}{(x - 1)^3}
$$

16.

$$
\begin{array}{r|rrrr}
-3 & 5 & 18 & 7 & -6 \\
   &   & -15 & -9 & 6 \\
\hline
   & 5 & 3 & -2 & 0
\end{array}
$$

$$(5x^3 + 18x^2 + 7x - 6) \div (x + 3) = 5x^2 + 3x - 2$$

18.

$$
\begin{array}{r|rrrr}
2 & 9 & -18 & -16 & 32 \\
  &   & 18 & 0 & -32 \\
\hline
  & 9 & 0 & -16 & 0
\end{array}
$$

$$(9x^3 - 16x - 18x^2 + 32) \div (x - 2) = 9x^2 - 16$$

20.

$$
\begin{array}{r|rrrr}
6 & 3 & -16 & 0 & -72 \\
  &   & 18 & 12 & 72 \\
\hline
  & 3 & 2 & 12 & 0
\end{array}
$$

$$(3x^3 - 16x^2 - 72) \div (x - 6) = 3x^2 + 2x + 12$$

22.

$$
\begin{array}{r|rrrr}
-2 & 5 & 0 & 6 & 8 \\
   &   & -10 & 20 & -52 \\
\hline
   & 5 & -10 & 26 & -44
\end{array}
$$

$$(5x^3 + 6x + 8) \div (x + 2) = 5x^2 - 10x + 26 - \frac{44}{x + 2}$$

24.

$$
\begin{array}{r|rrrrrr}
-3 & 1 & -13 & 0 & 0 & -120 & 80 \\
   &   & -3 & 48 & -144 & 432 & -936 \\
\hline
   & 1 & -16 & 48 & -144 & 312 & -856
\end{array}
$$

$$(x^5 - 13x^4 - 120x + 80) \div (x + 3)$$

$$= x^4 - 16x^3 + 48x^2 - 144x + 312 - \frac{856}{x + 3}$$

26.

$$
\begin{array}{r|rrrr}
-3 & 5 & 0 & 0 & 0 \\
   &   & -15 & 45 & -135 \\
\hline
   & 5 & -15 & 45 & -135
\end{array}
$$

$$5x^3 \div (x + 3) = 5x^2 - 15x + 45 - \frac{135}{x + 3}$$

28.

$$\begin{array}{r|rrrrr} -2 & -3 & 0 & 0 & 0 & 0 \\ & & 6 & -12 & 24 & -48 \\ \hline & -3 & 6 & -12 & 24 & -48 \end{array}$$

$-3x^4 \div (x+2)$

$\qquad = -3x^3 + 6x^2 - 12x + 24 - \dfrac{48}{x+2}$

30.

$$\begin{array}{r|rrrrr} 6 & -1 & 0 & 0 & 180 & 0 \\ & & -6 & -36 & -216 & -216 \\ \hline & -1 & -6 & -36 & -36 & -216 \end{array}$$

$(180x - x^4) \div (x - 6)$

$\qquad = -x^3 - 6x^2 - 36x - 36 - \dfrac{216}{x-6}$

32.

$$\begin{array}{r|rrrr} \frac{3}{2} & 3 & -4 & 0 & 5 \\ & & \frac{9}{2} & \frac{3}{4} & \frac{9}{8} \\ \hline & 3 & \frac{1}{2} & \frac{3}{4} & \frac{49}{8} \end{array}$$

Quotient: $(3x^3 - 4x^2 + 5) \div \left(x - \dfrac{3}{2}\right)$

$\qquad = 3x^2 + \dfrac{1}{2}x + \dfrac{3}{4} + \dfrac{49}{8x - 12}$

34.

$$\begin{array}{r|rrrr} -4 & 1 & 0 & -28 & -48 \\ & & -4 & 16 & 48 \\ \hline & 1 & -4 & -12 & 0 \end{array}$$

$x^3 - 28x - 48 = (x+4)(x^2 - 4x - 12)$

$\qquad = (x+4)(x+2)(x-6)$

36.

$$\begin{array}{r|rrrr} \frac{2}{3} & 48 & -80 & 41 & -6 \\ & & 32 & -32 & 6 \\ \hline & 48 & -48 & 9 & 0 \end{array}$$

$48x^3 - 80x^2 + 41x - 6 = (x - \frac{2}{3})(48x^2 - 48x + 9)$

$\qquad = (3x - 2)(4x - 3)(4x - 1)$

38.

$$\begin{array}{r|rrrr} \sqrt{2} & 1 & 2 & -2 & -4 \\ & & \sqrt{2} & 2\sqrt{2}+2 & 4 \\ \hline & 1 & 2+\sqrt{2} & 2\sqrt{2} & 0 \end{array}$$

$x^3 + 2x^2 - 2x - 4 = (x - \sqrt{2})(x^2 + (2 + \sqrt{2})x + 2\sqrt{2})$

$\qquad = (x - \sqrt{2})(x + \sqrt{2})(x + 2)$

40. $f(x) = \frac{1}{3}(15x^4 + 10x^3 - 6x^2 + 17x + 14), \quad k = -\frac{2}{3}$

$$\begin{array}{r|rrrrr} -\frac{2}{3} & 5 & \frac{10}{3} & -2 & \frac{17}{3} & \frac{14}{3} \\ & & -\frac{10}{3} & 0 & \frac{4}{3} & -\frac{14}{3} \\ \hline & 5 & 0 & -2 & 7 & 0 \end{array}$$

$f(x) = (x + \frac{2}{3})(5x^3 - 2x + 7)$

$f(-\frac{2}{3}) = (0)(\frac{185}{27}) = 0$

42. $f(x) = 4x^3 - 6x^2 - 12x - 4,\quad k = 1 - \sqrt{3}$

$$
\begin{array}{r|rrrr}
1-\sqrt{3} & 4 & -6 & -12 & -4 \\
 & & 4-4\sqrt{3} & 10-2\sqrt{3} & 4 \\
\hline
 & 4 & -2-4\sqrt{3} & -2-2\sqrt{3} & 0
\end{array}
$$

$f(x) = (x - 1 + \sqrt{3})[4x^2 - (2 + 4\sqrt{3})x - (2 + 2\sqrt{3})]$

$f(1 - \sqrt{3}) = (0)(24 - 12\sqrt{3}) = 0$

44. $g(x) = x^6 - 4x^4 + 3x^2 + 2$

(a)
$$
\begin{array}{r|rrrrrrr}
2 & 1 & 0 & -4 & 0 & 3 & 0 & 2 \\
 & & 2 & 4 & 0 & 0 & 6 & 12 \\
\hline
 & 1 & 2 & 0 & 0 & 3 & 6 & 14 \;=g(2)
\end{array}
$$

(b)
$$
\begin{array}{r|rrrrrrr}
-4 & 1 & 0 & -4 & 0 & 3 & 0 & 2 \\
 & & -4 & 16 & -48 & 192 & -780 & 3120 \\
\hline
 & 1 & -4 & 12 & -48 & 195 & -780 & 3122 \;=g(-4)
\end{array}
$$

(c)
$$
\begin{array}{r|rrrrrrr}
3 & 1 & 0 & -4 & 0 & 3 & 0 & 2 \\
 & & 3 & 9 & 15 & 45 & 144 & 432 \\
\hline
 & 1 & 3 & 5 & 15 & 48 & 144 & 434 \;=g(3)
\end{array}
$$

(d)
$$
\begin{array}{r|rrrrrrr}
-1 & 1 & 0 & -4 & 0 & 3 & 0 & 2 \\
 & & -1 & 1 & 3 & -3 & 0 & 0 \\
\hline
 & 1 & -1 & -3 & 3 & 0 & 0 & 2 \;=g(-1)
\end{array}
$$

46. $f(x) = 0.4x^4 - 1.6x^3 + 0.7x^2 - 2$

(a)
$$
\begin{array}{r|rrrrr}
1 & 0.4 & -1.6 & 0.7 & 0 & -2 \\
 & & 0.4 & -1.2 & -0.5 & -0.5 \\
\hline
 & 0.4 & -1.2 & -0.5 & -0.5 & -2.5 = f(1)
\end{array}
$$

48. $g(x) = x^3 + 3x^2$
Sign variations: 0, positive zeros: 0
$g(-x) = -x^3 + 3x^2$
Sign variations: 1, negative zeros: 1

50. $h(x) = 2x^4 - 3x + 2$
Sign variations: 2, positive zeros: 2 or 0
$h(-x) = 2x^4 + 3x + 2$
Sign variations: 0, negative zeros: 0

52. $f(x) = 4x^3 - 3x^2 + 2x - 1$
Sign variations: 3, positive zeros: 3 or 1
$f(-x) = -4x^3 - 3x^2 - 2x - 1$
Sign variations: 0, negative zeros: 0

54. $g(x) = 5x^5 + 10x$
Sign variations: 0, positive zeros: 0
$g(-x) = -5x^5 - 10x$
Sign variations: 0, negative zeros: 0

56. $f(x) = 3x^3 + 2x^2 + x + 3$
Sign variations: 0, positive zeros: 0
$f(-x) = -3x^3 + 2x^2 - x + 3$
Sign variations: 3, negative zeros: 3 or 1

58. $f(x) = -3x^3 + 20x^2 - 36x + 16$
Possible rational zeros: $\dfrac{\pm 1,\ \pm 2,\ \pm 4,\ \pm 8,\ \pm 16}{\pm 1,\ \pm 3} = \pm\dfrac{1}{3},\ \pm\dfrac{2}{3},\ \pm 1,\ \pm\dfrac{4}{3},\ \pm 2,\ \pm\dfrac{8}{3},\ \pm 4,\ \pm\dfrac{16}{3},\ \pm 8,\ \pm 16$

$\dfrac{2}{3},\ 2,\ 4$ on graph

60. $f(x) = 4x^3 - 12x^2 - x + 15$
Possible rational zeros: $\dfrac{\pm 1,\ \pm 3,\ \pm 5,\ \pm 15}{\pm 1,\ \pm 2,\ \pm 4} = \pm\dfrac{1}{4},\ \pm\dfrac{1}{2},\ \pm\dfrac{3}{4},\ \pm 1,$

$\pm\dfrac{5}{4},\ \pm\dfrac{3}{2},\ \pm\dfrac{5}{2},\ \pm 3,\ \pm\dfrac{15}{4},\ \pm 5,\ \pm\dfrac{15}{2},\ \pm 15$

$-1,\ \dfrac{3}{2},\ \dfrac{5}{2}$ on graph

62. $f(x) = 4x^4 - 17x^2 + 4$
(a) Possible rational zeros:

$\dfrac{\pm 1,\ \pm 2,\ \pm 4}{\pm 1,\ \pm 2,\ \pm 4} = \pm\dfrac{1}{4},\ \pm\dfrac{1}{2},\ \pm 1,\ \pm 2,\ \pm 4$

(c) $-2,\ -\dfrac{1}{2},\ \dfrac{1}{2},\ 2$ on graph

(b)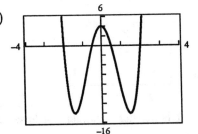

64. $f(x) = 6x^3 - x^2 - 13x + 8$

(a) $\pm\frac{1}{6}$, $\pm\frac{1}{3}$, $\pm\frac{1}{2}$, $\pm\frac{2}{3}$, ± 1, $\pm\frac{4}{3}$, ± 2, $\pm\frac{8}{3}$, ± 4, ± 8

(b)

(c) Real zeros: $1, \dfrac{-5 \pm \sqrt{217}}{12}$

66. $f(x) = 2x^3 + 5x^2 - 21x - 10$

(a) $\pm\frac{1}{2}$, ± 1, ± 2, $\pm\frac{5}{2}$, ± 5, ± 10

(b)

(c) Real zeros: $\dfrac{5}{2}, \dfrac{-5 \pm \sqrt{17}}{2}$

68. $f(x) = 2x^3 - 3x^2 - 12x + 8$

(a)
$$
\begin{array}{r|rrrr}
2 & 2 & -3 & -12 & 8 \\
  &   & 4 & 2 & -20 \\
\hline
  & 2 & 1 & -10 & -12
\end{array}
$$

2 is neither an upper nor lower bound.

(b)
$$
\begin{array}{r|rrrr}
4 & 2 & -3 & -12 & 8 \\
  &   & 8 & 20 & 32 \\
\hline
  & 2 & 5 & 8 & 40
\end{array}
$$

4 is an upper bound.

(c)
$$
\begin{array}{r|rrrr}
-1 & 2 & -3 & -12 & 8 \\
   &   & -2 & 5 & 7 \\
\hline
   & 2 & -5 & -7 & 15
\end{array}
$$

−1 is neither an upper nor lower bound.

70. $f(x) = 2x^4 - 8x + 3$

(a)
$$
\begin{array}{r|rrrrr}
1 & 2 & 0 & 0 & -8 & 3 \\
  &   & 2 & 2 & 2 & -6 \\
\hline
  & 2 & 2 & 2 & -6 & -3
\end{array}
$$

1 is neither an upper nor lower bound.

(b)
$$
\begin{array}{r|rrrrr}
3 & 2 & 0 & 0 & -8 & 3 \\
  &   & 6 & 18 & 54 & 138 \\
\hline
  & 2 & 6 & 18 & 46 & 141
\end{array}
$$

3 is an upper bound.

(c)
$$
\begin{array}{r|rrrrr}
-4 & 2 & 0 & 0 & -8 & 3 \\
   &   & -8 & 32 & -128 & 544 \\
\hline
   & 2 & -8 & 32 & -136 & 547
\end{array}
$$

−4 is a lower bound.

72. Possible rational zeros: ± 1, ± 2, ± 3, ± 6

$$
\begin{array}{r|rrrr}
-1 & 1 & 0 & -7 & -6 \\
   &   & -1 & 1 & 6 \\
\hline
   & 1 & -1 & -6 & 0
\end{array}
$$

$$
\begin{aligned}
x^3 - 7x - 6 &= (x+1)(x^2 - x - 6) \\
&= (x+1)(x-3)(x+2)
\end{aligned}
$$

The zeros are -2, -1 and 3.

74. Possible rational zeros:
± 1, ± 2, ± 3, ± 4, ± 6, ± 12

$$
\begin{array}{r|rrrr}
1 & 1 & -9 & 20 & -12 \\
  &   & 1 & -8 & 12 \\
\hline
  & 1 & -8 & 12 & 0
\end{array}
$$

$$
\begin{aligned}
x^3 - 9x^2 + 20x - 12 &= (x-1)(x^2 - 8x + 12) \\
&= (x-1)(x-6)(x-2)
\end{aligned}
$$

The zeros are 1, 2, and 6.

76. Possible rational zeros: ± 1, ± 2, ± 4, ± 8

$$
\begin{array}{r|rrrr}
-2 & 1 & 6 & 12 & 8 \\
   &   & -2 & -8 & -8 \\
\hline
   & 1 & 4 & 4 & 0
\end{array}
$$

$$
\begin{aligned}
x^3 + 6x^2 + 12x + 8 &= (x+2)(x^2 + 4x + 4) \\
&= (x+2)^3
\end{aligned}
$$

The zero is -2.

78. Possible rational zeros: ± 1, ± 3, ± 9, ± 27

$$
\begin{array}{r|rrrr}
3 & 1 & -9 & 27 & -27 \\
  &   & 3 & -18 & 27 \\
\hline
  & 1 & -6 & 9 & 0
\end{array}
$$

$$
\begin{aligned}
x^3 - 9x^2 + 27x - 27 &= (x-3)(x^2 - 6x + 9) \\
&= (x-3)^3
\end{aligned}
$$

The zero is 3.

80. Possible rational zeros: $\pm \frac{1}{3}$, ± 1, ± 3, ± 9

$$
\begin{array}{r|rrrr}
3 & 3 & -19 & 33 & -9 \\
  &   & 9 & -30 & 9 \\
\hline
  & 3 & -10 & 3 & 0
\end{array}
$$

$$
\begin{aligned}
3x^3 - 19x^2 + 33x - 9 &= (x-3)(3x^2 - 10x + 3) \\
&= (x-3)(3x-1)(x-3)
\end{aligned}
$$

The zeros are $\frac{1}{3}$ and 3.

82. Possible rational zeros: $\pm 1, \pm 3, \pm 9, \pm \frac{1}{2},$
$$\pm \frac{3}{2}, \pm \frac{9}{2}, \pm \frac{1}{3}, \pm \frac{1}{4},$$
$$\pm \frac{3}{4}, \pm \frac{9}{4}, \pm \frac{1}{6}, \pm \frac{1}{12}$$

$$
\begin{array}{r|rrrr}
\frac{1}{3} & 12 & -4 & -27 & 9 \\
 & & 4 & 0 & -9 \\
\hline
 & 12 & 0 & -27 & 0
\end{array}
$$

$$12z^3 - 4z^2 - 27z + 9 = \left(z - \tfrac{1}{3}\right)\left(12z^2 - 27\right)$$
$$= 3\left(z - \tfrac{1}{3}\right)\left(4z^2 - 9\right)$$
$$= 3\left(z - \tfrac{1}{3}\right)(2z + 3)(2z - 3)$$

The zeros are $-\frac{3}{2}$, $\frac{1}{3}$, and $\frac{3}{2}$.

84. Possible rational zeros: $\pm 1, \pm 2, \pm 5, \pm 10,$
$$\pm \tfrac{1}{3}, \pm \tfrac{2}{3}, \pm \tfrac{5}{3}, \pm \tfrac{10}{3}$$

$$
\begin{array}{r|rrrr}
\frac{2}{3} & 3 & -2 & 15 & -10 \\
 & & 2 & 0 & 10 \\
\hline
 & 3 & 0 & 15 & 0
\end{array}
$$

$$3x^3 - 2x^2 + 15x - 10 = \left(x - \tfrac{2}{3}\right)\left(3x^2 + 15\right)$$
The zero is $\frac{2}{3}$.

86. $P(t) = t^4 - 7t^2 + 12$
$$= (t^2 - 4)(t^2 - 3)$$
$$= (t + 2)(t - 2)(t + \sqrt{3})(t - \sqrt{3})$$

The zeros are ± 2 and $\pm \sqrt{3}$.

88. $f(x) = x^3 - \frac{3}{2}x^2 - \frac{23}{2}x + 6$
$$= \tfrac{1}{2}(2x^3 - 3x^2 - 23x + 12)$$

Possible rational zeros:
$\pm \frac{1}{2}, \pm 1, \pm \frac{3}{2}, \pm 2, \pm 3, \pm 4, \pm 6, \pm 12$
$x = -3, \frac{1}{2}, 4$

90. $f(z) = z^3 + \frac{11}{6}z^2 - \frac{1}{2}z - \frac{1}{3}$
$$= \tfrac{1}{6}(6z^3 + 11z^2 - 3z - 2)$$

Possible rational zeros:
$\pm \frac{1}{6}, \pm \frac{1}{3}, \pm \frac{1}{2}, \pm \frac{2}{3}, \pm 1, \pm 2$
$x = -2, -\frac{1}{3}, \frac{1}{2}$

92. $f(x) = x^3 - 2 = 0 \Rightarrow x^3 = 2 \Rightarrow x = \sqrt[3]{2}$
Zero rational zeros; one irrational zero
Matches (a)

94. $f(x) = x^3 - 2x = x(x^2 - 2) = x(x + \sqrt{2})(x - \sqrt{2})$
One rational zero; two irrational zeros
Matches (c)

96. $f(x) = x^3 - 3x - 1$

(a) ± 1

(b)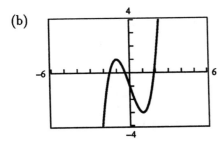

(c) -1.53, -0.35, 1.88

98. $f(x) = x^5 + x - 1$

(a) ± 1

(b)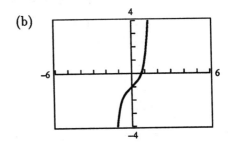

(c) 0.75

100. $4(x) + y = 108 \Rightarrow y = 108 - 4x$

$$V = x^2 y = x^2(108 - 4x)$$

$$= -4x^3 + 108x^2 = 11{,}664$$

$$4x^3 - 108x^2 + 11{,}664 = 0$$

$$x^3 - 27x^2 + 2916 = 0$$

$$(x + 9)(x - 18)^2 = 0$$

$$x = 18$$

$$y = 36$$

Dimensions: 18 in. × 18 in. × 36 in.

102. $P = -45x^3 + 2500x^2 - 275{,}000,\ 0 \le x \le 50$

$$800{,}000 = -45x^3 + 2500x^2 - 275{,}000$$

$$45x^3 - 2500x^2 + 1{,}075{,}000 = 0,\ [30, 40]$$

Using the methods of this section, we have $x \approx 32$ which corresponds to \$320,000 being spent on advertising.

104. During the year 1990 $(t = 10)$

106.

Iteration	a	c	b	$f(a)$	$f(c)$	$f(b)$	Error
1	3.0000	3.5000	4.0000	−20.0000	−2.0000	30.0000	0.5000
2	3.5000	3.7500	4.0000	−2.0000	12.0625	30.0000	0.2500
3	3.5000	3.6250	3.7500	−2.0000	4.5703	12.0625	0.1250
4	3.5000	3.5625	3.6250	−2.0000	1.1729	4.5703	0.0625
5	3.5000	3.5313	3.5625	−2.0000	−0.4413	1.1729	0.0313
6	3.5313	3.5469	3.5625	−0.4413	0.3588	1.1729	0.0156
7	3.5313	3.5391	3.5469	−0.4413	−0.0430	0.3588	0.0078
8	3.5391	3.5430	3.5469	−0.0430	0.1575	0.3588	0.0039
9	3.5391	3.5410	3.5430	−0.0430	0.0571	0.1575	0.0020

The zero of f is approximately 3.54.

108.

Iteration	a	c	b	$f(a)$	$f(c)$	$f(b)$	Error
1	0.0000	0.5000	1.0000	-4.0000	3.4375	18.0000	0.5000
2	0.0000	0.2500	0.5000	-4.0000	-1.4883	3.4375	0.2500
3	0.2500	0.3750	0.5000	-1.4883	0.7234	3.4375	0.1250
4	0.2500	0.3125	0.3750	-1.4883	-0.4509	0.7234	0.0625
5	0.3125	0.3438	0.3750	-0.4509	0.1199	0.7234	0.0313
6	0.3125	0.3281	0.3438	-0.4509	-0.1697	0.1199	0.0156
7	0.3281	0.3359	0.3438	-0.1697	-0.0259	0.1199	0.0078
8	0.3359	0.3398	0.3438	-0.0259	0.0467	0.1199	0.0039

The zero of f is approximately 0.34.

3.4   Complex Zeros and the Fundamental Theorem of Algebra

2. $f(x) = x^2 - x + 56$

Zeros: $x = \dfrac{1 \pm \sqrt{223}\,i}{2}$

$f(x) = \left(x - \dfrac{1 - \sqrt{223}\,i}{2}\right)\left(x - \dfrac{1 + \sqrt{223}\,i}{2}\right)$

4. $g(x) = x^2 + 10x + 23$

Zeros: $x = \dfrac{-10 \pm \sqrt{8}}{2} = -5 \pm \sqrt{2}$

$g(x) = (x + 5 + \sqrt{2})(x + 5 - \sqrt{2})$

6. $f(y) = y^4 - 625$

Zeros: $x = \pm 5,\ \pm 5i$

$f(y) = (y + 5)(y - 5)(y + 5i)(y - 5i)$

8. $h(x) = x^3 - 3x^2 + 4x - 2$

$$\begin{array}{r|rrrr}
1 & 1 & -3 & 4 & -2 \\
 &   & 1 & -2 & 2 \\
\hline
 & 1 & -2 & 2 & 0
\end{array}$$

Zeros: $x = 1,\ \dfrac{2 \pm \sqrt{4}\,i}{2} = 1 \pm i$

$h(x) = (x - 1)(x - 1 - i)(x - 1 + i)$

10. $f(x) = x^3 + 11x^2 + 39x + 29$

$$\begin{array}{r|rrrr}
-1 & 1 & 11 & 39 & 29 \\
 &   & -1 & -10 & -29 \\
\hline
 & 1 & 10 & 29 & 0
\end{array}$$

Zeros: $x = -1,\ \dfrac{-10 \pm \sqrt{16}\,i}{2} = -5 \pm 2i$

$f(x) = (x + 1)(x + 5 + 2i)(x + 5 - 2i)$

12. $f(s) = 2s^3 - 5s^2 + 12s - 5$

$$\begin{array}{r|rrrr}
\frac{1}{2} & 2 & -5 & 12 & -5 \\
 &   & 1 & -2 & 5 \\
\hline
 & 2 & -4 & 10 & 0
\end{array}$$

Zeros: $s = \dfrac{1}{2},\ \dfrac{4 \pm \sqrt{64}\,i}{4} = 1 \pm 2i$

$f(s) = (2s - 1)(s - 1 + 2i)(s - 1 - 2i)$

14. $g(x) = 3x^3 - 4x^2 + 8x + 8$

$$
-\frac{2}{3} \begin{array}{|rrrr} 3 & -4 & 8 & 8 \\ & -2 & 4 & -8 \\ \hline 3 & -6 & 12 & 0 \end{array}
$$

Zeros: $x = -\dfrac{2}{3}, \; \dfrac{6 \pm \sqrt{108}\,i}{6} = 1 \pm \sqrt{3}\,i$

$g(x) = (3x + 2)(x - 1 + \sqrt{3}\,i)(x - 1 - \sqrt{3}\,i)$

16. $f(x) = x^4 + 29x^2 + 100$

$\qquad = (x^2 + 25)(x^2 + 4)$

Zeros: $x = \pm 2i, \; \pm 5i$

$f(x) = (x + 2i)(x - 2i)(x + 5i)(x - 5i)$

18. $h(x) = x^4 + 6x^3 + 10x^2 + 6x + 9$

$$
-3 \begin{array}{|rrrrr} 1 & 6 & 10 & 6 & 9 \\ & -3 & -9 & -3 & -9 \\ \hline 1 & 3 & 1 & 3 & 0 \end{array}
$$

$$
-3 \begin{array}{|rrrr} 1 & 3 & 1 & 3 \\ & -3 & 0 & -3 \\ \hline 1 & 0 & 1 & 0 \end{array}
$$

Zeros: $x = -3, \; \pm i$

$h(x) = (x + 3)^2(x + i)(x - i)$

20. $g(x) = x^5 - 8x^4 + 28x^3 - 56x^2 + 64x - 32$

$$
2 \begin{array}{|rrrrrr} 1 & -8 & 28 & -56 & 64 & -32 \\ & 2 & -12 & 32 & -48 & 32 \\ \hline \end{array}
$$

$$
2 \begin{array}{|rrrrr} 1 & -6 & 16 & -24 & 16 & 0 \\ & 2 & -8 & 16 & -16 & \\ \hline \end{array}
$$

$$
2 \begin{array}{|rrrr} 1 & -4 & 8 & -8 & 0 \\ & 2 & -4 & 8 & \\ \hline 1 & -2 & 4 & 0 \end{array}
$$

Zeros: $x = 2, \; \dfrac{2 \pm \sqrt{12}\,i}{2} = 1 \pm \sqrt{3}\,i$

$g(x) = (x - 2)^3(x - 1 + \sqrt{3}\,i)(x - 1 - \sqrt{3}\,i)$

22. $f(x) = (x - 4)(x - 3i)(x + 3i)$

$\qquad = (x - 4)(x^2 + 9)$

$\qquad = x^3 - 4x^2 + 9x - 36$

24. $f(x) = (x - 6)(x + 5 - 2i)(x + 5 + 2i)$

$\qquad = (x - 6)(x^2 + 10x + 29)$

$\qquad = x^3 + 4x^2 - 31x - 174$

26. $f(x) = (x - 2)^3(x - 4i)(x + 4i)$

$\qquad = (x^3 - 6x^2 + 12x - 8)(x^2 + 16)$

$\qquad = x^5 - 6x^4 + 28x^3 - 104x^2 + 192x - 128$

28. $f(x) = (3x - 2)(x + 1)(x - 3 - \sqrt{2}\,i)(x - 3 + \sqrt{2}\,i)$

$\qquad = (3x^2 + x - 2)(x^2 - 6x + 11)$

$\qquad = 3x^4 - 17x^3 + 25x^2 + 23x - 22$

30. $f(x) = x^2(x-4)(x-1-i)(x-1+i)$
$= (x^3 - 4x^2)(x^2 - 2x + 2)$
$= x^5 - 6x^4 + 10x^3 - 8x^2$

32. $f(x) = x^4 - 2x^3 - 3x^2 + 12x - 18$
(a) $f(x) = (x^2 - 6)(x^2 - 2x + 3)$
(b) $f(x) = (x + \sqrt{6})(x - \sqrt{6})(x^2 - 2x + 3)$
(c) $f(x) = (x + \sqrt{6})(x - \sqrt{6})(x - 1 - \sqrt{2}\,i)(x - 1 + \sqrt{2}\,i)$

34. $f(x) = x^4 - 3x^3 - x^2 - 12x - 20$
(a) $f(x) = (x^2 + 4)(x^2 - 3x - 5)$
(b) $f(x) = (x^2 + 4)\left(x - \dfrac{3 + \sqrt{29}}{2}\right)\left(x - \dfrac{3 - \sqrt{29}}{2}\right)$
(c) $f(x) = (x + 2i)(x - 2i)\left(x - \dfrac{3 + \sqrt{29}}{2}\right)\left(x - \dfrac{3 - \sqrt{29}}{2}\right)$

36.

$$
\begin{array}{r|rrrr}
3i & 1 & 1 & 9 & 9 \\
 & & 3i & -9+3i & -9 \\
\hline
-3i & 1 & 1+3i & 3i & 0 \\
 & & -3i & -3i & \\
\hline
 & 1 & 1 & 0 & \\
\end{array}
$$

Zeros: $-1,\ \pm 3i$

38.

$$
\begin{array}{r|rrrr}
5+2i & 1 & -7 & -1 & 87 \\
 & & 5+2i & -14+6i & -87 \\
\hline
5-2i & 1 & -2+2i & -15+6i & 0 \\
 & & 5-2i & 15-6i & \\
\hline
 & 1 & 3 & 0 & \\
\end{array}
$$

Zeros: $-3,\ 5 \pm 2i$

40.

$$
\begin{array}{r|rrrr}
1-\sqrt{3}\,i & 3 & -4 & 8 & 8 \\
 & & 3-3\sqrt{3}\,i & -10-2\sqrt{3}\,i & -8 \\
\hline
1+\sqrt{3}\,i & 3 & -1-3\sqrt{3}\,i & -2-2\sqrt{3}\,i & 0 \\
 & & 3+3\sqrt{3}\,i & 2+2\sqrt{3}\,i & \\
\hline
 & 3 & 2 & 0 & \\
\end{array}
$$

Zeros: $-\dfrac{2}{3},\ 1 \pm \sqrt{3}\,i$

42.

$$
\begin{array}{c|cccc}
-1-3i & 1 & 4 & 14 & 20 \\
 & & -1-3i & -12-6i & -20 \\
\hline
-1+3i & 1 & 3-3i & 2-6i & \\
 & & -1+3i & -2-6i & \\
\hline
\end{array}
$$

Zeros: $-1 \pm 3i, \ -2$

44.

$$
\begin{array}{c|cccc}
\dfrac{-2+\sqrt{2}\,i}{5} & 25 & -55 & -54 & -18 \\
 & & -10+5\sqrt{2}i & 24-15\sqrt{2}i & 18 \\
\hline
\dfrac{-2-\sqrt{2}\,i}{5} & 25 & -65+5\sqrt{2}i & -30-15\sqrt{2}i & 0 \\
 & & -10-5\sqrt{2}i & 30+15\sqrt{2}i & \\
\hline
 & 25 & -75 & 0 &
\end{array}
$$

Zeros: $3, \ -\dfrac{2}{5} \pm \dfrac{\sqrt{2}}{5}i$

46. $P = R - C = xp - C$

$\qquad = x(140 - 0.0001x) - (80x + 150{,}000)$

$\qquad = -0.0001x^2 + 60x - 150{,}000$

$\qquad = 9{,}000{,}000$

Thus, $0 = 0.0001x^2 - 60x + 9{,}150{,}000$

$x = \dfrac{60 \pm \sqrt{-60}}{0.0002} = 300{,}000 \pm 10{,}000\sqrt{15}i.$

Since the zeros are both complex, it is not possible to determine a price p that would yield a profit of 9 million dollars.

48. $f(x) = (x - (a + bi))(x - (a - bi))$

$\qquad = ((x - a) - bi)((x - a) + bi)$

$\qquad = (x - a)^2 + b^2$

$\qquad = x^2 - 2ax + a^2 + b^2$

3.5 Rational Functions and Asymptotes

2. $f(x) = \dfrac{5x}{x-2}$

(a)

x	1	1.5	1.9	1.99	1.999
$f(x)$	−5	−15	−95	−995	−9995

x	3	2.5	2.1	2.01	2.001
$f(x)$	15	25	105	1005	10005

x	3	5	10	100	1000
$f(x)$	15	8.3333	6.25	5.1020	5.0100

(b) Vertical asymptote: $x = 2$
Horizontal asymptote: $y = 5$

(c) Domain: $x \neq 2$

4. $f(x) = \dfrac{3}{|x-2|}$

(a)

x	1	1.5	1.9	1.99	1.999
$f(x)$	3	6	30	300	3000

x	3	2.5	2.1	2.01	2.001
$f(x)$	3	6	30	300	3000

x	3	5	10	100	1000
$f(x)$	3	1	0.375	0.0306	0.0030

(b) Vertical asymptote: $x = 2$
Horizontal asymptote: $y = 0$

(c) Domain: $x \neq 2$

6. $f(x) = \dfrac{4x}{x^2 - 4}$

(a)

x	1	1.5	1.9	1.99	1.999
$f(x)$	−1.3333	−3.4286	−19.4872	−199.499	−1999.5

x	3	2.5	2.1	2.01	2.001
$f(x)$	2.4	4.4444	20.4878	200.4988	2000.5

x	3	5	10	100	1000
$f(x)$	2.4	0.9524	0.4167	0.0400	0.004

(b) Vertical asymptote: $x = \pm 2$
Horizontal asymptote: $y = 0$

(c) Domain: $x \neq \pm 2$

8. $f(x) = \dfrac{4}{(x-2)^3}$

Domain: all real numbers except 2
Vertical asymptote: $x = 2$
Horizontal asymptote: $y = 0$

10. $f(x) = \dfrac{1-5x}{1+2x} = \dfrac{-5x+1}{2x+1}$

Domain: all real numbers except $-\frac{1}{2}$
Vertical asymptote: $x = -\frac{1}{2}$
Horizontal asymptote: $y = -\frac{5}{2}$

12. $f(x) = \dfrac{2x^2}{x+1} = 2x - 2 + \dfrac{2}{x+1}$

Domain: all real numbers except -1
Vertical asymptote: $x = -1$
Slant asymptote: $y = 2x - 2$

14. $f(x) = \dfrac{3x^2 + x - 5}{x^2 + 1} = 3 + \dfrac{x-8}{x^2+1}$

Domain: all real numbers
Horizontal asymptote: $y = 3$

16. $f(x) = \dfrac{1}{x-4}$

Vertical asymptote: $x = 4$
Horizontal asymptote: $y = 0$
y-intercept: $\left(0, -\frac{1}{4}\right)$
Graph (e)

```
RANGE
Xmin=-1
Xmax=8
Xscl=1
Ymin=-3
Ymax=3
Yscl=1
```

18. $f(x) = \dfrac{1-2x}{x}$

Vertical asymptote: $x = 0$
Horizontal asymptote: $y = -2$
x-intercept: $\left(\frac{1}{2}, 0\right)$
Graph (b)

```
RANGE
Xmin=-3
Xmax=3
Xscl=1
Ymin=-5
Ymax=1
Yscl=1
```

20. $f(x) = -\dfrac{x+2}{x+1}$

Vertical asymptote: $x = -1$
Horizontal asymptote: $y = -1$
x-intercept: $(-2, 0)$
y-intercept: $(0, -2)$
Graph (d)

```
RANGE
Xmin=-4
Xmax=2
Xscl=1
Ymin=-3
Ymax=1
Yscl=1
```

22. $f(x) = \dfrac{x^2(x-3)}{x^2 - 3x}$, $g(x) = x$

(a) Domain of f: $x \neq 0, 3$
   Domain of g: all x

(b) f has no vertical asymptotes.

(c)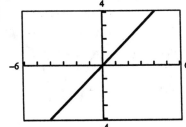

(d) Some viewing rectangles will show the difference in domains. For instance, on the TI-81, use the viewing rectangle $[-4.5, 5] \times [-4.5, 5]$.

24. $f(x) = \dfrac{2x-8}{x^2 - 9x + 20} = \dfrac{2(x-4)}{(x-4)(x-5)}$

$g(x) = \dfrac{2}{x-5}$

(a) Domain of f: $x \neq 4, 5$
   Domain of g: $x \neq 5$

(b) f has a vertical asymptote at $x = 5$.

(c)

(d) Some viewing rectangles will show the difference in domains. For instance, on the TI-81, use the viewing rectangle $[0, 9.5] \times [-4.5, 5]$.

26. (a) $C = \dfrac{25,000(15)}{100 - 15} \approx 4411.76$

The cost would be \$4411.76.

(b) $C = \dfrac{25,000(50)}{100 - 50} = 25,000$

The cost would be \$25,000.

(c) $C = \dfrac{25,000(90)}{100 - 90} = 225,000$

The cost would be \$225,000.

(d) No. The model is undefined for $p = 100$.

28. The moth will become satiated at the horizontal asymptote:

$$y = \frac{1.568}{6.360} \approx 0.2465 \text{ mg}$$

30. (a) $f(x) = 4 - \dfrac{1}{x}$ approaches 4 as x increases.

    (i) $f(x)$ is less than 4 when x is positive.

    (ii) $f(x)$ is greater than 4 when x is negative.

(b) $f(x) = 2 + \dfrac{1}{x - 3}$ approaches 2 as x increases.

    (i) $f(x)$ is greater than 2 when x is positive.

    (ii) $f(x)$ is less than 2 when x is negative.

(c) $f(x) = \dfrac{2x - 1}{x - 3}$ approaches 2 as x increases.

    (i) $f(x)$ is greater than 2 when x is positive.

    (ii) $f(x)$ is less than 2 when x is negative.

(d) $f(x) = \dfrac{2x - 1}{x^2 + 1}$ approaches 0 as x increases.

    (i) $f(x)$ is greater than 0 when x is positive.

    (ii) $f(x)$ is less than 0 when x is negative.

3.6   Graphs of Rational Functions

2. $g(x) = f(x - 1) = \dfrac{1}{x - 1}$

Horizontal shift one unit right

4. $g(x) = f(x + 2) = \dfrac{1}{x + 2}$

Horizontal shift two units left

6. $g(x) = -f(x) = -\dfrac{4}{x^2}$

Reflection in the x-axis

8. $g(x) = \frac{1}{4}f(x) = \frac{1}{x^2}$

Each y-value is multiplied by $\frac{1}{4}$.

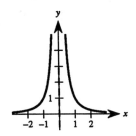

10. $g(x) = f(x) + 1 = \frac{8}{x^3} + 1$

Vertical shift one unit upward

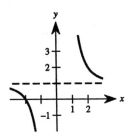

12. $g(x) = \frac{1}{8}f(x) = \frac{1}{x^3}$

Each y-value is multiplied by $\frac{1}{8}$.

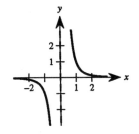

14. $f(x) = \frac{1}{x-3}$

y-intercept: $\left(0, -\frac{1}{3}\right)$
Vertical asymptote: $x = 3$
Horizontal asymptote: $y = 0$

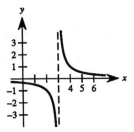

16. $g(x) = \frac{1}{3-x} = -\frac{1}{x-3}$

y-intercept: $\left(0, \frac{1}{3}\right)$
Vertical asymptote: $x = 3$
Horizontal asymptote: $y = 0$

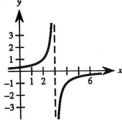

18. $P(x) = \frac{1-3x}{1-x} = \frac{3x-1}{x-1}$

x-intercept: $\left(\frac{1}{3}, 0\right)$
y-intercept: $(0, 1)$
Vertical asymptote: $x = 1$
Horizontal asymptote: $y = 3$

20. $f(t) = \frac{1-2t}{t} = -\frac{2t-1}{t}$

t-intercept: $\left(\frac{1}{2}, 0\right)$
Vertical asymptote: $t = 0$
Horizontal asymptote:
$y = -2$

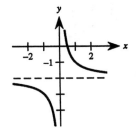

22. $f(x) = 2 - \frac{3}{x^2}$

x-intercepts:
$\left(-\frac{\sqrt{6}}{2}, 0\right), \left(\frac{\sqrt{6}}{2}, 0\right)$
Vertical asymptote: $x = 0$
Horizontal asymptote: $y = 2$
y-axis symmetry

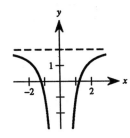

24. $g(x) = \frac{x}{x^2 - 9}$

Intercept: $(0, 0)$
Vertical asymptotes: $x = \pm 3$
Horizontal asymptote: $y = 0$
Origin symmetry

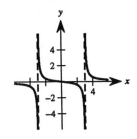

26. $f(x) = -\dfrac{1}{(x-2)^2}$

y-intercept: $\left(0, -\frac{1}{4}\right)$
Vertical asymptote: $x = 2$
Horizontal asymptote: $y = 0$

28. $f(x) = \dfrac{2x}{x^2 + x - 2} = \dfrac{2x}{(x+2)(x-1)}$

Intercept: $(0, 0)$
Vertical asymptotes: $x = -2, 1$
Horizontal asymptote: $y = 0$

30. $f(x) = \dfrac{3-x}{2-x}$

Domain: all real numbers $x \neq 2$
Range: all real numbers $y \neq 1$

32. $h(x) = \dfrac{1}{x-3} + 1$

Domain: all real numbers $x \neq 3$
Range: all real numbers $y \neq 1$

34. $g(x) = -\dfrac{x}{(x-2)^2}$

Domain: all real numbers $x \neq 2$
Range: $y < 0.125$

36. $f(x) = 5\left(\dfrac{1}{x-4} - \dfrac{1}{x+2}\right)$

Domain: all real numbers $x \neq 4, -2$
Range: $(0, \infty) \cup (-\infty, -3.33)$

38. $f(x) = \dfrac{1-x^2}{x} = -x + \dfrac{1}{x}$

x-intercepts: $(-1, 0)$, $(1, 0)$
Vertical asymptote: $x = 0$
Slant asymptote: $y = -x$
Origin symmetry

40. $h(x) = \dfrac{x^2}{x-1} = x + 1 + \dfrac{1}{x-1}$

Intercept: $(0, 0)$
Vertical asymptote: $x = 1$
Slant asymptote: $y = x + 1$

42. $g(x) = \dfrac{x^3}{2x^2-8} = \dfrac{1}{2}x + \dfrac{4x}{2x^2-8}$

Intercept: $(0, 0)$
Vertical asymptotes: $x = \pm 2$
Slant asymptote: $y = \dfrac{1}{2}x$
Origin symmetry

44. $f(x) = \dfrac{2x^2-5x+5}{x-2} = 2x - 1 + \dfrac{3}{x-2}$

y-intercept: $\left(0, -\dfrac{5}{2}\right)$
Vertical asymptote: $x = 2$
Slant asymptote: $y = 2x - 1$

46. $f(x) = \dfrac{2x^2+x}{x+1} = 2x - 1 + \dfrac{1}{x+1}$

x-intercepts: $\left(-\dfrac{1}{2}, 0\right)$, $(0, 0)$
y-intercept: $(0, 0)$
Vertical asymptote: $x = -1$
Slant asymptote: $y = 2x - 1$

48. $g(x) = \dfrac{4|x-2|}{x+1}$

Horizontal asymptotes: $y = \pm 4$

50. $g(x) = \dfrac{3x^4 - 5x + 3}{x^4 + 1}$

52. There is no vertical asymptote since

$$\frac{x^2 + x - 2}{x - 1} = (x + 2), \quad x \neq 1.$$

54. $C(x) = x + \dfrac{16}{x}$

Relative maximum: $(-4, -8)$

Relative minimum: $(4, 8)$

56. (a) $xy = 500 \Rightarrow y = \dfrac{500}{x}$

(b) $x > 0$

(c) If $x = 30, \quad y = 16.67$ meters.

58. (a) $y - 0 = \dfrac{3 - 0}{2 - a}(x - a) \Rightarrow$

$$y = \frac{3(x - a)}{2 - a}, \quad 0 \leq x \leq a$$

(b) $A = \dfrac{1}{2}(\text{base})(\text{height})$

$$= \frac{1}{2}a\left(-\frac{3a}{2 - a}\right) = \frac{-3a^2}{2(2 - a)}, \quad 2 < a$$

(c) A is a minimum when $a = 4$.

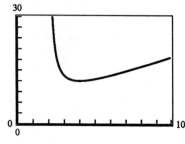

Review Exercises for Chapter 3

2. $y = (x-4)^2 - 4$

When $x = 0$, $y = (0-4)^2 - 4 = 12$.

When $y = 0$, $0 = (x-4)^2 - 4 \Rightarrow x = 2, 6$.

Vertex: $(4, -4)$

Intercept: $(0, 12)$, $(2, 0)$, $(6, 0)$

4. $y = 3x^2 - 12x + 11$

$= 3(x^2 - 4x + 4) + 11 - 12$

$= 3(x-2)^2 - 1$

When $x = 0$, $y = 3(0-2)^2 - 1 = 11$.

When $y = 0$, $0 = 3(x-2)^2 - 1 \Rightarrow x = 2 \pm \frac{1}{3}\sqrt{3}$.

Vertex: $(2, -1)$

Intercepts: $(0, 11), (2 \pm \frac{1}{3}\sqrt{3}, 0)$

6. $f(x) = a(x-2)^2 + 3$

$6 = a(-1-2)^2 + 3$

$3 = 9a$

$\frac{1}{3} = a$

$f(x) = \frac{1}{3}(x-2)^2 + 3$

8. $f(x) = x^2 + 8x + 10$

$= (x^2 + 8x + 16) + 10 - 16$

$= (x+4)^2 - 6$

Minimum: $(-4, -6)$

10. $h(x) = 3 + 4x - x^2$

$= -(x^2 - 4x + 4) + 3 + 4$

$= -(x-2)^2 + 7$

Maximum: $(2, 7)$

12. $h(x) = 4x^2 + 4x + 13$

$= 4\left(x^2 + x + \frac{1}{4}\right) + 13 - 1$

$= 4\left(x + \frac{1}{2}\right)^2 + 12$

Minimum: $\left(-\frac{1}{2}, 12\right)$

14. $f(x) = 4x^2 + 4x + 5$

$= 4\left(x^2 + x + \frac{1}{4}\right) + 5 - 1$

$= 4\left(x + \frac{1}{2}\right)^2 + 4$

Minimum: $\left(-\frac{1}{2},\ 4\right)$

16. $P = 2x + 2y = 200$

$x + y = 100$

$y = 100 - x$

$A = xy = x(100 - x)$

$= -(x^2 - 100x + 2500) + 2500$

$= -(x - 50)^2 + 2500$

A is maximum when $x = 50$ feet and
$y = 100 - 50 = 50$ feet.

18. $P = 230 + 20x - \frac{1}{2}x^2$

Using the zoom and trace features, $x = 20$
yields a maximum profit. Or, completing the
square,

$P = -\frac{1}{2}x^2 + 20x + 230$

$= -\frac{1}{2}(x^2 - 40x + 400) + 230 + 200$

$= -\frac{1}{2}(x - 20)^2 + 430$

Hence, $x = 20$ yields a maximum profit (of **430**).

20. $C = 20{,}000 - 120x + 0.055x^2$

$= 0.055\left(x^2 - \frac{24{,}000}{11} + \frac{12{,}000^2}{11^2}\right) + 20{,}000 - \frac{720{,}000}{11} = 0.055\left(x - \frac{12{,}000}{11}\right)^2 - \frac{500{,}000}{11}$

C is minimum when $x = \frac{12{,}000}{11}$ or **1091 units.**

22. $f(x) = \frac{1}{2}x^3 + 2x$

Since the degree is odd and the leading
coefficient is positive, the graph falls to the left
and rises to the right.

24. $h(x) = -x^5 - 7x^2 + 10x$

Since the degree is odd and the leading
coefficient is negative, the graph rises to the left
and falls to the right.

26. $f(x) = (x+1)^3$

28. $h(x) = -2x^3 - x^2 + x$
$$= -x(2x^2 + x - 1)$$
$$= x(2x - 1)(x + 1)$$

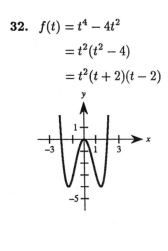

30. $f(x) = -x^3 + 3x - 2$
$$= (x-1)(-x^2 - x + 2)$$
$$= -(x-1)(x^2 + x - 2)$$
$$= -(x-1)(x+2)(x-1)$$
$$= -(x-1)^2(x+2)$$

32. $f(t) = t^4 - 4t^2$
$$= t^2(t^2 - 4)$$
$$= t^2(t+2)(t-2)$$

34.
$$
\begin{array}{r}
5x + 2 \\
x^2 - 3x + 1 \overline{\smash{)}\ 5x^3 - 13x^2 - x + 2} \\
\underline{5x^3 - 15x^2 + 5x} \\
2x^2 - 6x + 2 \\
\underline{2x^2 - 6x + 2} \\
0
\end{array}
$$

$$\frac{5x^3 - 13x^2 - x + 2}{x^2 - 3x + 1} = 5x + 2$$

36.
$$
\begin{array}{r}
3x^2 + 3 \\
x^2 - 1 \overline{\smash{)}\ 3x^4} \\
\underline{3x^4 - 3x^2} \\
3x^2 \\
\underline{3x^2 - 3} \\
3
\end{array}
$$

$$\frac{3x^4}{x^2 - 1} = 3x^2 + 3 + \frac{3}{x^2 - 1}$$

38.

$$
\begin{array}{r}
3x^2 + 5x + 8 \\
2x^2 - 1 \overline{\smash{\big)}\ 6x^4 + 10x^3 + 13x^2 - 5x + 2} \\
\underline{6x^4 - 3x^2} \\
10x^3 + 16x^2 - 5x \\
\underline{10x^3 - 5x} \\
16x^2 + 2 \\
\underline{16x^2 - 8} \\
10
\end{array}
$$

$$
\frac{6x^4 + 10x^3 + 13x^2 - 5x + 2}{2x^2 - 1} =
$$
$$
3x^2 + 5x + 8 + \frac{10}{2x^2 - 1}
$$

40.

$$
\begin{array}{r|rrrr}
\frac{1}{2} & 2 & 2 & -1 & 2 \\
& & 1 & \frac{3}{2} & \frac{1}{4} \\
\hline
& 2 & 3 & \frac{1}{2} & \frac{9}{4}
\end{array}
$$

$$
\frac{2x^3 + 2x - x + 2}{x - (1/2)} = 2x^2 + 3x + \frac{1}{2} + \frac{9/4}{x - (1/2)}
$$

42.

$$
\begin{array}{r|rrrr}
\frac{1}{3} + \frac{2}{3}i & 9 & -15 & 11 & -5 \\
& & 3+6i & -8-6i & 5 \\
\hline
& 9 & -12+6i & 3-6i & 0
\end{array}
$$

$$
\frac{9x^3 - 15x^2 + 11x - 5}{x - [(1/3) + (2/3)i]} = 9x^2 + (-12+6i)x + (3-6i)
$$

44. $f(x) = 20x^4 + 9x^3 - 14x^2 - 3x$

(a) $x = -1$ is a zero.

$$
\begin{array}{r|rrrrr}
-1 & 20 & 9 & -14 & -3 & 0 \\
& & -20 & 11 & 3 & 0 \\
\hline
& 20 & -11 & -3 & 0 & 0
\end{array}
$$

(c) $x = 0$ is a zero.

$$
\begin{array}{r|rrrrr}
0 & 20 & 9 & -14 & -3 & 0 \\
& & 0 & 0 & 0 & 0 \\
\hline
& 20 & 9 & -14 & -3 & 0
\end{array}
$$

(b) $x = \frac{3}{4}$ is a zero.

$$
\begin{array}{r|rrrrr}
\frac{3}{4} & 20 & 9 & -14 & -3 & 0 \\
& & 15 & 18 & 3 & 0 \\
\hline
& 20 & 24 & 4 & 0 & 0
\end{array}
$$

(d) $x = 1$ is not a zero.

$$
\begin{array}{r|rrrrr}
1 & 20 & 9 & -14 & -3 & 0 \\
& & 20 & 29 & 15 & 12 \\
\hline
& 20 & 29 & 15 & 12 & 12
\end{array}
$$

46. $f(x) = 3x^3 - 26x^2 + 364x - 232$

(a) $x = 4 - 10i$ is a zero.

$$
\begin{array}{r|rrrr}
4-10i & 3 & -26 & 364 & -232 \\
& & 12-30i & -356+20i & 232 \\
\hline
& 3 & -14-30i & 8+20i & 0
\end{array}
$$

(c) $x = \frac{2}{3}$ is a zero.

$$
\begin{array}{r|rrrr}
\frac{2}{3} & 3 & -26 & 364 & -232 \\
& & 2 & -16 & 232 \\
\hline
& 3 & -24 & 348 & 0
\end{array}
$$

(b) $x = 4$ is not a zero.

$$
\begin{array}{r|rrrr}
4 & 3 & -26 & 364 & -232 \\
& & 12 & -56 & 1232 \\
\hline
& 3 & -14 & 308 & 1000
\end{array}
$$

(d) $x = -1$ is not a zero.

$$
\begin{array}{r|rrrr}
-1 & 3 & -26 & 364 & -232 \\
& & -3 & 29 & -393 \\
\hline
& 3 & -29 & 393 & -625
\end{array}
$$

48. $h(x) = 5x^5 - 2x^4 - 45x + 18$

(a) 2 |

5	−2	0	0	−45	18
	10	16	32	64	38
5	8	16	32	19	56 $= h(2)$

(b) $\sqrt{3}$ |

5	−2	0	0	−45	18
	$5\sqrt{3}$	$15 - 2\sqrt{3}$	$-6 + 15\sqrt{3}$	$45 - 6\sqrt{3}$	-18
5	$-2 + 5\sqrt{3}$	$15 - 2\sqrt{3}$	$-6 + 15\sqrt{3}$	$-6\sqrt{3}$	0 $= h(\sqrt{3})$

50. $g(t) = 2t^5 - 5t^4 - 8t + 20$

(a) −4 |

2	−5	0	0	−8	20
	−8	52	−208	832	−3296
2	−13	52	−208	824	−3276 $= g(-4)$

(b) $\sqrt{2}$ |

2	−5	0	0	−8	20
	$2\sqrt{2}$	$4 - 5\sqrt{2}$	$-10 + 4\sqrt{2}$	$8 - 10\sqrt{2}$	−20
2	$-5 + 2\sqrt{2}$	$4 - 5\sqrt{2}$	$-10 + 4\sqrt{2}$	$-10\sqrt{2}$	0 $= g(\sqrt{2})$

52. $h(x) = -2x^5 + 4x^3 - 2x^2 + 5$
Sign variations; 3, positive zeros: 3 or 1
$h(-x) = 2x^5 - 4x^3 - 2x^2 + 5$
Sign variations; 2, negative zeros: 2 or 0

54. $f(x) = 3x^4 + 4x^3 - 5x^2 - 8$
Possible rational zeros:
$$\frac{\pm 1, \ \pm 2, \ \pm 4, \ \pm 8}{\pm 1, \ \pm 3} = \pm 1, \ \pm 2, \ \pm 4, \ \pm 8,$$
$$\pm \frac{1}{3}, \ \pm \frac{2}{3}, \ \pm \frac{4}{3}, \ \pm \frac{8}{3}$$

56. $f(x) = 10x^3 + 21x^2 - x - 6$

−2 |

10	21	−1	−6
	−20	−2	6
10	1	−3	0

Zeros: $-2, \ -\frac{3}{5}, \ \frac{1}{2}$

58. $f(x) = x^3 - 1.3x^2 - 1.7x + 0.6$

2 |

1	−1.3	−1.7	0.6
	2	1.4	−0.6
1	0.7	−0.3	0

Zeros: $-1, \ 0.3, \ 2$

60. $f(x) = 5x^4 + 126x^2 + 25$
$$= (5x^2 + 1)(x^2 + 25)$$

Zeros: $\pm \dfrac{\sqrt{5}}{5}i, \ \pm 5i$

62. $g(x) = x^3 - 3x^2 + 3x + 2$

(a) $\pm 1,\ \pm 2$

(b)

(c) -0.44

64. $f(x) = x^5 + 2x^3 - 3x - 20$

(a) $\pm 1,\ \pm 2,\ \pm 4,\ \pm 5,\ \pm 10,\ \pm 20$

(b)

(c) 1.72

66.
$$y = -0.00428x^2 + 1.442x - 3.136, \qquad 20 \le x \le 55$$
$$30 = -0.00428x^2 + 1.442x - 3.136$$
$$0.00428x^2 - 1.442x + 33.136 = 0$$

Using the zoom and trace features, we find that $x = 24.805$. Therefore, the average age of the groom is 30 when the age of the bride is 24.8 years.

68. $f(x) = \dfrac{2x^2 + 5x - 3}{x^2 + 2}$

Domain: All real numbers
Vertical asymptote: none
Horizontal asymptote: $y = 2$

70. $g(x) = \dfrac{1}{(x-3)^2}$

Domain: All real numbers except 3
Vertical asymptote: $x = 3$
Horizontal asymptote: $y = 0$

72. $f(x) = \dfrac{4}{x}$

Origin symmetry
Vertical asymptote: $x = 0$
Horizontal asymptote: $y = 0$

74. $g(x) = \dfrac{-2}{(x+3)^2}$

y-intercept: $\left(0,\ -\dfrac{2}{9}\right)$
Vertical asymptote: $x = -3$
Horizontal asymptote: $y = 0$

76. $f(x) = \dfrac{2x}{x^2 + 4}$

Intercept: $(0,\ 0)$
Origin symmetry
Horizontal asymptote: $y = 0$

78. $h(x) = \dfrac{4}{(x-1)^2}$

y-intercept: $(0,\ 4)$
Vertical asymptote: $x = 1$
Horizontal asymptote: $y = 0$

80. $y = \dfrac{2x^2}{x^2 - 4}$

Intercept: $(0,\ 0)$
y-axis symmetry
Vertical asymptotes:
    $x = 2$ and $x = -2$
Horizontal asymptote:
    $y = 2$

82. $y = \dfrac{5x}{x^2 - 4}$

84. $y = \dfrac{1}{x+3} + 2$

86. $\overline{C} = \dfrac{100,000 + 0.9x}{x}, \quad 0 < x$

$\quad \overline{C}(1,000) = \$100.90$

$\quad \overline{C}(10,000) = \$10.90$

$\quad \overline{C}(100,000) = \$1.90$

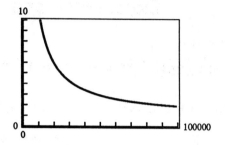

88. (a) When $t = 5$,

$$N = \frac{20(4 + 3(5))}{1 + 0.05(5)} = 304 \quad \text{or} \quad 304,000 \text{ fish.}$$

When $t = 10$,

$$N = \frac{20(4 + 3(10))}{1 + 0.05(10)} \approx 453.333 \quad \text{or} \quad 453,333 \text{ fish.}$$

When $t = 25$,

$$N = \frac{20(4 + 3(25))}{1 + 0.05(25)} \approx 702.222 \quad \text{or} \quad 702,222 \text{ fish.}$$

(b) $N = \dfrac{20(4 + 3t)}{1 + 0.05t} = \dfrac{60t + 80}{0.05t + 1}$

The limiting number of fish in the lake is determined by the horizontal asymptote.

$$N = \frac{60}{0.05} = 1200 \quad \text{or} \quad 1,200,000 \text{ fish}$$

90. The limiting amount of CO_2 uptake is determined by the horizontal asymptote.

$$y = \frac{18.47}{0.23} \approx 80.3 \text{ mg/dc}^2/\text{hr}$$

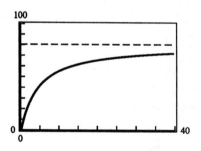

CHAPTER FOUR
Exponential and Logarithmic Functions

4.1 Exponential Functions and Their Graphs

2. $5000(2^{-1.5}) \approx 1767.767$

4. $(1.005)^{400} \approx 7.352$

6. $\sqrt[3]{4395} \approx 16.380$

8. $e^{1/2} \approx 1.649$

10. $e^{3.2} \approx 24.533$

12. $f(x) = -3^x$
y-intercept: $(0, -1)$
-3^x decreases as x increases.
Matches graph (e)

```
RANGE
Xmin=-3
Xmax=3
Xscl=1
Ymin=-5
Ymax=1
Yscl=1
```

14. $f(x) = -3^{-x}$
y-intercept: $(0, -1)$
-3^{-x} increases as x increases.
Matches graph (h)

```
RANGE
Xmin=-3
Xmax=3
Xscl=1
Ymin=-5
Ymax=1
Yscl=1
```

16. $f(x) = 3^x + 1$
y-intercept: $(0, 2)$
$3^x + 1$ increases as x increases.
Matches graph (a)

```
RANGE
Xmin=-3
Xmax=3
Xscl=1
Ymin=0
Ymax=4
Yscl=1
```

18. $f(x) = 3^{x-2}$
y-intercept: $\left(0, \frac{1}{9}\right)$
3^{x-2} increases as x increases.
Point: $(2, 1)$
Matches graph (c)

```
RANGE
Xmin=-2
Xmax=4
Xscl=1
Ymin=-1
Ymax=3
Yscl=1
```

20. $\left(\frac{1}{4}\right)^x < \left(\frac{1}{2}\right)^x$ for $x > 0$

22. $f(x) = \left(\frac{3}{2}\right)^x$

24. $h(x) = \left(\frac{3}{2}\right)^{-x}$

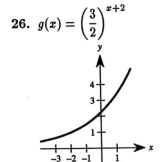

26. $g(x) = \left(\frac{3}{2}\right)^{x+2}$

28. $f(x) = \left(\frac{3}{2}\right)^{-x} + 2$

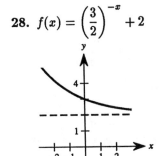

30. $f(x) = 4^{x+1} - 2$

32. $y = 1.08^{5x}$

34. $s(t) = 3e^{-0.2t}$

36. $h(x) = e^{x-2}$

38. (a) $f(x) = \dfrac{8}{1 + e^{-0.5x}}$

Asymptotes: $y = 0$, $y = 8$

(b) $g(x) = \dfrac{8}{1 + e^{-0.5/x}}$

Asymptotes: $y = 4$, $x = 0$

40. $y_1 = \left(1 + \dfrac{0.5}{x}\right)^x$

$y_2 = e^{0.5}$

The graph of y_1 approaches the graph of y_2 as x increases.

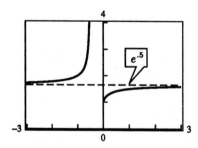

42. $P = \$1000, \quad r = 10\%, \quad t = 10$ years

Compounded n times per year: $A = 1000\left(1 + \dfrac{0.10}{n}\right)^{10n}$

Compounded continuously: $A = 1000e^{0.10(10)}$

n	1	2	4	12	365	Continuous compounding
A	\$2,593.74	\$2,653.30	\$2,685.06	\$2,707.04	\$2,717.91	\$2,718.28

44. $P = \$1000, \quad r = 10\%, \quad t = 40$ years

Compounded n times per year: $A = 1000\left(1 + \dfrac{0.10}{n}\right)^{40n}$

Compounded continuously: $A = 1000e^{0.10(40)}$

n	1	2	4	12	365	Continuous compounding
A	\$45,259.26	\$49,561.44	\$51,977.87	\$53,700.66	\$54,568.25	\$54,598.15

46. $P = 100{,}000e^{-0.12t}$

t	1	10	20	30	40	50
P	\$88,692.04	\$30,119.42	\$9,071.80	\$2,732.37	\$822.97	\$247.88

48. $P = 100{,}000\left(1 + \dfrac{0.07}{365}\right)^{-365t}$

t	1	10	20	30	40	50
P	\$93,240.01	\$49,661.86	\$24,663.01	\$12,248.11	\$6,082.64	\$3,020.75

50. $A = 5000e^{(0.075)(50)} \approx \$212{,}605.51$

52. (a) $x = 100$

$$p = 5000\left(1 - \frac{4}{4 + e^{-0.2}}\right) \approx \$849.53$$

(b) $x = 500$

$$p = 5000\left(1 - \frac{4}{4 + e^{-1}}\right) \approx \$421.12$$

The graph of p is a decreasing function with domain $(0, \infty)$ and range $(0, 5000)$.

54. (a) $P(5) = 2500e^{(0.0293)(5)} \approx \mathbf{2894}$

(b) $P(10) = 2500e^{(0.0293)(10)} \approx \mathbf{3551}$

(c) $P(20) = 2500e^{(0.0293)(20)} \approx \mathbf{4492}$

56. (a) When $t = 0$, $Q = 10 \left(\frac{1}{2}\right)^{0/5730}$

$$= 10 \text{ units.}$$

(b) When $t = 2000$, $Q = 10 \left(\frac{1}{2}\right)^{2000/5730}$

$$\approx 7.851 \text{ units.}$$

(c)

58. $C(10) = 19.95(1.05)^{10} \approx \$32.50$

60. (a) $f(u + v) = a^{u+v} = a^u \cdot a^v = f(u) \cdot f(v)$

(b) $f(2x) = a^{2x} = (a^x)^2 = [f(x)]^2$

$\boxed{4.2}$ Logarithmic Functions and Their Graphs

2. $\log_2\left(\frac{1}{8}\right) = \log_2 2^{-3} = -3$

4. $\log_{27} 9 = \log_{27} 27^{2/3} = \frac{2}{3}$

6. $\log_{10} 1000 = \log_{10} 10^3 = 3$

8. $\log_{10} 10 = \log_{10} 10^1 = 1$

10. $\ln 1 = \ln e^0 = 0$

12. $\log_a \frac{1}{a} = \log_a a^{-1} = -1$

14. $8^2 = 64$

$\qquad 2 = \log_8 64$

16. $9^{3/2} = 27$

$\qquad \frac{3}{2} = \log_9 27$

18. $10^{-3} = 0.001$

$\qquad -3 = \log_{10} 0.001$

20. $e^0 = 1$

$\qquad 0 = \ln 1$

22. $\log_{10}\left(\frac{4}{5}\right) \approx -0.097$

24. $\log_{10} 12.5 \approx 1.097$

26. $\ln(\sqrt{5} - 2) \approx -1.444$

28. $f(x) = 5^x$, $g(x) = \log_5 x$

30. $f(x) = 10^x$, $g(x) = \log_{10} x$

32. $f(x) = -\ln x$ falls to the right.
Vertical asymptote: $x = 0$
Intercept: $(1, 0)$
Graph (e)

```
RANGE
Xmin=-1
Xmax=4
Xscl=1
Ymin=-2
Ymax=2
Yscl=1
```

34. $f(x) = \ln(x - 1)$ rises to the right.
Vertical asymptote: $x = 1$
Intercept: $(2, 0)$
Graph (c)

```
RANGE
Xmin=-1
Xmax=4
Xscl=1
Ymin=-2
Ymax=2
Yscl=1
```

36. $f(x) = -\ln(-x)$ falls to the left.
Vertical asymptote: $x = 0$
Intercept: $(-1, 0)$
Graph (b)

```
RANGE
Xmin=-4
Xmax=1
Xscl=1
Ymin=-2
Ymax=2
Yscl=1
```

38. $f(x) = -\log_6(x + 2)$
Domain: $(-2, \infty)$
Vertical asymptote: $x = -2$
Intercept: $(-1, 0)$

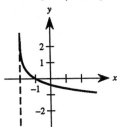

40. $y = \log_{10}\left(\dfrac{x}{5}\right)$
Domain: $(0, \infty)$
Vertical asymptote: $x = 0$
Intercept: $(5, 0)$

42. $g(x) = \ln(-x)$
Domain: $(-\infty, 0)$
Vertical asymptote: $x = 0$
Intercept: $(-1, 0)$

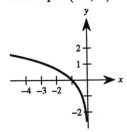

44. $g(x) = \dfrac{12 \ln x}{x}$
Increasing: $(0, 2.72)$
Decreasing: $(2.72, \infty)$
Relative maximum: $(2.72, 4.41)$

46. (a) False, y is not an exponential function of x. (y can never be 0.)
(b) True, y could be $\log_2 x$.
(c) True, x could be 2^y.
(d) False, y is not linear. (The points are not collinear.)

48. $t = \dfrac{10 \ln 2}{\ln 67 - \ln 50} \approx 23.68$ years

50. $t = \dfrac{\ln K}{0.095}$

(a)

K	1	2	4	6	8	10	12
t	0	7.3	14.6	18.9	21.9	24.2	26.2

(b)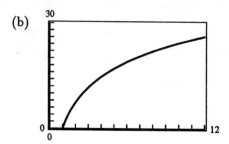

52. (a) $\dfrac{450}{30} = 15$ cubic feet per minute

(b) 380 cubic feet

(c) Total air space required $= 380(30) = 11{,}400$ cubic feet

Let $x =$ square feet of floor space and $h = 30$ feet.

$$V = xh$$
$$11{,}400 = x(30)$$
$$380 = x$$

The minimum number of square feet of floor space required when the ceiling height is 30 feet is 380 square feet.

54. $t = \dfrac{5.315}{-6.7968 + \ln x}, \; 1000 < x$

$\quad = \dfrac{5.315}{-6.7968 + \ln 1068.45}$

$t \approx 30$

The length of the mortgage will be about 30 years.

56. From Exercise 54, we know the length of the mortgage will be about 30 years. Thus, the total amount paid will be $12(30)(1068.45) = \$384{,}642.00$.

58. $\beta = 10 \log_{10} \left(\dfrac{10^{-4}}{10^{-16}} \right)$

$\quad = 10 \log_{10} \left(10^{12} \right) = 10 \cdot 12$

$\beta = 120$ decibels

60. (a) $(0, \infty)$

(b) $y = \log_{10} x$

$x = \log_{10} y$

$10^x = y$

$f^{-1}(x) = 10^x$

(c) Since $\log_{10} 1000 = 3$ and $\log_{10} 10{,}000 = 4$, the interval in which $f(x)$ will be found is $(3, 4)$.

(d) When $f(x)$ is negative, x is in the interval $(0, 1)$.

(e) $0 = \log_{10} 1$

$1 = \log_{10} 10$

$2 = \log_{10} 100$

$3 = \log_{10} 1000$

When $f(x)$ is increased by one unit, x is increased by a factor of 10.

(f) $\quad f(x_1) = 3n \qquad\qquad f(x_2) = n$

$\log_{10} x_1 = 3n \qquad\quad \log_{10} x_2 = n$

$x_1 = 10^{3n} \qquad\qquad x_2 = 10^n$

$x_1 : x_2 = 10^{3n} : 10^n = 10^{2n} : 1$

4.3 Properties of Logarithms

2. $\log_4 10 = \dfrac{\log_{10} 10}{\log_{10} 4} = \dfrac{1}{\log_{10} 4}$

4. $\ln 5 = \dfrac{\log_{10} 5}{\log_{10} e}$

6. $\log_4 10 = \dfrac{\ln 10}{\ln 4}$

8. $\log_{10} 5 = \dfrac{\ln 5}{\ln 10}$

10. $\log_7 4 = \dfrac{\log_{10} 4}{\log_{10} 7}$
$= \dfrac{\ln 4}{\ln 7}$
≈ 0.712

12. $\log_4(0.55) = \dfrac{\log_{10} 0.55}{\log_{10} 4}$
$= \dfrac{\ln 0.55}{\ln 4}$
≈ -0.431

14. $\log_{20} 125 = \dfrac{\log_{10} 125}{\log_{10} 20}$
$= \dfrac{\ln 125}{\ln 20}$
≈ 1.612

16. $\log_{1/3}(0.015) = \dfrac{\log_{10} 0.015}{\log_{10} \frac{1}{3}}$
$= \dfrac{\ln 0.015}{\ln \frac{1}{3}}$
≈ 3.823

18. $\log_{10} 10z = \log_{10} 10 + \log_{10} z$

20. $\log_{10} \dfrac{y}{2} = \log_{10} y - \log_{10} 2$

22. $\log_6 z^{-3} = -3 \log_6 z$

24. $\ln \sqrt[3]{t} = \ln t^{1/3} = \frac{1}{3} \ln t$

26. $\ln \dfrac{xy}{z} = \ln x + \ln y - \ln z$

28. $\ln \left(\dfrac{x^2 - 1}{x^3} \right) = \ln(x^2 - 1) - \ln x^3$
$= \ln[(x+1)(x-1)] - 3 \ln x$
$= \ln(x+1) + \ln(x-1) - 3 \ln x$

30. $\ln \sqrt{\dfrac{x^2}{y^3}} = \ln \left(\dfrac{x^2}{y^3} \right)^{1/2}$
$= \dfrac{1}{2} \ln \left(\dfrac{x^2}{y^3} \right)$
$= \dfrac{1}{2} \left(\ln x^2 - \ln y^3 \right)$
$= \dfrac{1}{2}(2 \ln x - 3 \ln y)$

32. $\ln \dfrac{x}{\sqrt{x^2 + 1}} = \ln x - \ln \sqrt{x^2 + 1}$
$= \ln x - \ln(x^2 + 1)^{1/2}$
$= \ln x - \dfrac{1}{2} \ln(x^2 + 1)$

34. $\ln \sqrt{x^2(x + 2)} = \ln[x^2(x + 2)]^{1/2}$
$= \frac{1}{2} \ln[x^2(x + 2)]$
$= \frac{1}{2} \ln x^2 + \frac{1}{2} \ln(x + 2)$
$= \ln x + \frac{1}{2} \ln(x + 2)$

36. $\log_b \dfrac{\sqrt{x}\, y^4}{z^4} = \log_b x^{1/2} + \log_b y^4 - \log_b z^4$
$= \dfrac{1}{2} \log_b x + 4 \log_b y - 4 \log_b z$

38. $\ln y + \ln z = \ln yz$

40. $\log_5 8 - \log_5 t = \log_5 \dfrac{8}{t}$

42. $-4\log_6 2x = \log_6 (2x)^{-4}$
$$= \log_6 \dfrac{1}{16x^4}$$

44. $\frac{3}{2}\log_7(z-2) = \log_7(z-2)^{3/2}$

46. $2\ln 8 + 5\ln z = \ln 8^2 + \ln z^5 = \ln 64z^5$

48. $3\ln x + 2\ln y - 4\ln z = \ln x^3 + \ln y^2 - \ln z^4 = \ln \dfrac{x^3 y^2}{z^4}$

50. $4[\ln z + \ln(z+5)] - 2\ln(z-5) = 4\ln z(z+5) - 2\ln(z-5) = \ln z^4(z+5)^4 - \ln(z-5)^2 = \ln \dfrac{z^4(z+5)^4}{(z-5)^2}$

52. $2[\ln x - \ln(x+1) - \ln(x-1)] = 2[\ln x - \ln(x+1)(x-1)] = 2\ln \dfrac{x}{x^2-1} = \ln \left(\dfrac{x}{x^2-1}\right)^2$

54. $\frac{1}{2}[\ln(x+1) + 2\ln(x-1)] + 3\ln x = \frac{1}{2}\ln(x+1)(x-1)^2 + 3\ln x$
$$= \ln \sqrt{(x^2-1)(x-1)} + \ln x^3$$
$$= \ln x^3 \sqrt{(x^2-1)(x-1)}$$

56. $\dfrac{3}{2}\ln 5t^6 - \dfrac{3}{4}\ln t^4 = \ln 5\sqrt{5}t^9 - \ln t^3 = \ln \dfrac{5\sqrt{5}t^9}{t^3} = \ln 5\sqrt{5}t^6$

58. $\log_b \frac{5}{3} = \log_b 5 - \log_b 3 \approx 0.8271 - 0.5646 = 0.2625$

60. $\log_b 18 = \log_b(2 \cdot 3^2) = \log_b 2 + 2\log_b 3 \approx 0.3562 + 2(0.5646) = 1.4854$

62. $\log_b \sqrt[3]{75} = \frac{1}{3}\log_b(3 \cdot 5^2) = \frac{1}{3}[2\log_b 5 + \log_b 3] \approx \frac{1}{3}[2(0.8271) + 0.5646] = 0.7396$

64. $\log_b(3b^2) = \log_b 3 + 2\log_b b \approx 0.5646 + 2 = 2.5646$

66. $\log_b 1 = 0$

68. $\log_6 \sqrt[3]{6} = \frac{1}{3}\log_6 6 = \frac{1}{3}(1) = \frac{1}{3}$

70. $\log_5 \left(\frac{1}{125}\right) = \log_5 5^{-3}$
$$= -3\log_5 5 = -3(1) = -3$$

72. $\ln \sqrt[4]{e^3} = \ln e^{3/4} = \frac{3}{4}\ln e = \frac{3}{4}(1) = \frac{3}{4}$

74. $\log_5 \frac{1}{15} = -\log_5(3 \cdot 5)$
$$= -\log_5 3 - \log_5 5$$
$$= -1 - \log_5 3$$

76. $\log_2(4^2 \cdot 3^4) = \log_2 4^2 + \log_2 3^4$
$$= \log_2 2^4 + \log_2 3^4$$
$$= 4\log_2 2 + 4\log_2 3$$
$$= 4(1) + 4\log_2 3$$
$$= 4 + 4\log_2 3$$

78. $\log_{10} \frac{9}{300} = \log_{10} 9 - \log_{10} 300$

$$= \log_{10} 9 - \log_{10}(3 \cdot 10^2)$$

$$= \log_{10} 9 - \log_{10} 3 - 2\log_{10} 10$$

$$= \log_{10}\left(\frac{9}{3}\right) - 2$$

$$= \log_{10} 3 - 2$$

80. $\ln \dfrac{6}{e^2} = \ln 6 - \ln e^2 = \ln 6 - 2$

82. $\ln 1 = \ln \frac{2}{2} = \ln 2 - \ln 2 = 0$

$\ln 2 \approx 0.6931$

$\ln 3 \approx 1.0986$

$\ln 4 = \ln 2^2 = 2\ln 2 \approx 2(0.6931) = 1.3862$

$\ln 5 \approx 1.6094$

$\ln 6 = \ln(2 \cdot 3) = \ln 2 + \ln 3 \approx 0.6931 + 1.0986 = 1.7917$

$\ln 8 = \ln 2^3 = 3\ln 2 \approx 3(0.6931) = 2.0793$

$\ln 9 = \ln 3^2 = 2\ln 3 \approx 2(1.0986) = 2.1972$

$\ln 10 = \ln(2 \cdot 5) = \ln 2 + \ln 5 \approx 0.6931 + 1.6094 = 2.3025$

$\ln 12 = \ln(2^2 \cdot 3) = 2\ln 2 + \ln 3 \approx 2(0.6931) + 1.0986 = 2.4848$

$\ln 15 = \ln(3 \cdot 5) = \ln 3 + \ln 5 \approx 1.0986 + 1.6094 = 2.7080$

$\ln 16 = \ln 2^4 = 4\ln 2 \approx 4(0.6931) = 2.7724$

$\ln 18 = \ln(2 \cdot 3^2) = \ln 2 + 2\ln 3 \approx 0.6931 + 2(1.0986) = 2.8903$

$\ln 20 = \ln(2^2 \cdot 5) = 2\ln 2 + \ln 5 \approx 2(0.6931) + 1.6094 = 2.9956$

84. Let $x = \log_b u$ and $y = \log_b v$, then $b^x = u$ and $b^y = v$.

$$\frac{u}{v} = \frac{b^x}{b^y} = b^{x-y}$$

$$\log_b \left(\frac{u}{v}\right) = \log_b(b^{x-y}) = x - y = \log_b u - \log_b v$$

4.4 Solving Exponential and Logarithmic Equations

2. $3^x = 243$

$3^x = 3^5$

$x = 5$

4. $8^x = 4$

$(2^3)^x = 2^2$

$2^{3x} = 2^2$

$3x = 2$

$x = \frac{2}{3}$

6. $3^{x-1} = 27$

$3^{x-1} = 3^3$

$x - 1 = 3$

$x = 4$

8. $\log_5 5x = 2$

$5x = 5^2$

$x = 5$

10. $\ln(2x - 1) = 0$
$2x - 1 = 1$
$2x = 2$
$x = 1$

12. $\ln e^{2x-1} = 2x - 1$

14. $-1 + \ln e^{2x} = 2x - 1$

16. $-8 + e^{\ln x^3} = x^3 - 8$

18. $4e^x = 91$
$e^x = \frac{91}{4}$
$x = \ln \frac{91}{4}$
≈ 3.125

20. $-14 + 3e^x = 11$
$3e^x = 25$
$e^x = \frac{25}{3}$
$x = \ln \frac{25}{3}$
≈ 2.120

22. $e^{2x} = 50$
$2x = \ln 50$
$x = \frac{1}{2} \ln 50$
≈ 1.956

24. $1000e^{-4x} = 75$
$e^{-4x} = \frac{3}{40}$
$-4x = \ln \frac{3}{40}$
$x = -\frac{1}{4} \ln \frac{3}{40}$
≈ 0.648

26. $e^{2x} - 5e^x + 6 = 0$
$(e^x - 2)(e^x - 3) = 0$

$e^x - 2 = 0$ or $e^x - 3 = 0$
$e^x = 2$ $\qquad e^x = 3$
$x = \ln 2$ $\qquad x = \ln 3$
≈ 0.693 $\qquad \approx 1.099$

28. $\frac{400}{1 + e^{-x}} = 200$
$1 + e^{-x} = 2$
$e^{-x} = 1$
$-x = \ln 1$
$x = 0$

30. $10^x = 570$
$x = \log_{10} 570$
≈ 2.756

32. $6^{5x} = 3000$
$\ln 6^{5x} = \ln 3000$
$5x \ln 6 = \ln 3000$
$x = \frac{\ln 3000}{5 \ln 6}$
≈ 0.894

34. $4^{-3t} = 0.10$
$\ln 4^{-3t} = \ln 0.10$
$-3t \ln 4 = \ln 0.10$
$t = -\frac{\ln 0.10}{3 \ln 4}$
≈ 0.554

36. $2^{3-x} = 565$
$\ln 2^{3-x} = \ln 565$
$(3 - x) \ln 2 = \ln 565$
$3 - x = \frac{\ln 565}{\ln 2}$
$x = 3 - \frac{\ln 565}{\ln 2}$
≈ -6.142

38. $6e^{1-x} = 25$

$\qquad f(x) = 6e^{1-x} - 25$

$\qquad x = -0.427$

40. $e^{0.125t} = 8$

$\qquad f(t) = e^{0.125t} - 8$

$\qquad t = 16.636$

42. $3(5^{x-1}) = 21$

$\qquad f(x) = 3(5^{x-1}) - 21$

$\qquad x = 2.209$

44. $\dfrac{3000}{2 + e^{2x}} = 2$

$\qquad f(x) = \dfrac{3000}{2 + e^{2x}} - 2$

$\qquad x = 3.656$

46. $\ln x = 2 \Rightarrow e^2 = x \Rightarrow x \approx 7.389$

48. $3\ln 5x = 10$

$\qquad \ln 5x = \frac{10}{3}$

$\qquad 5x = e^{10/3}$

$\qquad x = \frac{1}{5} \cdot e^{10/3}$

$\qquad \approx 5.606$

50. $\ln(x+1)^2 = 2$

$\qquad (x+1)^2 = e^2$

$\qquad x + 1 = \pm e$

$\qquad x = -1 \pm e$

$\qquad x = -1 + e \qquad x = -1 - e$

$\qquad \approx 1.718 \text{ or} \qquad \approx -3.718$

52. Using a graphing utility to solve
$y = \ln x + \ln(x+3) - 1 = 0$, we find that
$x = 0.729$. Or,

$\qquad \ln x + \ln(x+3) = 1$

$\qquad \ln x(x+3) = 1$

$\qquad x(x+3) = e^1$

$\qquad x^2 + 3x - e = 0$

$$x = \frac{-3 \pm \sqrt{3^2 - 4(1)(-e)}}{2(1)}$$

$$x = \frac{-3 \pm \sqrt{9 + 4e}}{2}$$

$$x = \frac{-3 + \sqrt{9 + 4e}}{2} \approx 0.729$$

$$x = \frac{-3 - \sqrt{9 + 4e}}{2}$$

$\qquad \approx -3.729 \text{ extraneous}$

54. $\log_{10} x^2 = 6$

$\qquad x^2 = 10^6$

$\qquad x = \pm 10^3$

$\qquad = \pm 1000$

56. $\log_4 x - \log_4 (x - 1) = \dfrac{1}{2}$

$\log_4 \dfrac{x}{x-1} = \dfrac{1}{2}$

$\dfrac{x}{x-1} = 4^{1/2}$

$\dfrac{x}{x-1} = 2$

$x = 2(x-1)$

$2 = x$

58. $\log_2 x + \log_2 (x + 2) = \log_2 (x + 6)$

$\log_2 x(x+2) = \log_2 (x + 6)$

$x(x+2) = (x+6)$

$x^2 + x - 6 = 0$

$(x+3)(x-2) = 0$

$x = -3,$ extraneous or

$x = 2$

60. $\ln(x+1) - \ln(x-2) = \ln x^2$

$\ln \dfrac{x+1}{x-2} = \ln x^2$

$\dfrac{x+1}{x-2} = x^2$

$x + 1 = x^3 - 2x^2$

$0 = x^3 - 2x^2 - x - 1$

Graphing the equation, you can see that the solution lies between 2 and 3. Using the zoom and trace features, $x \approx 2.547$.

62. $\ln 4x = 1$

$f(x) = \ln 4x - 1$

$x = 0.680$

64. $\log_{10} 8x - \log_{10}(1 + \sqrt{x}) = 2$

$f(x) = \log_{10} 8x - \log_{10}(1 + \sqrt{x}) - 2$

$x = 180.384$

66. $A = Pe^{rt}$

$2000 = 1000e^{0.12t}$

$2 = e^{0.12t}$

$\ln 2 = 0.12t$

$\dfrac{\ln 2}{0.12} = t$

$t \approx 5.8$ years

68. $A = Pe^{rt}$

$3000 = 1000e^{0.12t}$

$3 = e^{0.12t}$

$\ln 3 = 0.12t$

$\dfrac{\ln 3}{0.12} = t$

$t \approx 9.2$ years

70. (a)
$$600 = 5000\left(1 - \frac{4}{4 + e^{-0.002x}}\right)$$

$$-4400 = -\frac{20{,}000}{4 + e^{-0.002x}}$$

$$4 + e^{-0.002x} = \frac{50}{11}$$

$$e^{-0.002x} = \frac{6}{11}$$

$$-0.002x = \ln\frac{6}{11}$$

$$x = \frac{\ln\frac{6}{11}}{-0.002} \approx 303 \text{ units}$$

(b)
$$400 = 5000\left(1 - \frac{4}{4 + e^{-0.002x}}\right)$$

$$-4600 = -\frac{20{,}000}{4 + e^{-0.002x}}$$

$$4 + e^{-0.002x} = \frac{100}{23}$$

$$e^{-0.002x} = \frac{8}{23}$$

$$-0.002x = \ln\frac{8}{23}$$

$$x = \frac{\ln\frac{8}{23}}{-0.002} \approx 528 \text{ units}$$

72.
$$21 = 68 \cdot 10^{-0.04x}$$

$$\frac{21}{68} = 10^{-0.04x}$$

$$\log_{10}\frac{21}{68} = -0.04x$$

$$x = \frac{\log_{10}\frac{21}{68}}{-0.04}$$

$$\approx 12.8 \text{ inches}$$

74. $P = \dfrac{0.83}{1 + e^{-0.2n}}$

(a)

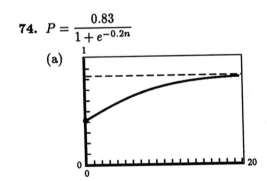

(b) $y = 0.83$ is the horizontal asymptote, which represents the limiting percentage of correct responses.

(c)
$$0.60 = \frac{0.83}{1 + e^{-0.2n}}$$

$$1 + e^{-0.2n} = \frac{0.83}{0.60}$$

$$e^{-0.2n} = \frac{0.23}{0.60}$$

$$-0.2n = \ln\frac{0.23}{0.60}$$

$$n = \frac{\ln\frac{0.23}{0.60}}{-0.2}$$

$$\approx 5 \text{ trials}$$

4.5 Applications of Exponential and Logarithmic Functions

	Initial investment	Annual % rate	Effective yield	Time to double	Amount after 10 yrs
2.	20,000	10.5%	11.07%	6.60 yrs	57,153.02
4.	10,000	13.86%	14.87%	5 yrs	39,988.23
6.	2000	4.4%	4.5%	15.75 yrs	3,105.41
8.	8986.58	8%	8.33%	8.66 yrs	20,000.00
10.	250	11.5%	12.19%	6.03 yrs	789.55

Formulas used in Exercises 2–10:

Initial investment: $P = Ae^{-rt}$

Annual % rate: $r = \dfrac{\ln(A/P)}{t}$ or $r = \ln(1 + \text{effective yield})$

Effective yield $= e^r - 1$

Time to double $= \dfrac{\ln 2}{r}$

Amount after ten years $= Pe^{10r}$

12. $500{,}000 = P\left(1 + \dfrac{0.12}{12}\right)^{(12)(40)}$

$500{,}000 = P(1.01)^{480}$

$P \approx \$4214.16$

14. $P = 1000$, $r = 10.5\%$

(a) $n = 1$

$t = \dfrac{\ln 2}{\ln(1 + 0.105)} \approx 6.942$ years

(b) $n = 12$

$t = \dfrac{\ln 2}{12 \ln\left(1 + \dfrac{0.105}{12}\right)} \approx 6.630$ years

(c) $n = 365$

$t = \dfrac{\ln 2}{365 \ln\left(1 + \dfrac{0.105}{365}\right)} \approx 6.602$ years

(d) Continuously

$t = \dfrac{\ln 2}{0.105} \approx 6.601$ years

16. $3P = P(1+r)^t$

$3 = (1+r)^t$

$\ln 3 = t \ln(1+r)$

$t = \dfrac{\ln 3}{\ln(1+r)}$

r	2%	4%	6%	8%	10%	12%
t	55.48	28.01	18.85	14.27	11.53	9.69

18. $A = 1 + 0.06t$

$A = \left(1 + \dfrac{0.055}{365}\right)^{365t}$

$5\frac{1}{2}\%$ interest compounded daily grows faster.

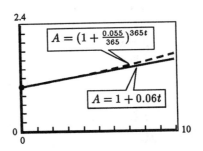

Isotope	Half-life	Initial quantity	Amt. after 1000 yrs	Amt. after 10,000 yrs
20. Ra^{226}	1620	2.30 grams	1.5 grams	0.03 grams
22. C^{14}	5730	3 grams	1.63 grams	0.89 grams
24. Pu^{230}	24,360	0.53 grams	0.52 grams	0.4 grams

Formula used in Exercises 20–24:

$Q(t) = Ce^{-(\ln 2/T)t}$

$Q(t)$ = amount, C = initial amount, T = half-life, t = time

26. $y = \frac{1}{2}e^{kt}$

$6 = \frac{1}{2}e^{4k}$

$12 = e^{4k}$

$4k = \ln 12$

$k = \frac{1}{4}\ln 12$

≈ 0.6212

28. $y = 5e^{kt}$

$1 = 5e^{3k}$

$\frac{1}{5} = e^{3k}$

$3k = \ln\frac{1}{5}$

$k = \frac{1}{3}\ln\frac{1}{5}$

≈ -0.5365

30. (a) If $t = 0$, $P = 240,360$.

    (b) $250,000 = 240,360e^{0.012t}$

$$0.012t = \ln\left(\frac{250,000}{240,360}\right)$$

$$t = \frac{1}{0.012}\ln\left(\frac{250,000}{240,360}\right)$$

$$\approx 3.27 \text{ years}$$

$$1990 + 3 = 1993$$

The city will have a population of 250,000 in the year 1993.

32. $100,250 = 140,500e^{k(-30)}$

$$-30k = \ln\left(\frac{100,250}{140,500}\right)$$

$$k = \left(-\frac{1}{30}\right)\ln\left(\frac{100,250}{140,500}\right) \approx 0.0113$$

$$P = 140,500e^{0.0113(10)} \approx 157,308$$

34. $P = Ce^{kt}$

$$2.30 = Ce^{k(0)} \Rightarrow C = 2.30$$

$$2.65 = 2.30Ce^{k(10)}$$

$$10k = \ln\frac{2.65}{2.30}$$

$$k = \frac{1}{10}\ln\frac{2.65}{2.30} \approx 0.0142$$

$$P = 2.30e^{0.0142t}$$

When $t = 20$, $P \approx 3.06$ million.

36. $N = 250e^{kt}$

$$280 = 250e^{10k}$$

$$\frac{28}{25} = e^{10k}$$

$$k = \frac{\ln 1.12}{10} \approx 0.0113$$

$$N = 250e^{0.0113t}$$

$$500 = 250e^{0.0113t}$$

$$t = \frac{\ln 2}{0.0113} \approx 61.34 \text{ hours}$$

38. $Q(t) = Ce^{kt}$

$$\frac{1}{2}C = Ce^{k(5730)}$$

$$5730k = \ln\frac{1}{2}$$

$$k = \frac{1}{5730}\ln\frac{1}{2} \approx -0.0001$$

$$0.15C = Ce^{-0.0001t}$$

$$-0.0001t = \ln 0.15$$

$$t = -\frac{1}{0.0001}\ln 0.15 \approx 18,971$$

40. $3000 = 4600e^{k(2)}$

$$\frac{3000}{4600} = e^{2k}$$

$$2k = \ln\frac{15}{23}$$

$$k = \frac{1}{2}\ln\frac{15}{23} \approx -0.2137$$

$$y = 4600e^{-0.2137(3)}$$

$$y \approx \$2,423$$

42. $N = 30(1 - e^{kt})$

    (a) $N = 19, \quad t = 20$

$$19 = 30(1 - e^{20k})$$

$$20k = \ln \frac{11}{30}$$

$$k \approx -0.0502$$

$$N = 30(1 - e^{-0.0502t})$$

    (b) $N = 25$

$$25 = 30(1 - e^{-0.0502t})$$

$$\frac{5}{30} = e^{-0.0502t}$$

$$t = -\frac{1}{0.0502} \ln \frac{5}{30} \approx 36 \text{ days}$$

44. $p = 0.193 e^{-(x-70)^2/14.32}$

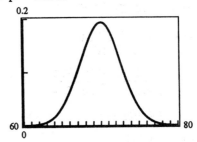

The average height is 70.0, the highest point on the graph.

46. $p(t) = \dfrac{1000}{1 + 9e^{-0.1656t}}$

    (a)

$p = 1000$ which represents the carrying capacity of the game preserve for this animal.

  (b) $p(5) = \dfrac{1000}{1 + 9e^{-0.828}} \approx 203$ animals

  (c) $500 = \dfrac{1000}{1 + 9e^{-0.1656t}}$

$$1 + 9e^{-0.1656t} = 2$$

$$e^{-0.1656t} = \frac{1}{9}$$

$$t = -\frac{1}{0.1656} \ln \frac{1}{9} \approx 13.3 \text{ months}$$

48. $S = \dfrac{500,000}{1 + 0.6e^{kt}}$

    (a) $300,000 = \dfrac{500,000}{1 + 0.6e^{2k}}$

$$1 + 0.6e^{2k} = \frac{5}{3}$$

$$e^{2k} = \frac{10}{9}$$

$$k = \frac{1}{2} \ln \frac{10}{9} \approx 0.0527$$

  (b) $S = \dfrac{500,000}{1 + 0.6e^{0.0527(5)}} \approx 280,759$ units

50. $R = \log_{10} I$

    (a) $8.1 = \log_{10} I$

$$10^{8.1} = I$$

$$I \approx 125,892,541$$

    (b) $6.7 = \log_{10} I$

$$10^{6.7} = I$$

$$I = 5,011,872$$

52. (a) $\beta(10^{-13}) = 10\log_{10}\dfrac{10^{-13}}{10^{-16}} = 10\log_{10}10^3 = 30$ decibels

(b) $\beta(10^{-7.5}) = 10\log_{10}\dfrac{10^{-7.5}}{10^{-16}} = 10\log_{10}10^{8.5} = 85$ decibels

(c) $\beta(10^{-7}) = 10\log_{10}\dfrac{10^{-7}}{10^{-16}} = 10\log_{10}10^9 = 90$ decibels

(d) $\beta(10^{-4.5}) = 10\log_{10}\dfrac{10^{-4.5}}{10^{-16}} = 10\log_{10}10^{11.5} = 115$ decibels

54. From Exercise 53, we have

$$\% \text{ decrease} = \frac{I_0 10^{88/10} - I_0 10^{72/10}}{I_0 10^{88/10}} \times 100 \approx 97\%.$$

56. $\text{pH} = -\log_{10}[\text{H}^+] = -\log_{10}[11.3 \times 10^{-6}] \approx 4.95$

58.
$$\text{pH} = -\log[\text{H}^+]$$
$$3.2 = -\log[\text{H}^+]$$
$$[\text{H}^+] = 10^{-3.2} \approx 6.3 \times 10^{-4}$$

60.
$$\text{pH} - 1 = -\log_{10}[\text{H}^+]$$
$$-(\text{pH} - 1) = \log_{10}[\text{H}^+]$$
$$10^{-(\text{pH}-1)} = [\text{H}^+]$$
$$10^{-\text{pH}+1} = [\text{H}^+]$$
$$10^{-\text{pH}} \cdot 10 = [\text{H}^+]$$

The hydrogen ion concentration is increased by a factor of 10.

62. (a)

64. The greater the rate of growth, the greater the value of k.

(b) $u = 80{,}000$ when $t = 16.73$ years.

Yes, it is possible.

4.6 | Exploring Data: **Nonlinear Models**

2.

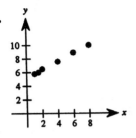

A linear model seems best.

4.

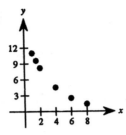

An exponential model seems best.

6.

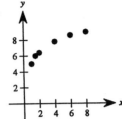

A logarithmic model seems best.

8. $y = 5.5441(1.3560)^x$

$= 5.5441e^{0.3045x}$

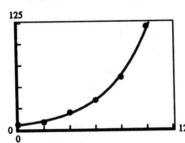

10. $y = 87.2618(0.9438)^x$

$= 87.2618e^{-0.05779x}$

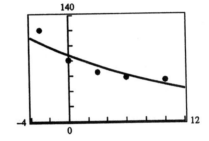

12. $y = 9.0275 + 2.5366 ln x$

14. $y = 20.0758 - 5.0269 ln x$

16. $y = 3.3969x^{1.6504}$

18. $y = 525.00794x^{0.2257}$

20. $y = 4.2154(0.0958)^x$

22. $y = 321.2589 + 193.2820 \ln t$

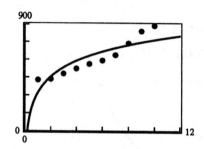

Review Exercises for Chapter 4

2. $f(x) = 2^{-x}$
Horizontal asymptote: $y = 0$
y-intercept: $(0, 1)$
As x increases, $f(x)$ decreases.
Matches (f)

```
RANGE
Xmin=-3
Xmax=4
Xscl=1
Ymin=-1
Ymax=4
Yscl=1
```

4. $f(x) = 2^x + 1$
Horizontal asymptote: $y = 1$
y-intercept: $(0, 2)$
As x increases, $f(x)$ increases.
Matches (b)

```
RANGE
Xmin=-4
Xmax=4
Xscl=1
Ymin=0
Ymax=6
Yscl=1
```

6. $f(x) = \log_2(x - 1)$
Vertical asymptote: $x = 1$
x-intercept: $(2, 0)$
As x increases, $f(x)$ increases.
Matches (e)

```
RANGE
Xmin=-1
Xmax=7
Xscl=1
Ymin=-4
Ymax=3
Yscl=1
```

8. $f(x) = 0.3^x$

10. $g(x) = 0.3^{-x}$

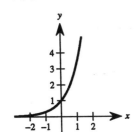

12. $h(x) = 2 - e^{-x/2}$

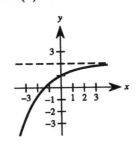

14. $s(t) = 4e^{-2/t}$, $t > 0$

16. $n = 1$: $A = 2000(1 + 0.12)^{30}$ $\approx \$59{,}919.84$

$n = 2$: $A = 2000\left(1 + \dfrac{0.12}{2}\right)^{60}$ $\approx \$65{,}975.38$

$n = 4$: $A = 2000\left(1 + \dfrac{0.12}{4}\right)^{120}$ $\approx \$69{,}421.97$

$n = 12$: $A = 2000\left(1 + \dfrac{0.12}{12}\right)^{360}$ $\approx \$71{,}899.28$

$n = 365$: $A = 2000\left(1 + \dfrac{0.12}{365}\right)^{10950}$ $\approx \$73{,}153.17$

Continuous: $A = 2000e^{(0.12)(30)} \approx \$73{,}196.47$

18. $A = P\left(1 + \dfrac{r}{n}\right)^{nt}$

$P = A\left(1 + \dfrac{r}{n}\right)^{-nt} = 200{,}000\left(1 + \dfrac{0.10}{12}\right)^{-12t}$

$t = 1$: $P = 200{,}000\left(1 + \dfrac{0.10}{12}\right)^{-12}$ $\approx \$181{,}042.49$

$t = 10$: $P = 200{,}000\left(1 + \dfrac{0.10}{12}\right)^{-120}$ $\approx \$73{,}881.39$

$t = 20$: $P = 200{,}000\left(1 + \dfrac{0.10}{12}\right)^{-240}$ $\approx \$27{,}292.30$

$t = 30$: $P = 200{,}000\left(1 + \dfrac{0.10}{12}\right)^{-360}$ $\approx \$10{,}081.97$

$t = 40$: $P = 200{,}000\left(1 + \dfrac{0.10}{12}\right)^{-480}$ $\approx \$3{,}724.35$

$t = 50$: $P = 200{,}000\left(1 + \dfrac{0.10}{12}\right)^{-600}$ $\approx \$1{,}375.80$

20. $V(2) = 14{,}000 \left(\dfrac{3}{4}\right)^2$

$\qquad = \$7875$

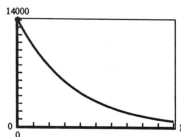

22. $P(t) = 1 - e^{-t/3}$

(a) $P(0.5) = 1 - e^{-0.5/3}$

$\qquad = 1 - e^{-1/6}$

$\qquad \approx 0.154$

(b) $P(2) = 1 - e^{-2/3} \approx 0.487$

(c) $P(5) = 1 - e^{-5/3} \approx 0.811$
   The algebraic solution is
   easier.

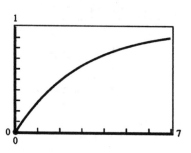

24. $C(10) = 69.95(1.05)^{10} \approx \$113.94$

26. $g(x) = \log_5 x$

28. $f(x) = \ln(x-3)$

30. $f(x) = \frac{1}{4} \ln x$

32. $25^{3/2} = 125$

$\quad \frac{3}{2} = \log_{25} 125$

34. $\log_9 3 = \log_9 9^{1/2}$

$\qquad = \frac{1}{2}$

36. $\log_4 \frac{1}{16} = \log_4 4^{-2}$

$\qquad = -2$

38. $\log_a \dfrac{1}{a} = \log_a a^{-1}$

$\qquad = -1$

40. $\ln e^{-3} = -3 \ln e$

$\qquad = -3(1)$

$\qquad = -3$

42. $\log_{1/2} 5 = \dfrac{\log_{10} 5}{\log_{10} 1/2}$

$\qquad = \dfrac{\ln 5}{\ln 1/2}$

$\qquad \approx -2.322$

44. $\log_3 0.28 = \dfrac{\log_{10} 0.28}{\log_{10} 3}$

$\qquad = \dfrac{\ln 0.28}{\ln 3}$

$\qquad \approx -1.159$

46. $\log_7 \dfrac{\sqrt{x}}{4} = \log_7 x^{1/2} - \log_7 4$

$\qquad = \dfrac{1}{2} \log_7 x - \log_7 4$

48. $\ln \left| \dfrac{x-1}{x+1} \right| = \ln |x-1| - \ln |x+1|$

50. $\ln \sqrt[5]{\dfrac{4x^2-1}{4x^2+1}} = \ln \left[\dfrac{(2x-1)(2x+1)}{4x^2+1}\right]^{1/5}$

$$= \frac{1}{5}\left[\ln(2x-1) + \ln(2x+1) - \ln(4x^2+1)\right]$$

$$= \frac{1}{5}\ln(2x-1) + \frac{1}{5}\ln(2x+1) - \frac{1}{5}\ln(4x^2+1)$$

52. $\log_6 y - 2\log_6 z = \log_6 y - \log_6 z^2 = \log_6 \dfrac{y}{z^2}$

54. $5\ln|x-2| - \ln|x+2| - 3\ln|x| = \ln|(x-2)^5| - [\ln|x+2| + \ln|x^3|] = \ln\left|\dfrac{(x-2)^5}{(x+2)x^3}\right|$

56. $3[\ln x - 2\ln(x^2+1)] + 2\ln 5 = 3[\ln x - \ln(x^2+1)^2] + \ln 5^2$

$$= 3\ln\frac{x}{(x^2+1)^2} + \ln 25 = \ln\frac{x^3}{(x^2+1)^6} + \ln 25 = \ln\frac{25x^3}{(x^2+1)^6}$$

58. True

$\log_b b^{2x} = 2x\log_b b$

$= 2x(1)$

$= 2x$

60. True

$e^{x-1} = e^x e^{-1}$

$= e^x \cdot \dfrac{1}{e}$

$= \dfrac{e^x}{e}$

Try graphing $y_1 = e^{x-1}$ and $y_2 = e^x/e$.

62. $\log_b \dfrac{25}{9} = \log_b \left(\dfrac{5}{3}\right)^2$

$= 2[\log_b 5 - \log_b 3]$

$\approx 2[0.8271 - 0.5646]$

$= 0.5250$

64. $\log_b 30 = \log_b(2 \cdot 3 \cdot 15)$

$= \log_b 2 + \log_b 3 + \log_b 5$

$\approx 0.3562 + 0.5646 + 0.8271$

$= 1.7479$

66. $t = 50\log_{10}\dfrac{18{,}000}{18{,}000 - 4000}$

≈ 5.46 minutes

68. $e^{3x} = 25$

$3x = \ln 25$

$x = \frac{1}{3}\ln 25 \approx 1.073$

70. $14e^{3x+2} = 560$

$e^{3x+2} = 40$

$3x + 2 = \ln 40$

$x = \frac{1}{3}(\ln 40 - 2)$

≈ 0.563

72. $e^{2x} - 6e^x + 8 = 0$

$(e^x - 4)(e^x - 2) = 0$

$e^x - 4 = 0$

$\quad x = \ln 4 \approx 1.386$

or

$e^x - 2 = 0$

$\quad x = \ln 2 \approx 0.693$

74. $2\ln 4x = 15$

$4x = e^{7.5}$

$x = \tfrac{1}{4}e^{7.5}$

≈ 452.011

76. $\ln\sqrt{x+1} = 2$

$\tfrac{1}{2}\ln(x+1) = 2$

$x + 1 = e^4$

$x = e^4 - 1$

≈ 53.598

78. $\tfrac{1}{2} = Ce^{k(0)}$

$\tfrac{1}{2} = Ce^0 = C$

$5 = \tfrac{1}{2}e^{k(5)}$

$10 = e^{5k}$

$5k = \ln 10$

$k = \tfrac{1}{5}\ln 10 \approx 0.4605$

$y = \tfrac{1}{2}e^{0.4605t}$

80. $2 = Ce^{k(0)}$

$2 = Ce^0 = C$

$1 = 2e^{k(5)}$

$\tfrac{1}{2} = e^{5k}$

$5k = \ln\tfrac{1}{2}$

$k = \tfrac{1}{5}\ln\tfrac{1}{2} \approx -0.1386$

$y = 2e^{-0.1386t}$

82. (a) $50 = \dfrac{157}{1 + 5.4e^{-0.12t}}$

$1 + 5.4e^{-0.12t} = \dfrac{157}{50}$

$5.4e^{-0.12t} = \dfrac{107}{50}$

$-0.12t = \ln\dfrac{107}{270}$

$t = -\dfrac{1}{0.12}\ln\dfrac{107}{270} \approx 7.7 \text{ weeks}$

(b) $75 = \dfrac{157}{1 + 5.4e^{-0.12t}}$

$1 + 5.4e^{-0.12t} = \dfrac{157}{75}$

$5.4e^{-0.12t} = \dfrac{82}{75}$

$-0.12t = \ln\dfrac{82}{405}$

$t = -\dfrac{1}{0.12}\ln\dfrac{82}{405} \approx 13.3 \text{ weeks}$

84. (a) $t = \dfrac{\ln 2}{r}$

$5 = \dfrac{\ln 2}{r}$

$r = \dfrac{\ln 2}{5}$

$\approx 0.1386 \text{ or } = 13.86\%$

(b) $A = 10,000e^{0.1386(1)}$

$\approx \$11,486.65$

(c) $A = Pe^{0.1386(1)}$

$\approx P(1.1487)$

$= P(1 + 0.1487)$

Effective yield ≈ 0.1487

$= 14.87\%$

86. $R = \log_{10}\left(\dfrac{I}{I_0}\right)$ where $I_0 = 1$

(a) $8.4 = \log_{10}\left(\dfrac{I}{1}\right)$

$10^{8.4} = I$

(b) $6.85 = \log_{10}\left(\dfrac{I}{1}\right)$

$10^{6.85} = I$

(c) $9.1 = \log_{10}\left(\dfrac{I}{1}\right)$

$10^{9.1} = I$

88. (a) $y = 2.3347 + 2.2893x$ correlation $= 0.95$

(b) $y = 1.5379 + 8.3734 ln x$ correlation $= 0.997$

(c)

The logarithmic model is better.

CHAPTER FIVE
Trigonometric Functions

5.1 Radian and Degree Measure

2. (a) $-\dfrac{\pi}{12}$ Quadrant IV

(b) $-\dfrac{11\pi}{9}$ Quadrant II

4. (a) 5.63   Quadrant IV

(b) -2.25 Quadrant III

6. (a) $-260°$ Quadrant II

(b) $-3.4°$ Quadrant IV

8. (a)

(b)

10. (a)

(b)

12. (a) $-\dfrac{9\pi}{4} + 4\pi = \dfrac{7\pi}{4}$

$-\dfrac{9\pi}{4} + 2\pi = -\dfrac{\pi}{4}$

(b) $-\dfrac{2\pi}{15} + 2\pi = \dfrac{28\pi}{15}$

$-\dfrac{2\pi}{15} - 2\pi = -\dfrac{32\pi}{15}$

14. (a) $-120° + 360° = 240°$

$-120° - 360° = -480°$

(b) $390° - 360° = 30°$

$390° - 720° = -330°$

16. (a) $-420° + 720° = 300°$

$-420° + 360° = -60°$

(b) $230° + 360° = 590°$

$230° - 360° = -130°$

18. (a) Complement: $\dfrac{\pi}{2} - 1 \approx 0.57$

Supplement: $\pi - 1 \approx 2.14$

(b) Complement: None

Supplement: $\pi - 2 \approx 1.14$

20. (a) Complement: $90° - 79° = 11°$

Supplement: $180° - 79° = 101°$

(b) Complement: None

Supplement: $180° - 150° = 30°$

22. (a) $-\dfrac{7\pi}{12} = -\dfrac{7\pi}{12}\left(\dfrac{180}{\pi}\right) = -105°$

(b) $\dfrac{\pi}{9} = \dfrac{\pi}{9}\left(\dfrac{180}{\pi}\right) = 20°$

24. (a) $\dfrac{11\pi}{6} = \dfrac{11\pi}{6}\left(\dfrac{180}{\pi}\right) = 330°$

(b) $\dfrac{34\pi}{15}\left(\dfrac{180}{\pi}\right) = 408°$

26. (a) $315° = 315\left(\dfrac{\pi}{180}\right) = \dfrac{7\pi}{4}$ radians

(b) $120° = 120\left(\dfrac{\pi}{180}\right) = \dfrac{2\pi}{3}$ radians

28. (a) $-270° = -270\left(\dfrac{\pi}{180}\right) = -\dfrac{3\pi}{2}$ radians

(b) $144° = 144\left(\dfrac{\pi}{180}\right) = \dfrac{4\pi}{5}$ radians

30. (a) $-216.35° = -216.35\left(\dfrac{\pi}{180}\right)$

≈ -3.776 radians

(b) $-48.27° = -48.27\left(\dfrac{\pi}{180}\right) \approx -0.842$ radians

32. (a) $-0.83° = -0.83\left(\dfrac{\pi}{180}\right)$

≈ -0.014 radians

(b) $345° = 345\left(\dfrac{\pi}{180}\right) \approx 6.021$ radians

34. (a) $\dfrac{15\pi}{8} = \dfrac{15\pi}{8}\left(\dfrac{180}{\pi}\right) \approx 337.500°$

(b) $6.5\pi = 6.5\pi\left(\dfrac{180}{\pi}\right) \approx 1170.000°$

36. (a) $-2 = -2\left(\dfrac{180}{\pi}\right) \approx -114.592°$

(b) $-0.57 = -0.57\left(\dfrac{180}{\pi}\right) \approx -32.659°$

38. (a) $-135°36'' = -135\dfrac{36}{3600} \approx -135.01°$

(b) $-408°16'25'' = -\left(408 + \dfrac{16}{60} + \dfrac{25}{3600}\right)$

$\approx -408.274°$

(b) $0.45 = 0.45\left(\dfrac{180}{\pi}\right)^{\circ}$

$\approx 25.7831°$

$= 25° + (0.7831)(60')$

$\approx 25°46' + (0.986)(60'')$

$= 25°46'59''$

40. (a) $-345.12° = -(345° + (0.12)(60'))$

$= -(345° + 7' + 0.2(60''))$

$= -345°7'12''$

42. (a) $-0.355 = -0.355 \left(\dfrac{180}{\pi}\right)^{\circ}$

$\approx -20.34°$

$= -(20° + (0.34)(60'))$

$\approx -(20°20' + 0.4(60''))$

$= -20°20'24''$

(b) $0.7865 = 0.7865 \left(\dfrac{180}{\pi}\right)^{\circ}$

$\approx 45.063°$

$= 45° + (0.063)(60')$

$\approx 45°3' + (0.78784)(60'')$

$= 45°3'47''$

44. $\theta = \dfrac{s}{r} = \dfrac{10}{16} = \dfrac{5}{8}$

46. $r = 80$ km, $s = 160$ km

$\theta = \dfrac{s}{r} = \dfrac{160 \text{ km}}{80 \text{ km}} = 2$

48. $\theta = 60° \left(\dfrac{\pi}{180°}\right) = \dfrac{\pi}{3}$

$s = r\theta = 9 \left(\dfrac{\pi}{3}\right) = 3\pi$ ft

50. $s = r\theta = 40 \left(\dfrac{3\pi}{4}\right) = 30\pi$ cm

52. $\theta = 47°36'32'' - 37°46'39'' = 9°49'53'' \approx 0.1716$

$s \approx 4000(0.1716) \approx 686.4$ miles

54. $\theta = 31°47' + 26°10' = 57°57' \approx 1.0114$

$s \approx 4000(1.0114) \approx 4045.7$ miles

56. $\theta = \dfrac{500}{4000} = 0.125$ radian $\approx 7.162°$

58. $\theta = \dfrac{s}{r} = \dfrac{12}{4} = 3$ radians $\approx 171.89°$

60. (a) 50 mi/hr $= 4400$ ft/min

Circumference $= 3\pi$ ft

$\dfrac{\text{Revolutions}}{\text{Minute}} = \dfrac{4400 \text{ ft/min}}{3\pi \text{ ft}} \approx 466.85$

(b) Angular speed $\approx (466.85 \text{ rot/min})(2\pi \text{ rad})$

≈ 2933.3 rad/min

62. $12 \dfrac{\text{Radians}}{\text{Second}} = \dfrac{100(2\pi)}{t}$

$t = \dfrac{50\pi}{3} \approx 52.36$ seconds

64. (a) Arc length of larger sprocket in feet:

$$s = r\theta$$

$$s = \frac{1}{3}(2\pi) = \frac{2\pi}{3} \text{ ft}$$

Therefore, the chain moves $2\pi/3$ feet as does the smaller rear sprocket. Thus, the angle, θ, of the smaller sprocket is

$$\theta = \frac{s}{r} = \frac{2\pi/3}{2} = \frac{\pi}{3} \text{ and the arc length of}$$
the tire is

$$s = \theta r$$

$$s = \left(\frac{\pi}{3}\right)(13)$$

$$s = \frac{13\pi}{3} \text{ ft.}$$

$$\text{Speed } = \frac{s}{t} = \frac{13\pi/3}{1 \text{ sec}}$$

$$= \frac{13\pi}{3} \text{ ft/sec} \approx 13.6 \text{ ft/sec}$$

(b) Speed mph:

$$\frac{13\pi}{3} \frac{\text{ft}}{\text{sec}} \times 3600 \frac{\text{sec}}{\text{hr}} \times \frac{1 \text{ mi}}{5280 \text{ ft}} = 9.3 \text{ mph}$$

5.2 The Trigonometric Functions and the Unit Circle

2. $t = \frac{\pi}{3} \Rightarrow \left(\frac{1}{2}, \frac{\sqrt{3}}{2}\right)$ **4.** $t = \frac{5\pi}{4} \Rightarrow$ $\left(-\frac{\sqrt{2}}{2}, -\frac{\sqrt{2}}{2}\right)$ **6.** $t = \frac{11\pi}{6} \Rightarrow$ $\left(\frac{\sqrt{3}}{2}, -\frac{1}{2}\right)$ **8.** $t = \pi \Rightarrow (-1, 0)$

10. $t = -\frac{\pi}{4}$ corresponds to the point $\left(\frac{\sqrt{2}}{2}, -\frac{\sqrt{2}}{2}\right)$.

$$\sin\left(-\frac{\pi}{4}\right) = y = -\frac{\sqrt{2}}{2}$$

$$\cos\left(-\frac{\pi}{4}\right) = x = \frac{\sqrt{2}}{2}$$

$$\tan\left(-\frac{\pi}{4}\right) = \frac{y}{x} = \frac{-\sqrt{2}/2}{\sqrt{2}/2} = -1$$

12. $t = -\frac{5\pi}{6}$ corresponds to the point $\left(-\frac{\sqrt{3}}{2}, -\frac{1}{2}\right)$.

$$\sin\left(-\frac{5\pi}{6}\right) = y = -\frac{1}{2}$$

$$\cos\left(-\frac{5\pi}{6}\right) = x = -\frac{\sqrt{3}}{2}$$

$$\tan\left(-\frac{5\pi}{6}\right) = \frac{y}{x} = \frac{-1/2}{-\sqrt{3}/2} = \frac{\sqrt{3}}{3}$$

14. $t = \frac{2}{3}\pi$ corresponds to the point $\left(-\frac{1}{2}, \frac{\sqrt{3}}{2}\right)$.

$$\sin\left(\frac{2}{3}\pi\right) = y = \frac{\sqrt{3}}{2}$$

$$\cos\left(\frac{2}{3}\pi\right) = x = -\frac{1}{2}$$

$$\tan\left(\frac{2}{3}\pi\right) = \frac{y}{x} = \frac{\sqrt{3}/2}{-1/2} = -\sqrt{3}$$

16. $t = \frac{7\pi}{4}$ corresponds to the point $\left(\frac{\sqrt{2}}{2}, -\frac{\sqrt{2}}{2}\right)$.

$$\sin\frac{7\pi}{4} = y = -\frac{\sqrt{2}}{2}$$

$$\cos\frac{7\pi}{4} = x = \frac{\sqrt{2}}{2}$$

$$\tan\frac{7\pi}{4} = \frac{y}{x} = \frac{-\sqrt{2}/2}{\sqrt{2}/2} = -1$$

18. $t = -\frac{2}{3}\pi$ corresponds to the point $\left(-\frac{1}{2}, -\frac{\sqrt{3}}{2}\right)$.

$$\sin\left(-\frac{2}{3}\pi\right) = y = -\frac{\sqrt{3}}{2}$$

$$\csc\left(-\frac{2}{3}\pi\right) = \frac{1}{y} = \frac{1}{-\sqrt{3}/2} = -\frac{2\sqrt{3}}{3}$$

$$\cos\left(-\frac{2}{3}\pi\right) = x = -\frac{1}{2}$$

$$\sec\left(-\frac{2}{3}\pi\right) = \frac{1}{x} = \frac{1}{-1/2} = -2$$

$$\tan\left(-\frac{2}{3}\pi\right) = \frac{y}{x} = \frac{-\sqrt{3}/2}{-1/2} = \sqrt{3}$$

$$\cot\left(-\frac{2}{3}\pi\right) = \frac{x}{y} = \frac{-1/2}{-\sqrt{3}/2} = \frac{\sqrt{3}}{3}$$

20. $t = \frac{3\pi}{2}$ corresponds to the point $(0, -1)$.

$$\sin\frac{3\pi}{2} = y = -1 \qquad \csc\frac{3\pi}{2} = \frac{1}{y} = \frac{1}{-1} = -1$$

$$\cos\frac{3\pi}{2} = x = 0 \qquad \sec\frac{3\pi}{2} \text{ is undefined.}$$

$$\tan\frac{3\pi}{2} \text{ is undefined.} \qquad \cot\frac{3\pi}{2} = \frac{x}{y} = \frac{0}{1} = 0$$

22. $t = -\frac{11\pi}{6}$ corresponds to the point $\left(\frac{\sqrt{3}}{2}, \frac{1}{2}\right)$.

$$\sin\left(-\frac{11\pi}{6}\right) = y = \frac{1}{2} \qquad \csc\left(-\frac{11\pi}{6}\right) = \frac{1}{y} = \frac{1}{1/2} = 2$$

$$\cos\left(-\frac{11\pi}{6}\right) = x = \frac{\sqrt{3}}{2} \qquad \sec\left(-\frac{11\pi}{6}\right) = \frac{1}{x} = \frac{1}{\sqrt{3}/2} = \frac{2\sqrt{3}}{3}$$

$$\tan\left(-\frac{11\pi}{6}\right) = \frac{y}{x} = \frac{1/2}{\sqrt{3}/2} = \frac{\sqrt{3}}{3} \qquad \cot\left(-\frac{11\pi}{6}\right) = \frac{x}{y} = \frac{\sqrt{3}/2}{1/2} = \sqrt{3}$$

24. $\cos 3\pi = \cos \pi = -1$

26. $\sin\frac{9\pi}{4} = \sin\frac{\pi}{4} = \frac{\sqrt{2}}{2}$

28. $\sin\left(-\frac{13\pi}{6}\right) = \sin\frac{11\pi}{6} = -\frac{1}{2}$

30. $\cos\left(-\frac{8\pi}{3}\right) = \cos\frac{4\pi}{3} = -\frac{1}{2}$

32. (a) $\sin t = -\sin(-t) = -\dfrac{2}{5}$

    (b) $\csc t = \dfrac{1}{\sin t}$

              $= -\dfrac{1}{\sin(-t)} = -\dfrac{5}{2}$

34. (a) $\cos(-t) = \cos t = -\dfrac{3}{4}$

    (b) $\sec(-t) = \sec t$

            $= \dfrac{1}{\cos t} = -\dfrac{4}{3}$

36. (a) $\cos(\pi - t) = -\cos t = -\frac{4}{5}$

    (b) $\cos(t + \pi) = -\cos t = -\frac{4}{5}$

38. $\tan \pi = 0$

40. $\cot 1 \approx 0.6421$

42. $\sec 1.8 \approx -4.4014$

44. $\sin(-0.9) \approx -0.7833$

46. (a)

      (b)

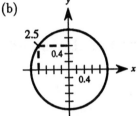

The approximate point corresponding to $t = 0.75$ radians is $(0.7, 0.7)$.

$$\sin 0.75 \approx 0.7$$

The approximate point corresponding to $t = 2.5$ is $(-0.8, 0.6)$.

$$\cos 2.5 \approx -0.8$$

48. (a)

      (b)

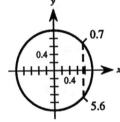

$\sin t = -0.75 = y$

$t \approx 4.0$ or $t \approx 5.4$

$\cos t = 0.75 = x$

$t \approx 0.7$ or $t \approx 5.6$

50. $y(t) = \frac{1}{4}e^{-t}\cos 6t$

    (a) $y(0) = \frac{1}{4}e^{-0}\cos(6 \cdot 0) = 0.2500$ ft

    (b) $y\left(\frac{1}{4}\right) = \frac{1}{4}e^{-1/4}\cos\left(6 \cdot \frac{1}{4}\right) \approx 0.0138$ ft

    (c) $y\left(\frac{1}{2}\right) = \frac{1}{4}e^{-1/2}\cos\left(6 \cdot \frac{1}{2}\right) \approx -0.1501$ ft

52.

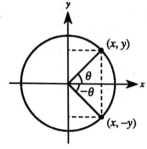

$$\cos \theta = x = \cos(-\theta)$$

$$\sec \theta = \frac{1}{x} = \sec(-\theta)$$

$$\sin \theta = y$$

$$\sin(-\theta) = -y = -\sin \theta$$

$$\sec \theta = \frac{1}{y}$$

$$\sec(-\theta) = -\frac{1}{y} = -\sec \theta$$

$$\tan \theta = \frac{y}{x}$$

$$\tan(-\theta) = \frac{-y}{x} = -\tan \theta$$

$$\cot \theta = \frac{x}{y}$$

$$\cot(-\theta) = \frac{x}{-y} = -\cot \theta$$

▌5.3▐ Trigonometric Functions and Right Triangles

2.

$$b = \sqrt{3^2 - 1^2} = 2\sqrt{2}$$

$$\sin \theta = \frac{1}{3} \qquad\qquad \csc \theta = 3$$

$$\cos \theta = \frac{2\sqrt{2}}{3} \qquad \sec \theta = \frac{3}{2\sqrt{2}} = \frac{3\sqrt{2}}{4}$$

$$\tan \theta = \frac{1}{2\sqrt{2}} = \frac{\sqrt{2}}{4} \qquad \cot \theta = 2\sqrt{2}$$

4.

$$a = \sqrt{13^2 - 5^2} = 12$$

$$\sin \theta = \frac{12}{13} \qquad \csc \theta = \frac{13}{12}$$

$$\cos \theta = \frac{5}{13} \qquad \sec \theta = \frac{13}{5}$$

$$\tan \theta = \frac{12}{5} \qquad \cot \theta = \frac{5}{12}$$

6.

$$c = \sqrt{2^2 + 1^2} = \sqrt{5}$$

$$\sin \theta = \frac{1}{\sqrt{5}} = \frac{\sqrt{5}}{5} \qquad \csc \theta = \frac{\sqrt{5}}{1} = \sqrt{5}$$

$$\cos \theta = \frac{2}{\sqrt{5}} = \frac{2\sqrt{5}}{5} \qquad \sec \theta = \frac{\sqrt{5}}{2}$$

$$\tan \theta = \frac{1}{2} \qquad\qquad \cot \theta = 2$$

8.

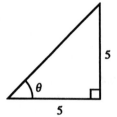

$$c = \sqrt{25 + 25} = 5\sqrt{2}$$

$$\sin \theta = \frac{5}{5\sqrt{2}} = \frac{\sqrt{2}}{2} \qquad \csc \theta = \frac{\sqrt{2}}{1} = \sqrt{2}$$

$$\cos \theta = \frac{5}{5\sqrt{2}} = \frac{\sqrt{2}}{2} \qquad \sec \theta = \frac{\sqrt{2}}{1} = \sqrt{2}$$

$$\tan \theta = \frac{5}{5} = 1 \qquad\qquad \cot \theta = \frac{1}{1} = 1$$

10. Given: $\cot \theta = 5$

$$1^2 + 5^2 = (\text{hyp})^2$$

$$\text{hyp} = \sqrt{26}$$

$$\sin \theta = \frac{\sqrt{26}}{26}$$

$$\cos \theta = \frac{5\sqrt{26}}{26}$$

$$\tan \theta = \frac{1}{5}$$

$$\sec \theta = \frac{\sqrt{26}}{5}$$

$$\csc \theta = \sqrt{26}$$

12. Given: $\cos \theta = \frac{5}{7}$

$$(\text{opp})^2 + 5^2 = 7^2$$

$$\text{opp} = 2\sqrt{6}$$

$$\sin \theta = \frac{2\sqrt{6}}{7}$$

$$\tan \theta = \frac{2\sqrt{6}}{5}$$

$$\cot \theta = \frac{5\sqrt{6}}{12}$$

$$\sec \theta = \frac{7}{5}$$

$$\csc \theta = \frac{7\sqrt{6}}{12}$$

14. Given: $\csc \theta = \frac{17}{4}$

$$4^2 + (\text{adj})^2 = 17^2$$

$$\text{adj} = \sqrt{273}$$

$$\sin \theta = \frac{4}{17}$$

$$\cos \theta = \frac{\sqrt{273}}{17}$$

$$\tan \theta = \frac{4\sqrt{273}}{273}$$

$$\cot \theta = \frac{\sqrt{273}}{4}$$

$$\sec \theta = \frac{17\sqrt{273}}{273}$$

16. Given: $\sin \theta = \frac{3}{8}$

$$3^2 + (\text{adj})^2 = 8^2$$

$$\text{adj} = \sqrt{55}$$

$$\cos \theta = \frac{\sqrt{55}}{8}$$

$$\tan \theta = \frac{3}{\sqrt{55}} = \frac{3\sqrt{55}}{55}$$

$$\csc \theta = \frac{8}{3}$$

$$\sec \theta = \frac{8}{\sqrt{55}} = \frac{8\sqrt{55}}{55}$$

$$\cot \theta = \frac{\sqrt{55}}{3}$$

18. $\sin 30° = \dfrac{1}{2}, \quad \tan 30° = \dfrac{\sqrt{3}}{3}$

    (a) $\csc 30° = \dfrac{1}{\sin 30°} = 2$

    (b) $\cot 60° = \tan 30° = \dfrac{\sqrt{3}}{3}$

    (c) $\cos 30° = \sqrt{1 - \sin^2 30°} = \dfrac{\sqrt{3}}{2}$

    (d) $\cot 30° = \dfrac{1}{\tan 30°} = \sqrt{3}$

20. $\sec \theta = 5, \quad \tan \theta = 2\sqrt{6}$

    (a) $\cos \theta = \dfrac{1}{\sec \theta} = \dfrac{1}{5}$

    (b) $\cot \theta = \dfrac{1}{\tan \theta} = \dfrac{\sqrt{6}}{12}$

    (c) $\cot(90° - \theta) = \tan \theta = 2\sqrt{6}$

    (d) $\sin \theta = \tan \theta \cos \theta = \dfrac{2\sqrt{6}}{5}$

22. (a) $\csc 30° = \dfrac{1}{\sin 30°} = 2$

    (b) $\sin \dfrac{\pi}{4} = \dfrac{\sqrt{2}}{2}$

24. (a) $\sin \dfrac{\pi}{3} = \dfrac{\sqrt{3}}{2}$

    (b) $\csc 45° = \dfrac{1}{\sin 45°} = \sqrt{2}$

26. (a) $\tan 23.5° \approx 0.4348$

    (b) $\cot 66.5° \approx 0.4348$

28. (a) $\cos 16°18' \approx 0.9598$

    (b) $\sin 73°56' \approx 0.9609$

30. (a) $\cos 4°50'15'' \approx 0.9964$

    (b) $\sec 4°50'15'' \approx 1.0036$

32. (a) $\sec 0.75 \approx 1.3667$

    (b) $\cos 0.75 \approx 0.7317$

34. (a) $\tan \dfrac{1}{2} \approx 0.5463$

    (b) $\cot\left(\dfrac{\pi}{2} - \dfrac{1}{2}\right) = \tan \dfrac{1}{2} \approx 0.5463$

36. (a) $\cos \theta = \dfrac{\sqrt{2}}{2} \Rightarrow \theta = 45° = \dfrac{\pi}{4}$

    (b) $\tan \theta = 1 \Rightarrow \theta = 45° = \dfrac{\pi}{4}$

38. (a) $\tan \theta = \sqrt{3} \Rightarrow \theta = 60° = \dfrac{\pi}{3}$

    (b) $\cos \theta = \dfrac{1}{2} \Rightarrow \theta = 60° = \dfrac{\pi}{3}$

40. (a) $\cot \theta = \dfrac{\sqrt{3}}{3} \Rightarrow \theta = 60° = \dfrac{\pi}{3}$

    (b) $\sec \theta = \sqrt{2} \Rightarrow \theta = 45° = \dfrac{\pi}{4}$

42. (a) $\cos \theta = 0.9848 \Rightarrow \theta \approx 10° \approx 0.175 \text{ radians}$

    (b) $\cos \theta = 0.8746 \Rightarrow \theta \approx 29° \approx 0.506 \text{ radians}$

44. (a) $\sin \theta = 0.3746 \Rightarrow \theta \approx 22° \approx 0.384 \text{ radians}$

    (b) $\cos \theta = 0.3746 \Rightarrow \theta \approx 68° \approx 1.187 \text{ radians}$

46. $\cos 60° = \dfrac{x}{10}$

    $x = 10 \cos 60° = 5$

48. $\csc 45° = \dfrac{r}{30}$

    $r = \dfrac{30}{\sin 45°} = 30\sqrt{2} \approx 42.4$

50. $\cot 20° = \dfrac{x}{30}$

    $x = \dfrac{30}{\tan 20°} \approx 82.4$

52. $\sec 75° = \dfrac{r}{20}$

    $r = \dfrac{20}{\cos 75°} \approx 77.3$

54. $\tan\theta = \dfrac{6}{3} = \dfrac{h}{135}$

$h = 270$ ft

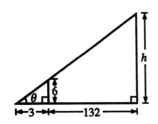

56. $\tan 50° = \dfrac{w}{100}$

$w = 100\tan 50° \approx 119$ ft

58. $\sin\theta = \dfrac{3\frac{1}{3}}{17\frac{1}{2}}$

$\theta \approx 11°$

60. $\tan 3° = \dfrac{h}{6}$

$h = 6\tan 3° \approx 0.314$

$d = 2 + 2h$

$\approx 2 + 2(0.314)$

$= 2.63$ in.

62. $\sin 75° \approx 0.97$

$\cos 75° \approx 0.27$

$\tan 75° \approx 3.73$

$\csc 75° \approx 1.04$

$\sec 75° \approx 3.86$

$\cot 75° \approx 0.27$

64. True, $\sec x = \csc(90° - x) \Rightarrow \sec 30° = \csc(90° - 30°) = \csc 60°$

66. True, $\cot^2\theta - \csc^2\theta = -1$, for all $\theta \Rightarrow \cot^2 10° - \csc^2 10° = -1$

68. False, $\tan[(0.8)^2] = \tan 0.64 \neq \tan^2(0.8) = (\tan 0.8)(\tan 0.8)$

5.4 Trigonometric Functions of Any Angle

2. (a) $(x, y) = (-12, -5)$

$r = \sqrt{144 + 25} = 13$

$\sin \theta = \dfrac{y}{r} = -\dfrac{5}{13}$ $\csc \theta = \dfrac{r}{y} = -\dfrac{13}{5}$

$\cos \theta = \dfrac{x}{r} = -\dfrac{12}{13}$ $\sec \theta = \dfrac{r}{x} = -\dfrac{13}{12}$

$\tan \theta = \dfrac{y}{x} = \dfrac{5}{12}$ $\cot \theta = \dfrac{x}{y} = \dfrac{12}{5}$

(b) $(x, y) = (1, -1)$

$r = \sqrt{1 + 1} = \sqrt{2}$

$\sin \theta = \dfrac{y}{r} = -\dfrac{\sqrt{2}}{2}$ $\csc \theta = \dfrac{r}{y} = -\sqrt{2}$

$\cos \theta = \dfrac{x}{r} = \dfrac{\sqrt{2}}{2}$ $\sec \theta = \dfrac{r}{x} = \sqrt{2}$

$\tan \theta = \dfrac{y}{x} = -1$ $\cot \theta = \dfrac{x}{y} = -1$

4. (a) $(x, y) = (3, -1)$

$r = \sqrt{9 + 1} = \sqrt{10}$

$\sin \theta = \dfrac{y}{r} = -\dfrac{\sqrt{10}}{10}$ $\csc \theta = \dfrac{r}{y} = -\sqrt{10}$

$\cos \theta = \dfrac{x}{r} = \dfrac{3\sqrt{10}}{10}$ $\sec \theta = \dfrac{r}{x} = \dfrac{\sqrt{10}}{3}$

$\tan \theta = \dfrac{y}{x} = -\dfrac{1}{3}$ $\cot \theta = \dfrac{x}{y} = -3$

(b) $(x, y) = (-2, 4)$

$r = \sqrt{4 + 16} = 2\sqrt{5}$

$\sin \theta = \dfrac{y}{r} = \dfrac{2\sqrt{5}}{5}$ $\csc \theta = \dfrac{r}{y} = \dfrac{\sqrt{5}}{2}$

$\cos \theta = \dfrac{x}{r} = -\dfrac{\sqrt{5}}{5}$ $\sec \theta = \dfrac{r}{x} = -\sqrt{5}$

$\tan \theta = \dfrac{y}{x} = \dfrac{4}{-2} = -2$ $\cot \theta = \dfrac{x}{y} = -\dfrac{1}{2}$

6. (a) $(x, y) = (8, 15)$

$r = \sqrt{64 + 225} = 17$

$\sin \theta = \dfrac{y}{r} = \dfrac{15}{17}$ $\csc \theta = \dfrac{r}{y} = \dfrac{17}{15}$

$\cos \theta = \dfrac{x}{r} = \dfrac{8}{17}$ $\sec \theta = \dfrac{r}{x} = \dfrac{17}{8}$

$\tan \theta = \dfrac{y}{x} = \dfrac{15}{8}$ $\cot \theta = \dfrac{x}{y} = \dfrac{8}{15}$

(b) $(x, y) = (-9, -40)$

$r = \sqrt{81 + 1600} = 41$

$\sin \theta = \dfrac{y}{r} = -\dfrac{40}{41}$ $\csc \theta = \dfrac{r}{y} = -\dfrac{41}{40}$

$\cos \theta = \dfrac{x}{r} = -\dfrac{9}{41}$ $\sec \theta = \dfrac{r}{x} = -\dfrac{41}{9}$

$\tan \theta = \dfrac{y}{x} = \dfrac{40}{9}$ $\cot \theta = \dfrac{x}{y} = \dfrac{9}{40}$

8. (a) $(x, y) = (-5, -2)$

$r = \sqrt{25 + 4} = \sqrt{29}$

$\sin \theta = \dfrac{y}{r} = -\dfrac{2\sqrt{29}}{29}$ $\csc \theta = \dfrac{r}{y} = -\dfrac{\sqrt{29}}{2}$

$\cos \theta = \dfrac{x}{r} = -\dfrac{5\sqrt{29}}{29}$ $\sec \theta = \dfrac{r}{x} = -\dfrac{\sqrt{29}}{5}$

$\tan \theta = \dfrac{y}{x} = \dfrac{2}{5}$ $\cot \theta = \dfrac{x}{y} = \dfrac{5}{2}$

(b) $(x, y) = \left(-\dfrac{3}{2}, 3\right)$

$r = \sqrt{\dfrac{9}{4} + 9} = \dfrac{3\sqrt{5}}{2}$

$\sin \theta = \dfrac{y}{r} = \dfrac{2\sqrt{5}}{5}$ $\csc \theta = \dfrac{r}{y} = \dfrac{\sqrt{5}}{2}$

$\cos \theta = \dfrac{x}{r} = -\dfrac{\sqrt{5}}{5}$ $\sec \theta = \dfrac{r}{x} = -\sqrt{5}$

$\tan \theta = \dfrac{y}{x} = -2$ $\cot \theta = \dfrac{x}{y} = -\dfrac{1}{2}$

10. $b_1 = 12, \quad c_1 = 13, \quad c_2 = 26$

(a) $a_1 = \sqrt{c_1{}^2 - b_1{}^2} = 5$

$a_2 = \dfrac{c_2}{c_1} a_1 = 2(5) = 10$

$b_2 = \dfrac{c_2}{c_1} b_1 = 2(12) = 24$

(b) $\sin \alpha_1 = \dfrac{a_1}{c_1} = \dfrac{5}{13} = \dfrac{a_2}{c_2} = \sin \alpha_2$

$\cos \alpha_1 = \dfrac{b_1}{c_1} = \dfrac{12}{13} = \dfrac{b_2}{c_2} = \cos \alpha_2$

$\tan \alpha_1 = \dfrac{a_1}{b_1} = \dfrac{5}{12} = \dfrac{a_2}{b_2} = \tan \alpha_2$

$\csc \alpha_1 = \dfrac{c_1}{a_1} = \dfrac{13}{5} = \dfrac{c_2}{a_2} = \csc \alpha_2$

$\sec \alpha_1 = \dfrac{c_1}{b_1} = \dfrac{13}{12} = \dfrac{c_2}{b_2} = \sec \alpha_2$

$\cot \alpha_1 = \dfrac{b_1}{a_1} = \dfrac{12}{5} = \dfrac{b_2}{a_2} = \cot \alpha_2$

12. $b_1 = 4, \quad a_2 = 4, \quad b_2 = 10$

(a) $c_2 = \sqrt{a_2{}^2 + b_2{}^2} = 2\sqrt{29}$

$a_1 = \dfrac{b_1}{b_2} a_2 = \dfrac{2}{5}(4) = \dfrac{8}{5}$

$c_1 = \dfrac{b_1}{b_2} c_2 = \dfrac{2}{5}(2\sqrt{29}) = \dfrac{4\sqrt{29}}{5}$

(b) $\sin \alpha_1 = \dfrac{a_1}{c_1} = \dfrac{2\sqrt{29}}{29} = \dfrac{a_2}{c_2} = \sin \alpha_2$

$\cos \alpha_1 = \dfrac{b_1}{c_1} = \dfrac{5\sqrt{29}}{29} = \dfrac{b_2}{c_2} = \cos \alpha_2$

$\tan \alpha_1 = \dfrac{a_1}{b_1} = \dfrac{2}{5} = \dfrac{a_2}{b_2} = \tan \alpha_2$

$\csc \alpha_1 = \dfrac{c_1}{a_1} = \dfrac{\sqrt{29}}{2} = \dfrac{c_2}{a_2} = \csc \alpha_2$

$\sec \alpha_1 = \dfrac{c_1}{b_1} = \dfrac{\sqrt{29}}{5} = \dfrac{c_2}{b_2} = \sec \alpha_2$

$\cot \alpha_1 = \dfrac{b_1}{a_1} = \dfrac{5}{2} = \dfrac{b_2}{a_2} = \cot \alpha_2$

14. (a) $\sin \theta > 0$ and $\cos \theta > 0$ in Quadrant I

(b) $\sin \theta < 0$ and $\cos \theta > 0$ in Quadrant IV

16. (a) $\sec \theta > 0$ and $\cot \theta < 0$ in Quadrant IV

(b) $\csc \theta < 0$ and $\tan \theta > 0$ in Quadrant III

18. $\cos \theta = \dfrac{x}{r} = \dfrac{-4}{5} \Rightarrow |y| = 3$

θ in Quadrant III $\Rightarrow y = -3$

$\sin \theta = \dfrac{y}{r} = -\dfrac{3}{5}$ $\qquad \csc \theta = \dfrac{r}{y} = -\dfrac{5}{3}$

$\cos \theta = \dfrac{x}{r} = -\dfrac{4}{5}$ $\qquad \sec \theta = \dfrac{r}{x} = -\dfrac{5}{4}$

$\tan \theta = \dfrac{y}{x} = \dfrac{3}{4}$ $\qquad \cot \theta = \dfrac{x}{y} = \dfrac{4}{3}$

20. $\cos \theta = \dfrac{x}{r} = \dfrac{8}{17} \Rightarrow |y| = 15$

$\tan \theta < 0 \Rightarrow y = -15$

$\sin \theta = \dfrac{y}{r} = -\dfrac{15}{17}$ $\qquad \csc \theta = \dfrac{r}{y} = -\dfrac{17}{15}$

$\cos \theta = \dfrac{x}{r} = \dfrac{8}{17}$ $\qquad \sec \theta = \dfrac{r}{x} = \dfrac{17}{8}$

$\tan \theta = \dfrac{y}{x} = -\dfrac{15}{8}$ $\qquad \cot \theta = \dfrac{x}{y} = -\dfrac{8}{15}$

22. $\cot \theta$ is undefined $\Rightarrow \theta = n\pi$

$\dfrac{\pi}{2} \leq \theta \leq \dfrac{3\pi}{2} \Rightarrow \theta = \pi, \quad y = 0, \quad x = -r$

$\sin \theta = \dfrac{y}{r} = \dfrac{0}{r} = 0$ $\qquad \csc \theta = \dfrac{r}{y}$ is undefined

$\cos \theta = \dfrac{x}{r} = \dfrac{-r}{r} = -1$ $\qquad \sec \theta = \dfrac{r}{x} = -1$

$\tan \theta = \dfrac{y}{x} = \dfrac{0}{x} = 0$ $\qquad \cot \theta = \dfrac{x}{y}$ is undefined

24. $\tan\theta$ is undefined $\Rightarrow \theta = n\pi + \dfrac{\pi}{2}$

$\pi \le \theta \le 2\pi \Rightarrow \theta = \dfrac{3\pi}{2}, \quad x = 0, \quad y = -r$

$\sin\theta = \dfrac{y}{r} = \dfrac{-r}{r} = -1 \qquad \csc\theta = \dfrac{r}{y} = -1$

$\cos\theta = \dfrac{x}{r} = \dfrac{0}{r} = 0 \qquad \sec\theta = \dfrac{r}{x}$ is undefined

$\tan\theta = \dfrac{y}{x}$ is undefined $\qquad \cot\theta = \dfrac{x}{y} = \dfrac{0}{y} = 0$

26. θ in Quadrant IV and $(x,\,y)$ satisfies $\left(x,\, -\frac{4}{3}x\right) \Rightarrow r^2 = x^2 + \frac{16}{9}x^2$ and $x > 0$. Thus, $r = \frac{5}{3}|x| = \frac{5}{3}x$.

$\sin\theta = \dfrac{y}{r} = -\dfrac{4}{5} \qquad \csc\theta = \dfrac{r}{y} = -\dfrac{5}{4}$

$\cos\theta = \dfrac{x}{r} = \dfrac{3}{5} \qquad \sec\theta = \dfrac{r}{x} = \dfrac{5}{3}$

$\tan\theta = \dfrac{y}{x} = -\dfrac{4}{3} \qquad \cot\theta = \dfrac{x}{y} = -\dfrac{3}{4}$

28. (a) $\theta = 309°$

$\theta' = 360° - 309° = 51°$

(b) $\theta = 226°$

$\theta' = 226° - 180° = 46°$

30. (a) $\theta = -145°$

$360° - 145° = 215°$ (coterminal angle)

$\theta' = 215° - 180° = 35°$

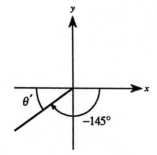

(b) $\theta = -239°$

$360° - 239° = 121°$ (coterminal angle)

$\theta' = 180° - 121° = 59°$

5.5 Graphs of Sine and Cosine

2. $y = 3\cos 3x$

Amplitude: $|3| = 3$

Period: $\dfrac{2\pi}{|3|} = \dfrac{2\pi}{3}$

```
RANGE
Xmin=0
Xmax=3π
Xscl=1
Ymin=-3
Ymax=3
Yscl=1
```

4. $y = -2\sin\dfrac{x}{3}$

Amplitude: $|-2| = 2$

Period: $\dfrac{2\pi}{|1/3|} = 6\pi$

```
RANGE
Xmin=0
Xmax=6π
Xscl=π
Ymin=-3
Ymax=3
Yscl=1
```

6. $y = \dfrac{5}{2}\cos\dfrac{\pi x}{2}$

Amplitude: $\left|\dfrac{5}{2}\right| = \dfrac{5}{2}$

Period: $\dfrac{2\pi}{|\pi/2|} = 4$

```
RANGE
Xmin=0
Xmax=8
Xscl=1
Ymin=-3
Ymax=3
Yscl=1
```

8. $y = -\dfrac{5}{2}\cos\dfrac{x}{4}$

Amplitude: $\left|-\dfrac{5}{2}\right| = \dfrac{5}{2}$

Period: $\dfrac{2\pi}{|1/4|} = 8\pi$

10. $y = \dfrac{2}{3}\cos\dfrac{\pi x}{10}$

Amplitude: $\left|\dfrac{2}{3}\right| = \dfrac{2}{3}$

Period: $\dfrac{2\pi}{|\pi/10|} = 20$

12. $f(x) = \cos x$

$g(x) = \cos(x + \pi)$

The graph of g is a horizontal shift to the left π units of the graph of f (a phase shift).

14. $f(x) = \sin 3x$

$g(x) = \sin(-3x) = -\sin(3x)$

The graph of g is a reflection in the x-axis of the graph of f.

16. $f(x) = \cos 4x$

$g(x) = -2 + \cos 4x$

The graph of g is a vertical shift downward of 2 units of the graph of f.

18. $f(x) = \sin x$

$g(x) = \sin\left(\dfrac{x}{3}\right)$

20. $f(x) = 2\cos 2x$

$g(x) = -\cos 4x$

22. $f(x) = 4 \sin \pi x$

$g(x) = 4 \sin \pi x - 3$

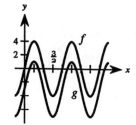

24. $f(x) = -\cos x$

$g(x) = -\cos(x - \pi)$

26. $f(x) = \sin x$

$g(x) = -\cos\left(x + \dfrac{\pi}{2}\right)$

28. $f(x) = \cos x$

$g(x) = -\cos(x - \pi)$

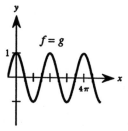

30. $y = -3 \cos 4x$

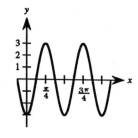

32. $y = \dfrac{3}{2} \sin \dfrac{\pi x}{4}$

34. $y = 10 \cos \dfrac{\pi x}{6}$

36. $y = 2 \cos x - 3$

38. $y = \dfrac{1}{2} \sin(x - \pi)$

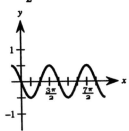

40. $y = 4 \cos \left(x + \dfrac{\pi}{4} \right)$

42. $y = -3 + 5 \cos \dfrac{\pi t}{12}$

44. $y = 4 \cos \left(x + \dfrac{\pi}{4} \right) + 4$

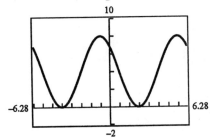

46. $y = -3 \cos(6x + \pi)$

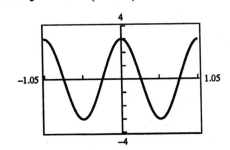

48. $y = -4 \sin \left(\dfrac{2}{3} x - \dfrac{\pi}{3} \right)$

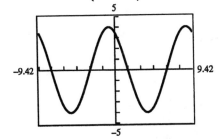

50. $y = 3\cos\left(\dfrac{\pi x}{2} + \dfrac{\pi}{2}\right) - 2$

52. $y = 5\sin(\pi - 2x) + 10$

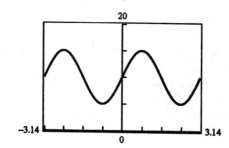

54. $y = \frac{1}{100}\sin(120\pi t)$

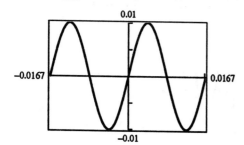

56. $f(x) = \cos x$

$\cos x = -1$ for $x = -\pi,\ \pi$

58.

$\sin x = \dfrac{\sqrt{3}}{2}$ for

$x = -\dfrac{5\pi}{3},\ -\dfrac{4\pi}{3},\ \dfrac{\pi}{3},\ \dfrac{2\pi}{3}$

60. $f(x) = a\cos x + d$

Amplitude is 3 and graph is reflected in **x-axis**: $a = -3$

Vertical shift upward 1 unit: $d = 1$

$f(x) = -3\cos x + 1$

62. $y = a\sin(bx - c)$

Amplitude: $a = \dfrac{1}{2}$

Period: $\pi = \dfrac{2\pi}{2} \Rightarrow b = 2$

Phase shift: $c = 0$

$y = \dfrac{1}{2}\sin 2x$

64. $y = a\sin(bx - c)$

Amplitude: $a = 2$

Period: $4 = \dfrac{2\pi}{b} \Rightarrow b = \dfrac{\pi}{2}$

Phase shift: $-1 = \dfrac{c}{b} \Rightarrow c = -\dfrac{\pi}{2}$

$y = 2\sin\left(\dfrac{\pi x}{2} + \dfrac{\pi}{2}\right)$

66. $f(x) = 2 \ln x$

$g(x) = a \cos x$

Graphing f and g on the same viewing rectangle (Xmin = 0, Xmax = 10, Ymin = −6, Ymax = 6), and changing the value of a, we find that $a = 4$ is the smallest integers value for which the graphs intersect more than once.

68. $v = 1.75 \sin \dfrac{\pi t}{2}$

(a) Period: $\dfrac{2\pi}{\pi/2} = 4$

(b) $\dfrac{60}{4} = 15$ cycles/min

(c)
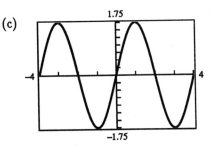

70. $P = 100 - 20 \cos \dfrac{5\pi t}{3}$

(a) Period: $\dfrac{2\pi}{|5\pi/3|} = \dfrac{6}{5}$ seconds

(b) Number of heartbeats per minute = $\dfrac{60 \text{ seconds}}{\frac{6}{5} \text{ seconds}} = 50$ heartbeats
(Number of cycles per minute)

(c)
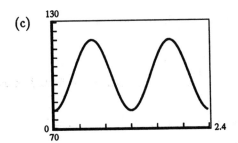

72. $S = 74.50 + 43.75 \sin \dfrac{\pi t}{6}$

74. $g(x) = f(2x)$

76. $g(x) = f(x) + 2$

The maximum sales occur in March ($t = 3$), and the minimum sales in September ($t = 9$).

5.6   Other Trigonometric Graphs

2. $y = \tan 3x$ matches graph (g)

```
RANGE
Xmin=-π/3
Xmax=π
Xscl=π/6
Ymin=-2
Ymax=2
Yscl=1
```

4. $y = 2 \csc \dfrac{x}{2}$ matches graph (a)

```
RANGE
Xmin=-π
Xmax=5π
Xscl=π
Ymin=-5
Ymax=5
Yscl=1
```

6. $y = \frac{1}{2} \sec \pi x$ matches graph (h)

```
RANGE
Xmin=-1
Xmax=3
Xscl=0.5
Ymin=-2
Ymax=2
Yscl=1
```

8. $y = -2 \csc 2\pi x$ matches graph (f)

```
RANGE
Xmin=-0.5
Xmax=2
Xscl=0.25
Ymin=-4
Ymax=4
Yscl=1
```

10. $y = \frac{1}{4} \tan x$

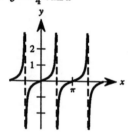

12. $y = -3 \tan \pi x$

14. $y = \frac{1}{4} \sec x$

16. $y = 2 \sec 4x$

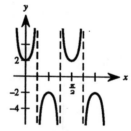

18. $y = -2\sec 4x + 2$

20. $y = \csc \dfrac{x}{3}$

22. $y = 3\cot \dfrac{\pi x}{2}$

24. $y = -\frac{1}{2}\tan x$

26. $y = \sec(x + \pi)$

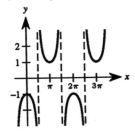

28. $y = \sec(\pi - x)$

30. $y = 2\cot\left(x + \dfrac{\pi}{2}\right)$

32. $y = -\tan 2x$

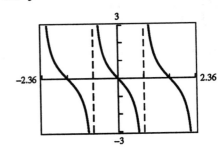

34. $y = \sec \pi x$

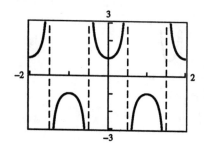

36. $y = -\csc(4x - \pi)$

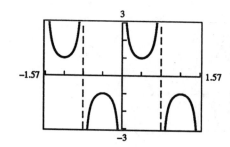

38. $y = \dfrac{1}{3} \sec \left(\dfrac{\pi x}{2} + \dfrac{\pi}{2} \right)$

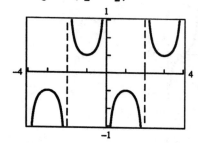

40. $y = 0.1 \tan \left(\dfrac{\pi}{4} x + \dfrac{\pi}{4} \right)$

42. $f(x) = \cot x$

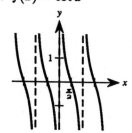

$\cot x = -\sqrt{3}$ for

$$x = -\frac{7\pi}{6}, \ -\frac{\pi}{6}, \ \frac{5\pi}{6}, \ \frac{11\pi}{6}$$

44. $f(x) = \csc x$

$\csc x = \sqrt{2}$ for

$$x = -\frac{7\pi}{4}, \ -\frac{5\pi}{4}, \ \frac{\pi}{4}, \ \frac{3\pi}{4}$$

46. The function $f(x) = \tan x$ is odd, since its graph is symmetric about the origin.

48. $f(x) = \tan\left(\dfrac{\pi x}{2}\right)$ and $g(x) = \dfrac{1}{2}\sec\left(\dfrac{\pi x}{2}\right)$, $-1 < x < 1$

(a)

(b) $f < g$ on the interval $(-1,\ 0.333)$.

50. $y = -3 + \cos x$

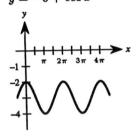

52. $y = 5 - \frac{1}{2}\sin 2\pi x$

54. $y = 2 + \tan(\pi x)$

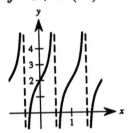

56. $y = \cos x + \cos 2x$
Period: 2π
Relative maxima: $(0,\ 2)$, $(\pi,\ 0)$
Relative minimum:
$(1.823,\ -1.125)$, $(4.460,\ -1.125)$

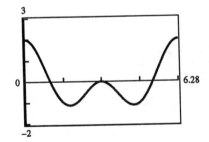

58. $y = 2\sin x + \cos 2x$

Period: 2π

Relative maxima: $(0.524, 1.5)$, $(2.618, 1.5)$

Relative minimum: $(1.571, 1)$, $(4.712, -3.0)$

60. $y = \sin x - \dfrac{1}{2}\sin\dfrac{x}{2}$

Period: 4π

Relative maxima:

    $(1.376, 0.664)$, $(8.017, 1.368)$

Relative minimum:

    $(4.550, -1.368)$, $(11.190, -0.664)$

62. $y = \cos x - \frac{1}{4}\cos 2x$

64. $y = \sin \pi x + \sin \dfrac{\pi x}{2}$

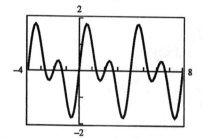

66. $y_1 = x + \cos x$

    $y_2 = x$

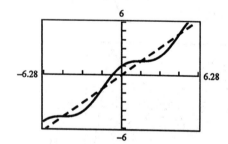

68. $y_1 = 2x - \sin x$

    $y_2 = 2x$

70. $y_1 = -\dfrac{s}{2} - \dfrac{1}{2}\sin\dfrac{\pi s}{y}$

$\quad\ y_2 = -\dfrac{s}{2}$

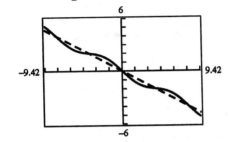

72. $y_1 = 4 - \dfrac{x^2}{16} - 4\cos\pi x$

$\quad\ y_2 = 4 - \dfrac{x^2}{16}$

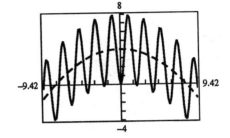

74. $y_1 = e^{-t}\cos t$

$\quad\ y_2 = e^{-t}$

The functional values approach 0 as x increases without bound.

76. $y_1 = 2^{-t^2/4}\sin t$

$\quad\ y_2 = 2^{-t^2/4}$

The functional values approach 0 as x increases without bound.

78. $y_1 = |x\sin x|$

$\quad\ y_2 = x$

$\quad\ y_3 = -x$

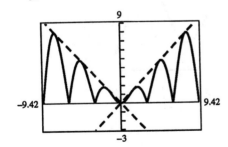

The functional values approach 0 as x approaches 0.

80. $y_1 = |x|\cos x$

$\quad\ y_2 = x$

$\quad\ y_3 = -x$

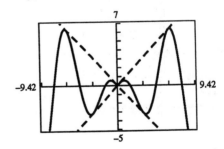

The functional values approach 0 as x approaches 0.

82. $y = \dfrac{4}{x} + \sin 2x,\ x > 0$

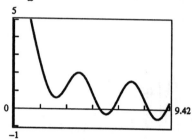

y increases without bound as x approaches 0.

84. $f(x) = \dfrac{1 - \cos x}{x}$

f approaches 0 as x approaches 0.

86. $h(x) = x \sin \dfrac{1}{x}$

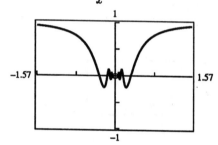

h approaches 0 as x approaches 0.

88. $f(x) = \sin x - \cos\left(x + \dfrac{\pi}{2}\right)$

$g(x) = 2 \sin x$

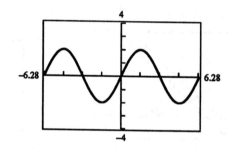

$f = g$ (The functions are equal.)

90. $f(x) = \cos^2\left(\dfrac{\pi x}{2}\right)$

$g(x) = \dfrac{1}{2}(1 + \cos \pi x)$

$f = g$ (The functions are equal.)

92. $f(x) = \sqrt{2x} \sin x,\ [0,\ 4\pi]$

94. $f(x) = \dfrac{1}{2} - \dfrac{4}{\pi^2}\left(\cos \pi x + \dfrac{1}{9}\cos 3\pi x\right)$, $[0, 2]$

96. $\cos x = \dfrac{100}{d}$

$d = \dfrac{100}{\cos x}$

$d = 100 \sec x$

98. $S = 25 + 2t + 20 \sin \dfrac{\pi t}{6}$

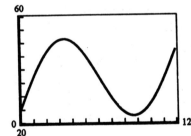

100. (a) Greatest difference during summer months
Smallest difference during winter months

(b) About one month

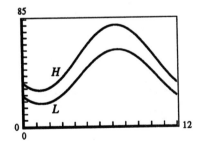

5.7 | Inverse Trigonometric Functions

2. $y = \arcsin 0 \Rightarrow \sin y = 0$ for $-\dfrac{\pi}{2} \le y \le \dfrac{\pi}{2} \Rightarrow y = 0$

4. $y = \arccos 0 \Rightarrow \cos y = 0$ for $0 \le y \le \pi \Rightarrow y = \dfrac{\pi}{2}$

6. $y = \arctan(-1) \Rightarrow \tan y = -1$ for $-\dfrac{\pi}{2} < y < \dfrac{\pi}{2} \Rightarrow y = -\dfrac{\pi}{4}$

8. $y = \arcsin\left(-\dfrac{\sqrt{2}}{2}\right) \Rightarrow \sin y = -\dfrac{\sqrt{2}}{2}$ for $-\dfrac{\pi}{2} \le y \le \dfrac{\pi}{2} \Rightarrow y = -\dfrac{\pi}{4}$

10. $y = \arctan(\sqrt{3}) \Rightarrow \tan y = \sqrt{3}$ for $-\dfrac{\pi}{2} < y < \dfrac{\pi}{2} \Rightarrow y = \dfrac{\pi}{3}$

12. $y = \arcsin\dfrac{\sqrt{2}}{2} \Rightarrow \sin y = \dfrac{\sqrt{2}}{2}$ for $-\dfrac{\pi}{2} \le y \le \dfrac{\pi}{2} \Rightarrow y = \dfrac{\pi}{4}$

14. $y = \arctan\left(-\dfrac{\sqrt{3}}{3}\right) \Rightarrow \tan y = -\dfrac{\sqrt{3}}{3}$ for $-\dfrac{\pi}{2} < y < \dfrac{\pi}{2} \Rightarrow y = -\dfrac{\pi}{6}$

16. $y = \arccos 1 \Rightarrow \cos y = 1$ for $0 \le y \le \pi \Rightarrow y = 0$

18. $\arcsin 0.45 \approx 0.47$

20. $\arccos(-0.8) \approx 2.50$

22. $\arctan 15 \approx 1.50$

24. $\arccos 0.26 \approx 1.31$

26. $\arcsin(-0.125) \approx -0.13$

28. $\arctan 2.8 \approx 1.23$

30. $f(x) = \sin x \quad -\dfrac{\pi}{2} < x < \dfrac{\pi}{2}$

32. $y = \tan(\arctan 25) = 25$

34. $\sin[\arcsin(-0.2)] = -0.2$

$g(x) = \arcsin x$

$y = x$

36. $\arccos\left(\cos\dfrac{7\pi}{2}\right) = \arccos 0$

$\qquad\qquad\qquad = \dfrac{\pi}{2}$

38. $y = \arcsin\dfrac{4}{5}$

$\sec y = \sec\left(\arcsin\dfrac{4}{5}\right)$

$\qquad = \dfrac{5}{3}$

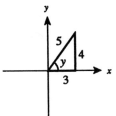

40. $y = \arccos\dfrac{\sqrt{5}}{5}$

$\sin y = \sin\left(\arccos\dfrac{\sqrt{5}}{5}\right)$

$\qquad = \dfrac{2\sqrt{5}}{5}$

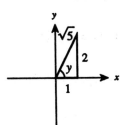

42. $y = \arctan\left(-\dfrac{5}{12}\right)$

$\csc y = \csc\left[\arctan\left(-\dfrac{5}{12}\right)\right]$

$\qquad = -\dfrac{13}{5}$

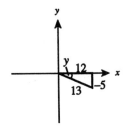

44. $y = \arcsin\left(-\dfrac{3}{4}\right)$

$\tan y = \tan\left[\arcsin\left(-\dfrac{3}{4}\right)\right]$

$= -\dfrac{3}{\sqrt{7}}$

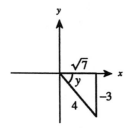

46. $f(x) = \tan\left(\arccos\dfrac{x}{2}\right)$

$g(x) = \dfrac{\sqrt{4 - x^2}}{x}$

Asymptote: $x = 0$

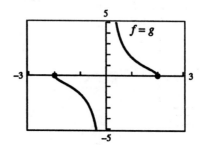

48. $y = \arctan x$

$\sin y = \sin(\arctan x)$

$= \dfrac{x}{\sqrt{x^2 + 1}}$

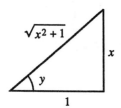

50. $y = \arctan 3x$

$\sec y = \sec(\arctan 3x)$

$= \sqrt{9x^2 + 1}$

52. $y = \arctan\dfrac{1}{x}$

$\cot y = \cot\left(\arctan\dfrac{1}{x}\right)$

$= x$

54. $y = \arcsin(x - 1)$

$\sec y = \sec[\arcsin(x - 1)]$

$= \dfrac{1}{\sqrt{2x - x^2}}$

56. $y = \arcsin\left(\dfrac{x-h}{r}\right)$

$\cos y = \cos\left[\arcsin\left(\dfrac{x-h}{r}\right)\right]$

$= \dfrac{\sqrt{r^2 - (x-h)^2}}{r}$

58. $\arcsin \dfrac{\sqrt{36-x^2}}{6} = \arccos \dfrac{x}{6}$

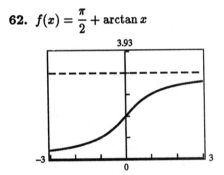

60. $\arccos \dfrac{x-2}{2} = \arctan\left(\dfrac{\sqrt{4x-x^2}}{x-2}\right)$

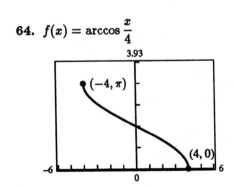

62. $f(x) = \dfrac{\pi}{2} + \arctan x$

64. $f(x) = \arccos \dfrac{x}{4}$

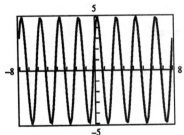

66. $w = \pi,\ A = 4,\ B = 3$

$f(t) = 4\cos \pi t + 3\sin \pi t$

$= \sqrt{16+9}\sin\left[\pi t + \arctan\left(\tfrac{3}{4}\right)\right]$

$= 5\sin(\pi t + 0.6435)$

68. $\theta = \arctan\left(\dfrac{s}{2000}\right)$

(a) $\theta = \arctan\left(\dfrac{1000}{2000}\right) \Rightarrow \tan\theta = \dfrac{1}{2} \Rightarrow \theta = 26.6°$

(b) $\theta = \arctan\left(\dfrac{4000}{2000}\right) \Rightarrow \tan\theta = 2 \Rightarrow \theta = 63.4°$

70. Area $= \arctan b - \arctan a$

(a) $a = 0, \quad b = 1$

Area $= \arctan(1) - \arctan(0) \approx 0.7854 - 0 = 0.7854$

(b) $a = -1, \quad b = 1$

Area $= \arctan(1) - \arctan(-1) \approx 0.7854 - (-0.7854) = 1.5708$

(c) $a = 0, \quad b = 3$

Area $= \arctan(3) - \arctan(0) \approx 1.2490 - 0 = 1.2490$

(d) $a = -1, \quad b = 3$

Area $= \arctan(3) - \arctan(-1) \approx 1.2490 - (-0.7854) = 2.0344$

72. $y = \text{arcsec}\, x$ if and only if $\sec y = x$ where $x \leq -1 \cup x \geq 1$ and $0 \leq y < \pi/2$ and $\pi/2 < y \leq \pi$. The domain of $y \, \text{arcsec}\, x$ is $(-\infty, -1] \cup [1, \infty)$ and the range is $[0, \pi/2) \cup (\pi/2, \pi]$.

74. (a) $y = \text{arcsec}\,\sqrt{2} \Rightarrow \sec y = \sqrt{2}$ and $0 \leq y < \dfrac{\pi}{2} \cup \dfrac{\pi}{2} < y \leq \pi \Rightarrow y = \dfrac{\pi}{4}$

(b) $y = \text{arcsec}\, 1 \Rightarrow \sec y = 1$ and $0 \leq y < \dfrac{\pi}{2} \cup \dfrac{\pi}{2} < y \leq \pi \Rightarrow y = 0$

(c) $y = \text{arccot}(-\sqrt{3}) \Rightarrow \cot y = -\sqrt{3}$ and $0 < y < \pi \Rightarrow y = \dfrac{5\pi}{6}$

(d) $y = \text{arccsc}\, 2 \Rightarrow \csc y = 2$ and $-\dfrac{\pi}{2} \leq y < 0 \cup 0 < y \leq \dfrac{\pi}{2} \Rightarrow y = \dfrac{\pi}{6}$

76.
$$y = \arctan(-x)$$
$$\tan y = -x, \quad -\frac{\pi}{2} < y < \frac{\pi}{2}$$
$$-\tan y = x$$
$$\tan(-y) = x, \quad -\frac{\pi}{2} < -y < \frac{\pi}{2}$$
$$\arctan(\tan(-y)) = \arctan x$$
$$-y = \arctan x$$
$$y = -\arctan x$$

78. $y_2 = \dfrac{\pi}{2} - y_1$

$$\arctan x + \arctan \frac{1}{x} = y_1 + y_2$$
$$= y_1 + \left(\frac{\pi}{2} - y_1\right)$$
$$= \frac{\pi}{2}$$

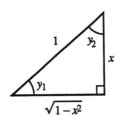

80. $\arcsin x = \arcsin \dfrac{x}{1}$

$$= \arctan \frac{x}{\sqrt{1-x^2}}$$

5.8 Applications of Trigonometry

2. Given: $B = 54°$, $c = 15$

$$\sin B = \frac{b}{c} \Rightarrow b = c \sin B = 15 \sin 54° \approx 12.135$$
$$\cos B = \frac{a}{c} \Rightarrow a = c \cos B = 15 \cos 54° \approx 8.8168$$

$$A = 90° - 54° = 36°$$

4. Given: $A = 8.4°$, $a = 40.5$

$$\tan A = \frac{a}{b} \Rightarrow b = \frac{a}{\tan A} = \frac{40.5}{\tan 8.4°} \approx 274.27$$
$$\sin A = \frac{a}{c} \Rightarrow c = \frac{a}{\sin A} = \frac{40.5}{\sin 8.4°} \approx 277.24$$

$$B = 90° - 8.4° = 81.6°$$

6. Given: $B = 65°12'$, $a = 14.2$

$$\tan B = \frac{b}{a} \Rightarrow b = a \tan B$$
$$= 14.2 \tan 65.2°$$
$$\approx 30.732$$
$$\cos B = \frac{a}{c} \Rightarrow c = \frac{a}{\cos B} = \frac{14.2}{\cos 65.2°} \approx 33.854$$

$$A = 90° - 65°12' = 24°48'$$

8. Given: $a = 25$, $c = 35$

$$b^2 = c^2 - a^2 \Rightarrow b = \sqrt{1225 - 625}$$
$$= 10\sqrt{6}$$
$$\approx 24.495$$
$$\sin A = \frac{a}{c} = \frac{25}{35} \Rightarrow A = \arcsin \frac{5}{7} \approx 45.585°$$

$$B \approx 90° - 45.585° = 44.415°$$

10. Given: $b = 1.32$, $c = 9.45$

$$a^2 = c^2 - b^2 \Rightarrow a = \sqrt{89.3025 - 1.7424}$$
$$= 9.3574$$

$$\cos A = \frac{b}{c} = \frac{1.32}{9.45} \Rightarrow A = \arccos \frac{44}{315} \approx 81.971°$$
$$B \approx 90° - 81.971° = 8.0295°$$

12. $\tan \theta = \dfrac{h}{\frac{1}{2}b} \Rightarrow h = \frac{1}{2} b \tan \theta$

$$h = \frac{1}{2} b \tan \theta$$

$$h = \frac{1}{2}(10) \tan 18° \approx 1.62 \text{ m}$$

14. $\tan 20° = \dfrac{600}{x} \Rightarrow x = \dfrac{600}{\tan 20°} \approx 1648.5 \text{ ft}$

16. $\tan 33° = \dfrac{h}{125}$

$$h = 125 \tan 33°$$
$$\approx 81.2 \text{ ft}$$

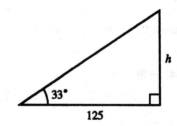

18. $\tan\theta = \dfrac{25/2}{52/3} = \dfrac{75}{104}$

$\theta = \arctan\left(\dfrac{75}{104}\right)$

≈ 0.6248 radians

$\approx 35.8°$

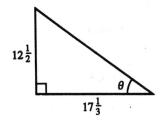

20. $\tan\theta = \dfrac{250}{2(5280)} \Rightarrow$

$\theta = \arctan\dfrac{250}{10,560}$

$\approx 1.356°$

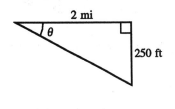

22. $\sin(12.5°) = \dfrac{a}{4}$

$a = 4\sin(12.5°)$

≈ 0.866 miles ≈ 4571 ft

24. $\tan 39°45' = \dfrac{H}{100}$

$H = 100\tan 39°45'$

$\tan 28° = \dfrac{h}{100}$

$h = 100\tan 28°$

Height of pole $= H - h$

$= 100\tan 39°45' - 100\tan 28°$

≈ 30 ft

26. $\sin 63° = \dfrac{a}{120} \Rightarrow a \approx 107$ nautical miles south

$\cos 63° = \dfrac{b}{120} \Rightarrow b \approx 54.5$ nautical miles west

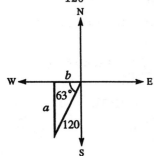

28. $\tan \theta = \dfrac{85}{120} \Rightarrow \theta = 35.3°$

Bearing: S 35.3° W

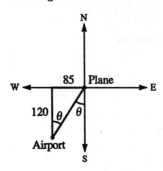

30. $\tan 14° = \dfrac{d}{x} \Rightarrow x = d \cot 14°$

$\tan 34° = \dfrac{d}{y} = \dfrac{d}{20 - x} = \dfrac{d}{20 - d \cot 14°}$

$\cot 34° = \dfrac{20 - d \cot 14°}{d}$

$d \cot 34° = 20 - d \cot 14°$

$d = \dfrac{20}{\cot 34° + \cot 14°} \approx 3.64$ miles

32. $\cot 55° = \dfrac{d}{30{,}000} \Rightarrow d \approx 21{,}006$ ft

$\cot 28° = \dfrac{D}{30{,}000} \Rightarrow D \approx 56{,}422$ ft

Distance between towns:

$D - d \approx 56{,}422 - 21{,}006$

$= 35{,}416$ ft ≈ 6.7 mi

34. $\cot 9° = \dfrac{d}{h} \Rightarrow d = h \cot 9°$

$\cot 3.5° = \dfrac{13 + d}{h} = \dfrac{13 + h \cot 9°}{h}$

$h \cot 3.5° = 13 + h \cot 9°$

$h = \dfrac{13}{\cot 3.5° - \cot 9°}$

≈ 1.3 miles ≈ 6840 ft

36. $\sin 30° = \dfrac{d}{25} \Rightarrow d = 12.5$

Length of side: $2d = 25$ inches

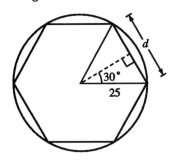

38. $c = \dfrac{20}{2} = 10$

$\sin 15° = \dfrac{a}{c}$

$a = c \sin 15° = 10 \sin 15° \approx 2.59$

Distance $= 2a \approx 2(2.59) = 5.18$ inches

40. $\tan\theta = \dfrac{12}{18}$

$\theta = \arctan \dfrac{2}{3} = 0.588 \text{ rad } \approx 33.7°$

$\cos\theta = \dfrac{18}{c}$

$c = \dfrac{18}{\cos\theta} \approx 21.6$

$f \approx \dfrac{21.6}{2} = 10.8$

$\phi \approx 90 - 33.7 = 56.3°$

$\sin\phi = \dfrac{6}{d}$

$d = \dfrac{6}{\sin\phi} \approx 7.2$

$g = \sqrt{10.8^2 + 7.2^2} \approx 12.98$

42. $d = \dfrac{1}{2}\cos 20\pi t$

(a) Maximum displacement = amplitude $= \dfrac{1}{2}$

(b) Frequency $= \dfrac{1}{\text{period}} = \dfrac{1}{1/10}$

$= 10$ cycles per unit of time

(c) $20\pi t = \dfrac{\pi}{2} \Rightarrow t = \dfrac{1}{40}$

44. $d = \dfrac{1}{64}\sin 729\pi t$

(a) Maximum displacement = amplitude $= \dfrac{1}{64}$

(b) Frequency $= \dfrac{1}{\text{period}} = \dfrac{1}{1/396}$

$= 396$ cycles per unit of time

(c) $792\pi t = \pi \Rightarrow t = \dfrac{1}{792}$

46. $d = a \cos bt$

Amplitude: $2a = 3.5 \Rightarrow a = 1.75 = \dfrac{7}{4}$ ft

Period: 10 seconds $= \dfrac{2\pi}{b} \Rightarrow b = \dfrac{\pi}{5}$

$d = \dfrac{7}{4} \cos \dfrac{\pi t}{5}$

Review Exercises for Chapter 5

2. $\theta = \dfrac{2\pi}{9}$

Coterminal angles: $\dfrac{20\pi}{9}, \ -\dfrac{16\pi}{9}$

4. $\theta = -405°$

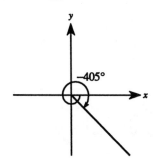

Coterminal angles: $315°, \ -45°$

6. $-234°50'' = -\left(234 + \frac{50}{3600}\right) \approx -234.014°$

8. $280°8'50'' = 280 + \frac{8}{60} + \frac{50}{3600} \approx 280.15°$

10. $25.1° = 25° + 0.1(60) = 25°6'$

12. $-327.85° = -[327° + (0.85)(60)] = -327°51'$

14. $\dfrac{-3\pi \text{ rad}}{5} = \dfrac{-3\pi \text{ rad}}{5} \cdot \dfrac{180°}{\pi \text{ rad}} = -108°$

16. $1.75 \text{ rad} = 1.75 \text{ rad} \cdot \dfrac{180°}{\pi \text{ rad}} \approx 100.27°$

18. $-16.5° = -16.5° \cdot \dfrac{\pi \text{ rad}}{180°} = \dfrac{-11\pi}{120} \text{ rad} \approx -0.2880 \text{ rad}$

20. $84°15' = 84.25° = 84.25° \cdot \dfrac{\pi \text{ rad}}{180°} = \dfrac{337\pi}{720} \text{ rad} \approx 1.4704 \text{ rad}$

22. $640°$ is in Quadrant IV and is coterminal to $280°$.

Reference angle $= 360° - 280° = 80°$

24. $\dfrac{17\pi}{3}$ is in Quadrant IV and is coterminal to $\dfrac{5\pi}{3}$.

Reference angle $= 2\pi - \dfrac{5\pi}{3} = \dfrac{\pi}{3}$

26. $x = x$, $y = 4x$, $r = \sqrt{x^2 + 16x^2} = \sqrt{17x^2} = x\sqrt{17}$

$$\sin \theta = \frac{4}{\sqrt{17}} = \frac{4\sqrt{17}}{17} \qquad \csc \theta = \frac{\sqrt{17}}{4}$$

$$\cos \theta = \frac{1}{\sqrt{17}} = \frac{\sqrt{17}}{17} \qquad \sec \theta = \sqrt{17}$$

$$\tan \theta = 4 \qquad\qquad\qquad \cot \theta = \frac{1}{4}$$

28. $x = 4$, $y = -8$, $r = \sqrt{16 + 64} = 4\sqrt{5}$

$$\sin \theta = \frac{y}{r} = -\frac{8}{4\sqrt{5}} = -\frac{2\sqrt{5}}{5} \qquad \csc \theta = -\frac{\sqrt{5}}{2}$$

$$\cos \theta = \frac{x}{r} = \frac{4}{4\sqrt{5}} = \frac{\sqrt{5}}{5} \qquad \sec \theta = \sqrt{5}$$

$$\tan \theta = \frac{y}{x} = -\frac{8}{4} = -2 \qquad \cot \theta = -\frac{1}{2}$$

30. $x = \frac{2}{3}$, $y = \frac{5}{2}$, $r = \sqrt{\frac{4}{9} + \frac{25}{4}} = \frac{\sqrt{241}}{6}$

$$\sin \theta = \frac{y}{r} = \frac{5/2}{\sqrt{241}/6} = \frac{15\sqrt{241}}{241} \qquad \csc \theta = \frac{\sqrt{241}}{15}$$

$$\cos \theta = \frac{x}{r} = \frac{2/3}{\sqrt{241}/6} = \frac{4\sqrt{241}}{241} \qquad \sec \theta = \frac{\sqrt{241}}{4}$$

$$\tan \theta = \frac{y}{x} = \frac{5/2}{2/3} = \frac{15}{4} \qquad\qquad \cot \theta = \frac{4}{15}$$

32. $\tan \theta = -\frac{12}{5}$, $\sin \theta > 0 \Rightarrow \theta$ is in Quadrant II.

$y = 12$, $x = -5$, $r = 13$

$$\sin \theta = \frac{y}{r} = \frac{12}{13} \qquad \csc \theta = \frac{13}{12}$$

$$\cos \theta = \frac{x}{r} = -\frac{5}{13} \qquad \sec \theta = -\frac{13}{5}$$

$$\tan \theta = \frac{y}{x} = -\frac{12}{5} \qquad \cot \theta = -\frac{5}{12}$$

34. $\cos \theta = -\frac{2}{5}$, $\sin \theta > 0 \Rightarrow \theta$ is in Quadrant II.

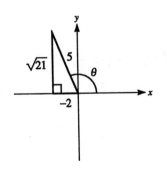

$x = -2$, $r = 5$, $y = \sqrt{21}$

$$\sin \theta = \frac{y}{r} = \frac{\sqrt{21}}{5} \qquad \csc \theta = \frac{5}{\sqrt{21}} = \frac{5\sqrt{21}}{21}$$

$$\cos \theta = \frac{x}{r} = -\frac{2}{5} \qquad \sec \theta = -\frac{5}{2}$$

$$\tan \theta = \frac{y}{x} = -\frac{\sqrt{21}}{2} \qquad \cot \theta = -\frac{2}{\sqrt{21}} = -\frac{2\sqrt{21}}{21}$$

36. $\sec \dfrac{\pi}{4} = \sqrt{2}$

38. $\cot\left(\dfrac{5\pi}{6}\right) = -\sqrt{3}$

40. $\csc 270° = -1$

42. $\csc 105° \approx 1.04$

44. $\sin\left(-\dfrac{\pi}{9}\right) \approx -0.34$

46. $\sec \theta$ is undefined $\Rightarrow \theta = \dfrac{\pi}{2},\ \dfrac{3\pi}{2}$ or $\theta = 90°,\ 270°$

48. $\tan \theta = \dfrac{\sqrt{3}}{3} \Rightarrow \theta$ is in Quadrant I or III.

Reference angle: $\dfrac{\pi}{6}$

$\theta = \dfrac{\pi}{6},\ \dfrac{7\pi}{6}$ or $\theta = 30°,\ 210°$

50. $\cot \theta = -1.5399,\ \theta$ is in Quadrant II or IV.

Reference angle: $33.0°$

$\theta = 147.0°\quad$ or $\quad 147° \cdot \dfrac{\pi}{180°} \approx 2.5656$ rad

$\theta = 327.0°\quad$ or $\quad 327° \cdot \dfrac{\pi}{180°} \approx 5.7072$ rad

52. $\csc \theta = 11.4737,\ \theta$ is in Quadrant I or II.

Reference angle: $5.0°$

$\theta = 5.0°$ or $5° \cdot \dfrac{\pi}{180°} \approx 0.873$ rad

$\theta = 175.0°$ or $175° \cdot \dfrac{\pi}{180°} \approx 3.054$ rad

54. $\tan\left(\arccos \dfrac{x}{2}\right) = \tan \theta$

$= \dfrac{\sqrt{4-x^2}}{x}$

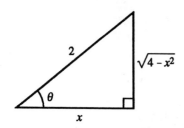

56. $\csc(\arcsin 10x) = \csc \theta = \dfrac{1}{10x}$

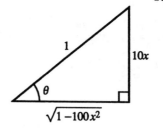

58. $y = -2 \sin \pi x$

60. $f(x) = 8\cos\left(-\dfrac{x}{4}\right)$

62. $f(x) = -\tan \dfrac{\pi x}{4}$

64. $g(t) = 3\cos(t + \pi)$

66. $h(t) = \sec\left(t - \dfrac{\pi}{4}\right)$

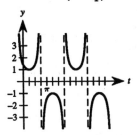

68. $f(t) = 3\csc\left(2t + \dfrac{\pi}{4}\right)$

70. $E(t) = 110\cos\left(120\pi t - \dfrac{\pi}{3}\right)$

72. $g(x) = 3\left(\sin \dfrac{\pi x}{3} + 1\right)$

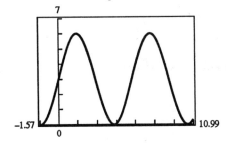

74. $y = 4 - \dfrac{x}{4} + \cos \pi x$

76. $f(t) = 2.5e^{-t/4}\sin 2\pi t$

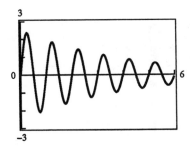

78. $y = 2 \arccos x$

80. $f(x) = \arccos(x - \pi)$

82. $g(x) = \sin(e^x)$

Not periodic

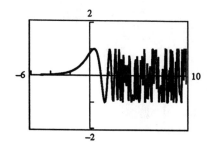

84. $h(x) = 4 \sin^2 x \cos^2 x$

The function is periodic with period $\dfrac{\pi}{2}$.

Relative Maximum:

(0.785, 1.00)

Relative Minimum: (0, 0)

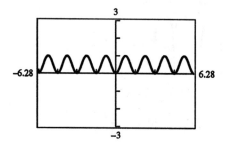

86. $\tan \theta = \dfrac{225}{105} \Rightarrow \theta = \arctan \dfrac{225}{105}$

$\Rightarrow \theta \approx 65.0°$

88.
$$\left.\begin{array}{l} \sin 48° = \dfrac{d_1}{650} \Rightarrow d_1 \approx 483 \\[2mm] \cos 25° = \dfrac{d_2}{810} \Rightarrow d_2 \approx 734 \end{array}\right\} d_1 + d_2 \approx 1217$$

$$\left.\begin{array}{l} \cos 48° = \dfrac{d_3}{650} \Rightarrow d_3 \approx 435 \\[2mm] \sin 25° = \dfrac{d_4}{810} \Rightarrow d_4 \approx 342 \end{array}\right\} d_3 - d_4 \approx 93$$

$$\tan \theta \approx \dfrac{93}{1217} \Rightarrow \theta \approx 4.4°$$

$$\sec 4.4° \approx \dfrac{D}{1217} \Rightarrow D \approx 1217 \sec 4.4° \approx 1221$$

The distance is 1221 miles and the bearing is
N 85.6° E.

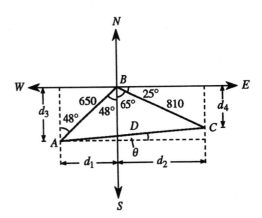

90. $\tan 14° = \dfrac{y}{35{,}000} \Rightarrow y = 35{,}000 \tan 14° \approx 8726.5$ feet

$\tan 58° = \dfrac{x + y}{35{,}000} \Rightarrow x + y = 35{,}000 \tan 58° \approx 56011.7$ feet

$x = 47{,}285.2$ feet ≈ 9.0 miles

Cumulative Test, Chapters 3—5

1. $f(x) = \dfrac{1}{4}\left(4x^2 - 12x + 17\right)$

$\qquad = x^2 - 3x + \dfrac{17}{4}$

$\qquad = x^2 - 3x + \dfrac{9}{4} + \dfrac{8}{4}$

$\qquad = \left(x - \dfrac{3}{2}\right)^2 + 2 \Rightarrow$ vertex is $\left(\dfrac{3}{2},\ 2\right)$.

2. $f(x) = a(x - h)^2 + k$

$\qquad = a(x - 0)^2 + 6$

$\qquad = ax^2 + 6$

Since $f(2) = 5$,

$\qquad 5 = a(2)^2 + 6 \Rightarrow a = -\frac{1}{4}$.

Hence, $f(x) = -\frac{1}{4}x^2 + 6$.

3. $f(x) = -\frac{2}{3}x^3 + 3x^2 - 2x + 1$

As $x \to \infty$, the graph goes downward to the right.

As $x \to -\infty$, the graph goes upward to the left.

4. $f(x) = (x + 4)(2x - 1)(x - 2)$

$\qquad = 2x^3 + 3x^2 - 18x + 8$

5. $f(t) = \frac{1}{4}t(t - 2)^2$

intercepts $t = 0,\ 2$

6.
$$
\begin{array}{r}
3x\ -2 \\
2x^2 + 1\ \overline{\smash{)}\ 6x^3 - 4x^2} \\
\underline{6x^3 + 3x} \\
-4x^2 - 3x \\
\underline{-4x^2 - 2} \\
-3x + 2
\end{array}
$$

Hence,

$$\frac{6x^3 - 4x^2}{2x^2 + 1} = 3x - 2 - \frac{3x - 2}{2x^2 + 1}.$$

7.

$$
\begin{array}{r|rrrr}
2 & 3 & 0 & -5 & 4 \\
  &   & 6 & 12 & 14 \\
\hline
  & 3 & 6 & 7 & 18
\end{array}
$$

Hence,

$$\frac{3x^3 - 5x + 4}{x - 2} = 3x^2 + 6x + 7 + \frac{18}{x - 2}.$$

8. $f(x) = 6x^3 - 25x^2 - 8x + 48$

From the graph of f and synthetic division, we see that the rational zeros of f are $-\frac{4}{3}$, $\frac{3}{2}$ and 4.

9. $g(t) = t^3 - 5t - 2$

Using a graphing utility, we see that $t = 2.4$ is a zero of g in the interval $[2, \ 3]$.

10. (a)

(b)

11. (a)

(b)

12. $\log_5 125 = \log_5 5^3$

$\qquad\qquad = 3 \log_5 5$

$\qquad\qquad = 3$

13. $2 \ln x - \frac{1}{2} \ln(x + 5) = \ln x^2 - \ln(x + 5)^{1/2}$

$$= \ln \left(\frac{x^2}{\sqrt{x + 5}} \right)$$

14. (a) $6e^{2x} = 72$

Using a graphing utility, we find that

$$y = 6e^{2x} - 72 = 0$$

for $x = 1.24$. Or,

$$6e^{2x} = 72$$

$$e^{2x} = 12$$

$$2x = \ln(12)$$

$$x = \frac{\ln(12)}{2}$$

$$\approx 1.24245.$$

(b) $\log_2 x + \log_2 5 = 6$

$$\log_2(5x) = 6$$

$$5x = 2^6$$

$$x = \frac{64}{5}$$

$$= 12.80$$

15. $A = Pe^{rt}$

$$= 2500e^{(0.075)(25)}$$

$$= 16{,}302.05$$

16. $\left(\dfrac{4\pi}{9}\,\text{rad}\right)\left(\dfrac{180}{\pi}\,\dfrac{\text{deg}}{\text{rad}}\right) = 80^\circ$

17. $(-120\,\text{deg})\left(\dfrac{\pi}{180}\,\dfrac{\text{rad}}{\text{deg}}\right) = -\dfrac{2\pi}{3}$

18.

$$\sin\theta = \frac{5}{13} \qquad \csc\theta = \frac{13}{5}$$

$$\cos\theta = \frac{12}{13} \qquad \sec\theta = \frac{13}{12}$$

$$\tan\theta = \frac{5}{12} \qquad \cot\theta = \frac{12}{5}$$

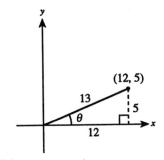

19. $\cos t = -\dfrac{2}{3}$

$$\sin^2 t = 1 - \cos^2 t = 1 - \frac{4}{9} = \frac{5}{9}$$

Since $\pi/2 < t < \pi$, $\sin t > 0$
and $\sin t = \sqrt{5}/3$. Hence,

$$\tan t = \frac{\sin t}{\cos t} = -\frac{\sqrt{5}}{2}.$$

20. $\sin(-1.25) \approx -0.9490$

21. $\sec\theta = \dfrac{1}{\cos\theta} = 2.125 \Rightarrow \cos\theta = 0.470588$

$$\Rightarrow \theta = 61.93° \text{ and } \theta = 360° - 61.93° = 298.07°$$

22. (a)

(b)

(c)

(d)

23. $f(x) = \sin x$

   (a) $g(x) = 10 + \sin x$ is 10 units above the graph of f.

   (b) $g(x) = \sin\left(\dfrac{\pi x}{2}\right)$ has period 4, whereas f has period 2π.

   (c) $g(x) = \sin\left(x + \dfrac{\pi}{4}\right)$ is $\dfrac{\pi}{4}$ units to the left of the graph of f.

   (d) $g(x) = -\sin x$ is a reflection in the x-axis of the graph of f.

24. Since the amplitude is 3, $a = 3$. Since the tic marks are spaced $\pi/4$ units apart, the period is π. Finally, the graph is shifted $\pi/4$ units to the right. Hence,

$$y = 3\sin\left(2\left(x - \frac{\pi}{4}\right)\right) = 3\sin\left(2x - \frac{\pi}{2}\right).$$

25. $f(x) = \sin 3x - 2\cos x$

   (a)

   (b) The period of $\cos x$ is 2π, and that of $\sin 3x$ is $2\pi/3$. Hence, the period of f is 2π.

   (c) Using the zoom and trace features, $x \approx 1.9$.

   (d) The maximum value is approximately $2.8 = f(2.7)$.

26. $f(t) = 2^{-t/2}\cos\left(\dfrac{\pi t}{2}\right)$

   (a)

   (b) No, the function is not periodic, because of the $2^{-t/2}$ factor.

   (c) For $t > 0.75$, $f(t) < 0.3$. (Graph f and $y = 0.3$ on the same viewing rectangle.)

27. (a) $y = \arcsin\dfrac{1}{2} \Leftrightarrow \sin y = \dfrac{1}{2}$,

        $-\dfrac{\pi}{2} \le y \le \dfrac{\pi}{2} \Leftrightarrow y = \dfrac{\pi}{6}$

   (b) $y = \arctan\sqrt{3} \Leftrightarrow \tan y = \sqrt{3}$,

        $-\dfrac{\pi}{2} < 0 < \dfrac{\pi}{2} \Leftrightarrow y = \dfrac{\pi}{3}$

28. $\sin\left(\arccos 2x\right)$

   Let $y = \arccos 2x \Rightarrow \cos y = 2x$.

   $\sin y = \dfrac{\sqrt{1 - 4x^2}}{1} = \sqrt{1 - 4x^2}$

29. $32°30' = 32.5°$

   $h = 600\tan(32.5°) = 382.24$ feet

CHAPTER SIX
Additional Topics in Trigonometry

6.1 | Applications of Fundamental Identities

2. $\tan x = \dfrac{\sqrt{3}}{3}$, $\cos x = -\dfrac{\sqrt{3}}{2} \Rightarrow x$ is in Quadrant III

$$\sin x = -\sqrt{1 - \cos^2 x} = -\sqrt{1 - \dfrac{3}{4}} = -\dfrac{1}{2}$$

$$\cot x = \dfrac{1}{\tan x} = \dfrac{3}{\sqrt{3}} = \sqrt{3}$$

$$\sec x = \dfrac{1}{\cos x} = -\dfrac{2}{\sqrt{3}} = -\dfrac{2\sqrt{3}}{3}$$

$$\csc x = \dfrac{1}{\sin x} = -2$$

4. $\csc \theta = \dfrac{5}{3}$, $\tan \theta = \dfrac{3}{4} \Rightarrow \theta$ is in Quadrant I

$$\sin \theta = \dfrac{1}{\csc \theta} = \dfrac{3}{5}$$

$$\cos \theta = \sqrt{1 - \sin^2 \theta} = \sqrt{1 - \dfrac{9}{25}} = \dfrac{4}{5}$$

$$\cot \theta = \dfrac{1}{\tan \theta} = \dfrac{4}{3}$$

$$\sec \theta = \dfrac{1}{\cos \theta} = \dfrac{5}{4}$$

6. $\cos \left(\dfrac{\pi}{2} - x \right) = \dfrac{3}{5} \Rightarrow \sin x = \dfrac{3}{5}$

$\sin x = \dfrac{3}{5}$, $\cos x = \dfrac{4}{5} \Rightarrow x$ is in Quadrant I

$$\tan x = \dfrac{\sin x}{\cos x} = \dfrac{3}{4}$$

$$\sec x = \dfrac{1}{\cos x} = \dfrac{5}{4}$$

$$\cot x = \dfrac{1}{\tan x} = \dfrac{4}{3}$$

$$\csc x = \dfrac{1}{\sin x} = \dfrac{5}{3}$$

8. $\sec \theta = -3$, $\tan \theta < 0 \Rightarrow \theta$ is in Quadrant II

$$\cos \theta = \dfrac{1}{\sec \theta} = -\dfrac{1}{3}$$

$$\sin \theta = \sqrt{1 - \cos^2 \theta} = \dfrac{2\sqrt{2}}{3}$$

$$\tan \theta = \dfrac{\sin \theta}{\cos \theta} = -2\sqrt{2}$$

$$\cot \theta = \dfrac{1}{\tan \theta} = -\dfrac{1}{2\sqrt{2}} = -\dfrac{\sqrt{2}}{4}$$

$$\csc \theta = \dfrac{1}{\sin \theta} = \dfrac{3\sqrt{2}}{4}$$

10. $\tan \theta$ is undefined, $\sin \theta > 0 \Rightarrow \theta = \dfrac{\pi}{2}$

$$\tan \theta = \dfrac{\sin \theta}{\cos \theta} \text{ is undefined} \Rightarrow \cos \theta = 0$$

$$\sin \theta = \sqrt{1 - \cos^2 \theta} = 1$$

$$\csc \theta = \dfrac{1}{\sin \theta} = 1$$

$$\sec \theta = \dfrac{1}{\cos \theta} \text{ is undefined.}$$

$$\cot \theta = \dfrac{\cos \theta}{\sin \theta} = 0$$

12. $\dfrac{\sin(-x)}{\cos(-x)} = \dfrac{-\sin x}{\cos x} = -\tan x$ $\Rightarrow$ The expression is matched with (e).

14. $\dfrac{1 - \cos^2 x}{\sin x} = \dfrac{\sin^2 x}{\sin x} = \sin x$ $\Rightarrow$ The expression is matched with (f).

16. $\dfrac{\sin[(\pi/2) - x]}{\cos[(\pi/2) - x]} = \dfrac{\cos x}{\sin x} = \cot x$ $\Rightarrow$ The expression is matched with (c).

18. $\cos^2 x(\sec^2 x - 1) = 1 - \cos^2 x = \sin^2 x$ $\Rightarrow$ The expression is matched with (c).

20. $\cot x \sec x = \dfrac{\cos x}{\sin x} \cdot \dfrac{1}{\cos x} = \dfrac{1}{\sin x} = \csc x$ $\Rightarrow$ The expression is matched with (a).

22. $\dfrac{\cos^2[(\pi/2) - x]}{\cos x} = \dfrac{\sin^2 x}{\cos x} = \sin x \cdot \dfrac{\sin x}{\cos x} = \sin x \tan x$ $\Rightarrow$ The expression is matched with (d).

24. $\sin\phi(\csc\phi - \sin\phi) = \sin\phi \csc\phi - \sin^2\phi$

$\qquad\qquad = \sin\phi \cdot \dfrac{1}{\sin\phi} - \sin^2\phi$

$\qquad\qquad = 1 - \sin^2\phi = \cos^2\phi$

26. $\sec\alpha \dfrac{\sin\alpha}{\tan\alpha} = \dfrac{1}{\cos\alpha}(\sin\alpha)\cot\alpha$

$\qquad\qquad = \dfrac{1}{\cos\alpha}(\sin\alpha)\left(\dfrac{\cos\alpha}{\sin\alpha}\right) = 1$

28. $\dfrac{\csc\theta}{\sec\theta} = \dfrac{1/\sin\theta}{1/\cos\theta}$

$\qquad = \dfrac{1}{\sin\theta} \cdot \dfrac{\cos\theta}{1} = \dfrac{\cos\theta}{\sin\theta} = \cot\theta$

30. $\dfrac{1}{\tan^2 x + 1} = \dfrac{1}{\sec^2 x} = \cos^2 x$

32. $\dfrac{\tan^2\theta}{\sec^2\theta} = \dfrac{\sin^2\theta / \cos^2\theta}{1/\cos^2\theta}$

$\qquad = \dfrac{\sin^2\theta}{\cos^2\theta} \cdot \cos^2\theta$

$\qquad = \sin^2\theta$

34. $\cot\left(\dfrac{\pi}{2} - x\right)\cos x = \tan x \cos x$

$\qquad\qquad = \left(\dfrac{\sin x}{\cos x}\right)\cos x$

$\qquad\qquad = \sin x$

36. $\cos t(1 + \tan^2 t) = \cos t(\sec^2 t)$

$\qquad\qquad = \dfrac{1}{\sec t} \cdot \sec^2 t$

$\qquad\qquad = \sec t$

38. $\sec^2 x \tan^2 x + \sec^2 x = \sec^2 x(\tan^2 x + 1)$

$\qquad\qquad = \sec^2 x \sec^2 x$

$\qquad\qquad = \sec^4 x$

40. $\dfrac{\sec^2 x - 1}{\sec x - 1} = \dfrac{(\sec x + 1)(\sec x - 1)}{\sec x - 1}$

$\qquad\qquad = \sec x + 1$

42. $1 - 2\cos^2 x + \cos^4 x = (1 - \cos^2 x)^2$

$\qquad\qquad = (\sin^2 x)^2$

$\qquad\qquad = \sin^4 x$

44. $\csc^3 x - \csc^2 x - \csc x + 1 = \csc^2 x(\csc x - 1) - (\csc x - 1)$

$\qquad\qquad = (\csc^2 x - 1)(\csc x - 1)$

$\qquad\qquad = \cot^2 x(\csc x - 1)$

46. $(\cot x + \csc x)(\cot x - \csc x) = \cot^2 x - \csc^2 x$
$$= -1$$

48. $(3 - 3\sin x)(3 + 3\sin x) = 9 - 9\sin x + 9\sin x - 9\sin^2 x$
$$= 9(1 - \sin^2 x)$$
$$= 9\cos^2 x$$

50. $\dfrac{1}{\sec x + 1} - \dfrac{1}{\sec x - 1} = \dfrac{\sec x - 1 - (\sec x + 1)}{(\sec x + 1)(\sec x - 1)}$
$$= \dfrac{-2}{\sec^2 x - 1}$$
$$= \dfrac{-2}{\tan^2 x}$$
$$= -2\cot^2 x$$

52. $\tan x - \dfrac{\sec^2 x}{\tan x} = \dfrac{\tan^2 x - \sec^2 x}{\tan x}$
$$= \dfrac{-1}{\tan x}$$
$$= -\cot x$$

54. $\dfrac{5}{\tan x + \sec x} \cdot \dfrac{\tan x - \sec x}{\tan x - \sec x} = \dfrac{5(\tan x - \sec x)}{\tan^2 x - \sec^2 x}$
$$= \dfrac{5(\tan x - \sec x)}{-1}$$
$$= -5(\tan x - \sec x)$$
$$= 5(\sec x - \tan x)$$

56. $\dfrac{\tan^2 x}{\csc x + 1} \cdot \dfrac{\csc x - 1}{\csc x - 1} = \dfrac{\tan^2 x(\csc x - 1)}{\csc^2 x - 1}$
$$= \dfrac{\tan^2 x(\csc x - 1)}{\cot^2 x}$$
$$= \tan^4 x(\csc x - 1)$$

58. $y_1 = \sec^2 x \csc x$

$y_2 = \sec^2 x + \csc x$

The graphs of y_1 and y_2 are different.
The equation is not an identity.

60. $y_1 = \sin x \tan x$

$y_2 = \csc x - \cos x$

The graphs of y_1 and y_2 are different.
The equation is not an identity.

62. Graphing $y_1 = \sin x \cdot \tan x + \cos x$ and $y_2 = \sec x$, we see that the graphs coincide. Hence, $\sin x \cdot \tan x + \cos x = \sec x$.

64. $\sqrt{16 - 4x^2} = \sqrt{16 - 4(2\sin\theta)^2}, \quad x = 2\sin\theta$

$$= \sqrt{16 - 16\sin^2\theta}$$
$$= \sqrt{16(1 - \sin^2\theta)}$$
$$= \sqrt{16\cos^2\theta}$$
$$= 4\cos\theta$$

66. $\sqrt{x^2 - 4} = \sqrt{(2\sec\theta)^2 - 4}, \quad x = 2\sec\theta$

$$= \sqrt{4\sec^2\theta - 4}$$
$$= \sqrt{4(\sec^2\theta - 1)}$$
$$= \sqrt{4\tan^2\theta}$$
$$= 2\tan\theta$$

68. $\sqrt{x^2 + 100} = \sqrt{(10\tan\theta)^2 + 100}, \quad x = 10\tan\theta$

$$= \sqrt{100\tan^2\theta + 100}$$
$$= \sqrt{100(\tan^2\theta + 1)}$$
$$= \sqrt{100\sec^2\theta}$$
$$= 10\sec\theta$$

70. $\sqrt{1 - e^{2x}} = \sqrt{1 - \sin^2\theta}, \quad e^x = \sin\theta$

$$= \sqrt{\cos^2\theta}$$
$$= \cos\theta$$

72. $\sqrt{(x^2 - 16)^3} = \sqrt{[(4\sec\theta)^2 - 16]^3}, \quad x = 4\sec\theta$

$$= \sqrt{(16\sec^2\theta - 16)^3}$$
$$= \sqrt{[16(\sec^2\theta - 1)]^3}$$
$$= \sqrt{[16(\tan^2\theta)]^3}$$
$$= \left(\sqrt{16\tan^2\theta}\right)^3$$
$$= (4\tan\theta)^3$$
$$= 64\tan^3\theta$$

74. $\sin\theta = -\sqrt{1 - \cos^2\theta}$ is valid for θ in Quadrant III, $\pi \le \theta \le \dfrac{3\pi}{2}$, and Quadrant IV, $\dfrac{3\pi}{2} \le \theta \le 2\pi$.

76. $\ln|\cot t| + \ln(1 + \tan^2 t) = \ln|(\cot t)(1 + \tan^2 t)|$

$$= \ln\left|\frac{1}{\sin t \cdot \cos t}\right|$$

$$= -\ln|\sin t \cdot \cos t|$$

78. False; $5\sec\theta = 5 \cdot \dfrac{1}{\cos\theta} = \dfrac{5}{\cos\theta} \neq \dfrac{1}{5\cos\theta}$

80. False; θ must be equal to ϕ (or differ by $2n\pi$).

82. $\tan^2\theta + 1 = \sec^2\theta$

   (a) $\theta = 346°$

      $(\tan 346°)^2 + 1 = (\sec 346°)^2$

          $0.0622 + 1 = 1.0622$

             $1.0622 = 1.0622$

   (b) $\theta = 3.1$

      $(\tan 3.1)^2 + 1 = (\sec 3.1)^2$

       $0.00173 + 1 = 1.00173$

          $1.00173 = 1.00173$

84. $\sin(-\theta) = -\sin\theta$

   (a) $\theta = 250°$

      $\sin(-250°) = -\sin(250°)$

         $0.9397 = -(-0.9397)$

         $0.9397 = 0.9397$

   (b) $\theta = \frac{1}{2}$

      $\sin\left(-\frac{1}{2}\right) = -\sin\left(\frac{1}{2}\right)$

      $-0.4794 = -(0.4794)$

      $-0.4794 = -0.4794$

86. $\sin\theta = \pm\sqrt{1 - \cos^2\theta}$

$$\tan\theta = \frac{\sin\theta}{\cos\theta} = \pm\frac{\sqrt{1 - \cos^2\theta}}{\cos\theta}$$

$$\cot\theta = \frac{\cos\theta}{\sin\theta} = \pm\frac{\cos\theta}{\sqrt{1 - \cos^2\theta}}$$

$$\sec\theta = \frac{1}{\cos\theta}$$

$$\csc\theta = \frac{1}{\sin\theta} = \pm\frac{1}{\sqrt{1 - \cos^2\theta}}$$

▮ 6.2 ▮ Verifying Trigonometric Identities

2. $\tan y \cot y = \tan y\left(\dfrac{1}{\tan y}\right) = 1$

4. $\cot^2 y(\sec^2 y - 1) = \cot^2 y \tan^2 y = 1$

6. $\cos^2\beta - \sin^2\beta = \cos^2\beta - (1 - \cos^2\beta)$

$$= 2\cos^2\beta - 1$$

8. $2 - \sec^2 z = 2 - (1 + \tan^2 z)$

$$= 1 - \tan^2 z$$

10. $\cos x + \sin x \tan x = \cos x + \sin x \left(\dfrac{\sin x}{\cos x} \right)$

$$= \frac{\cos^2 x + \sin^2 x}{\cos x}$$

$$= \frac{1}{\cos x}$$

$$= \sec x$$

12. $\cos t (\csc^2 t - 1) = \cos t \cot^2 t$

$$= \sin t \left(\frac{\cos t}{\sin t} \right) \cot^2 t$$

$$= \frac{1}{\csc t} \cot^3 t$$

$$= \frac{\cot^3 t}{\csc t}$$

14. $\dfrac{1}{\sin x} - \sin x = \dfrac{1 - \sin^2 x}{\sin x} = \dfrac{\cos^2 x}{\sin x}$

16. $\sec^6 x (\sec x \tan x) - \sec^4 x (\sec x \tan x) = \sec^4 x (\sec x \tan x)(\sec^2 x - 1)$

$$= \sec^4 x (\sec x \tan x) \tan^2 x$$

$$= \sec^5 x \tan^3 x$$

18. $\dfrac{\sec \theta - 1}{1 - \cos \theta} = \dfrac{\sec \theta - 1}{1 - (1/\sec \theta)} \cdot \dfrac{\sec \theta}{\sec \theta}$

$$= \frac{\sec \theta (\sec \theta - 1)}{\sec \theta - 1}$$

$$= \sec \theta$$

20. $\dfrac{\sec x + \tan x}{\sec x - \tan x} = \dfrac{\sec x + \tan x}{\sec x - \tan x} \cdot \dfrac{\sec x + \tan x}{\sec x + \tan x}$

$$= \frac{(\sec x + \tan x)^2}{\sec^2 x - \tan^2 x}$$

$$= \frac{(\sec x + \tan x)^2}{1}$$

$$= (\sec x + \tan x)^2$$

22. $\dfrac{1}{\sin x} - \dfrac{1}{\csc x} = \dfrac{\csc x - \sin x}{\sin x \csc x}$

$$= \frac{\csc x - \sin x}{1}$$

$$= \csc x - \sin x$$

24. $\cos x - \dfrac{\cos x}{1 - \tan x} = \dfrac{\cos x (1 - \tan x) - \cos x}{1 - \tan x}$

$$= \frac{- \cos x \tan x}{1 - \tan x}$$

$$= \frac{- \cos x (\sin x / \cos x)}{1 - (\sin x / \cos x)} \cdot \frac{\cos x}{\cos x}$$

$$= \frac{- \sin x \cos x}{\cos x - \sin x}$$

$$= \frac{\sin x \cos x}{\sin x - \cos x}$$

26. $\csc x (\csc x - \sin x) + \dfrac{\sin x - \cos x}{\sin x} + \cot x = \csc^2 x - \csc x \sin x + 1 - \dfrac{\cos x}{\sin x} + \cot x$

$$= \csc^2 x - 1 + 1 - \cot x + \cot x$$

$$= \csc^2 x$$

28. $4 \tan^4 x + \tan^2 x - 3 = (\tan^2 x + 1)(4 \tan^2 x - 3)$

$$= \sec^2 x (4 \tan^2 x - 3)$$

30. $\csc^4 \theta - \cot^4 \theta = (\csc^2 \theta - \cot^2 \theta)(\csc^2 \theta + \cot^2 \theta)$

$\qquad\qquad = \csc^2 \theta + \cot^2 \theta$

$\qquad\qquad = \csc^2 \theta + (\csc^2 \theta - 1)$

$\qquad\qquad = 2\csc^2 \theta - 1$

32. $\dfrac{\cot \alpha}{\csc \alpha - 1} \cdot \dfrac{\csc \alpha + 1}{\csc \alpha + 1} = \dfrac{\cot \alpha(\csc \alpha + 1)}{\csc^2 \alpha - 1}$

$\qquad\qquad\qquad = \dfrac{\cot \alpha(\csc \alpha + 1)}{\cot^2 \alpha}$

$\qquad\qquad\qquad = \dfrac{\csc \alpha + 1}{\cot \alpha}$

34. $\dfrac{\cos[(\pi/2) - x]}{\sin[(\pi/2) - x]} = \dfrac{\sin x}{\cos x} = \tan x$

36. $(1 + \sin y)[1 + \sin(-y)] = (1 + \sin y)(1 - \sin y)$

$\qquad\qquad\qquad = 1 - \sin^2 y$

$\qquad\qquad\qquad = \cos^2 y$

38. $\dfrac{1 + \sec(-\theta)}{\sin(-\theta) + \tan(-\theta)} = \dfrac{1 + \sec \theta}{-\sin \theta - \tan \theta}$

$\qquad\qquad\qquad = -\dfrac{1 + \sec \theta}{\sin \theta + \tan \theta}$

$\qquad\qquad\qquad = -\dfrac{1 + \sec \theta}{\sin \theta[1 + (1/\cos \theta)]}$

$\qquad\qquad\qquad = -\dfrac{1 + \sec \theta}{\sin \theta(1 + \sec \theta)}$

$\qquad\qquad\qquad = -\dfrac{1}{\sin \theta}$

$\qquad\qquad\qquad = -\csc \theta$

40. $\sec^2 y - \cot^2 \left(\dfrac{\pi}{2} - y\right) = \sec^2 y - \tan^2 y = 1$

42. $\sqrt{\dfrac{1 - \cos \theta}{1 + \cos \theta}} = \sqrt{\dfrac{1 - \cos \theta}{1 + \cos \theta} \cdot \dfrac{1 - \cos \theta}{1 - \cos \theta}}$

$\qquad\qquad = \sqrt{\dfrac{(1 - \cos \theta)^2}{1 - \cos^2 \theta}}$

$\qquad\qquad = \sqrt{\dfrac{(1 - \cos \theta)^2}{\sin^2 \theta}}$

$\qquad\qquad = \dfrac{1 - \cos \theta}{|\sin \theta|}$

44. $\dfrac{\tan x + \tan y}{1 - \tan x \tan y} = \dfrac{\dfrac{1}{\cot x} + \dfrac{1}{\cot y}}{1 - \dfrac{1}{\cot x} \cdot \dfrac{1}{\cot y}} \cdot \dfrac{\cot x \cot y}{\cot x \cot y} = \dfrac{\cot y + \cot x}{\cot x \cot y - 1}$

46. $\dfrac{\cos x - \cos y}{\sin x + \sin y} + \dfrac{\sin x - \sin y}{\cos x + \cos y} = \dfrac{(\cos x - \cos y)(\cos x + \cos y) + (\sin x - \sin y)(\sin x + \sin y)}{(\sin x + \sin y)(\cos x + \cos y)}$

$$= \frac{\cos^2 x - \cos^2 y + \sin^2 x - \sin^2 y}{(\sin x + \sin y)(\cos x + \cos y)}$$

$$= \frac{(\cos^2 x + \sin^2 x) - (\cos^2 y + \sin^2 y)}{(\sin x + \sin y)(\cos x + \cos y)}$$

$$= 0$$

48. $\ln |\sec \theta| = \ln \left| \dfrac{1}{\cos \theta} \right|$

$$= \ln 1 - \ln |\cos \theta|$$

$$= -\ln |\cos \theta|$$

50. $-\ln |\sec \theta + \tan \theta| = \ln \dfrac{1}{|\sec \theta + \tan \theta|}$

$$= \ln \left| \frac{1}{\sec \theta + \tan \theta} \cdot \frac{\sec \theta - \tan \theta}{\sec \theta - \tan \theta} \right|$$

$$= \ln \left| \frac{\sec \theta - \tan \theta}{\sec^2 \theta - \tan^2 \theta} \right|$$

$$= \ln |\sec \theta - \tan \theta|$$

52. True Identity: $\tan \theta = \pm\sqrt{\sec^2 \theta - 1}$

$\tan \theta = \sqrt{\sec^2 \theta - 1}$ is not true for $\pi/2 < \theta < \pi$ or $3\pi/2 < \theta < 2\pi$.

For instance, let $\theta = \dfrac{3\pi}{4}$.

54. $\sqrt{\sin^2 x + \cos^2 x} \neq \sin x + \cos x$

The left side is 1 for any x, but the right side is not necessarily 1.

For instance, let $x = \dfrac{\pi}{4}$.

56. $\cos x - \csc x \cot = \cos x - \dfrac{1}{\sin x} \dfrac{\cos x}{\sin x}$

$$= \cos x \left(1 - \frac{1}{\sin^2 x} \right)$$

$$= \cos x(1 - \csc^2 x)$$

$$= -\cos x(\csc^2 x - 1)$$

$$= -\cos x \cot^2 x$$

6.3 Solving Trigonometric Equations

2. $\csc x - 2 = 0$

    (a) $\csc \dfrac{\pi}{6} - 2 = 2 - 2 = 0$

    (b) $\csc \dfrac{5\pi}{6} - 2 = 2 - 2 = 0$

4. $2\cos^2 4x - 1 = 0$

    (a) $2\left[\cos 4\left(\dfrac{\pi}{16}\right)\right]^2 - 1 = 2\cos^2 \dfrac{\pi}{4} - 1$

$$= 2\left(\dfrac{1}{\sqrt{2}}\right)^2 - 1$$

$$= 0$$

    (b) $2\left[\cos 4\left(\dfrac{3\pi}{16}\right)\right]^2 - 1 = 2\cos^2 \dfrac{3\pi}{4} - 1$

$$= 2\left(-\dfrac{1}{\sqrt{2}}\right)^2 - 1$$

$$= 0$$

6. $\sec^4 x - 4\sec^2 x = 0$

    (a) $\sec^4 \dfrac{2\pi}{3} - 4\sec^2 \dfrac{2\pi}{3} = (-2)^4 - 4(-2)^2$

$$= 16 - 16$$

$$= 0$$

    (b) $\sec^4 \dfrac{5\pi}{3} - 4\sec^2 \dfrac{5\pi}{3} = 2^4 - 4(2)^2$

$$= 16 - 16$$

$$= 0$$

8. $2\sin x - 1 = 0$

$$2\sin x = 1$$

$$\sin x = \dfrac{1}{2}$$

$$x = \dfrac{\pi}{6} + 2n\pi$$

$$\text{or } x = \dfrac{5\pi}{6} + 2n\pi$$

10. $\tan x + 1 = 0$

$$\tan x = -1$$

$$x = \dfrac{3\pi}{4} + n\pi$$

$$\text{or } x = \dfrac{7\pi}{4} + n\pi$$

12. $\tan^2 x = 3$

$$\tan x = \pm\sqrt{3}$$

$$x = \dfrac{\pi}{3} + n\pi$$

$$\text{or } x = \dfrac{2\pi}{3} + n\pi$$

14. $\csc^2 x - 2 = 0$

$$\csc x = \pm\sqrt{2}$$

$$x = \dfrac{\pi}{4} + \dfrac{n\pi}{2}$$

16. $\cos x(2\cos x + 1) = 0$

$$\cos x = 0 \qquad \text{or} \qquad 2\cos x + 1 = 0$$

$$x = \dfrac{\pi}{2} + n\pi \qquad\qquad \cos x = -\dfrac{1}{2}$$

$$x = \dfrac{2\pi}{3} + 2n\pi$$

$$\text{or } x = \dfrac{4\pi}{3} + 2n\pi$$

18. $4\sin^2 x - 3 = 0$

$$\sin x = \pm\frac{\sqrt{3}}{2}$$

$$x = \frac{\pi}{3} + n\pi$$

$$\text{or } x = \frac{2\pi}{3} + n\pi$$

20. $(3\tan^2 x - 1)(\tan^2 x - 3) = 0$

$3\tan^2 x - 1 = 0$ $\qquad$ or $\quad \tan^2 x - 3 = 0$

$\tan x = \pm\dfrac{1}{\sqrt{3}}$ $\qquad\qquad$ $\tan x = \pm\sqrt{3}$

$x = \dfrac{\pi}{6} + n\pi$ $\qquad\qquad$ $x = \dfrac{\pi}{3} + n\pi$

or $x = \dfrac{5\pi}{6} + n\pi$ $\qquad$ or $x = \dfrac{2\pi}{3} + n\pi$

22. $\qquad \sec^2 x - \sec x - 2 = 0$

$(\sec x - 2)(\sec x + 1) = 0$

$\sec x - 2 = 0 \qquad$ or $\quad \sec x + 1 = 0$

$\sec x = 2 \qquad\qquad\quad \sec x = -1$

$x = \dfrac{\pi}{3}, \dfrac{5\pi}{3} \qquad\qquad\quad x = \pi$

24. $\qquad 3\tan^3 x - \tan x = 0$

$\tan x(3\tan^2 x - 1) = 0$

$\tan x = 0 \qquad$ or $\quad 3\tan^2 x - 1 = 0$

$x = 0,\ \pi$

$\qquad\qquad\qquad \tan x = \pm\dfrac{\sqrt{3}}{3}$

$\qquad\qquad x = \dfrac{\pi}{6}, \dfrac{5\pi}{6}, \dfrac{7\pi}{6}, \dfrac{11\pi}{6}$

26. $\qquad 2\sin^2 x = 2 + \cos x$

$2 - 2\cos^2 x = 2 + \cos x$

$2\cos^2 x + \cos x = 0$

$\cos x(2\cos x + 1) = 0$

$\cos x = 0 \qquad$ or $\quad 2\cos x + 1 = 0$

$x = \dfrac{\pi}{2}, \dfrac{3\pi}{2} \qquad\qquad 2\cos x = -1$

$\qquad\qquad\qquad\qquad \cos x = -\dfrac{1}{2}$

$\qquad\qquad\qquad\qquad x = \dfrac{2\pi}{3}, \dfrac{4\pi}{3}$

28. $\qquad \csc x + \cot x = 1$

$$\frac{1}{\sin x} + \frac{\cos x}{\sin x} = 1$$

$$1 + \cos x = \sin x$$

$$(1 + \cos x)^2 = \sin^2 x$$

$$1 + 2\cos x + \cos^2 x = 1 - \cos^2 x$$

$$2\cos^2 x + 2\cos x = 0$$

$$\cos x(\cos x + 1) = 0$$

$\cos x = 0 \qquad\qquad$ or $\quad \cos x = -1$

$x = \dfrac{\pi}{2}, \dfrac{3\pi}{2} \qquad\qquad\qquad x = \pi$

$(3\pi/2 \text{ is extraneous.}) \qquad (\pi \text{ is extraneous.})$

30. $\tan 3x = 1$

$$3x = \frac{\pi}{4} + 2n\pi \qquad \text{or} \qquad 3x = \frac{5\pi}{4} + 2n\pi$$

$$x = \frac{\pi}{12} + \frac{2n\pi}{3} \qquad\qquad\qquad x = \frac{5\pi}{12} + \frac{2n\pi}{3}$$

$$x = \frac{\pi}{12}, \frac{9\pi}{12}, \frac{17\pi}{12} \qquad\qquad x = \frac{5\pi}{12}, \frac{13\pi}{12}, \frac{21\pi}{12}$$

32. $\sec 4x = 2$

$$4x = \frac{\pi}{3} + 2n\pi \qquad\qquad \text{or} \qquad 4x = \frac{5\pi}{3} + 2n\pi$$

$$x = \frac{\pi}{12} + \frac{n\pi}{2} \qquad\qquad\qquad x = \frac{5\pi}{12} + \frac{n\pi}{2}$$

$$x = \frac{\pi}{12}, \frac{7\pi}{12}, \frac{13\pi}{12}, \frac{19\pi}{12} \qquad x = \frac{5\pi}{12}, \frac{11\pi}{12}, \frac{17\pi}{12}, \frac{23\pi}{12}$$

34.
$$y^2 + y - 20 = 0 \qquad\qquad\qquad \sin^2 x + \sin x - 20 = 0$$

$$(y+5)(y-4) = 0 \qquad\qquad\qquad (\sin x + 5)(\sin x - 4) = 0$$

$$y + 5 = 0 \quad \text{or} \quad y - 4 = 0 \qquad\quad \sin x = -5 \quad \text{or} \quad \sin x = 4$$

$$y = -5 \qquad\qquad y = 4 \qquad\quad \text{(No solution)} \qquad \text{(No solution)}$$

36. $2y^2 + 6y - 1 = 0 \qquad\qquad 2\cos^2 x + 6\cos x - 1 = 0$

$$y = \frac{-6 \pm \sqrt{36 + 8}}{4}$$

$$= \frac{-6 \pm \sqrt{44}}{4} \qquad\qquad \cos x = \frac{-6 \pm \sqrt{36 + 8}}{4}$$

$$= \frac{-6 \pm 2\sqrt{11}}{4} \qquad\qquad\qquad = \frac{-6 \pm \sqrt{44}}{4}$$

$$= \frac{-3 \pm \sqrt{11}}{2} \qquad\qquad\qquad = \frac{-3 \pm \sqrt{11}}{2} \approx 0.1583, -3.1583$$

$$\cos x \approx 0.1583 \qquad\qquad \text{or} \qquad \cos x \approx -3.1583$$

$$x \approx 1.4112, 4.8714 \qquad\qquad \text{(No solution)}$$

38. The graph of $y = 4\sin^3 x - 2\sin x - 1$ has two x-intercepts: $x = 1.086$ and $x = 2.056$.

40. The graph of $y = \dfrac{\cos x \cot x}{1 - \sin x} - 3$ has two x-intercepts: $x = 0.524$ and $x = 2.618$.

42. The graph of $y = x \cos x - 1$ has one x-intercept: $x = 4.917$.

44. The graph of $y = \csc^2 x + 0.5 \cot x - 5$ has four x-intercepts: $x = 0.515$, $x = 2.726$, $x = 3.657$, and $x = 5.868$.

46. The graph of $y = 12 \cos^2 x + 5 \cos x - 3$ has four x-intercepts: $x = 0.723$, $x = 1.231$, $x = 5.052$, and $x = 5.561$.

48. The graph of $y = 3 \tan^2 x + 4 \tan x - 4$ has four x-intercepts: $x = 0.588$, $x = 2.034$, $x = 3.730$, and $x = 5.561$.

50. The graph of $y = 4 \cos^2 x - 4 \cos x - 1$ has two intercepts: $x = 1.779$ and $x = 4.504$.

52. (a) $f(x) = 2 \sin x + \cos 2x$

   Maxima at $(0.524, 1.5)$, $(2.620, 1.5)$

   Minima at $(1.571, 1.0)$, $(4.712, -3.0)$

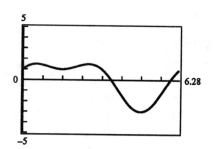

   (b) The graph of $y = 2 \cos x - 4 \sin x \cos x$ has the same x-intercepts: 0.524, 2.620, 1.571, 4.712.

54. (a) The equation $\sin \dfrac{1}{x} = 0$ has an infinite number of solutions in the interval $[-1, 1]$. You can see the oscillating behavior by using Xmin $= -0.1$ and Xmax $= 0.1$. In fact, any x of the form $x = \dfrac{1}{n\pi}$ yields a solution of $\sin \dfrac{1}{x} = 0$.

   (b) Yes, the greatest solution is $x = 0.318$, as can be seen by graphing y on the interval $0.2 \le x \le 20$.

56. $S = 74.50 + 43.75 \sin \dfrac{\pi t}{6}$, Graphing S and $y = 100$ in the same viewing rectangle, we see that $S > y$ during February, March and April; or, we can make a table of values.

y	1	2	3	4	5	6	7	8	9	10	11	12
S	96.4	112.4	118.3	112.4	96.4	74.5	52.6	36.6	30.8	36.6	52.6	74.5

Sales exceed 100,000 units during February, March, and April.

58. $y = 1.56e^{-0.22t} \cos 4.9t$

By graphing $y = 1$ and $y = -1$, we see that the displacement does not exceed 1 when $t > 1.958$.

60. $r = \frac{1}{32}v_0{}^2 \sin 2\theta$, $r = 3000$ ft, $v_0 = 1200$

$$3000 = \frac{1}{32}(1200)^2 \sin 2\theta$$

$$\sin 2\theta \approx 0.06667$$

$2\theta \approx 0.06672$ $\qquad$ or $\quad 2\theta \approx \pi - 0.06672 \approx 3.07487$

$\theta \approx 0.03336 \approx 1.9°$ $\qquad$ $\theta \approx 1.53744 \approx 88.1°$

62. $f(x) = e^{-x} = x$

Solving $e^{-x} - x = 0$, we see that $x = 0.567$.

64. $f(x) = \dfrac{5x + 2 - x^3}{5} = x$

Solving $\dfrac{5x + 2 - x^3}{5} - x = 0$, we see that 1.260.

■ 6.4 Sum and Difference Formulas

2. $\sin 15° = \sin(45° - 30°) = \sin 45° \cos 30° - \sin 30° \cos 45°$

$$= \frac{\sqrt{2}}{2} \cdot \frac{\sqrt{3}}{2} - \frac{1}{2} \cdot \frac{\sqrt{2}}{2} = \frac{\sqrt{2}}{4}(\sqrt{3} - 1)$$

$\cos 15° = \cos(45° - 30°) = \cos 45° \cos 30° + \sin 45° \sin 30°$

$$= \frac{\sqrt{2}}{2} \cdot \frac{\sqrt{3}}{2} + \frac{\sqrt{2}}{2} \cdot \frac{1}{2} = \frac{\sqrt{2}}{4}(\sqrt{3} + 1)$$

$$\tan 15° = \tan(45° - 30°) = \frac{\tan 45° - \tan 30°}{1 + \tan 45° \tan 30°} = \frac{1 - (\sqrt{3}/3)}{1 + (\sqrt{3}/3)} = 2 - \sqrt{3}$$

4. $\sin 165° = \sin(135° + 30°) = \sin 135° \cos 30° + \sin 30° \cos 135°$

$$= \sin 45° \cos 30° - \sin 30° \cos 45° = \frac{\sqrt{2}}{2} \cdot \frac{\sqrt{3}}{2} - \frac{1}{2} \cdot \frac{\sqrt{2}}{2} = \frac{\sqrt{2}}{4}(\sqrt{3} - 1)$$

$\cos 165° = \cos(135° + 30°) = \cos 135° \cos 30° - \sin 135° \sin 30°$

$$= -\cos 45° \cos 30° - \sin 45° \sin 30° = -\frac{\sqrt{2}}{2} \cdot \frac{\sqrt{3}}{2} - \frac{\sqrt{2}}{2} \cdot \frac{1}{2} = -\frac{\sqrt{2}}{4}(\sqrt{3} + 1)$$

$$\tan 165° = \tan(135° + 30°) = \frac{\tan 135° + \tan 30°}{1 - \tan 135° \tan 30°}$$

$$= \frac{-\tan 45° + \tan 30°}{1 + \tan 45° \tan 30°} = \frac{-1 + (\sqrt{3}/3)}{1 + (\sqrt{3}/3)} = -2 + \sqrt{3}$$

6. $\sin 255° = \sin(300° - 45°) = \sin 300° \cos 45° - \sin 45° \cos 300°$

$$= -\sin 60° \cos 45° - \sin 45° \cos 60° = -\frac{\sqrt{3}}{2} \cdot \frac{\sqrt{2}}{2} - \frac{\sqrt{2}}{2} \cdot \frac{1}{2} = -\frac{\sqrt{2}}{4}(\sqrt{3} + 1)$$

$\cos 255° = \cos(300° - 45°) = \cos 300° \cos 45° + \sin 300° \sin 45°$

$$= \cos 60° \cos 45° - \sin 60° \sin 45° = \frac{1}{2} \cdot \frac{\sqrt{2}}{2} - \frac{\sqrt{3}}{2} \cdot \frac{\sqrt{2}}{2} = \frac{\sqrt{2}}{4}(1 - \sqrt{3})$$

$\tan 255° = \tan(300° - 45°) = \dfrac{\tan 300° - \tan 45°}{1 + \tan 300° \tan 45°}$

$$= \frac{-\tan 60° - \tan 45°}{1 - \tan 60° \tan 45°} = \frac{-\sqrt{3} - 1}{1 - \sqrt{3}} = 2 + \sqrt{3}$$

8. $\sin \dfrac{7\pi}{12} = \sin\left(\dfrac{\pi}{3} + \dfrac{\pi}{4}\right) = \sin \dfrac{\pi}{3} \cos \dfrac{\pi}{4} + \sin \dfrac{\pi}{4} \cos \dfrac{\pi}{3}$

$$= \frac{\sqrt{3}}{2} \cdot \frac{\sqrt{2}}{2} + \frac{\sqrt{2}}{2} \cdot \frac{1}{2} = \frac{\sqrt{2}}{4}(\sqrt{3} + 1)$$

$\cos \dfrac{7\pi}{12} = \cos\left(\dfrac{\pi}{3} + \dfrac{\pi}{4}\right) = \cos \dfrac{\pi}{3} \cos \dfrac{\pi}{4} - \sin \dfrac{\pi}{3} \sin \dfrac{\pi}{4}$

$$= \frac{1}{2} \cdot \frac{\sqrt{2}}{2} - \frac{\sqrt{3}}{2} \cdot \frac{\sqrt{2}}{2} = \frac{\sqrt{2}}{4}(1 - \sqrt{3})$$

$\tan \dfrac{7\pi}{12} = \tan\left(\dfrac{\pi}{3} + \dfrac{\pi}{4}\right) = \dfrac{\tan(\pi/3) + \tan(\pi/4)}{1 - \tan(\pi/3)\tan(\pi/4)}$

$$= \frac{\sqrt{3} + 1}{1 - \sqrt{3}} = -2 - \sqrt{3}$$

10. $\sin\left(-\dfrac{\pi}{12}\right) = \sin\left(\dfrac{\pi}{6} - \dfrac{\pi}{4}\right) = \sin \dfrac{\pi}{6} \dfrac{\cos}{\pi} 4 - \dfrac{\sin}{\pi} 4 \dfrac{\cos}{\pi} 6$

$$= \frac{1}{2} \cdot \frac{\sqrt{2}}{2} - \frac{\sqrt{2}}{2} \cdot \frac{\sqrt{3}}{2} = \frac{\sqrt{2}}{4}(1 - \sqrt{3})$$

$\cos\left(-\dfrac{\pi}{12}\right) = \cos\left(\dfrac{\pi}{6} - \dfrac{\pi}{4}\right) = \cos \dfrac{\pi}{6} \cos \dfrac{\pi}{4} + \sin \dfrac{\pi}{6} \sin \dfrac{\pi}{4}$

$$= \frac{\sqrt{3}}{2} \cdot \frac{\sqrt{2}}{2} + \frac{1}{2} \cdot \frac{\sqrt{2}}{2} = \frac{\sqrt{2}}{4}(\sqrt{3} + 1)$$

$\tan\left(-\dfrac{\pi}{12}\right) = \tan\left(\dfrac{\pi}{6} - \dfrac{\pi}{4}\right) = \dfrac{\tan(\pi/6) - \tan(\pi/4)}{1 + \tan(\pi/6)\tan(\pi/4)}$

$$= \frac{(\sqrt{3}/3) - 1}{1 + (\sqrt{3}/3)} = -2 + \sqrt{3}$$

12. $\sin 140° \cos 50° + \cos 140° \sin 50° = \sin(140° + 50°) = \sin 190°$

14. $\cos 20° \cos 30° + \sin 20° \sin 30° = \cos(30° - 20°) = \cos 10°$

16. $\dfrac{\tan 140° - \tan 60°}{1 + \tan 140° \tan 60°} = \tan(140° - 60°) = \tan 80°$

18. $\cos \dfrac{\pi}{7} \cos \dfrac{\pi}{5} - \sin \dfrac{\pi}{7} \sin \dfrac{\pi}{5} = \cos\left(\dfrac{\pi}{7} + \dfrac{\pi}{5}\right) = \cos \dfrac{12\pi}{35}$

20. $\cos 3x \cos 2y + \sin 3x \sin 2y = \cos(3x - 2y)$

Figures for Exercises 22–24

22. $\cos(v - u) = \cos v \cos u + \sin v \sin u$

$$= \left(-\tfrac{3}{5}\right)\left(\tfrac{12}{13}\right) + \left(\tfrac{4}{5}\right)\left(\tfrac{5}{13}\right)$$

$$= \tfrac{-36}{65} + \tfrac{20}{65}$$

$$= -\tfrac{16}{65}$$

24. $\sin(u - v) = \sin u \cos v - \cos u \sin v$

$$= \tfrac{-15}{65} - \tfrac{48}{65}$$

$$= -\tfrac{63}{65}$$

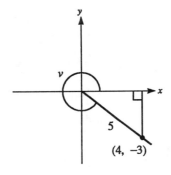

Figures for Exercises 26–28

26. $\sin(u + v) = \sin u \cos v + \cos u \sin v$

$$= \left(\tfrac{7}{25}\right)\left(\tfrac{4}{5}\right) + \left(-\tfrac{24}{25}\right)\left(-\tfrac{3}{5}\right)$$

$$= \tfrac{28}{125} + \tfrac{72}{125}$$

$$= \tfrac{100}{125}$$

$$= \tfrac{4}{5}$$

28. $\cos(u - v) = \cos u \cos v + \sin u \sin v$

$$= \left(-\tfrac{24}{25}\right)\left(\tfrac{4}{5}\right) + \left(\tfrac{7}{25}\right)\left(-\tfrac{3}{5}\right)$$

$$= \tfrac{-96}{125} - \tfrac{21}{125}$$

$$= -\tfrac{117}{125}$$

30. $\sin(3\pi - x) = \sin 3\pi \cos x - \sin x \cos 3\pi = (0)(\cos x) - (\sin x)(-1) = \sin x$

32. $\cos\left(\dfrac{5\pi}{4} - x\right) = \cos\dfrac{5\pi}{4}\cos x + \sin\dfrac{5\pi}{4}\sin x = -\dfrac{\sqrt{2}}{2}(\cos x + \sin x)$

34. $\tan\left(\dfrac{\pi}{4} - \theta\right) = \dfrac{\tan(\pi/4) - \tan\theta}{1 + \tan(\pi/4)\tan\theta} = \dfrac{1 - \tan\theta}{1 + \tan\theta}$

36. $\sin(x+y)\sin(x-y) = (\sin x \cos y + \sin y \cos x)(\sin x \cos y - \sin y \cos x)$

$$= \sin^2 x \cos^2 y - \sin^2 y \cos^2 x$$

$$= \sin^2 x (1 - \sin^2 y) - \sin^2 y \cos^2 x$$

$$= \sin^2 x - \sin^2 x \sin^2 y - \sin^2 y \cos^2 x$$

$$= \sin^2 x - \sin^2 y (\sin^2 x + \cos^2 x)$$

$$= \sin^2 x - \sin^2 y$$

38. $\cos(x+y) + \cos(x-y) = \cos x \cos y - \sin x \sin y + \cos x \cos y + \sin x \sin y$

$$= 2 \cos x \cos y$$

40. $\sin(n\pi + \theta) = \sin n\pi \cos \theta + \sin \theta \cos n\pi$

$$= (0)(\cos \theta) + (\sin \theta)(-1)^n$$

$$= (-1)^n (\sin \theta), \text{ where } n \text{ is an integer}$$

42. $C = \arctan \dfrac{a}{b} \Rightarrow \sin C = \dfrac{a}{\sqrt{a^2 + b^2}}, \quad \cos C = \dfrac{b}{\sqrt{a^2 + b^2}}$

$$\sqrt{a^2 + b^2} \cos(B\theta - C) = \sqrt{a^2 + b^2} \left(\cos B\theta \cdot \frac{b}{\sqrt{a^2 + b^2}} + \sin B\theta \cdot \frac{a}{\sqrt{a^2 + b^2}} \right)$$

$$= b \cos B\theta + a \sin B\theta$$

$$= a \sin B\theta + b \cos B\theta$$

44. $y_1 = \cos(\pi + x)$

$y_2 = -\cos x$

The graphs of y_1 and y_2 coincide.

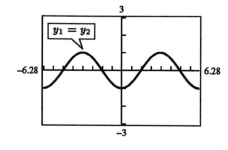

46. $y_1 = \tan(\pi + \theta)$

$y_2 = \tan \theta$

The graphs of y_1 and y_2 coincide.

48. $3\sin 2\theta + 4\cos 2\theta$

$a = 3, \quad b = 4, \quad B = 2$

(a) $C = \arctan \dfrac{b}{a} = \arctan \dfrac{4}{3} \approx 0.9273$

$$3\sin 2\theta + 4\cos 2\theta = \sqrt{a^2 + b^2}\,\sin(B\theta + C)$$
$$\approx 5\sin(2\theta + 0.9273)$$

(b) $C = \arctan \dfrac{a}{b} = \arctan \dfrac{3}{4} \approx 0.6435$

$$3\sin 2\theta + 4\cos 2\theta = \sqrt{a^2 + b^2}\,\cos(B\theta - C)$$
$$\approx 5\cos(2\theta - 0.6435)$$

50. $\sin 2\theta - \cos 2\theta$

$a = 1, \quad b = -1, \quad B = 2$

(a) $C = \arctan \dfrac{b}{a} = \arctan(-1) = -\dfrac{\pi}{4}$

$$\sin 2\theta - \cos \theta = \sqrt{a^2 + b^2}\,\sin(B\theta + C)$$
$$= \sqrt{2}\,\sin\left(2\theta - \dfrac{\pi}{4}\right)$$

(b) $C = \arctan \dfrac{a}{b} = \arctan(-1) = -\dfrac{\pi}{4}$

$$\sin 2\theta - \cos 2\theta = \sqrt{a^2 + b^2}\,\cos(B\theta - C)$$
$$= \sqrt{2}\,\cos\left(2\theta + \dfrac{\pi}{4}\right)$$

52. $C = \arctan \dfrac{a}{b} = -\dfrac{3\pi}{4} \Rightarrow a = b, \quad a < 0, \quad b < 0$

$\sqrt{a^2 + b^2} = 5 \Rightarrow a = b = \dfrac{-5\sqrt{2}}{2}$

$B = 1$

$5\cos\left(\theta + \dfrac{3\pi}{4}\right) = -\dfrac{5\sqrt{2}}{2}\sin\theta - \dfrac{5\sqrt{2}}{2}\cos\theta$

54. $\sin(\arctan 2x - \arccos x) = \sin(\arctan 2x)\cos(\arccos x) - \sin(\arccos x)\cos(\arctan 2x)$

$$= \left(\dfrac{2x}{\sqrt{1 + 4x^2}}\right)(x) - \left(\sqrt{1 - x^2}\right)\left(\dfrac{1}{\sqrt{1 + 4x^2}}\right)$$
$$= \dfrac{2x^2}{\sqrt{1 + 4x^2}} - \dfrac{\sqrt{1 - x^2}}{\sqrt{1 + 4x^2}}$$
$$= \dfrac{2x^2 - \sqrt{1 - x^2}}{\sqrt{1 + 4x^2}}$$

$\theta = \arctan 2x$

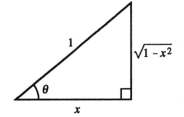

$\theta = \arccos x$

56.

$$\sin\left(x+\frac{\pi}{6}\right) - \sin\left(x-\frac{\pi}{6}\right) = \frac{1}{2}$$

$$\sin x \cos\frac{\pi}{6} + \cos x \sin\frac{\pi}{6} - \left(\sin x \cos\frac{\pi}{6} - \cos x \sin\frac{\pi}{6}\right) = \frac{1}{2}$$

$$2\cos x(0.5) = \frac{1}{2}$$

$$\cos x = \frac{1}{2}$$

$$x = \frac{\pi}{3},\ \frac{5\pi}{3}$$

58.

$$\tan(x+\pi) - \cos\left(x+\frac{\pi}{2}\right) = 0$$

$$\frac{\tan x + \tan\pi}{1 - \tan x \tan\pi} - \left(\cos x \cos\frac{\pi}{2} - \sin x \sin\frac{\pi}{2}\right) = 0$$

$$\frac{\tan x + 0}{1 - \tan x(0)} - (\cos x)(0) + (\sin x)(1) = 0$$

$$\tan x + \sin x = 0$$

$$\frac{\sin x}{\cos x} = -\sin x$$

$$\sin x = -\sin x \cos x$$

$$\sin x(1 + \cos x) = 0$$

$$\sin x = 0 \qquad \text{or} \qquad \cos x = -1$$

$$x = 0,\ \pi \qquad\qquad x = \pi$$

60. $\tan(x+\pi) + 2\sin(x+\pi) = 0$

Graphing $y = \tan(x+\pi) + 2\sin(x+\pi)$, we see that there are four x-intercepts in the interval $[0,\ 2\pi)$: $x = 0,\ 1.047,\ 3.142$ and 5.236.

62. $y = \dfrac{1}{3}\sin 2t + \dfrac{1}{4}\cos 2t$

(a) $a = \dfrac{1}{3}, \quad b = \dfrac{1}{4}, \quad B = 2$

$\quad C = \arctan \dfrac{b}{a} = \arctan \dfrac{3}{4} \approx 0.6435$

$\quad y \approx \sqrt{\left(\dfrac{1}{3}\right)^2 + \left(\dfrac{1}{4}\right)^2}\, \sin(2t + 0.6435)$

$\quad\quad = \dfrac{5}{12}\sin(2t + 0.6435)$

(b)
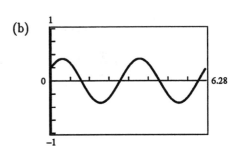

(c) Amplitude: $\dfrac{5}{12}$

(d) Frequency: $= \dfrac{1}{\text{period}} = \dfrac{b}{2\pi} = \dfrac{2}{2\pi} = \dfrac{1}{\pi}$

64. $\tan(u+v) = \dfrac{\sin(u+v)}{\cos(u+v)} = \dfrac{\sin u \cos v + \cos u \sin v}{\cos u \cos v - \sin u \sin v} = \dfrac{\dfrac{\sin u \cos v}{\cos u \cos v} + \dfrac{\cos u \sin v}{\cos u \cos v}}{\dfrac{\cos u \cos v}{\cos u \cos v} - \dfrac{\sin u \sin v}{\cos u \cos v}} = \dfrac{\tan u + \tan v}{1 - \tan u \tan v}$

6.5 Multiple–Angle and Product–to–Sum Formulas

2. Graphing the function on the interval $[0,\ 2\pi)$, we find the following zeros: 1.571, 3.665, 4.712, 5.760

Algebraically:

$\quad \sin 2x + \cos x = 0$

$\quad 2\sin x \cos x + \cos x = 0$

$\quad \cos x(2\sin x + 1) = 0$

$\cos x = 0 \qquad\quad \text{or}\quad 2\sin x + 1 = 0$

$x = \dfrac{\pi}{2},\ \dfrac{3\pi}{2} \qquad\qquad \sin x = -\dfrac{1}{2}$

$\qquad\qquad\qquad\qquad\quad x = \dfrac{7\pi}{6},\ \dfrac{11\pi}{6}$

$x = \dfrac{\pi}{2},\ \dfrac{7\pi}{6},\ \dfrac{3\pi}{2},\ \dfrac{11\pi}{6}$

4. Graphing the function on the interval $[0, 2\pi)$, we find the following zeros: 0.785, 1.571, 2.356, 3.927, 4.712, 5.498

Algebraically:

$$\sin 2x \sin x = \cos x$$

$$2\sin x \cos x \sin x = \cos x$$

$$2\sin^2 x \cos x - \cos x = 0$$

$$\cos x(2\sin^2 x - 1) = 0$$

$$\cos x = 0 \qquad \text{or} \quad 2\sin^2 x - 1 = 0$$

$$x = \frac{\pi}{2}, \frac{3\pi}{2} \qquad\qquad \sin x = \pm\frac{\sqrt{2}}{2}$$

$$x = \frac{\pi}{4}, \frac{3\pi}{4}, \frac{5\pi}{4}, \frac{7\pi}{4}$$

6. Graphing the function on the interval $[0, 2\pi)$, we find the following zeros: 1.571, 3.665, 5.760.

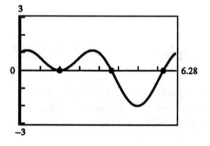

Algebraically:

$$\cos 2x + \sin x = 0$$

$$1 - 2\sin^2 x + \sin x = 0$$

$$2\sin^2 x - \sin x - 1 = 0$$

$$(2\sin x + 1)(\sin x - 1) = 0$$

$$2\sin x + 1 = 0 \qquad \text{or} \quad \sin x = 1$$

$$\sin x = -\frac{1}{2} \qquad\qquad x = \frac{\pi}{2}$$

$$x = \frac{7\pi}{6}, \frac{11\pi}{6}$$

$$x = \frac{\pi}{2}, \frac{7\pi}{6}, \frac{11\pi}{6}$$

8. Graphing the function on the interval $[0,\ 2\pi)$, we find the following zeros: 0.524, 1.571, 2.618, 4.712

Algebraically:

$$\tan 2x - 2\cos x = 0$$

$$\frac{2\tan x}{1 - \tan^2 x} = 2\cos x$$

$$2\tan x = 2\cos x(1 - \tan^2 x)$$

$$2\tan x = 2\cos x - 2\cos x \tan^2 x$$

$$2\tan x = 2\cos x - 2\cos x \frac{\sin^2 x}{\cos^2 x}$$

$$2\tan x = 2\cos x - 2\frac{\sin^2 x}{\cos x}$$

$$\tan x = \cos x - \frac{\sin^2 x}{\cos x}$$

$$\frac{\sin x}{\cos x} = \cos x - \frac{\sin^2 x}{\cos x}$$

$$\frac{\sin x}{\cos x} + \frac{\sin^2 x}{\cos x} - \cos x = 0$$

$$\frac{\sin x + \sin^2 x - \cos^2 x}{\cos x} = 0$$

$$\frac{1}{\cos x}\left[\sin x + \sin^2 x - (1 - \sin^2 x)\right] = 0$$

$$\sec x[2\sin^2 x + \sin x - 1] = 0$$

$$\sec x(2\sin x - 1)(\sin x + 1) = 0$$

$$\sec x = 0 \qquad \text{or} \quad 2\sin x - 1 = 0 \qquad \text{or} \quad \sin x + 1 = 0$$

$$x = \frac{\pi}{2},\ \frac{3\pi}{2} \qquad\qquad \sin x = \frac{1}{2} \qquad\qquad \sin x = -1$$

$$x = \frac{\pi}{6},\ \frac{5\pi}{6} \qquad\qquad x = \frac{3\pi}{2}$$

$$x = \frac{\pi}{6},\ \frac{\pi}{2},\ \frac{5\pi}{6},\ \frac{3\pi}{2}$$

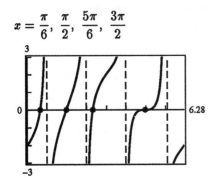

10. Graphing the function on the interval
$[0, \ 2\pi)$, we find the following zeros:
0.0, 0.785, 1.571, 2.356, 3.142, 3.927, 4.712, 5.498.

Algebraically:

$$(\sin 2s + \cos 2s)^2 = 1$$

$$\sin^2 s2 + 2\sin 2s \cos 2s + \cos^2 2s = 1$$

$$2\sin 2s \cos 2s = 0$$

$$\sin 4s = 0$$

$$4s = n\pi$$

$$s = \frac{n\pi}{4}$$

$$s = 0, \ \frac{\pi}{4}, \ \frac{\pi}{2},$$

$$\frac{3\pi}{4}, \ \pi, \ \frac{5\pi}{4},$$

$$\frac{3\pi}{2}, \ \frac{7\pi}{4}$$

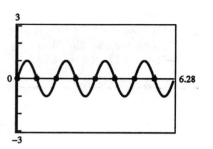

12. $g(x) = 4\sin x \cos x + 2$

$\qquad = 2(2\sin x \cos x) + 2$

$\qquad = 2\sin 2x + 2$

Relative maxima: $(0.785, 4), (3.927, 4)$

Relative minima: $(2.356, 0), (5.498, 0)$

14. $f(x) = (\cos x + \sin x)(\cos x - \sin x)$

$\qquad = \cos^2 x - \sin^2 x$

$\qquad = \cos 2x$

Relative maxima: $(0, 1), (3.142, 1)$

Relative minima: $(1.571, -1), (4.712, -1)$

16. $\sin 2u = 2 \sin u \cos u$

$$= 2 \cdot \frac{\sqrt{5}}{3}\left(-\frac{2}{3}\right) = -\frac{4\sqrt{5}}{9}$$

$$\cos 2u = \cos^2 u - \sin^2 u = \frac{4}{9} - \frac{5}{9} = -\frac{1}{9}$$

$$\tan 2u = \frac{2 \tan u}{1 - \tan^2 u} = \frac{2(-\sqrt{5}/2)}{1 - (5/4)} = 4\sqrt{5}$$

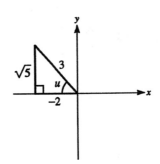

18. $\sin 2u = 2 \sin u \cos u$

$$= 2\left(-\frac{1}{\sqrt{17}}\right)\left(\frac{4}{\sqrt{17}}\right) = -\frac{8}{17}$$

$$\cos 2u = \cos^2 u - \sin^2 u$$

$$= \left(\frac{4}{\sqrt{17}}\right)^2 - \left(-\frac{1}{\sqrt{17}}\right)^2 = \frac{15}{17}$$

$$\tan 2u = \frac{2 \tan u}{1 - \tan^2 u} = \frac{2(-1/4)}{1 - (-1/4)^2} = -\frac{8}{15}$$

20. $\sin 2u = 2 \sin u \cos u$

$$= 2 \cdot \frac{1}{3}\left(-\frac{2\sqrt{2}}{3}\right) = -\frac{4\sqrt{2}}{9}$$

$$\cos 2u = \cos^2 u - \sin^2 u$$

$$= \left(-\frac{2\sqrt{2}}{3}\right)^2 - \left(\frac{1}{3}\right)^2 = \frac{7}{9}$$

$$\tan 2u = \frac{2 \tan u}{1 - \tan^2 u} = \frac{2(-\sqrt{2}/4)}{1 - (-\sqrt{2}/4)^2} = -\frac{4\sqrt{2}}{7}$$

22. $\sin^4 x = (\sin^2 x)(\sin^2 x)$

$$= \left(\frac{1 - \cos 2x}{2}\right)\left(\frac{1 - \cos 2x}{2}\right)$$

$$= \frac{1 - 2 \cos 2x + \cos^2 2x}{4}$$

$$= \frac{1 - 2 \cos 2x + \left(\dfrac{1 + \cos 4x}{2}\right)}{4}$$

$$= \frac{2 - 4 \cos 2x + 1 + \cos 4x}{8}$$

$$= \frac{1}{8}(3 - 4 \cos 2x + \cos 4x)$$

24. $\cos^6 x = (\cos^2 x)^3 = \left(\dfrac{1+\cos 2x}{2}\right)^3 = \dfrac{1}{8}(1 + 3\cos 2x + 3\cos^2 2x + \cos^3 2x)$

$$= \dfrac{1}{8}\left[1 + 3\cos 2x + 3 \cdot \dfrac{1+\cos 4x}{2} + \dfrac{3\cos 2x + \cos 6x}{4}\right]$$

$$= \dfrac{1}{32}(10 + 15\cos 2x + 6\cos 4x + \cos 6x)$$

26. $\sin^4 x \cos^2 x = \sin^2 x \sin^2 x \cos^2 x$

$$= \left(\dfrac{1-\cos 2x}{2}\right)\left(\dfrac{1-\cos 2x}{2}\right)\left(\dfrac{1+\cos 2x}{2}\right)$$

$$= \dfrac{1}{8}(1-\cos 2x)(1-\cos^2 2x)$$

$$= \dfrac{1}{8}(1 - \cos 2x - \cos^2 2x + \cos^3 2x)$$

$$= \dfrac{1}{8}\left[1 - \cos 2x - \left(\dfrac{1+\cos 4x}{2}\right) + \cos 2x\left(\dfrac{1+\cos 4x}{2}\right)\right]$$

$$= \dfrac{1}{16}[2 - 2\cos 2x - 1 - \cos 4x + \cos 2x + \cos 2x \cos 4x]$$

$$= \dfrac{1}{16}\left[1 - \cos 2x - \cos 4x + \dfrac{1}{2}\cos 2x + \dfrac{1}{2}\cos 6x\right]$$

$$= \dfrac{1}{32}[2 - 2\cos 2x - 2\cos 4x + \cos 2x + \cos 6x]$$

$$= \dfrac{1}{32}[2 - \cos 2x - 2\cos 4x + \cos 6x]$$

28. $\sin 165° = \sin\left(\dfrac{1}{2} \cdot 330°\right) = \sqrt{\dfrac{1-\cos 330°}{2}} = \sqrt{\dfrac{1-(\sqrt{3}/2)}{2}} = \dfrac{1}{2}\sqrt{2-\sqrt{3}}$

$\cos 165° = \cos\left(\dfrac{1}{2} \cdot 330°\right) = -\sqrt{\dfrac{1+\cos 330°}{2}} = -\sqrt{\dfrac{1+(\sqrt{3}/2)}{2}} = -\dfrac{1}{2}\sqrt{2+\sqrt{3}}$

$\tan 165° = \tan\left(\dfrac{1}{2} \cdot 330°\right) = \dfrac{\sin 330°}{1+\cos 330°} = \dfrac{-1/2}{1+(\sqrt{3}/2)} = \dfrac{-1}{2+\sqrt{3}} = \sqrt{3} - 2$

30. $\sin 67°30' = \sin\left(\dfrac{1}{2} \cdot 135°\right) = \sqrt{\dfrac{1-\cos 135°}{2}} = \sqrt{\dfrac{1+(\sqrt{2}/2)}{2}} = \dfrac{1}{2}\sqrt{2+\sqrt{2}}$

$\cos 67°30' = \cos\left(\dfrac{1}{2} \cdot 135°\right) = \sqrt{\dfrac{1+\cos 135°}{2}} = \sqrt{\dfrac{1-(\sqrt{2}/2)}{2}} = \dfrac{1}{2}\sqrt{2-\sqrt{2}}$

$\tan 67°30' = \tan\left(\dfrac{1}{2} \cdot 135°\right) = \dfrac{\sin 135°}{1+\cos 135°} = \dfrac{\sqrt{2}/2}{1-(\sqrt{2}/2)} = 1+\sqrt{2}$

32. $\sin\dfrac{\pi}{12} = \sin\left[\dfrac{1}{2}\left(\dfrac{\pi}{6}\right)\right] = \sqrt{\dfrac{1-\cos(\pi/6)}{2}} = \sqrt{\dfrac{1-(\sqrt{3}/2)}{2}} = \dfrac{1}{2}\sqrt{2-\sqrt{3}}$

$\cos\dfrac{\pi}{12} = \cos\left[\dfrac{1}{2}\left(\dfrac{\pi}{6}\right)\right] = \sqrt{\dfrac{1+\cos(\pi/6)}{2}} = \dfrac{1}{2}\sqrt{2+\sqrt{3}}$

$\tan\dfrac{\pi}{12} = \tan\left[\dfrac{1}{2}\left(\dfrac{\pi}{6}\right)\right] = \dfrac{\sin(\pi/6)}{1+\cos(\pi/6)} = \dfrac{1/2}{1+(\sqrt{3}/2)} = 2-\sqrt{3}$

34. $\sin\left(\dfrac{u}{2}\right) = \sqrt{\dfrac{1-\cos u}{2}} = \sqrt{\dfrac{1-(3/5)}{2}} = \dfrac{\sqrt{5}}{5}$

$\cos\left(\dfrac{u}{2}\right) = \sqrt{\dfrac{1+\cos u}{2}} = \sqrt{\dfrac{1+(3/5)}{2}} = \dfrac{2\sqrt{5}}{5}$

$\tan\left(\dfrac{u}{2}\right) = \dfrac{\sin u}{1+\cos u} = \dfrac{4/5}{1+(3/5)} = \dfrac{1}{2}$

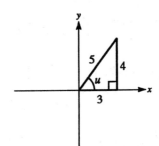

36. $\sin\left(\dfrac{u}{2}\right) = \sqrt{\dfrac{1-\cos u}{2}} = \sqrt{\dfrac{1+(3/\sqrt{10})}{2}} = \sqrt{\dfrac{10+3\sqrt{10}}{20}}$

$\cos\left(\dfrac{u}{2}\right) = -\sqrt{\dfrac{1+\cos u}{2}} = -\sqrt{\dfrac{1-(3/\sqrt{10})}{2}} = -\sqrt{\dfrac{10-3\sqrt{10}}{20}}$

$\tan\left(\dfrac{u}{2}\right) = \dfrac{1-\cos u}{\sin u} = \dfrac{1+(3/\sqrt{10})}{-1/\sqrt{10}} = -3-\sqrt{10}$

38. $\sin\left(\dfrac{u}{2}\right) = \sqrt{\dfrac{1-\cos u}{2}} = \sqrt{\dfrac{1+(2/7)}{2}} = \dfrac{3\sqrt{14}}{14}$

$\cos\left(\dfrac{u}{2}\right) = \sqrt{\dfrac{1+\cos u}{2}} = \sqrt{\dfrac{1-(2/7)}{2}} = \dfrac{\sqrt{70}}{14}$

$\tan\left(\dfrac{u}{2}\right) = \dfrac{1-\cos u}{\sin u} = \dfrac{1+(2/7)}{3(\sqrt{5}/7)} = \dfrac{3\sqrt{5}}{5}$

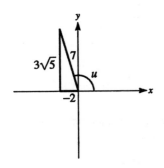

40. $\sqrt{\dfrac{1 + \cos 4x}{2}} = |\cos 2x|$

42. $-\sqrt{\dfrac{1 - \cos(x - 1)}{2}} = -\left|\sin\left(\dfrac{x - 1}{2}\right)\right|$

44. Graphing the function on the interval $[0,\ 2\pi)$, we find the following zeros: 0.0, 1.047, 5.236.

Algebraically:

$$\sin\frac{x}{2} + \cos x = 1$$

$$\pm\sqrt{\frac{1 - \cos x}{2}} = 1 - \cos x$$

$$\frac{1 - \cos x}{2} = 1 - 2\cos x + \cos^2 x$$

$$1 - \cos x = 2 - 4\cos x + 2\cos^2 x$$

$$2\cos^2 x - 3\cos x + 1 = 0$$

$$(2\cos x - 1)(\cos x - 1) = 0$$

$$2\cos x - 1 = 0 \qquad \text{or } \cos x - 1 = 0$$
$$\cos x = \frac{1}{2} \qquad\qquad \cos x = 1$$
$$x = \frac{\pi}{3},\ \frac{5\pi}{3} \qquad\qquad x = 0$$

0, $\pi/3$, and $5\pi/3$ are all solutions to the equation.

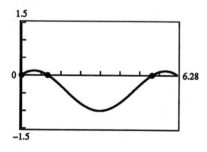

46. Graphing the function on the interval $[0,\ 2\pi)$, we find the following zeros: 0.0, 1.571, 4.712.

Algebraically:

$$\tan\frac{x}{2} - \sin x = 0$$

$$\frac{1 - \cos x}{\sin x} = \sin x$$

$$1 - \cos x = \sin^2 x$$

$$1 - \cos x = 1 - \cos^2 x$$

$$\cos^2 x - \cos x = 0$$

$$\cos x(\cos x - 1) = 0$$

$$\cos x = 0 \qquad \text{or } \cos x - 1 = 0$$
$$x = \frac{\pi}{2},\ \frac{3\pi}{2} \qquad\qquad \cos x = 1$$
$$\qquad\qquad\qquad x = 0$$

0, $\pi/2$, and $3\pi/2$ are all solutions to the equation.

48. $4\sin\dfrac{\pi}{3}\cos\dfrac{5\pi}{6} = 4 \cdot \dfrac{1}{2}\left[\sin\left(\dfrac{\pi}{3} + \dfrac{5\pi}{6}\right) + \sin\left(\dfrac{\pi}{3} - \dfrac{5\pi}{6}\right)\right]$

$\qquad = 2\left[\sin\dfrac{7\pi}{6} + \sin\left(-\dfrac{\pi}{2}\right)\right] = 2\left(\sin\dfrac{7\pi}{6} - \sin\dfrac{\pi}{2}\right)$

50. $3\sin 2\alpha \sin 3\alpha = 3 \cdot \frac{1}{2}[\cos(2\alpha - 3\alpha) - \cos(2\alpha + 3\alpha)] = \frac{3}{2}[\cos(-\alpha) - \cos 5\alpha] = \frac{3}{2}(\cos\alpha - \cos 5\alpha)$

52. $\cos 2\theta \cos 4\theta = \frac{1}{2}[\cos(2\theta - 4\theta) + \cos(2\theta + 4\theta)] = \frac{1}{2}[\cos(-2\theta) + \cos 6\theta] = \frac{1}{2}(\cos 2\theta + \cos 6\theta)$

54. $\sin(x + y)\cos(x - y) = \frac{1}{2}(\sin 2x + \sin 2y)$

56. $10\cos 75° \cos 15° = 10 \cdot \frac{1}{2}(\cos 60° + \cos 90°)$

58. $\cos 120° + \cos 30° = 2\cos\left(\dfrac{120° + 30°}{2}\right)\cos\left(\dfrac{120° - 30°}{2}\right) = 2\cos 75° \cos 45°$

60. $\sin 5\theta - \sin 3\theta = 2\cos\left(\dfrac{5\theta + 3\theta}{2}\right)\sin\left(\dfrac{5\theta - 3\theta}{2}\right) = 2\cos 4\theta \sin \theta$

62. $\sin x + \sin 5x = 2\sin\left(\dfrac{x + 5x}{2}\right)\cos\left(\dfrac{x - 5x}{2}\right) = 2\sin 3x \cos(-2x) = 2\sin 3x \cos 2x$

64. $\cos\left(\theta + \dfrac{\pi}{2}\right) - \cos\left(\theta - \dfrac{\pi}{2}\right) = -2\sin\left(\dfrac{\theta + (\pi/2) + \theta - (\pi/2)}{2}\right)$

$\sin\left(\dfrac{\theta + (\pi/2) - \theta + (\pi/2)}{2}\right) = -2\sin 2\theta \sin\dfrac{\pi}{2}$

66. $\sin\left(x + \dfrac{\pi}{2}\right) + \sin\left(x - \dfrac{\pi}{2}\right) = 2\sin\left(\dfrac{x + (\pi/2) + x - (\pi/2)}{2}\right)\cos\left(\dfrac{x + (\pi/2) - x + (\pi/2)}{2}\right)$

$= 2\sin x \cos\dfrac{\pi}{2} = 0$

68. Graphing the function on the interval $[0,\ 2\pi)$, we find the following zeros: 0.0, 0.785, 1.571, 2.356, 3.142, 3.927, 4.712, 5.498.

Algebraically:

$$\cos 2x - \cos 6x = 0$$

$$-2\sin 4x \sin(-2x) = 0$$

$$2\sin 4x \sin 2x = 0$$

$\sin 4x = 0$	or $\sin 2x = 0$
$4x = n\pi$	$2x = n\pi$
$x = \dfrac{n\pi}{4}$	$x = \dfrac{n\pi}{2}$
$x = 0,\ \dfrac{\pi}{4},\ \dfrac{\pi}{2},\ \dfrac{3\pi}{4},\ \pi,\ \dfrac{5\pi}{4},\ \dfrac{3\pi}{2},\ \dfrac{7\pi}{4}$	$x = 0,\ \dfrac{\pi}{2},\ \pi,\ \dfrac{3\pi}{2}$

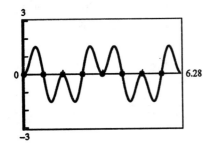

70. Graphing the function on the interval $[0, 2\pi)$, we find the following zeros: 0.0, 3.142.

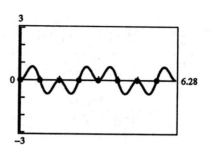

Algebraically:

$$\sin^2 3x - \sin^2 x = 0$$

$$(\sin 3x + \sin x)(\sin 3x - \sin x) = 0$$

$$(2\sin 2x \cos x)(2\cos 2x \sin x) = 0$$

$$\sin 2x = 0 \Rightarrow x = 0, \frac{\pi}{2}, \pi, \frac{3\pi}{2} \quad \text{or}$$

$$\cos x = 0 \Rightarrow x = \frac{\pi}{2}, \frac{3\pi}{2} \quad \text{or}$$

$$\cos 2x = 0 \Rightarrow x = \frac{\pi}{4}, \frac{3\pi}{4}, \frac{5\pi}{4}, \frac{7\pi}{4} \quad \text{or}$$

$$\sin x = 0 \Rightarrow x = 0, \pi$$

72.
$$\sec 2\theta = \frac{1}{\cos 2\theta} = \frac{1}{\cos^2 \theta - \sin^2 \theta}$$
$$= \frac{1/\cos^2 \theta}{1 - (\sin^2 \theta / \cos^2 \theta)}$$
$$= \frac{\sec^2 \theta}{1 - \tan^2 \theta}$$
$$= \frac{\sec^2 \theta}{1 - (\sec^2 \theta - 1)}$$
$$= \frac{\sec^2 \theta}{2 - \sec^2 \theta}$$

74.
$$\cos^4 x - \sin^4 x = (\cos^2 x - \sin^2 x)(\cos^2 x + \sin^2 x)$$
$$= (\cos 2x)(1)$$
$$= \cos 2x$$

76.
$$\sin\left(\frac{\alpha}{3}\right)\cos\left(\frac{\alpha}{3}\right) = \frac{1}{2}\left[2\left(\sin\left(\frac{\alpha}{3}\right)\cos\left(\frac{\alpha}{3}\right)\right)\right]$$
$$= \frac{1}{2}\sin\frac{2\alpha}{3}$$

78.
$$\sin 4\beta = 2\sin 2\beta \cos 2\beta$$
$$= 2[2\sin\beta\cos\beta(\cos^2\beta - \sin^2\beta)]$$
$$= 2[2\sin\beta\cos\beta(1 - \sin^2\beta - \sin^2\beta)]$$
$$= 4\sin\beta\cos\beta(1 - 2\sin^2\beta)$$

80.
$$\frac{\cos 3\beta}{\cos\beta} = \frac{\cos^3\beta - 3\sin^2\beta\cos\beta}{\cos\beta}$$
$$= \cos^2\beta - 3\sin^2\beta$$
$$= 1 - \sin^2\beta - 3\sin^2\beta$$
$$= 1 - 4\sin^2\beta$$

82.
$$\tan\frac{u}{2} = \frac{1 - \cos u}{\sin u}$$
$$= \frac{1}{\sin u} - \frac{\cos u}{\sin u}$$
$$= \csc u - \cot u$$

84. $\dfrac{\cos 3x - \cos x}{\sin 3x - \sin x} = \dfrac{-2\sin\left(\dfrac{3x+x}{2}\right)\sin\left(\dfrac{3x-x}{2}\right)}{2\cos\left(\dfrac{3x+x}{2}\right)\sin\left(\dfrac{3x-x}{2}\right)}$

$\qquad\qquad\qquad = \dfrac{-2\sin 2x \sin x}{2\cos 2x \sin x}$

$\qquad\qquad\qquad = -\tan 2x$

86. $\dfrac{\sin x \pm \sin y}{\cos x + \cos y} = \dfrac{2\sin\left(\dfrac{x\pm y}{2}\right)\cos\left(\dfrac{x\mp y}{2}\right)}{2\cos\left(\dfrac{x+y}{2}\right)\cos\left(\dfrac{x-y}{2}\right)}$

$\qquad\qquad\qquad = \tan\left(\dfrac{x\pm y}{2}\right)$

88. $\sin\left(\dfrac{\pi}{6}+x\right) + \sin\left(\dfrac{\pi}{6}-x\right) = 2\sin\dfrac{\pi}{6}\cos x$

$\qquad\qquad\qquad\qquad\qquad = 2 \cdot \dfrac{1}{2}\cos x$

$\qquad\qquad\qquad\qquad\qquad = \cos x$

90. $\cos^2 x = \dfrac{1 + \cos 2x}{2} = \dfrac{1}{2} + \dfrac{\cos 2x}{2}$

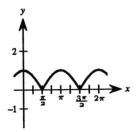

92. $\cos(2\arccos x) = \cos^2(\arccos x) - \sin^2(\arccos x)$

$\qquad\qquad\qquad = x^2 - (1 - x^2) = 2x^2 - 1$

94. $\frac{1}{2}[\cos(u-v) + \cos(u+v)]$

$\qquad = \frac{1}{2}[(\cos u \cos v + \sin u \sin v)$

$\qquad\qquad + (\cos u \cos v - \sin u \sin v)]$

$\qquad = \frac{1}{2}(2\cos u \cos v)$

$\qquad = \cos u \cos v$

96. $r = \frac{1}{32}v_0{}^2 \sin 2\theta$

$\qquad r = \frac{1}{32}v_0{}^2\, 2\sin\theta\cos\theta$

$\qquad r = \frac{1}{16}v_0{}^2 \sin\theta\cos\theta$

6.6 Law of Sines

2. Given: $B = 45°$, $C = 120°$, $c = 20$

$$A = 180° - B - C = 15°$$

$$a = \frac{c}{\sin C}(\sin A) = \frac{20 \sin 15°}{\sin 120°} \approx 5.977$$

$$b = \frac{c}{\sin C}(\sin B) = \frac{20 \sin 45°}{\sin 120°} \approx 16.33$$

4. Given: $B = 10°$, $C = 135°$, $c = 60$

$$A = 180° - B - C = 35°$$

$$a = \frac{c}{\sin C}(\sin A) = \frac{60}{\sin 135°}(\sin 35°) \approx 48.67$$

$$c = \frac{c}{\sin C}(\sin B) = \frac{60}{\sin 135°}(\sin 10°) \approx 14.73$$

6. Given: $A = 60°$, $a = 9$, $c = 10$

$$\sin C = \frac{c \sin A}{a} = \frac{10 \sin 60°}{9} \approx 0.9623 \Rightarrow C \approx 74.2° \text{ or } C \approx 105.8°$$

Case 1

$$C \approx 74.2°$$

$$B \approx 180° - 60° - 74.2° = 45.8°$$

$$b = \frac{a}{\sin A}(\sin B) \approx \frac{9}{\sin 60°}(\sin 45.8°) \approx 7.449$$

Case 2

$$C \approx 105.8°$$

$$B \approx 180° - 60° - 105.8° = 14.2°$$

$$b = \frac{a}{\sin A}(\sin B) \approx \frac{9}{\sin 60°}(\sin 14.2°) \approx 2.551$$

8. Given: $A = 24.3°$, $C = 54.6°$, $c = 2.68$

$$B = 180° - A - C = 180° - 24.3° - 54.6° = 101.1°$$

$$a = \frac{c}{\sin C}(\sin A) = \frac{2.68}{\sin 54.6°}(\sin 24.3°) \approx 1.353$$

$$b = \frac{c}{\sin C}(\sin B) = \frac{2.68}{\sin 54.6°}(\sin 101.1°) \approx 3.226$$

10. Given: $A = 5°40'$, $B = 8°15'$, $b = 4.8$

$$C = 180° - A - B = 180° - 5°40' - 8°15' = 166°5'$$

$$a = \frac{b}{\sin B}(\sin A) = \frac{4.8}{\sin 8°15'}(\sin 5°40') \approx 3.303$$

$$c = \frac{b}{\sin B}(\sin C) = \frac{4.8}{\sin 8°15'}(\sin 166°5') \approx 8.045$$

12. Given: $C = 85°20'$, $a = 35$, $c = 50$

$$\sin A = \frac{a \sin C}{c} = \frac{35 \sin 85°20'}{50} \approx 0.6977 \Rightarrow A \approx 44.24°$$

$$B = 180° - A - C \approx 180° - 44.2° - 85.33° \approx 50.43°$$

$$b = \frac{c}{\sin C}(\sin B) \approx \frac{50}{\sin 85°20'}(\sin 50.43°) \approx 38.66$$

14. Given: $A = 100°$, $a = 125$, $c = 10$

$$\sin C = \frac{c \sin A}{a}$$

$$= \frac{10 \sin 100°}{125} \approx 0.0788 \Rightarrow C = 4.52°$$

$$B = 180° - A - C$$

$$\approx 180° - 100° - 4.52° = 75.5°$$

$$b = \frac{a}{\sin A}(\sin B)$$

$$\approx \frac{125}{\sin 100°}(\sin 75.5°) \approx 122.9$$

16. Given: $B = 2°45'$, $b = 6.2$, $c = 5.8$

$$\sin C = \frac{c \sin B}{b} = \frac{5.8 \sin 2°45'}{6.2}$$

$$\approx 0.04488 \Rightarrow C = 2.57°$$

$$A = 180° - B - C$$

$$\approx 180° - 2.75° - 2.57° = 174.68°$$

$$a = \frac{b}{\sin B}(\sin A)$$

$$\approx \frac{6.2}{\sin 2°45'}(\sin 174.68°) \approx 11.99$$

18. Given: $a = 11.4$, $b = 12.8$, $A = 58°$

$$\sin B = \frac{b \sin A}{a} = \frac{12.8 \sin 58°}{11.4} \approx 0.95219 \Rightarrow B \approx 72.2° \text{ or } B \approx 107.8°$$

<u>Case 1</u>

$B \approx 72.2°$

$C \approx 180° - 72.2° - 58° = 49.8°$

$c \approx \dfrac{11.4}{\sin 58°}(\sin 49.8°) \approx 10.27$

<u>Case 2</u>

$B \approx 107.8°$

$C \approx 180° - 107.8° - 58° = 14.2°$

$c \approx \dfrac{11.4}{\sin 58°}(\sin 14.2°) \approx 3.30$

20. Given: $a = 42$, $b = 50$, $A = 58°$

$h = 50 \sin 58° \approx 42.40$

Since $a < h$, no triangle is formed.

22. Given: $a = 125$, $b = 100$, $A = 110°$

$$\sin B = \frac{b \sin A}{a} = \frac{100 \sin 110°}{125}$$

$$\approx 0.7518 \Rightarrow B \approx 48.74°$$

$$C \approx 180° - 110° - 48.74° = 21.3°$$

$$c = \frac{a}{\sin A}(\sin C) \approx \frac{125}{\sin 110°}(\sin 21.3°) \approx 48.23$$

24. Given: $A = 60°$, $a = 10$

(a) One solution if $b \leq 10$ or $b = \dfrac{10}{\sin 60°}$

(b) Two solutions if $10 < b < \dfrac{10}{\sin 60°}$

(c) No solution if $b > \dfrac{10}{\sin 60°}$

26. Area $= \frac{1}{2}ac \sin B$

$= \frac{1}{2}(105)(64) \sin 72°30'$

≈ 3204

28. Area $= \frac{1}{2}bc \sin A$

$= \frac{1}{2}(4.5)(22) \sin 5°15'$

≈ 4.529

30. Area $= \frac{1}{2}ab \sin C$

$= \frac{1}{2}(16)(20) \sin 84°30'$

≈ 159.3

32. $A = 180° - 22°50' - 96° = 61°10'$

$h = \dfrac{100}{\sin 61°10'}(\sin 22°50') \approx 44.3$ ft

34. (a) $A = B \Rightarrow 2A = 180° - 40° \Rightarrow A = 70°$

$r = \dfrac{3000}{\sin 40°}(\sin 70°) \approx 4385.71$ ft

(b) $s \approx \left(\dfrac{40\pi}{180}\right) 4385.71 \approx 3061.80$ ft

36. $A = 51°$, $B = 112°$, $c = 2.5$

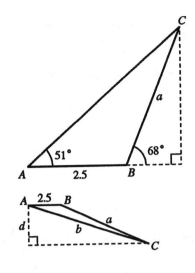

$C = 180° - 51° - 112° = 17°$

$a = \dfrac{c}{\sin C}(\sin A) = \dfrac{2.5}{\sin 17°}(\sin 51°) \approx 6.65$

$h \approx 6.65 \sin 68° \approx 6.2$ mi

38. $A = 18°$, $B = 90° + 66° = 156°$, $c = 2.5$

$C = 180° - 18° - 156° = 6°$

$b = \dfrac{c}{\sin C}(\sin B) = \dfrac{2.5}{\sin 6°}(\sin 156°) \approx 9.73$

$d \approx 9.73 \sin 18° \approx 3$ miles

40.

$\dfrac{6}{\sin 5°} = \dfrac{1.5}{\sin C} \Rightarrow \sin C = \dfrac{1.5 \sin 5°}{6}$

$\sin C \approx 0.02179$

$C \approx 1.2485°$

$B \approx 180° - 5° - 1.2485°$

$\qquad = 173.7515°$

$\dfrac{x}{\sin B°} = \dfrac{6}{\sin 5°} \Rightarrow \dfrac{6 \sin 173.7515°}{\sin 5°}$

$x \approx 7.493$ in.

$(6 + 1.5) - 7.493 \approx 0.007$ in.

42. $\alpha = 180 - (\phi + 180 - \theta) = \theta - \phi$

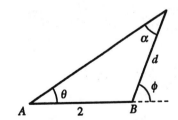

$\dfrac{d}{\sin \phi} = \dfrac{2}{\sin \alpha}$

$d = \dfrac{2 \sin \phi}{\sin(\theta - \phi)}$

6.7 | Law of Cosines

2. Given: $a = 9$, $b = 4$, $c = 10$

$$\cos C = \frac{a^2 + b^2 - c^2}{2ab} = \frac{81 + 16 - 100}{2(9)(4)} \approx -0.0417 \Rightarrow C \approx 92.39°$$

$$\sin A = \frac{a \sin C}{c} \approx \frac{9 \sin 92.4°}{10} \approx 0.8992 \Rightarrow A \approx 64.06°$$

$$B \approx 180° - 92.39° - 64.06° = 23.55°$$

4. Given: $C = 110°$, $a = 10$, $b = 4.5$

$$c^2 = a^2 + b^2 - 2ab \cos C$$

$$\approx 100 + 20.25 - 2(10)(4.5)(-0.3420) \approx 151.03 \Rightarrow c \approx 12.29$$

$$\sin A = \frac{a \sin C}{c} \approx \frac{10(0.9397)}{12.29} \approx 0.7646 \Rightarrow A \approx 49.9°$$

$$B \approx 180° - 110° - 49.9° = 20.1°$$

6. Given: $a = 55$, $b = 25$, $c = 72$

$$\cos C = \frac{a^2 + b^2 - c^2}{2ab} = \frac{55^2 + 25^2 - 72^2}{2(55)(25)} \approx -0.5578 \Rightarrow C \approx 123.91°$$

$$\sin A = \frac{a \sin C}{c} = \frac{55(0.8300)}{72} \approx 0.6340 \Rightarrow A \approx 39.35°$$

$$B \approx 180° - 123.91° - 39.35° = 16.74°$$

8. Given: $a = 1.42$, $b = 0.75$, $c = 1.25$

$$\cos A = \frac{b^2 + c^2 - a^2}{2bc} = \frac{(0.75)^2 + (1.25)^2 - (1.42)^2}{2(0.75)(1.25)} \approx 0.0579 \Rightarrow A \approx 86.7°$$

$$\sin B = \frac{b \sin A}{a} \approx \frac{0.75(0.9983)}{1.42} \approx 0.5273 \Rightarrow B \approx 31.8°$$

$$C \approx 180° - 86.7° - 31.8° = 61.5°$$

10. Given: $A = 55°$, $b = 3$, $c = 10$

$$a^2 = b^2 + c^2 - 2bc \cos A = 9 + 100 - 60 \cos 55° \approx 74.59 \Rightarrow a \approx 8.64$$

$$\sin B = \frac{b \sin A}{a} \approx \frac{3 \sin 55°}{8.64} \approx 0.2844 \Rightarrow B \approx 16.52°$$

$$C \approx 180° - 55° - 16.52° = 108.48°$$

12. Given: $B = 75°20'$, $a = 6.2$, $c = 9.5$

$$b^2 = a^2 + c^2 - 2ac\cos B \approx (6.2)^2 + (9.5)^2 - 2(6.2)(9.5)(0.2532) \approx 98.86 \Rightarrow b \approx 9.943$$

$$\sin A = \frac{a\sin B}{b} = \frac{6.2(0.9674)}{9.943} = 0.6032 \Rightarrow A = 37.1°$$

$$C \approx 180° - 75.33° - 37.1° = 67.57°$$

14. Given: $C = 15°$, $a = 6.25$, $b = 2.15$

$$c^2 = a^2 + b^2 - 2ab\cos C = (6.25)^2 + (2.15)^2 - 2(6.25)(2.15)(0.9659) \approx 17.73 \Rightarrow c \approx 4.210$$

$$\cos A = \frac{b^2 + c^2 - a^2}{2bc} = \frac{(2.15)^2 + (4.21)^2 - (6.25)^2}{2(2.15)(4.21)} = -0.9233 \Rightarrow A = 157.4°$$

$$B \approx 180° - 15° - 157.4° = 7.6°$$

16.

18.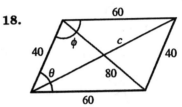

$$c^2 = 25^2 + 35^2 - 2(25)(35)\cos 120°$$

$$c \approx 52.20$$

$$2\theta = 360° - 2(120°)$$

$$\theta = 60°$$

$$d^2 = 25^2 + 35^2 - 2(25)(35)\cos 60°$$

$$d \approx 31.22$$

$$\cos\theta = \frac{40^2 + 60^2 - 80^2}{2(40)(60)}$$

$$\theta \approx 104.5°$$

$$2\phi \approx 360 - 2(104.5°)$$

$$\phi = 75.5°$$

$$c^2 = 40^2 + 60^2 - 2(40)(60)\cos 75.5°$$

$$c \approx 63.25$$

20.

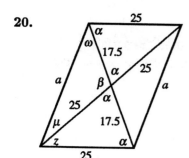

$$\cos \alpha = \frac{25^2 + 17.5^2 - 25^2}{2(25)(17.5)}$$

$$\alpha \approx 69.512°$$

$$\beta = 180 - \alpha \approx 110.488°$$

$$a^2 = 17.5^2 + 25^2 - 2(17.5)(25) \cos 110.488°$$

$$a \approx 35.18$$

$$z = 180 - 2\alpha \approx 40.976$$

$$\cos \mu = \frac{25^2 + 35.18^2 - 17.5^2}{2(25)(35.18)}$$

$$\mu \approx 27.771°$$

$$\theta = \mu + z \approx 68.7°$$

$$\omega = 180° - \mu - \beta \approx 41.741°$$

$$\phi = \omega + \alpha \approx 111.3°$$

22. $a = 2.5, \ b = 10.2, \ c = 9 \Rightarrow s = \dfrac{2.5 + 10.2 + 9}{2} = 10.85$

Area $= \sqrt{10.85(8.35)(0.65)(1.85)} \approx 10.44$

24. $a = 75.4, \ b = 52, \ c = 52 \Rightarrow s = \dfrac{75.4 + 52 + 52}{2} = 89.7$

Area $= \sqrt{89.7(14.3)(37.7)(37.7)} \approx 1350$

26. $a = 4.25, \ b = 1.55, \ c = 3.00 \Rightarrow s = \dfrac{4.25 + 1.55 + 3.00}{2} = 4.4$

Area $= \sqrt{4.4(0.15)(2.85)(1.4)} \approx 1.623$

28. $d^2 = 4^2 + 6^2 - 2(4)(6) \cos 30°$

$d \approx 3.23$

$a = 4, \ b = 6, \ c = 3.23 \Rightarrow s = \dfrac{a + b + c}{2} \approx 6.62$

Area of triangle $= \sqrt{s(s - a)(s - b)(s - c)} \approx 6.0 \ \text{ft}^2$

Area of parallelogram $= 2 \times$ area of triangle $\approx 2(60) = 12.0 \ \text{yd}^2$

30. Given: $a = 540$, $c = 675$, $B = 180° - 75° + 32° = 137°$

The angle for the bearing from C to A is $C + 32°$. We wish to find that bearing and side b.

$$b^2 = a^2 + c^2 - 2ac \cos B = 540^2 + 675^2 - 2(540)(675)(-0.7314) \approx 1,280,000 \Rightarrow b \approx 1132$$

$$\sin C = \frac{c \sin B}{b} \approx \frac{675(0.6820)}{1132} \approx 0.4067 \Rightarrow C \approx 24°$$

Distance from C to A: 1132 miles

Bearing from C to A: S 56° W

32. The angles at the base of the tower are 98° and 82°. The longer guy wire, g_1, is given by

$$g_1{}^2 = 75^2 + 100^2 - 2(75)(100) \cos 98° \approx 17,700 \Rightarrow g_1 \approx 133.1 \text{ ft.}$$

The shorter wire, g_2, is given by

$$g_2{}^2 = 75^2 + 100^2 - 2(75)(100) \cos 82° \approx 13,540 \Rightarrow g_2 \approx 116.4 \text{ ft.}$$

34. $\cos A = \dfrac{375^2 + 250^2 - 300^2}{2(375)(250)}$

$\approx 0.6033 \Rightarrow A \approx 52.9°$

$\sin C = \dfrac{250(0.7975)}{300}$

$\approx 0.6646 \Rightarrow C \approx 41.6°$

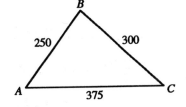

36. $a = \sqrt{35^2 + 20^2 - 2(35)(20) \cos 42°}$

$\approx \sqrt{584.60}$

$\approx 24 \text{ mi}$

38. $d^2 = 10^2 + 7^2 - 2(10)(7) \cos \theta$

$\theta = \arccos \left[\dfrac{10^2 + 7^2 - d^2}{2(10)(7)} \right]$

$s = \dfrac{360° - \theta}{360°}(2\pi r) = \dfrac{(360° - \theta)\pi}{45}$

d (inches)	9	10	11	12	13	14	15	16
θ (degrees)	60.9°	69.5°	78.5°	88.0°	98.2°	109.6°	122.9°	139.8°
s (inches)	20.88	20.28	19.65	18.99	18.28	17.48	16.55	15.37

40. Bearing from P to M: N θ E

Bearing from P to A: N ϕ E

Since M is due west of A, it follows that

$$\theta = M - 90° \text{ and } \phi = 90° - A.$$

$$\cos M = \frac{(6.5)^2 + (8.5)^2 - (14.5)^2}{2(6.5)(8.5)} \approx -0.8665 \Rightarrow M \approx 150°$$

$$\cos A = \frac{(6.5)^2 + (14.5)^2 - (8.5)^2}{2(6.5)(14.5)} \approx 0.9562 \Rightarrow A \approx 17°$$

Bearing from Phoenix to Minneapolis: N 60° E

Bearing from Phoenix to Albany: N 73° E

42. $x^2 = 330^2 + 420^2 - 2(330)(420)\cos 9°$

$x = 107.3$ feet

44. (a) Working with $\triangle OBC$, we have

$$\cos \alpha = \frac{a/2}{R}.$$

This implies that

$$2R = \frac{a}{\cos \alpha}.$$

Since we know that

$$\frac{a}{\sin A} = \frac{b}{\sin B} = \frac{c}{\sin C},$$

we can complete the proof by showing that

$$\cos \alpha = \sin A.$$

The solution of the system

$$A + B + C = 180°$$
$$\alpha - C + A = \beta$$
$$\alpha + \beta = B$$

is $\alpha = 90° - A$. Therefore,

$$2R = \frac{a}{\cos \alpha} = \frac{a}{\cos(90° - A)} = \frac{a}{\sin A}.$$

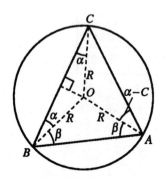

(b) By Heron's Formula, the area of the triangle is

$$\text{Area} = \sqrt{s(s-a)(s-b)(s-c)}.$$

We can also find the area by dividing the area into six triangles and using the fact that the area is $\frac{1}{2}$ the base times the height. Using the figure as given, we have

$$\text{Area} = \tfrac{1}{2}xr + \tfrac{1}{2}xr + \tfrac{1}{2}yr + \tfrac{1}{2}yr + \tfrac{1}{2}zr + \tfrac{1}{2}zr$$
$$= r(x + y + z)$$
$$= rs.$$

Therefore, $rs = \sqrt{s(s-a)(s-b)(s-c)} \Rightarrow$

$$r = \sqrt{\frac{(s-a)(s-b)(s-c)}{s}}.$$

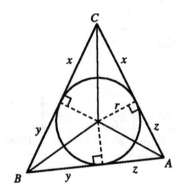

46. Given: $a = 200$ ft, $b = 250$ ft, $c = 325$ ft

$$s = \frac{200 + 250 + 325}{2} \approx 387.5$$

Radius of the inscribed circle:
$$r = \sqrt{\frac{(s-a)(s-b)(s-c)}{s}} =$$
$$\sqrt{\frac{(187.5)(137.5)(62.5)}{387.5}} \approx 64.5 \text{ ft}$$

Circumference of an inscribed circle:
$$C = 2\pi r \approx 2\pi(64.5) \approx 405 \text{ ft}$$

48.
$$-\cos A = \frac{a^2 - (b^2 + c^2)}{2bc}$$
$$1 - \cos A = \frac{a^2 - (b^2 - 2bc + c^2)}{2bc}$$
$$\frac{1}{2}bc(1 - \cos A) = \frac{a^2 - (b-c)^2}{4}$$
$$= \left(\frac{a-b+c}{2}\right)\left(\frac{a+b-c}{2}\right)$$

6.8 DeMoivre's Theorem and nth Roots

2. $|-5| = \sqrt{5^2 + 0^2}$
$$= \sqrt{25} = 5$$

4. $|5 - 12i| = \sqrt{5^2 + (-12)^2}$
$$= \sqrt{169} = 13$$

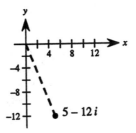

6. $|-8 + 3i| = \sqrt{(-8)^2 + (3)^2}$
$$= \sqrt{73}$$

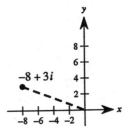

8. $z = 2$
$$r = \sqrt{2^2 + 0^2} = \sqrt{4} = 2$$
$$\tan\theta = \frac{0}{2} = 0 \Rightarrow \theta = 0$$
$$z = 2(\cos 0 + i\sin 0)$$

10. $z = -2 + 2\sqrt{3}\,i$
$$r = \sqrt{(-2)^2 + (2\sqrt{3})^2} = \sqrt{16} = 4$$
$$\tan\theta = \frac{2\sqrt{3}}{-2} = -\sqrt{3} \Rightarrow \theta = \frac{2\pi}{3}$$
$$z = 4\left(\cos\frac{2\pi}{3} + i\sin\frac{2\pi}{3}\right)$$

12. $z = -2 - 2i$

$r = \sqrt{(-2)^2 + (-2)^2} = \sqrt{8} = 2\sqrt{2}$

$\tan\theta = \dfrac{-2}{-2} = 1 \Rightarrow \theta = \dfrac{5\pi}{4}$

$z = 2\sqrt{2}\left(\cos\dfrac{5\pi}{4} + i\sin\dfrac{5\pi}{4}\right)$

14. $z = -1 + \sqrt{3}\,i$

$r = \sqrt{(-1)^2 + (\sqrt{3})^2} = \sqrt{4} = 2$

$\tan\theta = \dfrac{\sqrt{3}}{-1} = -\sqrt{3} \Rightarrow \theta = \dfrac{2\pi}{3}$

$z = 2\left(\cos\dfrac{2\pi}{3} + i\sin\dfrac{2\pi}{3}\right)$

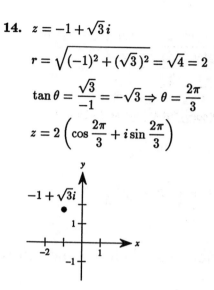

16. $z = \dfrac{5}{2}(\sqrt{3} - i)$

$r = \sqrt{\left(\dfrac{5}{2}\sqrt{3}\right)^2 + \left(\dfrac{5}{2}(-1)\right)^2}$

$= \sqrt{\dfrac{100}{4}} = \sqrt{25} = 5$

$\tan\theta = \dfrac{-1}{\sqrt{3}} = \dfrac{-\sqrt{3}}{3} \Rightarrow \theta = \dfrac{11\pi}{6}$

$z = 5\left(\cos\dfrac{11\pi}{6} + i\sin\dfrac{11\pi}{6}\right)$

18. $z = 4 + 0i$

$r = \sqrt{4^2 + 0^2} = \sqrt{16} = 4$

$\tan\theta = \dfrac{0}{4} = 0 \Rightarrow \theta = 0$

$z = 4(\cos 0 + i\sin 0)$

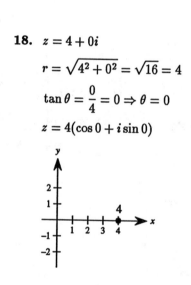

20. $z = 3 - i$

$r = \sqrt{(3)^2 + (-1)^2} = \sqrt{10}$

$\tan \theta = \dfrac{-1}{3} \Rightarrow \theta \approx -18.4°$

$z \approx \sqrt{10}(\cos(-18.4°) + i \sin(-18.4°))$

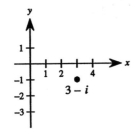

22. $z = 0 - 2i$

$r = \sqrt{0^2 + (-2)^2} = \sqrt{4} = 2$

$\tan \theta = \dfrac{-2}{0}, \quad \text{undefined} \Rightarrow \theta = \dfrac{3\pi}{2}$

$z = 2\left(\cos \dfrac{3\pi}{2} + i \sin \dfrac{3\pi}{2}\right)$

24. $z = 2\sqrt{2} - i$

$r = \sqrt{(2\sqrt{2})^2 + (-1)^2} = \sqrt{9} = 3$

$\tan \theta = \dfrac{-1}{2\sqrt{2}} = -\dfrac{\sqrt{2}}{4} \Rightarrow \theta \approx (-19.5°)$

$z \approx 3(\cos(-19.5°) + i \sin(-19.5°))$

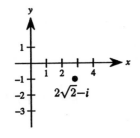

26. $z = 1 + 3i$

$r = \sqrt{1^2 + 3^2} = \sqrt{10}$

$\tan \theta = \dfrac{3}{1} = 3 \Rightarrow \theta \approx 71.6°$

$z \approx \sqrt{10}(\cos 71.6° + i \sin 71.6°)$

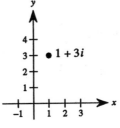

28. $5(\cos 135° + i \sin 135°) = 5\left[-\dfrac{\sqrt{2}}{2} + i\left(\dfrac{\sqrt{2}}{2}\right)\right]$

$= -\dfrac{5\sqrt{2}}{2} + \dfrac{5\sqrt{2}}{2}i$

30. $\dfrac{3}{4}(\cos 315° + i \sin 315°) = \dfrac{3}{4}\left[\dfrac{\sqrt{2}}{2} + i\left(-\dfrac{\sqrt{2}}{2}\right)\right]$

$= \dfrac{3\sqrt{2}}{8} - \dfrac{3\sqrt{2}}{8}i$

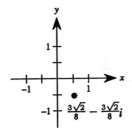

32. $8\left(\cos\dfrac{\pi}{12} + i\sin\dfrac{\pi}{12}\right) \approx 8(0.9659 + 0.2588i)$

$$\approx 7.7274 + 2.0706i$$

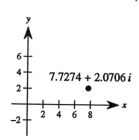

34. $7(\cos 0° + i\sin 0°) = 7$

36. $6[\cos(230°30') + i\sin(230°30')] \approx -3.816 - 4.630i$

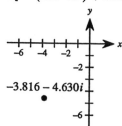

38. $\left[\dfrac{3}{2}\left(\cos\dfrac{\pi}{2} + i\sin\dfrac{\pi}{2}\right)\right]\left[6\left(\cos\dfrac{\pi}{4} + i\sin\dfrac{\pi}{4}\right)\right] = \left(\dfrac{3}{2}\right)(6)\left[\cos\left(\dfrac{\pi}{2} + \dfrac{\pi}{4}\right) + i\sin\left(\dfrac{\pi}{2} + \dfrac{\pi}{4}\right)\right]$

$$= 9\left(\cos\dfrac{3\pi}{4} + i\sin\dfrac{3\pi}{4}\right)$$

40. $[0.5(\cos 100° + i\sin 100°)][0.8(\cos 300° + i\sin 300°)] = (0.5)(0.8)[\cos(100° + 300°) + i\sin(100° + 300°)]$

$$= 0.4(\cos 400° + i\sin 400°)$$

$$= 0.4(\cos 40° + i\sin 40°)$$

42. $(\cos 5° + i\sin 5°)(\cos 20° + i\sin 20°) = \cos(5° + 20°) + i\sin(5° + 20°) = \cos 25° + i\sin 25°$

44. $\dfrac{\cos 40° + i\sin 40°}{\cos 10° + i\sin 10°} = \cos(40° - 10°) + i\sin(40° - 10°) = \cos 30° + i\sin 30°$

46. $\dfrac{5[\cos(4.3) + i\sin(4.3)]}{4[\cos(2.1) + i\sin(2.1)]} = \dfrac{5}{4}[\cos(4.3 - 2.1) + i\sin(4.3 - 2.1)] = \dfrac{5}{4}[\cos(2.2) + i\sin(2.2)]$

48. $\dfrac{9(\cos 20° + i\sin 20°)}{5(\cos 75° + i\sin 75°)} = \dfrac{9}{5}[\cos(20° - 75°) + i\sin(20° - 75°)] = \dfrac{9}{5}[\cos(-55°) + i\sin(-55°)]$

$$= \dfrac{9}{5}[\cos 305° + i\sin 305°]$$

50. (a) $\sqrt{3} + i = 2(\cos 30° + i\sin 30°)$

$\quad\quad\ 1 + i = \sqrt{2}(\cos 45° + i\sin 45°)$

(b) $(\sqrt{3}+i)(1+i) = [2(\cos 30° + i\sin 30°)][\sqrt{2}(\cos 45° + i\sin 45°)]$

$$= 2\sqrt{2}(\cos 75° + i\sin 75°)$$

$$= 2\sqrt{2}\left[\left(\frac{\sqrt{6}-\sqrt{2}}{4}\right) + \left(\frac{\sqrt{6}+\sqrt{2}}{4}\right)i\right]$$

$$= (\sqrt{3}-1) + (\sqrt{3}+1)i$$

(c) $(\sqrt{3}+i)(1+i) = \sqrt{3} + (\sqrt{3}+1)i + i^2 = (\sqrt{3}-1) + (\sqrt{3}+1)i$

52. (a) $\quad 3+4i = 5(\cos 53.13° + i\sin 53.13°)$

$\quad\quad 1 - \sqrt{3}\,i = 2(\cos 300° + i\sin 300°)$

(b) $\dfrac{3+4i}{1-\sqrt{3}\,i} = \dfrac{5(\cos 53.13° + i\sin 53.13°)}{2(\cos 300° + i\sin 300°)} = 2.5[\cos(-246.9°) + i\sin(-246.9°)]$

$$= 2.5(\cos 113.1° + i\sin 113.1°)$$

$$\approx -0.9808 + 2.299i$$

(c) $\dfrac{3+4i}{1-\sqrt{3}\,i} = \dfrac{3+4i}{1-\sqrt{3}\,i} \cdot \dfrac{1+\sqrt{3}\,i}{1+\sqrt{3}\,i}$

$$= \frac{3 + (4+3\sqrt{3})i + 4\sqrt{3}\,i^2}{1+3}$$

$$= \frac{3 - 4\sqrt{3}}{4} + \frac{4 + 3\sqrt{3}}{4}i \approx -0.9821 + 2.299i$$

54. (a) $\quad\quad 4i = 4(\cos 90° + i\sin 90°)$

$\quad\quad -4 + 2i = 2\sqrt{5}(\cos 153.4° + i\sin 153.4°)$

(b) $\dfrac{4i}{-4+2i} = \dfrac{4(\cos 90° + i\sin 90°)}{2\sqrt{5}(\cos 153.4° + i\sin 153.4°)}$

$$= \frac{2\sqrt{5}}{5}(\cos 296.6° + i\sin 296.6°)$$

$$\approx 0.400 - 0.800i$$

(c) $\dfrac{4i}{-4+2i} = \dfrac{4i}{-4+2i} \cdot \dfrac{-4-2i}{-4-2i} = \dfrac{8-16i}{20} = \dfrac{2}{5} - \dfrac{4}{5}i = 0.400 - 0.800i$

56. $z = r(\cos\theta + i\sin\theta)$

$\overline{z} = r(\cos\theta - i\sin\theta)$

$\quad = r[\cos(-\theta) + i\sin(-\theta)]$

58. $z = r(\cos\theta + i\sin\theta)$

$-z = -r(\cos\theta + i\sin\theta)$

$\quad = r(-\cos\theta - i\sin\theta)$

$\quad = r[\cos(\theta + \pi) + i\sin(\theta + \pi)]$

60. Let $z = x + iy$ such that:

$$\tan\frac{\pi}{6} = \frac{y}{x}$$

62. $(2 + 2i)^6 = \left[2\sqrt{2}\left(\cos\dfrac{\pi}{4} + i\sin\dfrac{\pi}{4}\right)\right]^6$

$\quad = (2\sqrt{2})^6\left(\cos\dfrac{6\pi}{4} + i\sin\dfrac{6\pi}{4}\right)$

$\quad = 512\left(\cos\dfrac{3\pi}{2} + i\sin\dfrac{3\pi}{2}\right)$

$\quad = -512i$

64. $(1 - i)^{12} = \left[\sqrt{2}\left(\cos\dfrac{7\pi}{4} + i\sin\dfrac{7\pi}{4}\right)\right]^{12}$

$\quad = (\sqrt{2})^{12}(\cos 21\pi + i\sin 21\pi)$

$\quad = 64(\cos\pi + i\sin\pi)$

$\quad = 64(-1)$

$\quad = -64$

66. $4(1 - \sqrt{3}\,i)^3 = 4\left[2\left(\cos\dfrac{5\pi}{3} + i\sin\dfrac{5\pi}{3}\right)\right]^3$

$\quad = 4[2^3(\cos 5\pi + i\sin 5\pi)]$

$\quad = 32(-1)$

$\quad = -32$

68. $[3(\cos 150° + i\sin 150°)]^4 = 3^4(\cos 600° + i\sin 600°)$

$\quad = 81(\cos 240° + i\sin 240°)$

$\quad = 81(-\cos 60° - i\sin 60°)$

$\quad = -\dfrac{81}{2} - \dfrac{81\sqrt{3}}{2}i$

70. $\left[2\left(\cos\dfrac{\pi}{2} + i\sin\dfrac{\pi}{2}\right)\right]^8 = 2^8(\cos 4\pi + i\sin 4\pi)$

$\quad = 256(\cos 0 + i\sin 0) = 256$

72. $(\cos 0 + i\sin 0)^{20} = \cos 0 + i\sin 0 = 1$

74. (a) Square roots of $16(\cos 60° + i \sin 60°)$:

$$\sqrt{16}\left[\cos\left(\frac{60° + 360°k}{2}\right) + i \sin\left(\frac{60° + 360°k}{2}\right)\right], \quad k = 0,\, 1$$

$4(\cos 30° + i \sin 30°)$

$4(\cos 210° + i \sin 210°)$

(b)

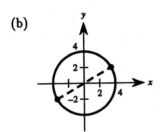

(c) $4\left(\dfrac{\sqrt{3}}{2} + \dfrac{1}{2}i\right) = 2\sqrt{3} + 2i$

$4\left(-\dfrac{\sqrt{3}}{2} - \dfrac{1}{2}i\right) = -2\sqrt{3} - 2i$

76. (a) Fifth roots of $32\left(\cos\dfrac{5\pi}{6} + i \sin\dfrac{5\pi}{6}\right)$:

$$\sqrt[5]{32}\left[\cos\left(\frac{(5\pi/6) + 2k\pi}{5}\right) + i \sin\left(\frac{(5\pi/6) + 2k\pi}{5}\right)\right],$$

$$k = 0,\, 1,\, 2,\, 3,\, 4$$

$2\left(\cos\dfrac{\pi}{6} + i \sin\dfrac{\pi}{6}\right)$

$2\left(\cos\dfrac{17\pi}{30} + i \sin\dfrac{17\pi}{30}\right)$

$2\left(\cos\dfrac{29\pi}{30} + i \sin\dfrac{29\pi}{30}\right)$

$2\left(\cos\dfrac{41\pi}{30} + i \sin\dfrac{41\pi}{30}\right)$

$2\left(\cos\dfrac{53\pi}{30} + i \sin\dfrac{53\pi}{30}\right)$

(b)

(c) $1.732 + i,\ -0.4158 + 1.956i,\ -1.989 + 0.2091i,$

$-0.8135 - 1.827i,\ 1.486 - 1.338i$

78. (a) Fourth roots of $625i = 625\left(\cos\dfrac{\pi}{2} + i\sin\dfrac{\pi}{2}\right)$:

$$\sqrt[4]{625}\left[\cos\left(\frac{(\pi/2) + 2k\pi}{4}\right) + i\sin\left(\frac{(\pi/2) + 2k\pi}{4}\right)\right],\ k = 0,\ 1,\ 2,\ 3$$

$$5\left(\cos\frac{\pi}{8} + i\sin\frac{\pi}{8}\right)$$

$$5\left(\cos\frac{5\pi}{8} + i\sin\frac{5\pi}{8}\right)$$

$$5\left(\cos\frac{9\pi}{8} + i\sin\frac{9\pi}{8}\right)$$

$$5\left(\cos\frac{13\pi}{8} + i\sin\frac{13\pi}{8}\right)$$

(b)

(c) $4.619 + 1.913i,\ -1.913 + 4.619i,$

$-4.619 - 1.913i,\ 1.913 - 4.619i$

80. (a) Cube roots of $-8\sqrt{2}(1 - i) = 8\left(\cos\dfrac{3\pi}{4} + i\sin\dfrac{3\pi}{4}\right)$:

$$\sqrt[3]{8}\left[\cos\left(\frac{(3\pi/4) + 2k\pi}{3}\right) + i\sin\left(\frac{(3\pi/4) + 2k\pi}{3}\right)\right],\ k = 0,\ 1,\ 2$$

$$2\left(\cos\frac{\pi}{4} + i\sin\frac{\pi}{4}\right)$$

$$2\left(\cos\frac{11\pi}{12} + i\sin\frac{11\pi}{12}\right)$$

$$2\left(\cos\frac{19\pi}{12} + i\sin\frac{19\pi}{12}\right)$$

(b)

(c) $1.414 + 1.414i,\ -1.932 + 0.5176i,\ 0.5176 - 1.9319i$

82. (a) Fourth roots of $i = \cos \dfrac{\pi}{2} + i \sin \dfrac{\pi}{2}$:

$$\sqrt[4]{1}\left[\cos\left(\frac{(\pi/2)+2k\pi}{4}\right)+i\sin\left(\frac{(\pi/2)+2k\pi}{4}\right)\right],\ k=0,\ 1,\ 2,\ 3$$

$$\cos\frac{\pi}{8}+i\sin\frac{\pi}{8}$$

$$\cos\frac{5\pi}{8}+i\sin\frac{5\pi}{8}$$

$$\cos\frac{9\pi}{8}+i\sin\frac{9\pi}{8}$$

$$\cos\frac{13\pi}{8}+i\sin\frac{13\pi}{8}$$

(b)

(c) $0.9239 + 0.3827i,\ -0.3827 + 0.9239i,$

$-0.9239 - 0.3827i,\ 0.3827 - 0.9239i$

84. (a) Cube roots of $1000 = 1000(\cos 0 + i \sin 0)$:

$$\sqrt[3]{1000}\left(\cos\frac{2k\pi}{3}+i\sin\frac{2k\pi}{3}\right),\ k=0,\ 1,\ 2$$

$$10(\cos 0 + i \sin 0)$$

$$10\left(\cos\frac{2\pi}{3}+i\sin\frac{2\pi}{3}\right)$$

$$10\left(\cos\frac{4\pi}{3}+i\sin\frac{4\pi}{3}\right)$$

(b)

(c) $10,\ -5+5\sqrt{3}\,i,\ -5-5\sqrt{3}\,i$

86. $x^3 + 1 = 0$

$x^3 = -1$

The solutions are the cube roots of $-1 = \cos \pi + i \sin \pi$:

$$\cos\left(\frac{\pi + 2k\pi}{3}\right) + i \sin\left(\frac{\pi + 2k\pi}{3}\right), \; k = 0, 1, 2$$

$$\cos\frac{\pi}{3} + i \sin\frac{\pi}{3} = \frac{1}{2} + \frac{\sqrt{3}}{2}i$$

$$\cos\pi + i\sin\pi = -1$$

$$\cos\frac{5\pi}{3} + i\sin\frac{5\pi}{3} = \frac{1}{2} - \frac{\sqrt{3}}{2}i$$

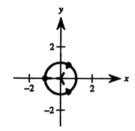

88. $x^4 - 81 = 0$

$x^4 = 81$

The solutions are the fourth roots of 81:

$$\sqrt[4]{81}\left(\cos\frac{0 + 2\pi k}{4} + i\sin\frac{0 + 2\pi k}{4}\right), \; k = 0, 1, 2, 3$$

$$3(\cos 0 + i \sin 0) = 3$$

$$3\left(\cos\frac{\pi}{2} + i\sin\frac{\pi}{2}\right) = 3i$$

$$3\left(\cos\pi + i\sin\pi\right) = -3$$

$$3\left(\cos\frac{3\pi}{2} + i\sin\frac{3\pi}{2}\right) = -3i$$

90. $x^6 - 64i = 0$

$\qquad x^6 = 64i$

The solutions are the sixth roots of $64i$:

$$\sqrt[6]{64}\left[\cos\left(\frac{(\pi/2)+2k\pi}{6}\right) + i\sin\left(\frac{(\pi/2)+2k\pi}{6}\right)\right],$$

$\qquad k = 0,\ 1,\ 2,\ 3,\ 4,\ 5$

$$2\left(\cos\frac{\pi}{12} + i\sin\frac{\pi}{12}\right) \approx 1.932 + 0.5176i$$

$$2\left(\cos\frac{5\pi}{12} + i\sin\frac{5\pi}{12}\right) \approx 0.5176 + 1.932i$$

$$2\left(\cos\frac{3\pi}{4} + i\sin\frac{3\pi}{4}\right) \approx -1.414 + 1.414i$$

$$2\left(\cos\frac{13\pi}{12} + i\sin\frac{13\pi}{12}\right) \approx -1.932 - 0.5176i$$

$$2\left(\cos\frac{17\pi}{12} + i\sin\frac{17\pi}{12}\right) \approx -0.5176 - 1.932i$$

$$2\left(\cos\frac{7\pi}{4} + i\sin\frac{7\pi}{4}\right) \approx 1.414 - 1.414i$$

92. $x^4 + (1+i) = 0$

$\qquad x^4 = -1 - i = \sqrt{2}\,(\cos 225° + i\sin 225°)$

The solutions are the fourth roots of $-1 - i$:

$$\sqrt[4]{\sqrt{2}}\left[\cos\left(\frac{225° + 360°k}{4}\right) + i\sin\left(\frac{225° + 360°k}{4}\right)\right],$$

$\qquad k = 0,\ 1,\ 2,\ 3$

$$\sqrt[4]{\sqrt{2}}\,(\cos 56.25° + i\sin 56.25°) \approx 0.6059 + 0.9067i$$

$$\sqrt[4]{\sqrt{2}}\,(\cos 146.25° + i\sin 146.25°) \approx -0.9067 + 0.6059i$$

$$\sqrt[4]{\sqrt{2}}\,(\cos 236.25° + i\sin 236.25°) \approx -0.6059 - 0.9067i$$

$$\sqrt[4]{\sqrt{2}}\,(\cos 326.25° + i\sin 326.25°) \approx 0.9067 - 0.6059i$$

Review Exercises for Chapter 6

2. $\dfrac{\sin 2\alpha}{\cos^2 \alpha - \sin^2 \alpha} = \dfrac{\sin 2\alpha}{\cos 2\alpha} = \tan 2\alpha$

4. $\dfrac{\sin^3 \beta + \cos^3 \beta}{\sin \beta + \cos \beta} = \dfrac{(\sin \beta + \cos \beta)(\sin^2 \beta - \sin \beta \cos \beta + \cos^2 \beta)}{\sin \beta + \cos \beta}$

$$= 1 - \sin \beta \cos \beta$$

$$= 1 - \frac{1}{2} \sin 2\beta$$

6. $\dfrac{\sin \theta}{1 + \cos \theta} + \dfrac{1 + \cos \theta}{\sin \theta} = \dfrac{\sin^2 \theta + (1 + \cos \theta)^2}{\sin \theta (1 + \cos \theta)}$

$$= \dfrac{\sin^2 \theta + 1 + 2\cos \theta + \cos^2 \theta}{\sin \theta (1 + \cos \theta)}$$

$$= \dfrac{2}{\sin \theta}$$

$$= 2 \csc \theta$$

8. $\dfrac{2 \tan(x + 1)}{1 - \tan^2(x + 1)} = \tan[2(x + 1)]$

$$= \tan(2x + 2)$$

10. $\sqrt{\dfrac{1 - \cos^2 x}{1 + \cos x}} = \sqrt{1 - \cos x} = \sqrt{2}\left|\sin \dfrac{x}{2}\right|$

12. $\cos x(\tan^2 x + 1) = \cos x \sec^2 x$

$$= \dfrac{1}{\sec x} \sec^2 x$$

$$= \sec x$$

14. $\sin^3 \theta + \sin \theta \cos^2 \theta = \sin \theta (\sin^2 \theta + \cos^2 \theta)$

$$= \sin \theta$$

16. $\cos^3 x \sin^2 x = \cos x(\cos^2 x) \sin^2 x$

$$= \cos x(1 - \sin^2 x) \sin^2 x$$

$$= (\sin^2 x - \sin^4 x) \cos x$$

18. Using a product-sum formula, we have

$$\sin 3x \cos 2x = \tfrac{1}{2}(\sin 5x + \sin x).$$

20. $\sqrt{1 - \cos x} = \sqrt{(1 - \cos x)\dfrac{1 + \cos x}{1 + \cos x}}$

$$= \sqrt{\dfrac{\sin^2 x}{1 + \cos x}}$$

$$= \dfrac{|\sin x|}{\sqrt{1 + \cos x}}$$

22. $\cos\left(x + \dfrac{\pi}{2}\right) = \cos x \cos \dfrac{\pi}{2} - \sin x \sin \dfrac{\pi}{2}$

$$= (\cos x)(0) - (\sin x)(1)$$

$$= -\sin x$$

24. $\sin(\pi - x) = \sin \pi \cos x - \sin x \cos \pi$

$$= (0)(\cos x) - (\sin x)(-1)$$

$$= \sin x$$

26. $\dfrac{2\cos 3x}{\sin 4x - \sin 2x} = \dfrac{2\cos 3x}{2\cos 3x \sin x} = \dfrac{1}{\sin x} = \csc x$

28. $1 - \cos 2x = 1 - (\cos^2 x - \sin^2 x)$

$\qquad\qquad = (1 - \cos^2 x) + \sin^2 x$

$\qquad\qquad = 2\sin^2 x$

30. $\dfrac{\sin(\alpha + \beta)}{\cos \alpha \cos \beta} = \dfrac{\sin \alpha \cos \beta + \sin \beta \cos \alpha}{\cos \alpha \cos \beta}$

$\qquad\qquad = \dfrac{\sin \alpha}{\cos \alpha} + \dfrac{\sin \beta}{\cos \beta}$

$\qquad\qquad = \tan \alpha + \tan \beta$

32. $y_1 = \sin 4x$

$y_2 = 8\cos^3 x \sin x - 4\cos x \sin x$

The graphs of y_1 and y_2 coincide.

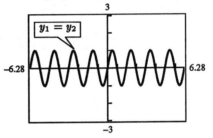

34. $y_1 = \cos^2 5x - \cos^2 x$

$y_2 = -\sin 4x \sin 6x$

The graphs of y_1 and y_2 coincide.

36. $y_1 = \sin 2x + \sin 4x - \sin 6x$

$y_2 = 4\sin x \sin 2x \sin 3x$

The graphs of y_1 and y_2 coincide.

38. $\cos(285°) = \cos(225° + 60°)$

$= \cos 225° \cos 60° - \sin 225° \sin 60°$

$= \left(-\dfrac{\sqrt{2}}{2}\right)\left(\dfrac{1}{2}\right) - \left(-\dfrac{\sqrt{2}}{2}\right)\left(\dfrac{\sqrt{3}}{2}\right)$

$= \dfrac{\sqrt{2}}{4}(\sqrt{3} - 1)$

40. $\sin\dfrac{3\pi}{8} = \sqrt{\dfrac{1 - \cos(3\pi/4)}{2}}$

$= \sin\left[\dfrac{1}{2}\left(\dfrac{3\pi}{4}\right)\right]$

$= \sqrt{\dfrac{1 - \cos(3\pi/4)}{2}}$

$= \sqrt{\dfrac{1 + (\sqrt{2}/2)}{2}}$

$= \dfrac{\sqrt{2 + \sqrt{2}}}{2}$

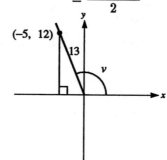

Figures for Exercises 42–46

42. $\tan(u + v) = \dfrac{\tan u + \tan v}{1 - \tan u \tan v}$

$= \dfrac{-(3/\sqrt{7}) - (12/5)}{1 - (-3/\sqrt{7})(-12/5)}$

$= \dfrac{15 + 12\sqrt{7}}{36 - 5\sqrt{7}}$

44. $\sin 2v = 2 \sin v \cos v$

$= 2\left(\dfrac{12}{13}\right)\left(-\dfrac{5}{13}\right)$

$= -\dfrac{120}{169}$

46. $\tan 2v = \dfrac{\sin 2v}{\cos 2v}$

$= \dfrac{2 \sin v \cos v}{2 \cos^2 v - 1}$

$= \dfrac{2(12/13)(-5/13)}{2(-5/13)^2 - 1}$

$= \dfrac{-120/169}{(50/169) - (169/169)}$

$= \dfrac{120}{119}$

48. $\sin(x + y) = \sin x + \sin y. False.$

$\sin(x + y) = \sin x \cos y + \cos x \sin y$

50. $4 \sin 45° \cos 15° = 1 + \sqrt{3}.$ True.

$4 \sin 45° \cos 15° = 4\left(\dfrac{1}{2}[\sin(45° + 15°) + \sin(45° - 15°)]\right)$

$= 2[\sin 60° + \sin 30°] = 2\left[\dfrac{\sqrt{3}}{2} + \dfrac{1}{2}\right] = 2\left(\dfrac{\sqrt{3} + 1}{2}\right) = 1 + \sqrt{3}$

52. Graphing the function on the interval $[0, 2\pi)$, we find the following zeros: $1.047, 5.236$

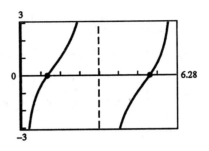

    Algebraically:

$$\csc x - 2\cot x = 0$$

$$\frac{1}{\sin x} - \frac{2\cos x}{\sin x} = 0$$

$$1 - 2\cos x = 0$$

$$\cos x = \frac{1}{2}$$

$$x = \frac{\pi}{3},\ \frac{5\pi}{3}$$

54. Graphing the function on the interval $[0, 2\pi)$, we find the following zeros: $1.571, 4.712$

    Algebraically:

$$\cos 4x - 7\cos 2x = 8$$

$$2\cos^2 2x - 1 - 7\cos 2x = 8$$

$$2\cos^2 2x - 7\cos 2x - 9 = 0$$

$$(2\cos 2x - 9)(\cos 2x + 1) = 0$$

$$2\cos 2x - 9 = 0 \quad \text{or} \quad \cos 2x + 1 = 0$$

$$\cos 2x = \frac{9}{2} \qquad\qquad \cos 2x = -1$$

$$\text{(No solution)} \qquad\qquad 2x = \pi + 2n\pi$$

$$x = \frac{\pi}{2} + n\pi$$

$$x = \frac{\pi}{2},\ \frac{3\pi}{2}$$

56. Graphing the function on the interval $[0, \ 2\pi)$, we find the following zeros: 0.0, 0.524, 1.571, 2.618, 3.142, 3.665, 4.712, 5.760

Algebraically:

$$\sin 4x - \sin 2x = 0$$

$$2 \cos 3x \sin x = 0$$

$\cos 3x = 0$ or $\sin x = 0$

$$3x = \frac{\pi}{2} + n\pi \qquad \qquad x = 0, \ \pi$$

$$x = \frac{\pi}{6} + \frac{n\pi}{3}$$

$$x = \frac{\pi}{6}, \ \frac{\pi}{2}, \ \frac{5\pi}{6},$$

$$\frac{7\pi}{6}, \ \frac{3\pi}{2}, \ \frac{11\pi}{6}$$

58. Graphing the function on the interval $[0, \ 2\pi)$, we find the following zeros: 0.0, 4.189

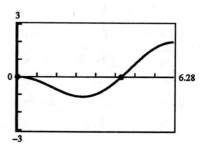

Algebraically:

$$\cos x = \cos \frac{x}{2}$$

$$\cos^2 x = \cos^2 \frac{x}{2}$$

$$\cos^2 x - \cos^2 \frac{x}{2} = 0$$

$$\left(\cos x + \cos \frac{x}{2} \right) \left(\cos x - \cos \frac{x}{2} \right) = 0$$

$$2 \cos \frac{3x}{4} \cos \frac{x}{4} \left(-2 \sin \frac{3x}{4} \sin \frac{x}{4} \right) = 0$$

$$-4 \cos \frac{3x}{4} \cos \frac{x}{4} \sin \frac{3x}{4} \sin \frac{x}{4} = 0$$

$\cos \dfrac{3x}{4} = 0 \Rightarrow x = \dfrac{4\pi}{3} \left(x = \dfrac{2\pi}{3} \text{ is extraneous.} \right)$

or $\cos \dfrac{x}{4} = 0 \Rightarrow$ (No solution in $[0, \ 2\pi)$)

or $\sin \dfrac{3x}{4} = 0$ or $\sin \dfrac{x}{4} = 0$

$\qquad \qquad x = 0, \ \dfrac{4\pi}{3} \qquad \qquad x = 0$

60. Graphing the function on the interval $[0, 2\pi)$, we find the following zeros:
1.047, 2.094, 3.142, 4.189, 5.236

Algebraically:

$$\sin s + \sin 3s + \sin 5s = 0$$

$$2\sin 3s \cos 2s + \sin 3s = 0$$

$$\sin 3s(2\cos 2s + 1) = 0$$

$\sin 3s = 0$ or $2\cos 2s + 1 = 0$

$3s = n\pi$ $\cos 2s = -\dfrac{1}{2}$

$s = \dfrac{n\pi}{3}$ $2s = \dfrac{2\pi}{3} + 2n\pi$

$s = 0, \dfrac{\pi}{3}, \dfrac{2\pi}{3},$ or $2s = \dfrac{4\pi}{3} + 2n\pi$

$\pi, \dfrac{4\pi}{3}, \dfrac{5\pi}{3}$ $s = \dfrac{\pi}{3}, \dfrac{2\pi}{3}, \dfrac{4\pi}{3}, \dfrac{5\pi}{3}$

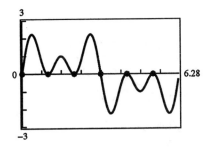

62. $\sin\left(x + \dfrac{\pi}{4}\right) - \sin\left(x - \dfrac{\pi}{4}\right) = 2\cos x \sin \dfrac{\pi}{4}$

$$= \sqrt{2}\cos x$$

64. $\cos \dfrac{x}{2} \cos \dfrac{x}{4} = \dfrac{1}{2}\left[\cos\left(\dfrac{x}{2} - \dfrac{x}{4}\right) + \cos\left(\dfrac{x}{2} + \dfrac{x}{4}\right)\right]$

$$= \dfrac{1}{2}\left(\cos \dfrac{x}{4} + \cos \dfrac{3x}{4}\right)$$

66. $\sin(2\arctan x) = \sin 2\theta$

$$= 2\sin\theta\cos\theta$$

$$= 2\left(\dfrac{x}{\sqrt{x^2+1}}\right)\left(\dfrac{1}{\sqrt{x^2+1}}\right)$$

$$= \dfrac{2x}{x^2+1}$$

68. $r = \frac{1}{32} v_0^2 \sin 2\theta$

$100 = \frac{1}{32}(80)^2 \sin 2\theta$

$\sin 2\theta = 0.5$

$2\theta = 30°$ or $2\theta = 180° - 30° = 150°$

$\theta = 15°$ $\qquad \theta = 75°$

70. Volume of the trough, V, will be the area of the isosceles triangle, A, times the length, l, of the trough.

$V = A \cdot l$

(a) $\quad A = \frac{1}{2}bh$

$\cos \dfrac{\theta}{2} = \dfrac{h}{1.5} \Rightarrow h = 1.5 \cos \dfrac{\theta}{2}$

$\sin \dfrac{\theta}{2} = \dfrac{(1/2)b}{1.5} \Rightarrow \dfrac{1}{2}b = 1.5 \sin \dfrac{\theta}{2}$

$A = 1.5 \sin \dfrac{\theta}{2} 1.5 \cos \dfrac{\theta}{2}$

$A = 1.5^2 \sin \dfrac{\theta}{2} \cos \dfrac{\theta}{2}$

$A = 2.25 \sin \dfrac{\theta}{2} \cos \dfrac{\theta}{2}$ square feet

$V = (2.25)(16) \sin \dfrac{\theta}{2} \cos \dfrac{\theta}{2}$ cubic feet

$V = 36 \sin \dfrac{\theta}{2} \cos \dfrac{\theta}{2}$ cubic feet

(b) $\quad V = 36 \sin \dfrac{\theta}{2} \cos \dfrac{\theta}{2}$

$V = 18 \left(2 \sin \dfrac{\theta}{2} \cos \dfrac{\theta}{2} \right)$

$V = 18 \sin \theta$ cubic feet

Maximum when $\theta = \pi/2$.

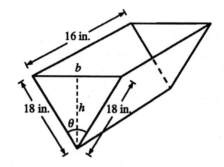

72. Given: $a = 6$, $b = 9$, $c = 45°$

$c^2 = a^2 + b^2 - 2ab \cos 45° \approx 36 + 81 - 2(6)(9)(0.7071) \approx 40.63 \Rightarrow c \approx 6.374$

$\cos B = \dfrac{a^2 + c^2 - b^2}{2ac} \approx \dfrac{36 + 40.63 - 81}{2(6)(6.374)} \approx -0.0571 \Rightarrow B \approx 93.3°, \; A \approx 41.7°$

74. Given: $B = 110°$, $C = 30°$, $c = 10.5$

$A = 180° - 110° - 30° = 40°$

$a = \dfrac{c \sin A}{\sin C} = \dfrac{10.5(\sin 40°)}{\sin 30°} \approx \dfrac{10.5(0.6428)}{0.5} \approx 13.5$

$b = \dfrac{c \sin B}{\sin C} = \dfrac{10.5(\sin 110°)}{\sin 30°} \approx \dfrac{10.5(0.9397)}{0.5} \approx 19.7$

76. Given: $a = 80$, $b = 60$, $c = 100$

$$\cos C = \frac{a^2 + b^2 - c^2}{2ab} = \frac{6400 + 3600 - 10000}{2(80)(60)} = 0 \Rightarrow C = 90°$$

$$\sin A = \frac{80}{100} = 0.8 \Rightarrow A \approx 53.1°$$

$$\sin B = \frac{60}{100} = 0.6 \Rightarrow B \approx 36.9°$$

78. Given: $A = 130°$, $a = 50$, $b = 30$

$$\sin B = \frac{b \sin A}{a} = \frac{30 \sin 130°}{50} \approx \frac{30(0.7660)}{50} \approx 0.4596 \Rightarrow B \approx 27.4°$$

$$C \approx 180° - 130° - 27.4° = 22.6°$$

$$c^2 = a^2 + b^2 - 2ab \cos C \approx 50^2 + 30^2 - 2(50)(30)(0.9232) \approx 630.4 \Rightarrow c \approx 25.11$$

80. Given: $C = 50°$, $a = 25$, $c = 22$

$$\sin A = \frac{a \sin C}{c} = \frac{25 \sin 50°}{22} \approx \frac{25(0.7660)}{22} \approx 0.8705 \Rightarrow A \approx 60.5° \text{ or } 119.5°$$

Case 1:   Case 2:

$A \approx 60.5°$ $A \approx 119.5°$

$B \approx 180° - 50° - 60.5° = 69.5°$ $B \approx 180° - 50° - 119.5° = 10.5°$

$b = \dfrac{c \sin B}{\sin C} \approx \dfrac{22(0.9367)}{0.7660} \approx 26.90$ $b = \dfrac{c \sin B}{\sin C} \approx 5.234$

82. Given: $B = 150°$, $a = 64$, $b = 10$

$$\sin A = \frac{a \sin B}{b} = \frac{64 \sin 150°}{10} \approx 3.2 \Rightarrow \text{no triangle formed}$$

84. Given: $a = 2.5$, $b = 15.0$, $c = 4.5$

Since $a + c < b$, a triangle is not formed.

86. Given: $B = 90°$, $a = 5$, $c = 12$

$$b = \sqrt{12^2 + 5^2} = \sqrt{169} = 13$$

$$A = \arctan \tfrac{5}{12} \approx 22.6°$$

$$C = \arctan \tfrac{12}{5} \approx 67.4°$$

88. $a = 15$, $b = 8$, $c = 10$

$$s = \frac{15 + 8 + 10}{2} = 16.5$$

$$\text{Area} = \sqrt{16.5(1.5)(8.5)(6.5)} \approx 36.98$$

90. $B = 80°$, $a = 4$, $c = 8$

$$\text{Area} = \tfrac{1}{2}ac \sin B = \tfrac{1}{2}(4)(8)(0.9848) = 15.76$$

92. $b^2 = a^2 + c^2 - 2ac \cos B = 325^2 + 450^2 - 2(325)(450) \cos 115° \approx 431{,}741 \Rightarrow b \approx 657 \text{ meters}$

94.
$$\frac{a}{\sin 75°} = \frac{400}{\sin 37.5°}$$
$$a = \frac{400 \sin 75°}{\sin 37.5°} \approx 634.7 \text{ ft}$$
$$\sin 67.5° = \frac{w}{a}$$
$$w = 634.7 \sin 67.5° \approx 586.4 \text{ ft}$$

96. $a^2 = 5^2 + 8^2 - 2(5)(8) \cos 152°$
$$\approx 159.6 \Rightarrow a \approx 12.63 \text{ ft}$$
$$b^2 = 5^2 + 8^2 - 2(5)(8) \cos 28°$$
$$\approx 18.36 \Rightarrow b \approx 4.285 \text{ ft}$$

98. $r = 6, \quad \theta = 150°$
$$-3\sqrt{3} + 3i = 6(\cos 150° + i \sin 150°)$$

100. $r = 7, \quad \theta = 180°$
$$-7 = 7(\cos 180° + i \sin 180°)$$

102. $24(\cos 330° + i \sin 330°) = 24\left(\dfrac{\sqrt{3}}{2} - \dfrac{1}{2}i\right)$
$$= 12\sqrt{3} - 12i$$

104. $8\left(\cos \dfrac{5\pi}{6} + i \sin \dfrac{5\pi}{6}\right) = 8\left(-\dfrac{\sqrt{3}}{2} + \dfrac{1}{2}i\right)$
$$= -4\sqrt{3} + 4i$$

106. (a) $z_1 = 2\sqrt{3} - 2i = 4(\cos 330° + i \sin 330°)$
$$z_2 = -10i = 10(\cos 270° + i \sin 270°)$$

    (b) $z_1 z_2 = [4(\cos 330° + i \sin 330°)][10(\cos 270° + i \sin 270°)]$
$$= 40(\cos 600° + i \sin 600°)$$
$$= 40(\cos 240° + i \sin 240°)$$
$$\approx -20.00 - 34.64i$$
$$\frac{z_1}{z_2} = \frac{4(\cos 330° + i \sin 330°)}{10(\cos 270° + i \sin 270°)}$$
$$= 0.4(\cos 60° + i \sin 60°)$$
$$\approx 0.4(0.5 + 0.8660i)$$
$$\approx 0.2 + 0.3464i$$

108. (a) $z_1 = 5i = 5(\cos 90° + i \sin 90°)$

$\qquad z_2 = 2 - 2i = 2\sqrt{2}(\cos 315° + i \sin 315°)$

(b) $z_1 z_2 = [5(\cos 90° + i \sin 90°)][2\sqrt{2}(\cos 315° + i \sin 315°)]$

$\qquad = 10\sqrt{2}(\cos 405° + i \sin 405°)$

$\qquad = 10\sqrt{2}(\cos 45° + i \sin 45°)$

$\qquad = 10 + 10i$

$\dfrac{z_1}{z_2} = \dfrac{5(\cos 90° + i \sin 90°)}{2\sqrt{2}(\cos 315° + i \sin 315°)}$

$\qquad = \dfrac{5\sqrt{2}}{4}[\cos(-225°) + i \sin(-225°)]$

$\qquad = \dfrac{5\sqrt{2}}{4}(\cos 135° + i \sin 135°)$

$\qquad = -\dfrac{5}{4} + \dfrac{5}{4}i$

110. $\left[2\left(\cos \dfrac{4\pi}{15} + i \sin \dfrac{4\pi}{15}\right)\right]^5 = 2^5\left(\cos \dfrac{4\pi}{3} + i \sin \dfrac{4\pi}{3}\right)$

$\qquad = 32\left(-\dfrac{1}{2} - \dfrac{\sqrt{3}}{2}i\right)$

$\qquad = -16 - 16\sqrt{3}\,i$

112. $(1 - i)^8 = [\sqrt{2}(\cos 315° + i \sin 315°)]^8$

$\qquad = 16(\cos 2520° + i \sin 2520°)$

$\qquad = 16(\cos 0° + i \sin 0°)$

$\qquad = 16$

114. Fourth roots of $256 = 256(\cos 0 + i \sin 0)$:

$\qquad \sqrt[4]{256}\left(\cos \dfrac{2\pi k}{4} + i \sin \dfrac{2\pi k}{4}\right)$, $\quad k = 0,\ 1,\ 2,\ 3$

$\qquad 4(\cos 0 + i \sin 0) = 4$

$\qquad 4(\cos \pi + i \sin \pi) = -4$

$\qquad 4\left(\cos \dfrac{\pi}{2} + i \sin \dfrac{\pi}{2}\right) = 4i$

$\qquad 4\left(\cos \dfrac{3\pi}{2} + i \sin \dfrac{3\pi}{2}\right) = -4i$

116. Fourth roots of $-1 + i = \sqrt{2}\left(\cos\dfrac{3\pi}{4} + i\sin\dfrac{3\pi}{4}\right)$:

$$(\sqrt{2})^{1/4}\left(\cos\dfrac{(3\pi/4) + 2k\pi}{4} + i\sin\dfrac{(3\pi/4) + 2k\pi}{4}\right), \ k = 0, \ 1, \ 2, \ 3$$

$$(\sqrt{2})^{1/4}\left(\cos\dfrac{3\pi}{16} + i\sin\dfrac{3\pi}{16}\right) \approx 0.9067 + 0.6059i$$

$$(\sqrt{2})^{1/4}\left(\cos\dfrac{11\pi}{16} + i\sin\dfrac{11\pi}{16}\right) \approx -0.6059 + 0.9067i$$

$$(\sqrt{2})^{1/4}\left(\cos\dfrac{19\pi}{16} + i\sin\dfrac{19\pi}{16}\right) \approx -0.9067 - 0.6059i$$

$$(\sqrt{2})^{1/4}\left(\cos\dfrac{27\pi}{16} + i\sin\dfrac{27\pi}{16}\right) \approx 0.6059 - 0.9067i$$

118. $x^5 - 32 = 0$

$$x^5 = 32$$

$$32 = 32(\cos 0 + i\sin 0)$$

$$\sqrt[5]{32} = \sqrt[5]{32}\left[\cos\left(0 + \dfrac{2\pi k}{5}\right) + i\sin\left(\dfrac{0 + 2\pi k}{5}\right)\right],$$

$$k = 0, \ 1, \ 2, \ 3, \ 4$$

$$2(\cos 0 + i\sin 0) = 2$$

$$2\left(\cos\dfrac{2\pi}{5} + i\sin\dfrac{2\pi}{5}\right) = 0.6180 + 1.9021i$$

$$2\left(\cos\dfrac{4\pi}{5} + i\sin\dfrac{4\pi}{5}\right) = -1.6180 + 1.1756i$$

$$2\left(\cos\dfrac{6\pi}{5} + i\sin\dfrac{6\pi}{5}\right) = -1.6180 - 1.1756i$$

$$2\left(\cos\dfrac{8\pi}{5} + i\sin\dfrac{8\pi}{5}\right) = 0.6180 - 1.9021i$$

120. $x^3 + 8i = 0$

$$x^3 = -8i$$

$$-8i = 8\left(\cos\frac{3\pi}{2} + i\sin\frac{3\pi}{2}\right)$$

$$\sqrt[3]{-8i} = \sqrt[3]{8}\left[\cos\left(\frac{(3\pi/2)+2\pi k}{3}\right) + i\sin\left(\frac{(3\pi/2)+2\pi k}{3}\right)\right]$$

$$k = 0,\ 1,\ 2$$

$$2\left(\cos\frac{\pi}{2} + i\sin\frac{\pi}{2}\right) = 2i$$

$$2\left(\cos\frac{7\pi}{6} + i\sin\frac{7\pi}{6}\right) = -\sqrt{3} - i$$

$$2\left(\cos\frac{11\pi}{6} + i\sin\frac{11\pi}{6}\right) = \sqrt{3} - i$$

CHAPTER SEVEN
Linear Models and Systems of Equations

7.1 Solving Systems of Equations Algebraically and Graphically

2. $x - y = -5$ Equation 1

$x + 2y = 4$ Equation 2

Solve for x in Equation 1: $x = y - 5$
Substitute for x in Equation 2: $(y - 5) + 2y = 4$
Solve for y: $3y = 9$ $\Rightarrow$ $y = 3$
Backsubstitute $y = 3$: $x = y - 5 = 3 - 5 = -2$
Solution: $(-2, 3)$

4. $3x - y = -2$ Equation 1

$x^3 - y = 0$ Equation 2

Solve for y in Equation 1: $y = 3x + 2$
Substitute for y in Equation 2: $x^3 - (3x + 2) = 0$
Solve for x: $x^3 - 3x - 2 = 0$ $\Rightarrow$ $(x - 2)(x + 1)^2 = 0$ $\Rightarrow$ $x = 2, -1$
Backsubstitute $x = 2$: $y = 3x + 2 = 6 + 2 = 8$
Backsubstitute $x = -1$: $y = 3x + 2 = -3 + 2 = -1$
Solutions: $(-1, -1)$, $(2, 8)$

6. $x - y = 0$ Equation 1

$x^3 - 5x + y = 0$ Equation 2

Solve for y in Equation 1: $y = x$
Substitute for y in Equation 2: $x^3 - 5x + x = 0$
Solve for x: $x^3 - 4x = 0$ $\Rightarrow$ $x(x^2 - 4) = 0$ $\Rightarrow$ $x = 0, \pm 2$
Backsubstitute $x = 2$: $y = x = 2$
Backsubstitute $x = -2$: $y = x = -2$
Backsubstitute $x = 0$: $y = x = 0$
Solutions: $(2, 2)$, $(-2, -2)$, $(0, 0)$

8. $y = -x^2 + 1$ Equation 1

$y = \quad x^4 - 2x^2 + 1$ Equation 2

Substitute for y in Equation 1: $x^2 + (x^4 - 2x^2 + 1) = 1$

Solve for x: $x^4 - x^2 = 0$ $\Rightarrow$ $x^2(x^2 - 1) = 0$ $\Rightarrow$ $x = 0, \pm 1$

Backsubstitute $x = 0$: $y = x^4 - 2x^2 + 1 = 1$

Backsubstitute $x = 1$: $y = x^4 - 2x^2 + 1 = 0$

Backsubstitute $x = -1$: $y = x^4 - 2x^2 + 1 = 0$

Solutions: $(0, 1)$, $(1, 0)$, $(-1, 0)$

10. $y = x^3 - 3x^2 + 3$ Equation 1

$y = \quad - 2x + 3$ Equation 2

Substitute for y in Equation 2: $-2x + 3 = x^3 - 3x^2 + 3$

Solve for x: $x^3 - 3x^2 + 2x = 0$ $\Rightarrow$ $x(x - 2)(x - 1) = 0$ $\Rightarrow$ $x = 0, 2, 1$

Backsubstitute $x = 0$: $y = x^3 - 3x^2 + 3 = 3$

Backsubstitute $x = 2$: $y = x^3 - 3x^2 + 3 = 8 - 12 + 3 = -1$

Backsubstitute $x = 1$: $y = x^3 - 3x^2 + 3 = 1 - 3 + 3 = 1$

Solutions: $(0, 3)$, $(2, -1)$, $(1, 1)$

12. $x + 2y = \quad 1$ Equation 1

$5x - 4y = -23$ Equation 2

Solve for x in Equation 1: $x = 1 - 2y$

Substitute for x in Equation 2:

$5(1 - 2y) - 4y = -23$

Solve for y: $-14y = -28$ $\Rightarrow$ $y = 2$

Backsubstitute $y = 2$: $x = 1 - 2y = 1 - 4 = -3$

Solution: $(-3, 2)$

14. $6x - 3y - 4 = 0$ Equation 1

$x + 2y - 4 = 0$ Equation 2

Solve for x in Equation 2: $x = 4 - 2y$

Substitute for x in Equation 1: $6(4 - 2y) - 3y - 4 = 0$

Solve for y: $24 - 12y - 3y - 4 = 0$ $\Rightarrow$ $-15y = -20$ $\Rightarrow$ $y = \frac{4}{3}$

Backsubstitute $y = \frac{4}{3}$: $x = 4 - 2y = 4 - 2\left(\frac{4}{3}\right) = \frac{4}{3}$

Solution: $\left(\frac{4}{3}, \frac{4}{3}\right)$

16. $1.5x + 0.8y = 2.3$ Equation 1

$0.3x - 0.2y = 0.1$ Equation 2

Solve for y in Equation 2: $y = 1.5x - 0.5$

Substitute for y in Equation 1: $1.5x + 0.8(1.5x - 0.5) = 2.3$

Solve for x: $1.5x + 1.2x - 0.4 = 2.3$ $\Rightarrow$ $2.7x = 2.7$ $\Rightarrow$ $x = 1$

Backsubstitute $x = 1$: $y = 1.5x - 0.5 = 1.5 - 0.5 = 1$

Solution: $(1, 1)$

18. $\frac{1}{2}x + \frac{3}{4}y = 10$ Equation 1

    $\frac{3}{2}x - \ y = 4$ Equation 2

Solve for y in Equation 2: $y = \frac{3}{2}x - 4$

Substitute for y in Equation 1: $\frac{1}{2}x + \frac{3}{4}\left(\frac{3}{2}x - 4\right) = 10$

Solve for x: $\frac{1}{2}x + \frac{9}{8}x - 3 = 10$ $\Rightarrow$ $\frac{13}{8}x = 13$ $\Rightarrow x = 8$

Backsubstitute $x = 8$: $y = \frac{3}{2}x - 4 = \frac{3}{2}(8) - 4 = 12 - 4 = 8$

Solution: $(8, 8)$

20. $x - 2y = 0$ Equation 1

    $3x - \ y = 0$ Equation 2

Solve for x in Equation 1: $x = 2y$

Substitute for x in Equation 2: $3(2y) - y = 0$

Solve for y: $6y - y = 0$ $\Rightarrow$ $5y = 0$ $\Rightarrow$ $y = 0$

Backsubstitute $y = 0$: $x = 2y = 0$

Solution: $(0, 0)$

22. $x + y = 4$ Equation 1

    $x^2 - y = 2$ Equation 2

Solve for y in Equation 1: $y = 4 - x$

Substitute for y in Equation 2: $x^2 - (4 - x) = 2$

Solve for x: $x^2 + x - 6 = 0$ $\Rightarrow$ $(x + 3)(x - 2) = 0$ $\Rightarrow$ $x = -3, 2$

Backsubstitute $x = -3$: $y = 4 - x = 4 - (-3) = 7$

Backsubstitute $x = 2$: $y = 4 - x = 4 - (2) = 2$

Solutions: $(-3, \ 7), \ (2, \ 2)$

24. $x^2 + y^2 = 25$ Equation 1

    $2x + y = 10$ Equation 2

Solve for y in Equation 2: $y = 10 - 2x$

Substitute for y in Equation 1: $x^2 + (10 - 2x)^2 = 25$

Solve for x: $x^2 + 100 - 40x + 4x^2 - 25 = 0 \Rightarrow 5x^2 - 40x + 75 = 0 \Rightarrow (x - 3)(x - 5) = 0 \Rightarrow x = 3, 5$

Backsubstitute $x = 3$: $y = 10 - 2x = 10 - 2(3) = 4$

Backsubstitute $x = 5$: $y = 10 - 2x = 10 - 2(5) = 0$

Solutions: $(3, \ 4), \ (5, \ 0)$

26. $y = x^3 - 2x^2 + x - 1$ Equation 1

$y = -x^2 + 3x - 1$ Equation 2

Substitute for y in Equation 1: $-x^2 + 3x - 1 = x^3 - 2x^2 + x - 1$

Solve for x: $x^3 - x^2 - 2x = 0$ $\Rightarrow$ $x(x-2)(x+1) = 0$ $\Rightarrow$ $x = 0,\ 2,\ -1$

Backsubstitute $x = 0$: $y = -x^2 + 3x - 1 = -1$

Backsubstitute $x = 2$: $y = -x^2 + 3x - 1 = -2^2 + 3(2) - 1 = -4 + 6 - 1 = 1$

Backsubstitute $x = -1$: $y = -x^2 + 3x - 1 = -(-1)^2 + 3(-1) - 1 = -1 - 3 - 1 = -5$

Solutions: $(0, -1),\ (2, 1),\ (-1, -5)$

28. $x^2 + y = 4 \Rightarrow y = -x^2 + 4$

$e^x - y = 0 \Rightarrow y = e^x$

Graphing the two functions, you can see that there are two points of intersection. Using the zoom and trace features, you will find that the solutions are $(1.06, 2.88)$ and $(-1.96, 0.14)$.

30. $x - 2y = 1$ Equation 1

$y = \sqrt{x-1}$ Equation 2

Substitute for y in Equation 1: $x - 2\sqrt{x-1} = 1$

Solve for x: $2\sqrt{x-1} = x - 1 \Rightarrow 4(x-1) = x^2 - 2x + 1$

$$\Rightarrow x^2 - 6x + 5 = 0$$

$$\Rightarrow (x-5)(x-1) = 0$$

$$\Rightarrow x = 5,\ 1$$

Backsubstitute $x = 5$: $y = \sqrt{x-1} = \sqrt{5-1} = 2$

Backsubstitute $x = 1$: $y = \sqrt{x-1} = \sqrt{1-1} = 0$

Solution: $(5, 2),\ (1, 0)$

32. $x - y + 3 = 0$ $\Rightarrow$ $y = x + 3$

$x^2 - 4x + 7 = y$ $\Rightarrow$ $(x-2)^2 + 3 = y$

Solutions: $(1, 4),\ (4, 7)$

34. $3x - 2y = 0$ $\Rightarrow$ $y = \frac{3}{2}x$

$x^2 - y^2 = 4$

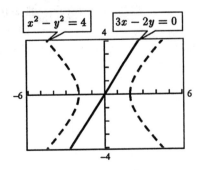

No points of intersection

36. $x^2 + y^2 = 8$

$\quad\quad y = x^2$

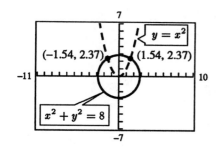

Solution:

$$\left(\pm\sqrt{\frac{-1+\sqrt{33}}{2}}, \ \frac{-1+\sqrt{33}}{2}\right) = (\pm 1.54, \ 2.37)$$

40. $x - y = 3 \quad\Rightarrow\quad y = x - 3$

$\quad\quad x - y^2 = 1$

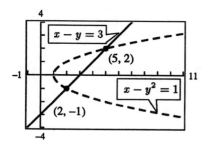

Solutions: $(2, -1)$, $(5, 2)$

Another way to solve this problem is to reverse the roles of x and y. Let $X = y$ and $Y = x$ to obtain $Y = X = 3$ and $Y = X^2 + 1$. Then $X^2 + 1 = X + 3 \Rightarrow X = -1, 2$ and $Y = 2, 5$. Returning to the original variables, the solutions are $(2, -1)$ and $(5, 2)$.

38. $x + 2y = 8 \quad\quad \Rightarrow \quad y = -\frac{1}{2}x + 4$

$\quad\quad y = \log_2 x$

Solution: $(4, 2)$

42. $x^2 + y^2 = 4$

$\quad\quad 2x^2 - y = 2$

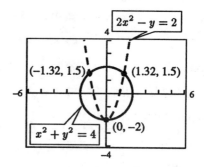

Solutions: $\left(\pm\dfrac{\sqrt{7}}{2}, \ \dfrac{3}{2}\right) = (\pm 1.32, \ 1.5), \ (0, -2)$

44. $C = 5.5\sqrt{x} + 10,000, \quad R = 3.29x$

$$R = C$$
$$3.29x = 5.5\sqrt{x} + 10,000$$
$$3.29x - 10,000 = 5.5\sqrt{x}$$
$$10.8241x^2 - 65,800x + 100,000,000 = 30.25x$$
$$10.8241x^2 - 65,830.25x + 100,000,000 = 0$$
$$x \approx 3133.7 \text{ units}$$

In order for the revenue to break even with the cost, 3134 units must be sold.

46. $C = 0.08x + 50,000, \quad R = 0.25x$

$$R = C$$
$$0.25x = 0.08x + 50,000$$
$$0.17x = 50,000$$
$$x \approx 294,117.6$$

In order for the revenue to break even with the cost, 294,118 units must be sold.

48. $C = 21.60x + 5000, \quad R = 34.10x$

$$R = C$$
$$34.10x = 21.60x + 5000$$
$$12.5x = 5000$$
$$x = 400 \text{ units}$$

50. $x + y = 18,000 \quad \Rightarrow \quad y = 18,000 - x$

$$0.0775x + 0.0825y = 1455$$
$$0.0775x + 0.0825(18,000 - x) = 1455$$
$$-0.005x + 1485 = 1455$$
$$0.005x = 30$$
$$x = \$6000 \text{ at } 7.75\%$$
$$y = 18,000 - 6000$$
$$= \$12,000 \text{ at } 8.25\%$$

52. $0.02x + 15,000 = 0.01x + 20,000$

$$0.01x = 5000$$
$$x = \$500,000$$

To make the second offer better, you would have to sell more than $500,000 per year.

54. (a)

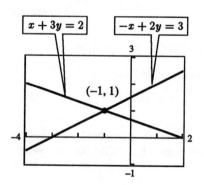

(b)
$$1.45 + 0.00014x^2 = (2.388 - 0.007x)^2$$
$$1.45 + 0.00014x^2 = 5.702544 - 0.033432x + 0.000049x^2$$
$$0.000091x^2 + 0.033432x - 4.252544 = 0$$

$$x \approx 100 \text{ bushels per day}$$

When $x = 100$, $p = 1.45 + 0.00014(100)^2 = \$2.85$ per bushel.

56. $A = \frac{1}{2}bh$

$1 = \frac{1}{2}a^2$

$a^2 = 2$

$a = \sqrt{2}$

The dimensions are $\sqrt{2} \times \sqrt{2} \times 2$.

58. (a) $y = 2^x$ and $y = x^2$ intersect at $(-0.77, 0.59)$, $(2, 4)$, and $(4, 16)$.

(b) $y = 10^x$ and $y = x^{10}$ intersect at $(-0.83, -0.15)$, $(1.37, 23.51)$, and $(10, 10^{10})$.

7.2 Systems of Linear Equations in Two Variables

2.

$$
\begin{array}{rcl}
x + 3y &=& 2 \\
-x + 2y &=& 3 \\
\hline
5y &=& 5 \\
y &=& 1 \\
x &=& -1
\end{array}
$$

Solution: $(-1, 1)$

4. $2x - y = 2 \Rightarrow 6x - 3y = 6$

$4x + 3y = 24 \Rightarrow \underline{4x + 3y = 24}$

$\qquad\qquad\qquad 10x \qquad = 30$

$\qquad\qquad\qquad\qquad x = 3$

$\qquad\qquad\qquad\qquad y = 4$

Solution: (3, 4)

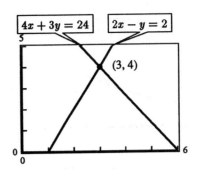

6. $3x + 2y = 2 \Rightarrow -6x - 4y = -4$

$6x + 4y = 14 \Rightarrow \underline{6x + 4y = 14}$

$\qquad\qquad\qquad\qquad\quad 0 = 10$

Inconsistent; no solution

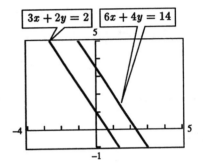

8. $x - 2y = 5$

$\underline{6x + 2y = 7}$

$7x \qquad = 12$

$\qquad x = \frac{12}{7}$

$\qquad y = -\frac{23}{14}$

Solution: $\left(\frac{12}{7}, -\frac{23}{14}\right)$

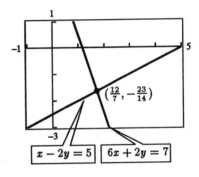

10. $5x + 3y = 18 \Rightarrow 35x + 21y = 126$

$2x - 7y = -1 \Rightarrow \underline{6x - 21y = -3}$

$\qquad\qquad\qquad\qquad 41x \qquad = 123$

$\qquad\qquad\qquad\qquad\quad x = 3$

$\qquad\qquad\qquad\qquad\quad y = 1$

Solution: (3, 1)

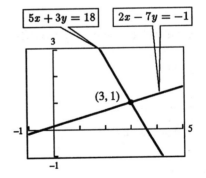

12. $3x - 5y = 2$

$\underline{2x + 5y = 13}$

$5x \quad\quad = 15$

$x = 3$

$y = \frac{7}{5}$

Solution: $\left(3, \frac{7}{5}\right)$

14. $x + 7y = 12 \Rightarrow -3x - 21y = -36$

$3x - 5y = 10 \Rightarrow \underline{\quad 3x - 5y = \quad 10 \quad}$

$-26y = -26$

$y = \quad 1$

$x = \quad 5$

Solution: $(5, 1)$

16. $8r + 16s = 20 \Rightarrow -16r - 32s = -40$

$16r + 50s = 55 \Rightarrow \underline{\quad 16r + 50s = \quad 55 \quad}$

$18s = \quad 15$

$s = \quad \frac{5}{6}$

$r = \quad \frac{5}{6}$

Solution: $\left(\frac{5}{6}, \frac{5}{6}\right)$

18. $5u + 6v = 24 \Rightarrow \quad 15u + 18v = \quad 72$

$3u + 5v = 18 \Rightarrow \underline{-15u - 25v = -90}$

$-7v = -18$

$v = \quad \frac{18}{7}$

$u = \quad \frac{12}{7}$

Solution: $\left(\frac{12}{7}, \frac{18}{7}\right)$

20. $1.8x + 1.2y = 4 \Rightarrow \quad 18x + 12y = 40$

$9x + \quad 6y = 3 \Rightarrow \underline{-18x - 12y = -6}$

$0 = 34$

Inconsistent; no solution

22. $\frac{2}{3}x + \frac{1}{6}y = \frac{2}{3} \Rightarrow \quad 4x + y = \quad 4$

$4x + \quad y = 4 \Rightarrow \underline{-4x - y = -4}$

$0 = \quad 0$

Solution: All points (x, y) lying on the line $4x + y = 4$

24. $\dfrac{x - 1}{2} + \dfrac{y + 2}{3} = 4 \Rightarrow 3(x - 1) + 2(y + 2) = 24 \Rightarrow 3x + 2y = 23$

$x - 2y = 5 \Rightarrow$

$\underline{\quad x - 2y = \quad 5 \quad}$

$4x \quad\quad = 28$

$x = 7$

$y = 1$

Solution: $(7, 1)$

26. $0.02x - 0.05y = -0.19 \Rightarrow 2x - 5y = -19 \Rightarrow -6x + 15y = \quad 57$

$0.03x + 0.04y = \quad 0.52 \Rightarrow 3x + 4y = \quad 52 \Rightarrow \underline{\quad 6x + \quad 8y = 104}$

$23y = 161$

$y = \quad 7$

$x = \quad 8$

Solution: $(8, 7)$

28. $0.2x - 0.5y = -27.8 \Rightarrow 2x - 5y = -278 \Rightarrow \quad 8x - 20y = -1112$

$0.3x + 0.4y = \quad 68.7 \Rightarrow 3x + 4y = \quad 687 \Rightarrow \quad \underline{15x + 20y = \quad 3435}$

$$23x \quad\quad = \quad 2323$$

$$x = \quad 101$$

$$y = \quad 96$$

Solution: $(101, 96)$

30. $3b + 3m = 7 \Rightarrow -3b - 3m = -7$

$3b + 5m = 3 \Rightarrow \quad \underline{3b + 5m = \quad 3}$

$$2m = -4$$

$$m = -2$$

$$b = \tfrac{13}{3}$$

Solution: $\left(\tfrac{13}{3}, -2\right)$

32. $25x - 24y = \quad 0 \Rightarrow \quad 25x - 24y = \quad 0$

$13x - 12y = 120 \Rightarrow \quad \underline{-26x + 24y = -240}$

$$-x \quad\quad = -240$$

$$x = \quad 240$$

$$y = \quad 250$$

Solution: $(240, 250)$

It is necessary to change the scale on the axis to see the point of intersection of the lines.

34. $2x + \tfrac{3}{2}y = 2000 \Rightarrow \quad 4x + 3y = 4000$

$\quad y - \quad x = \quad 50 \Rightarrow \quad \underline{-4x + 4y = \quad 200}$

$$7y = 4200$$

$$y = \quad 600$$

$$x = \quad 550$$

Solution:
The speed of the first plane is 550 mph.
The speed of the second plane is 600 mph.

36. Let $x =$ amount of 87 octane

$y =$ amount of 92 octane

$$x + y = 500$$

$$87x + 92y = 500(89)$$

$$87x + 92(500 - x) = 500(89) \Rightarrow$$

$$x = 300, \; y = 200$$

38. $x + \quad y = 32{,}000 \Rightarrow -575x - 575y = -18{,}400{,}000$

$0.0575x + 0.0625y = \quad 1930 \Rightarrow \quad \underline{575x + 625y = \quad 19{,}300{,}000}$

$$50y = \quad 900{,}000$$

$$y = \quad 18{,}000$$

$$x = \quad 14{,}000$$

Solution: \$14,000 is invested in 5.75% bond and \$18,000 is invested in 6.25% bond.

40. $x + \quad y = \quad 240 \Rightarrow -6695x - 6695y = -1{,}606{,}800$

$66.95x + 84.95y = 17{,}652 \Rightarrow \quad \underline{6695x + 8495y = \quad 1{,}765{,}200}$

$$1800y = \quad 158{,}400$$

$$y = \quad 88$$

$$x = \quad 152$$

Solution: 152 pairs of \$66.95 shoes and 88 pairs of \$84.95 shoes

42. Supply = Demand

$$10 + \tfrac{7}{3}x = 60 - x$$
$$\tfrac{10}{3}x = 50$$
$$x = 15$$
$$p = 45$$

Equilibrium point: (15, 45)

44. Supply = Demand

$$25 + 0.1x = 100 - 0.05x$$
$$0.15x = 75$$
$$x = 500$$
$$p = 75$$

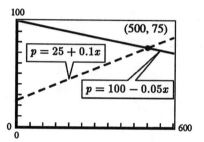

Equilibrium point: (500, 75)

46. Supply = Demand

$$225 + 0.0005x = 400 - 0.0002x$$
$$0.0007x = 175$$
$$x = 250,000$$
$$p = 350$$

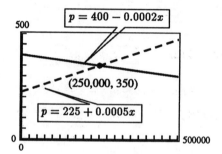

Equilibrium point: (250,000, 350)

48. $x + y = 1600 \Rightarrow -x - y = -1600$

$\quad\quad x = 4y \Rightarrow \underline{\quad x - 4y = \quad\quad 0}$

$$-5y = -1600$$
$$y = 320$$
$$x = 1280$$

Solution: One company hauled 320 tons and the other company hauled 1280 tons.

50. $5b + 10a = 11.7 \Rightarrow -10b - 20a = -23.4$

$\quad 10b + 30a = 25.6 \Rightarrow \underline{\quad 10b + 30a = \quad 25.6}$

$$10a = 2.2$$
$$a = 0.22$$
$$b = 1.90$$

$$y = ax + b$$
$$y = 0.22x + 1.90$$

52. $6b + 15a = 23.6 \Rightarrow -30b - 75a = -118$

$15b + 55a = 48.8 \Rightarrow \underline{30b + 110a = 97.6}$

$35a = -20.4$

$a = -\frac{20.4}{35}$

≈ -0.583

$b = \frac{566}{105}$

≈ 5.390

$y = ax + b$

$y = -0.583x + 5.390$

54. $n = 4, \qquad \sum_{i=1}^{4} x_i = 0, \qquad \sum_{i=1}^{4} y_i = 4,$

$\sum_{i=1}^{4} x_i^2 = 20, \qquad \sum_{i=1}^{4} x_i y_i = 6$

$4b + 0 = 4$

$4b = 4$

$b = 1$

$0 + 20a = 6$

$a = \frac{6}{20}$

$a = \frac{3}{10}$

$y = ax + b$

$y = \frac{3}{10}x + 1$

56. $n = 8, \qquad \sum_{i=1}^{8} x_i = 28, \qquad \sum_{i=1}^{8} y_i = 8,$

$\sum_{i=1}^{8} x_i^2 = 116, \qquad \sum_{i=1}^{8} x_i y_i = 37$

$8b + 28a = 8 \Rightarrow -56b - 196a = -56$

$28b + 116a = 37 \Rightarrow \underline{56b + 232a = 74}$

$36a = 18$

$a = \frac{1}{2}$

$b = -\frac{3}{4}$

$y = ax + b$

$y = \frac{1}{2}x - \frac{3}{4}$

58. $(1.0, 32)$, $(1.5, 41)$, $(2.0, 48)$, $(2.5, 53)$

$$n = 4, \qquad \sum_{i=1}^{4} x_i = 7, \qquad \sum_{i=1}^{4} y_i = 174,$$

$$\sum_{i=1}^{4} x_i^2 = 13.5, \qquad \sum_{i=1}^{4} x_i y_i = 322$$

$$4b + 7a = 174 \Rightarrow 28b + 49a = 1218$$
$$7b + 13.5a = 322 \Rightarrow \underline{-28b - 54a = -1288}$$
$$-5a = -70$$
$$a = 14$$
$$b = 19$$

$y = ax + b$

$y = 14x + 19$

When $x = 1.6$, $y = 14(1.6) + 19 = 41.4$ bushels per acre.

60. Since $x = 8$ and $y = -2$, one possible system of linear equations is

$$2x - y = 18$$
$$-3x + 4y = -32.$$

7.3 Systems of Linear Equations in More Than Two Variables

2.
$$x + y + z = 2$$
$$-x + 3y + 2z = 8$$
$$4x + y = 4$$

$$x + y + z = 2$$
$$4y + 3z = 10$$
$$-3y - 4z = -4$$

$$x + y + z = 2$$
$$12y + 9z = 30$$
$$-12y - 16z = -16$$

$$x + y + z = 2$$
$$12y + 9z = 30$$
$$-7z = 14$$

$$-7z = 14 \Rightarrow z = -2$$
$$12y + 9(-2) = 30 \Rightarrow y = 4$$
$$x + 4 - 2 = 2 \Rightarrow x = 0$$

Solution: $(0, 4, -2)$

4.
$$2x + 2z = 2$$
$$5x + 3y = 4$$
$$3y - 4z = 4$$

$$x + 3y - 4z = 0$$
$$2x + 2z = 2$$
$$3y - 4z = 4$$

$$x + 3y - 4z = 0$$
$$-6y + 10z = 2$$
$$3y - 4z = 4$$

$$x + 3y - 4z = 0$$
$$-6y + 10z = 2$$
$$z = 5$$

$$z = 5$$
$$-6y + 10(5) = 2 \Rightarrow y = 8$$
$$x + 3(8) - 4(5) = 0 \Rightarrow x = -4$$

Solution: $(-4, 8, 5)$

6.
$$2x + 4y + z = -4$$
$$2x - 4y + 6z = 13$$
$$4x - 2y + z = 6$$

$$2x + 4y + z = -4$$
$$-8y + 5z = 17$$
$$-10y - z = 14$$

$$2x + 4y + z = -4$$
$$-40y + 25z = 85$$
$$-40y - 4z = 56$$

$$2x + 4y + z = -4$$
$$-40y + 25z = 85$$
$$-29z = -29$$

$$-29z = -29 \Rightarrow z = 1$$
$$-40y + 25(1) = 85 \Rightarrow y = -\frac{3}{2}$$
$$2x + 4\left(-\frac{3}{2}\right) + 1 = -4 \Rightarrow x = \frac{1}{2}$$

Solution: $\left(\frac{1}{2}, -\frac{3}{2}, 1\right)$

8.
$$x - 11y + 4z = 3$$
$$5x - 3y + 2z = 3$$
$$2x + 4y - z = 7$$

$$x - 11y + 4z = 3$$
$$52y - 18z = -12$$
$$26y - 9z = 1$$

$$x - 11y + 4z = 3$$
$$52y - 18z = -12$$
$$0 = 7$$

Inconsistent; no solution

10. $2x + y + 3z = 1$

$2x + 6y + 8z = 3$

$6x + 8y + 18z = 5$

$2x + y + 3z = 1$

$5y + 5z = 2$

$5y + 9z = 2$

$2x + y + 3z = 1$

$5y + 5z = 2$

$4z = 0$

$4z = 0 \quad \Rightarrow \quad z = 0$

$5y + 5(0) = 2 \quad \Rightarrow \quad y = \frac{2}{5}$

$2x + \frac{2}{5} + 3(0) = 1 \quad \Rightarrow \quad x = \frac{3}{10}$

Solution: $\left(\frac{3}{10}, \frac{2}{5}, 0\right)$

12. $2x + y - 3z = 4$

$4x \quad + 2z = 10$

$-2x + 3y - 13z = -8$

$2x + y - 3z = 4$

$-2y + 8z = 2$

$4y - 16z = -4$

$2x + y - 3z = 4$

$y - 4z = -1$

$2x \quad + z = 5$

$y - 4z = -1$

$z = a$

$y = 4a - 1$

$x = -\frac{1}{2}a + \frac{5}{2}$

Solution: $\left(-\frac{1}{2}a + \frac{5}{2}, \, 4a - 1, \, a\right)$

14. $x + 2y - z = 5$

$4x - y + 5z = 11$

$5x - 8y + 13z = 7$

$x + 2y - z = 5$

$-9y + 9z = -9$

$-18y + 18z = -18$

$x + 2y - z = 5$

$y - z = 1$

$x \quad + z = 3$

$y - z = 1$

$z = a$

$y = a + 1$

$x = -a + 3$

Solution: $(-a + 3, \, a + 1, \, a)$

16. $x - 3y + 2z = 18$

$5x - 13y + 12z = 80$

$x - 3y + 2z = 18$

$2y + 2z = -10$

$x - 3y + 2z = 18$

$y + z = -5$

$x \quad + 5z = 3$

$y + z = -5$

$z = a$

$y = -a - 5$

$x = -5a + 3$

Solution: $(-5a + 3, \, -a - 5, \, a)$

18.
$$2x + 3y + 3z = 7$$
$$4x + 18y + 15z = 44$$

$$2x + 3y + 3z = 7$$
$$12y + 9z = 30$$

$$8x + 12y + 12z = 28$$
$$12y + 9z = 30$$

$$8x \quad\; + 3z = -2$$
$$4y + 3z = 10$$

$$z = a$$
$$y = -\tfrac{3}{4}a + \tfrac{5}{2}$$
$$x = -\tfrac{3}{8}a - \tfrac{1}{4}$$

Solution: $\left(-\tfrac{3}{8}a - \tfrac{1}{4},\ -\tfrac{3}{4}a + \tfrac{5}{2},\ a\right)$

20.
$$x + y + z + w = 6$$
$$2x + 3y \quad\;\; - w = 0$$
$$-3x + 4y + z + 2w = 4$$
$$x + 2y - z + w = 0$$

$$x + y + z + w = 6$$
$$y - 2z - 3w = -12$$
$$7y + 4z + 5w = 22$$
$$y - 2z \quad\;\; = -6$$

$$x + y + z + w = 6$$
$$y - 2z - 3w = -12$$
$$18z + 26w = 106$$
$$3w = 6$$

$$3w = 6 \quad \Rightarrow \quad w = 2$$
$$18z + 26(2) = 106 \quad \Rightarrow \quad z = 3$$
$$y - 2(3) - 3(2) = -12 \quad \Rightarrow \quad y = 0$$
$$x + 0 + 3 + 2 = 6 \quad \Rightarrow \quad x = 1$$

Solution: $(1, 0, 3, 2)$

22.
$$x - y - 5z = -3$$
$$3x - 2y - 6z = -4$$
$$-3x + 2y + 6z = 1$$

$$x - y - 5z = -3$$
$$y + 9z = 5$$
$$-y - 9z = -8$$

$$x - y - 5z = -3$$
$$y + 9z = 5$$
$$0 = -3$$

Inconsistent; no solution

24.
$$2x + 3y \quad\;\; = 0$$
$$4x + 3y - z = 0$$
$$8x + 3y + 3z = 0$$

$$2x + 3y \quad\;\; = 0$$
$$-3y - z = 0$$
$$-9y + 3z = 0$$

$$2x + 3y \quad\;\; = 0$$
$$-3y - z = 0$$
$$6z = 0$$

$$6z = 0 \quad \Rightarrow \quad z = 0$$
$$-3y - 0 = 0 \quad \Rightarrow \quad y = 0$$
$$2x + 3(0) = 0 \quad \Rightarrow \quad x = 0$$

Solution: $(0, 0, 0)$

26. $12x + 5y + z = 0$
$12x + 4y - z = 0$

$12x + 5y + z = 0$
$- y - 2z = 0$

$12x \qquad - 9z = 0$
$- y - 2z = 0$

$z = a$

$y = -2a$

$x = \frac{3}{4}a$

Solution: $\left(\frac{3}{4}a,\ -2a,\ a\right)$

28. $y = ax^2 + bx + c$
Passing through $(0, 5)$, $(1, 6)$, and $(2, 5)$

$5 =$	c	Equation 1
$6 = a + b + c$		Equation 2
$5 = 4a + 2b + c$		Equation 3

Solution: $a = -1$, $b = 2$, $c = 5$
The equation of the parabola is $y = -x^2 + 2x + 5$.

30. $y = ax^2 + bx + c$
Passing through $(1, 2)$, $(2, 1)$, and $(3, -4)$

$2 = a + b + c$	Equation 1
$1 = 4a + 2b + c$	Equation 2
$-4 = 9a + 3b + c$	Equation 3

Solution: $a = -2$, $b = 5$, $c = -1$
The equation of the parabola is
$y = -2x^2 + 5x - 1$.

32. $x^2 + y^2 + Dx + Ey + F = 0$
Passing through $(0, 0)$, $(0, 6)$, and $(-3, 3)$

$F = 0$	Equation 1
$36 + 6E + F = 0$	Equation 2
$18 - 3D + 3E + F = 0$	Equation 3

Solution: $D = 0$, $E = -6$, $F = 0$
The equation is $x^2 + y^2 - 6y = 0$.

34. $x^2 + y^2 + Dx + Ey + F = 0$
Passing through $(0, 0)$, $(0, 2)$, and $(3, 0)$

$F = 0$	Equation 1
$4 + 2E + F = 0$	Equation 2
$9 + 3D + F = 0$	Equation 3

Solution: $D = -3$, $E = -2$, $F = 0$
The equation is $x^2 + y^2 - 3x - 2y = 0$.

36. $s = \frac{1}{2}at^2 + v_0 t + s_0$
$(t,\ s) = (1,\ 48),\ (2,\ 64),\ (3,\ 48)$

$48 = \frac{1}{2}a + v_0 + s_0$	Equation 1
$64 = 2a + 2v_0 + s_0$	Equation 2
$48 = \frac{9}{2}a + 3v_0 + s_0$	Equation 3

$s = \frac{1}{2}(-32)t^2 + 64t + 0$

$s = -16t^2 + 64t$

38. $s = \frac{1}{2}at^2 + v_0 t + s_0$

$(t,\ s) = (2,\ 132),\ (3,\ 100),\ (4,\ 36)$

$132 = 2a + 2v_0 + s_0$ Equation 1

$100 = \frac{9}{2}a + 3v_0 + s_0$ Equation 2

$36 = 8a + 4v_0 + s_0$ Equation 3

$s = \frac{1}{2}(-32)t^2 + 48t + 100$

$s = -16t^2 + 48t + 100$

40. $0.05x + 0.07y + 0.08z = 1520$

$$x = \frac{y}{2}$$

$$y = z - 1500$$

$$0.05\left(\frac{y}{2}\right) + 0.07y + 0.08(y + 1500) = 1520$$

$$0.175y = 1400$$

$$y = 8000$$

$$x = \frac{y}{2} = 4000,\ \ z = y + 1500 = 9500$$

Solution: $x = \$4000$ at 5%

$y = \$8000$ at 7%

$z = \$9500$ at 8%

42. $\quad\quad\quad x + y + z = 800{,}000$

$0.08x + 0.09y + 0.10z = 67{,}000$

$$x = 5z$$

$y + \quad 6z = 800{,}000 \ \Rightarrow \quad 5y + 30z = \quad 4{,}000{,}000$

$0.09y + 0.50z = \ 67{,}000 \ \Rightarrow \ \underline{-5.4y - 30z = -4{,}020{,}000}$

$\quad\quad\quad\quad\quad\quad\quad\quad -0.4y \quad\quad = \quad -20{,}000$

$\quad\quad\quad\quad\quad\quad\quad\quad\quad\quad y = \quad 50{,}000$

$z = \dfrac{800{,}000 - 50{,}000}{6} = 125{,}000,\ \ x = 800{,}000 - 50{,}000 - 125{,}000 = 625{,}000$

Solution: $x = \$625{,}000$ at 8%

$y = \$50{,}000$ at 9%

$z = \$125{,}000$ at 10%

44.
$$C + M + B + G = 500{,}000$$
$$0.09C + 0.05M + 0.12B + 0.14G = 0.10(500{,}000)$$
$$B + G = \tfrac{1}{4}(500{,}000)$$

This system has infinitely many solutions.

Let $G = s$, then $B = 125{,}000 - s$
$$M = \tfrac{1}{2}s - 31{,}250$$
$$C = 406{,}250 - \tfrac{1}{2}s$$

Solution: $406{,}250 - \tfrac{1}{2}s$ in certificates of deposit
$\tfrac{1}{2}s - 31{,}250$ in municipal bonds
$125{,}000 - x$ in blue-chip stocks
s in growth stocks

46. (a) To use as little of the 50% solution as possible, the chemist should use no 10% solution.
$$x(0.20) + (10 - x)(0.50) = 10(0.25)$$
$$0.20x + 5 - 0.50x = 2.5$$
$$-0.30x = -2.5$$
$$x = 8\tfrac{1}{3} \text{ liters of 20\% solution}$$
$$10 - x = 1\tfrac{2}{3} \text{ liters of 50\% solution}$$

(b) To use as much 50% solution as possible, the chemist should use no 20% solution.
$$x(0.10) + (10 - x)0.50 = 10(0.25)$$
$$0.10x + 5 - 0.50x = 2.5$$
$$-0.40x = -2.5$$
$$x = 6\tfrac{1}{4} \text{ liters of 10\% solution}$$
$$10 - x = 3\tfrac{3}{4} \text{ liters of 50\% solution}$$

(c) To use 2 liters of 50% solution we let $x =$ the number of liters at 10% and $y =$ the number of liters at 20%.
$$0.10x + 0.20y + 2(0.50) = 10(0.25) \qquad \text{Equation 1}$$
$$x + y = 8 \qquad \text{Equation 2}$$

Solution: $y = 7$ liters of 20% solution
$x = 1$ liter of 10% solution

48.
$$I_1 - I_2 + I_3 = 0$$
$$3I_1 + 2I_2 \qquad = 7$$
$$2I_2 + 4I_3 = 8$$

$$I_1 - I_2 + I_3 = 0$$
$$5I_2 - 3I_3 = 7$$
$$2I_2 + 4I_3 = 8$$

$$I_1 - I_2 + I_3 = 0$$
$$10I_2 - 6I_3 = 14$$
$$10I_2 + 20I_3 = 40$$

$$I_1 - I_2 + I_3 = 0$$
$$10I_2 - 6I_3 = 14$$
$$26I_3 = 26$$

$$26I_3 = 26 \quad \Rightarrow \quad I_3 = 1$$
$$10I_2 - 6(1) = 14 \quad \Rightarrow \quad I_2 = 2$$
$$I_1 - 2 + 1 = 0 \quad \Rightarrow \quad I_1 = 1$$

Solution: $I_1 = 1$ ampere
$I_2 = 2$ amperes
$I_3 = 1$ ampere

50.
$$t_1 - 2t_2 \qquad = 0$$
$$t_1 \qquad - 2a = 128$$
$$t_2 + 2a = 64$$

$$t_1 - 2t_2 \qquad = 0$$
$$2t_2 - 2a = 128$$
$$t_2 + 2a = 64$$

$$t_1 - 2t_2 \qquad = 0$$
$$2t_2 - 2a = 128$$
$$3a = 0$$

$$3a = 0 \quad \Rightarrow \quad a = 0$$
$$2t_2 - 2(0) = 128 \quad \Rightarrow \quad t_2 = 64$$
$$t_1 - 2(64) = 0 \quad \Rightarrow \quad t_1 = 128$$

Solution: $a = 0$ ft/sec^2
$t_1 = 128$ lb
$t_2 = 64$ lb

52.
$$5c + 10a = 8$$
$$10b \qquad = 12$$
$$10c + 34a = 22$$

$$5c + 10a = 8$$
$$10b \qquad = 12$$
$$14a = 6$$

$$14a = 8 \quad \Rightarrow \quad a = \tfrac{3}{7}$$
$$10b = 12 \quad \Rightarrow \quad b = \tfrac{6}{5}$$
$$5c + 10\left(\tfrac{3}{7}\right) = 8 \quad \Rightarrow \quad c = \tfrac{26}{35}$$

$$y = ax^2 + bx + c$$
$$y = \tfrac{3}{7}x^2 + \tfrac{6}{5}x + \tfrac{26}{35}$$

54.
$$4c + 6b + 14a = 25$$
$$6c + 14b + 36a = 21$$
$$14c + 36b + 98a = 33$$

$$84c + 126b + 294a = 525$$
$$84c + 196b + 504a = 294$$
$$84c + 216b + 588a = 198$$

$$84c + 126b + 294a = 525$$
$$70b + 210a = -231$$
$$90b + 294a = -327$$

$$84c + 126b + 294a = 525$$
$$10b + 30a = -33$$
$$30b + 98a = -109$$

$$84c + 126b + 294a = 525$$
$$10b + 30a = -33$$
$$8a = -10$$

$$8a = -10 \quad \Rightarrow \quad a = -\frac{5}{4}$$
$$10b + 30\left(-\frac{5}{4}\right) = -33 \quad \Rightarrow \quad b = \frac{9}{20}$$
$$84c + 126\left(\frac{9}{20}\right) + 294\left(-\frac{5}{4}\right) \quad \Rightarrow \quad c = \frac{199}{20}$$

$$y = ax^2 + bx + c$$
$$y = -\frac{5}{4}x^2 + \frac{9}{20}x + \frac{199}{20}$$

56. (a) and (b)

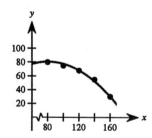

(b) (80, 80), (100, 75), (120, 68), (140, 55), (160, 30)

$$n = 5$$

$$\sum_{i=1}^{5} x_i = 600 \qquad \sum_{i=1}^{5} y_i = 308$$

$$\sum_{i=1}^{5} x_i{}^2 = 76{,}000 \qquad \sum_{i=1}^{5} x_i y_i = 34{,}560$$

$$\sum_{i=1}^{5} x_i{}^3 = 10{,}080{,}000 \qquad \sum_{i=1}^{5} x_i{}^2 y_i = 4{,}087{,}200$$

$$\sum_{i=1}^{5} x_i{}^4 = 1{,}387{,}840{,}000$$

–CONTINUED ON NEXT PAGE–

56. –CONTINUED–

$$5c + 600b + 76{,}000a = 308$$
$$600c + 76{,}000b + 10{,}080{,}000a = 34{,}560$$
$$76{,}000c + 10{,}080{,}000b + 1{,}387{,}840{,}000a = 4{,}087{,}200$$

Solving this system, we get $a \approx -0.008$, $b \approx 1.371$, and $c \approx 21.886$.

$$y = ax^2 + bx + c$$
$$y = -0.008x^2 + 1.371x + 21.886$$

58. Since $x = -\frac{3}{2}$, $y = 4$, and $z = -7$, one possible system of linear equations is

$$2x + 3y - 4z = 37$$
$$4x - y + z = -17$$
$$-2x + 3y + 2z = 1.$$

7.4 Partial Fractions

2. $\dfrac{1}{4x^2 - 9} = \dfrac{A}{2x + 3} + \dfrac{B}{2x - 3}$

$1 = A(2x - 3) + B(2x + 3)$

Let $x = -\frac{3}{2}$: $1 = -6A \Rightarrow A = -\frac{1}{6}$.
Let $x = \frac{3}{2}$: $1 = 6B \Rightarrow B = \frac{1}{6}$.

$$\frac{1}{4x^2 - 9} = \frac{1}{6}\left[\frac{1}{2x - 3} - \frac{1}{2x + 3}\right]$$

4. $\dfrac{3}{x^2 - 3x} = \dfrac{A}{x - 3} + \dfrac{B}{x}$

$3 = Ax + B(x - 3)$

Let $x = 3$: $3 = 3A \Rightarrow A = 1$.
Let $x = 0$: $3 = -3B \Rightarrow B = -1$.

$$\frac{3}{x^2 - 3x} = \frac{1}{x - 3} - \frac{1}{x}$$

6. $\dfrac{5}{x^2 + x - 6} = \dfrac{A}{x + 3} + \dfrac{B}{x - 2}$

$5 = A(x - 2) + B(x + 3)$

Let $x = -3$: $5 = -5A \Rightarrow A = -1$.
Let $x = 2$: $5 = 5B \Rightarrow B = 1$.

$$\frac{5}{x^2 + x - 6} = \frac{1}{x - 2} - \frac{1}{x + 3}$$

8. $\dfrac{x + 1}{x^2 + 4x + 3} = \dfrac{x + 1}{(x + 3)(x + 1)}$

$$= \frac{1}{x + 3}, \quad x \neq -1$$

10. $\dfrac{3x^2 - 7x - 2}{x^3 - x} = \dfrac{A}{x} + \dfrac{B}{x+1} + \dfrac{C}{x-1}$

$3x^2 - 7x - 2 = A(x^2 - 1) + Bx(x - 1)$
$\qquad\qquad\qquad + Cx(x + 1)$

Let $x = 0$: $-2 = -A \Rightarrow A = 2$.
Let $x = -1$: $8 = 2B \Rightarrow B = 4$.
Let $x = 1$: $-6 = 2C \Rightarrow C = -3$.

$\dfrac{3x^2 - 7x - 2}{x^3 - x} = \dfrac{2}{x} + \dfrac{4}{x+1} - \dfrac{3}{x-1}$

12. $\dfrac{x+2}{x(x-4)} = \dfrac{A}{x} + \dfrac{B}{x-4}$

$x + 2 = A(x - 4) + Bx$

Let $x = 0$: $2 = -4A \Rightarrow A = -\frac{1}{2}$.
Let $x = 4$: $6 = 4B \Rightarrow B = \frac{3}{2}$.

$\dfrac{x+2}{x(x-4)} = \dfrac{1}{2}\left[\dfrac{3}{x-4} - \dfrac{1}{x}\right]$

14. $\dfrac{2x-3}{(x-1)^2} = \dfrac{A}{x-1} + \dfrac{B}{(x-1)^2}$

$2x - 3 = A(x - 1) + B$

Let $x = 1$: $-1 = B$.
Let $x = 0$: $-3 = -A + B$

$\qquad\qquad -3 = -A - 1$

$\qquad\qquad 2 = A$

$\dfrac{2x-3}{(x-1)^2} = \dfrac{2}{x-1} - \dfrac{1}{(x-1)^2}$

16. $\dfrac{4x^2 - 1}{2x(x+1)^2} = \dfrac{A}{2x} + \dfrac{B}{x+1} + \dfrac{C}{(x+1)^2}$

$4x^2 - 1 = A(x+1)^2 + 2Bx(x+1) + 2Cx$

Let $x = 0$: $-1 = A$.
Let $x = -1$: $3 = -2C \Rightarrow C = -\frac{3}{2}$.
Let $x = 1$: $3 = 4A + 4B + 2C$

$\qquad\qquad 3 = -4 + 4B - 3$

$\qquad\qquad \frac{5}{2} = B$

$\dfrac{4x^2 - 1}{2x(x+1)^2} = \dfrac{1}{2}\left[-\dfrac{1}{x} + \dfrac{5}{x+1} - \dfrac{3}{(x+1)^2}\right]$

18. $\dfrac{6x^2 + 1}{x^2(x-1)^3} = \dfrac{A}{x} + \dfrac{B}{x^2} + \dfrac{C}{x-1} + \dfrac{D}{(x-1)^2} + \dfrac{E}{(x-1)^3}$

$6x^2 + 1 = Ax(x-1)^3 + B(x-1)^3 + Cx^2(x-1)^2 + Dx^2(x-1) + Ex^2$

Let $x = 0$: $1 = -B \Rightarrow B = -1$.
Let $x = 1$: $7 = E$.

Substitute B and E into the equation, expand the binomials, collect like terms, and equate the coefficients of like terms.

$x^3 - 4x^2 + 3x = (A + C)x^4 - (3A + 2C - D)x^3 + (3A + C - D)x^2 - Ax$

$\qquad -A = 3 \Rightarrow A = -3$

$\qquad A + C = 0 \Rightarrow C = 3$

$\qquad -3A - 2C + D = 1$

$\qquad 9 - 6 + D = 1 \Rightarrow D = -2$

$\dfrac{6x^2 + 1}{x^2(x-1)^3} = -\dfrac{3}{x} - \dfrac{1}{x^2} + \dfrac{3}{x-1} - \dfrac{2}{(x-1)^2} + \dfrac{7}{(x-1)^3}$

20. $\dfrac{x}{(x-1)(x^2+x+1)} = \dfrac{A}{x-1} + \dfrac{Bx+C}{x^2+x+1}$

$x = A(x^2+x+1) + (Bx+C)(x-1)$

Let $x = 1$: $1 = 3A \Rightarrow A = \frac{1}{3}$.

$x = \dfrac{1}{3}(x^2+x+1) + (Bx+C)(x-1)$

$= \dfrac{1}{3}x^2 + \dfrac{1}{3}x + \dfrac{1}{3} + Bx^2 - Bx + Cx - C = \left(\dfrac{1}{3}+B\right)x^2 + \left(\dfrac{1}{3}-B+C\right)x + \dfrac{1}{3} - C$

Equating coefficients of like powers:

$0 = \dfrac{1}{3} + B \Rightarrow B = -\dfrac{1}{3}$

$0 = \dfrac{1}{3} - C \Rightarrow C = \dfrac{1}{3}$

$\dfrac{x}{(x-1)(x^2+x+1)} = \dfrac{1}{3}\left[\dfrac{1}{x-1} - \dfrac{x-1}{x^2+x+1}\right]$

22. $\dfrac{2x^2+x+8}{(x^2+4)^2} = \dfrac{Ax+B}{x^2+4} + \dfrac{Cx+D}{(x^2+4)^2}$

$2x^2+x+8 = (Ax+B)(x^2+4) + Cx + D$

$2x^2+x+8 = Ax^3 + Bx^2 + (4A+C)x + (4B+D)$

Equating coefficients of like powers:

$0 = A$

$2 = B$

$1 = 4A + C \Rightarrow C = 1$

$8 = 4B + D \Rightarrow D = 0$

$\dfrac{2x^2+x+8}{(x^2+4)^2} = \dfrac{2}{x^2+4} + \dfrac{x}{(x^2+4)^2}$

24. $\dfrac{x^2-4x+7}{(x+1)(x^2-2x+3)} = \dfrac{A}{x+1} + \dfrac{Bx+C}{x^2-2x+3}$

$x^2-4x+7 = A(x^2-2x+3) + Bx(x+1)$

$\qquad\qquad\qquad + C(x+1)$

Let $x = -1$: $12 = 6A \Rightarrow A = 2$.

Let $x = 0$: $7 = 3A + C \Rightarrow C = 1$.

Let $x = 1$: $4 = 2A + 2B + 2C$

$\qquad 4 = 4 + 2B + 2 \Rightarrow B = -1$

$\dfrac{x^2-4x+7}{(x+1)(x^2-2x+3)} = \dfrac{2}{x+1} - \dfrac{x-1}{x^2-2x+3}$

26. $\dfrac{x^3}{(x+2)^2(x-2)^2} = \dfrac{A}{x+2} + \dfrac{B}{(x+2)^2} + \dfrac{C}{x-2} + \dfrac{D}{(x-2)^2}$

$x^3 = A(x+2)(x-2)^2 + B(x-2)^2 + C(x+2)^2(x-2) + D(x+2)^2$

Let $x = -2$: $-8 = 16B \Rightarrow B = -\frac{1}{2}$.

Let $x = 2$: $8 = 16D \Rightarrow D = \frac{1}{2}$.

$$x^3 = A(x+2)(x-2)^2 - \frac{1}{2}(x-2)^2 + C(x+2)^2(x-2) + \frac{1}{2}(x+2)^2$$

$$x^3 - 4x = (A+C)x^3 + (-2A+2C)x^2 + (-4A-4C)x + (8A-8C)$$

Equating coefficients of like powers:

$0 = -2A + 2C \Rightarrow A = C$

$1 = A + C$

$1 = 2A \Rightarrow A = \dfrac{1}{2} \Rightarrow C = \dfrac{1}{2}$

$$\frac{x^3}{(x+2)^2(x-2)^2} = \frac{1}{2}\left[\frac{1}{x+2} - \frac{1}{(x+2)^2} + \frac{1}{x-2} + \frac{1}{(x-2)^2}\right]$$

28. $\dfrac{x+1}{x^3+x} = \dfrac{A}{x} + \dfrac{Bx+C}{x^2+1}$

$x + 1 = A(x^2+1) + Bx^2 + Cx$

Let $x = 0$: $1 = A$.

$x + 1 = x^2 + 1 + Bx^2 + Cx$

$x + 1 = (1+B)x^2 + Cx + 1$

Equating coefficients of like powers:

$0 = 1 + B \Rightarrow B = -1$

$1 = C$

$\dfrac{x+1}{x^3+x} = \dfrac{1}{x} - \dfrac{x-1}{x^2+1}$

32. $\dfrac{x^2-x}{x^2+x+1} = 1 - \dfrac{2x+1}{x^2+x+1}$

30. $\dfrac{x^3-x+3}{x^2+x-2} = x - 1 + \dfrac{2x+1}{(x+2)(x-1)}$

$\dfrac{2x+1}{(x+2)(x-1)} = \dfrac{A}{x+2} + \dfrac{B}{x-1}$

$2x + 1 = A(x-1) + B(x+2)$

Let $x = -2$: $-3 = -3A \Rightarrow A = 1$.

Let $x = 1$: $3 = 3B \Rightarrow B = 1$.

$\dfrac{x^3-x+3}{x^2+x-2} = x - 1 + \dfrac{1}{x+2} + \dfrac{1}{x-1}$

34. $\dfrac{1}{x(x+a)} = \dfrac{A}{x} + \dfrac{B}{x+a}$, a is a constant

$1 = A(x+a) + Bx$

Let $x = 0$: $1 = aA \Rightarrow A = \dfrac{1}{a}$.

Let $x = -a$: $1 = -aB \Rightarrow B = -\dfrac{1}{a}$.

$\dfrac{1}{x(x+a)} = \dfrac{1}{a}\left[\dfrac{1}{x} - \dfrac{1}{x+a}\right]$

36. $\dfrac{1}{(x+1)(a-x)} = \dfrac{A}{x+1} + \dfrac{B}{a-x}$

$1 = A(a-x) + B(x+1)$

Let $x = -1$: $1 = A(a+1) \Rightarrow A = \dfrac{1}{a+1}$.

Let $x = a$: $1 = B(a+1) \Rightarrow B = \dfrac{1}{a+1}$.

$\dfrac{1}{(x+1)(a-x)} = \dfrac{1}{a+1}\left[\dfrac{1}{x+1} + \dfrac{1}{a-x}\right]$

7.5 Systems of Inequalities

2. The graph of $y \le 2$ is the half-plane below the line $y = 2$ together with the line $y = 2$. Graph (h)

4. The graph of $2x - y \ge -2$ is the half-plane below the line $2x - y = -2$ together with the line. Graph (c)

6. The graph of $(x-2)^2 + (y-3)^2 > 4$ is the exterior of the circle centered at $(2, 3)$ with radius 2. Graph (g)

8. The graph of $y \le 4 - x^2$ is the interior region of a parabola with vertex $(0, 4)$ and opening downward together with the graph of $y = 4 - x^2$. Graph (d)

10. $x \le 4$

12. $y \le 3$

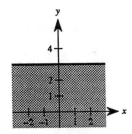

14. $y > 2x - 4$

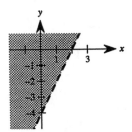

16. $5x + 3y \ge -15$

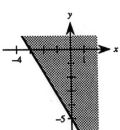

18. $y^2 - x < 0$

20. $y < \ln x$

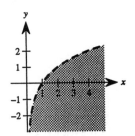

22. $3x + 2y < 6$

$x > 0$

$y > 0$

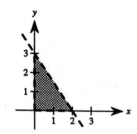

24. $2x + y \geq 2$

$x \leq 2$

$y \leq 1$

26. $x - 7y > -36$

$5x + 2y > 5$

$6x - 5y > 6$

28. $x - 2y < -6$

$5x - 3y > -9$

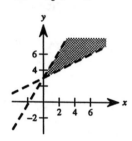

30. $x - y^2 > 0$

$x - y < 2$

32. $x^2 + y^2 \leq 25$

$4x - 3y \leq 0$

34. $x < 2y - y^2$

$0 < x + y$

36. $y < -x^2 + 2x + 3$

$y > x^2 - 4x + 3$

38. $y \geq x^4 - 2x^2 + 1$

$y \leq 1 - x^2$

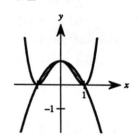

40. $y \le e^{-x^2/2}$

$y \ge 0$

$-2 \le x \le 2$

42. $0 \le y \le 4$

$4x - y \ge 0$

$4x - y \le 16$

44. $y \ge 0$

$y \le x + 1$

$y \le -x + 1$

46. $x^2 + y^2 \le 16$

$x \ge 0$

$y \ge x$

48. $x =$ number of model A

$y =$ number of model B

$x \ge 2y$

$8x + 12y \le 200$

$x \ge 4$

$y \ge 2$

50. $x =$ number of \$15 tickets

$y =$ number of \$25 tickets

$x + y \ge 15,000$

$x \ge 8,000$

$y \ge 4,000$

$15x + 25y \ge 275,000$

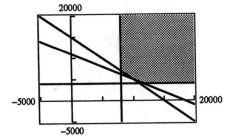

52. $x =$ ounces of food X

$y =$ ounces of food Y

Calcium: $20x + 10y \geq 300$

Iron: $15x + 10y \geq 150$

Vitamin B: $10x + 20y \geq 200$

$x \geq 0$

$y \geq 0$

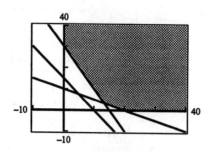

54. Demand $=$ Supply

$60 - x = 10 + \frac{7}{3}x$

$50 = \frac{10}{3}x$

$15 = x$

$45 = p$

Point of equilibrium: $(15, 45)$

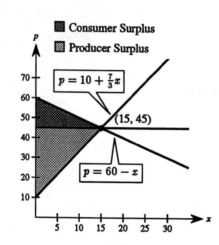

The consumer surplus is the area of the triangle bounded by

$p \leq 60 - x$

$p \geq 45$

$x \geq 0.$

Consumer surplus $= \frac{1}{2}$(base)(height)

$= \frac{1}{2}(15)(15)$

$= \frac{225}{2}$

The producer surplus is the area of the triangle bounded by

$p \geq 10 + \frac{7}{3}x$

$p \leq 45$

$x \geq 0.$

Producer surplus $= \frac{1}{2}$(base)(height)

$= \frac{1}{2}(15)(35)$

$= \frac{525}{2}$

56.   Demand $=$ Supply

$$100 - 0.05x = 25 + 0.1x$$

$$75 = 0.15x$$

$$500 = x$$

$$75 = p$$

Point of equilibrium: $(500, 75)$

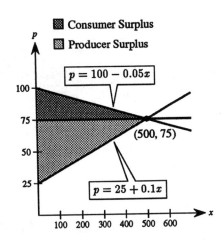

The consumer surplus is the area of the triangle bounded by

$$p \leq 100 - 0.05x$$

$$p \geq 75$$

$$x \geq 0.$$

Consumer surplus $= \dfrac{1}{2}$ (base)(height)

$$= \dfrac{1}{2}(500)(25)$$

$$= 6250$$

The producer surplus is the area of the triangle bounded by

$$p \geq 25 + 0.1x$$

$$p \leq 75$$

$$x \geq 0.$$

Producer surplus $= \dfrac{1}{2}$ (base)(height)

$$= \dfrac{1}{2}(500)(50)$$

$$= 12{,}500$$

58.

$$\text{Demand} = \text{Supply}$$

$$400 - 0.0002x = 225 + 0.0005x$$

$$175 = 0.0007x$$

$$250,000 = x$$

$$350 = p$$

Point of equilibrium: $(250,000, \ 350)$

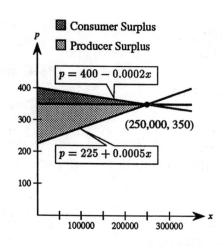

The consumer surplus is the area of the triangle bounded by

$$p \leq 400 - 0.0002x$$

$$p \geq 350$$

$$x \geq 0.$$

Consumer surplus $= \frac{1}{2}(\text{base})(\text{height}) = \frac{1}{2}(250,000)(50) = 6,250,000$

The producer surplus is the area of the triangle bounded by

$$p \geq 225 + 0.0005x$$

$$p \leq 350$$

$$x \geq 0.$$

Producer surplus $= \frac{1}{2}(\text{base})(\text{height}) = \frac{1}{2}(250,000)(125) = 15,625,000$

7.6 Linear Programming

2. $z = 2x + 8y$

At $(0, 4)$: $z = 2(0) + 8(4) = 32$
At $(0, 0)$: $z = 2(0) + 8(0) = 0$
At $(2, 0)$: $z = 2(2) + 8(0) = 4$
The maximum value is 32 at $(0, 4)$.
The minimum value is 0 at $(0, 0)$.

4. $z = 7x + 3y$

At $(0, 4)$: $z = 7(0) + 3(4) = 12$
At $(0, 0)$: $z = 7(0) + 3(0) = 0$
At $(2, 0)$: $z = 7(2) + 3(0) = 14$
The maximum value is 14 at $(2, 0)$.
The minimum value is 0 at $(0, 0)$.

6. $z = 4x + 3y$

At $(0, 4)$: $z = 4(0) + 3(4) = 12$
At $(3, 0)$: $z = 4(3) + 3(0) = 12$
At $(5, 3)$: $z = 4(5) + 3(3) = 29$
At $(0, 2)$: $z = 4(0) + 3(2) = 6$
The maximum value is 29 at $(5, 3)$.
The minimum value is 6 at $(0, 2)$.

8. $z = x + 6y$

At $(0, 4)$: $z = 0 + 6(4) = 24$
At $(3, 0)$: $z = 3 + 6(0) = 3$
At $(5, 3)$: $z = 5 + 6(3) = 23$
At $(0, 2)$: $z = 0 + 6(2) = 12$
The maximum value is 24 at $(0, 4)$.
The minimum value is 3 at $(3, 0)$.

10. $z = 50z + 35y$

At $(0, 800)$: $z = 50(0) + 35(800) = 28,000$
At $(900, 0)$: $z = 50(900) + 35(0) = 45,000$
At $(675, 0)$: $z = 50(675) + 35(0) = 33,750$
At $(0, 600)$: $z = 50(0) + 35(600) = 21,000$
The maximum value is 45,000 at $(900, 0)$.
The minimum value is 21,000 at $(0, 600)$.

12. $z = 16x + 18y$

At $(0, 800)$: $z = 16(0) + 18(9800) = 14,000$
At $(900, 0)$: $z = 16(900) + 18(0) = 14,000$
At $(675, 0)$: $z = 16(675) + 18(0) = 10,800$
At $(0, 600)$: $z = 16(0) + 18(600) = 10,800$
The maximum value is 14,000 at any point along the line segment connecting $(0, 800)$ and $(900, 0)$.

The minimum value is 10,800 at any point along the line segment connecting $(675, 0)$ and $(0, 600)$.

14. $z = 7x + 8y$

At $(0, 8)$: $z = 7(0) + 8(8) = 64$
At $(4, 0)$: $z = 7(4) + 8(0) = 28$
At $(0, 0)$: $z = 7(0) + 8(0) = 0$
The maximum value is 64 at $(0, 8)$.
The minimum value is 0 at $(0, 0)$.

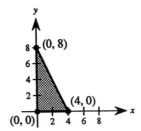

16. $z = 7x + 2y$

At $(0, 8)$: $z = 7(0) + 2(8) = 16$
At $(4, 0)$: $z = 7(4) + 2(0) = 28$
At $(0, 0)$: $z = 7(0) + 2(0) = 0$
The maximum value is 28 at $(4, 0)$.
The minimum value is 0 at $(0, 0)$.

18. $z = 4x + 5y$

At $(0, 0)$: $z = 4(0) + 5(0) = 0$
At $(5, 0)$: $z = 4(5) + 5(0) = 20$
At $(4, 1)$: $z = 4(4) + 5(1) = 21$
At $(0, 3)$: $z = 4(0) + 5(3) = 15$
The maximum value is 21 at $(4, 1)$.
The minimum value is 0 at $(0, 0)$.

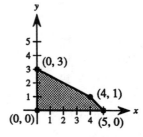

20. $z = 2x + y$

At $(0, 0)$: $z = 2(0) - 0 = 0$
At $(5, 0)$: $z = 2(5) - 0 = 10$
At $(4, 1)$: $z = 2(4) - 1 = 7$
At $(0, 3)$: $z = 2(0) - 3 = -3$
The maximum value is 10 at $(5, 0)$.
The minimum value is -3 at $(0, 3)$.

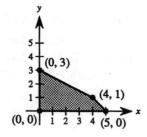

22. $z = x$

At $(0, 0)$: $z = 0$
At $(12, 0)$: $z = 12$
At $(10, 8)$: $z = 10$
At $(6, 16)$: $z = 6$
At $(0, 20)$: $z = 0$
The maximum value is 12 at $(12, 0)$.

The minimum value is 0 at any point along the
line segment connecting $(0, 0)$ and $(0, 20)$.

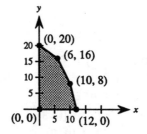

24. $z = y$

At $(0, 0)$: $z = 0$
At $(12, 0)$: $z = 0$
At $(10, 8)$: $z = 8$
At $(6, 16)$: $z = 16$
At $(0, 20)$: $z = 20$

The maximum value is 20 at $(0, 20)$.

The minimum value is 0 at any point along the line segment connecting $(0, 0)$ and $(12, 0)$.

26. $z = 5x + y$

At $(0, 10)$: $z = 5(0) + (10) = 10$
At $(3, 6)$: $z = 5(3) + (6) = 21$
At $(5, 0)$: $z = 5(5) + (0) = 25$
At $(0, 0)$: $z = 5(0) + (0) = 0$

The maximum value is 25 at $(5, 0)$.

28. $z = 3x + y$

At $(0, 10)$: $z = 3(0) + (10) = 10$
At $(3, 6)$: $z = 3(3) + (6) = 15$
At $(5, 0)$: $z = 3(5) + (0) = 15$
At $(0, 0)$: $z = 3()) + (0) = 0$

The maximum value is 15 at any point along the line segment connecting $(3, 6)$ or $(5, 1)$.

30. $z = 2x + 4y$

At $(0, 5)$: $z = 2(0) + 4(5) = 20$
At $(4, 4)$: $z = 2(4) + 4(4) = 24$
At $(5, 3)$: $z = 2(5) + 4(3) = 22$
At $(7, 0)$: $z = 2(7) + 4(0) = 14$
At $(0, 0)$: $z = 2(0) + 4(0) = 0$

The maximum value is 24 at $(4, 4)$.

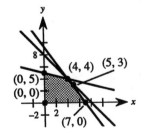

32. $z = 4z + y$

At $(0, 5)$: $z = 4(0) + (5) = 5$
At $(4, 4)$: $z = 4(4) + (4) = 20$
At $(5, 3)$: $z = 4(5) + (3) = 23$
At $(7, 0)$: $z = 4(7) + (0) = 28$
At $(0, 0)$: $z = 4(0) + (0) = 0$

The maximum value is 28 at $(7, 0)$.

34. x = number of acres for crop A

y = number of acres for crop B

Constraints: $x + y \leq 150$

$x + 2y \leq 240$

$\frac{3}{10}x + \frac{1}{10}y \leq 30$

$x \leq 0$

$y \leq 0$

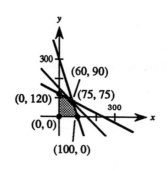

Objective function: $P = 140x + 235y$

Vertices: $(0, 120)$, $(60, 90)$, $(75, 75)$, $(100, 0)$, $(0, 0)$

At $(0, 120)$: $P = 140(0) + 235(120) = 28{,}200$
At $(60, 90)$: $P = 140(60) + 235(90) = 29{,}550$
At $(75, 75)$: $P = 140(75) + 235(75) = 28{,}125$
At $(0, 0)$: $P = 140(0) + 235(0) = 0$

At maximum profit of \$29,550 occurs when 60 acres are used for crop A and 90 acres are used for crop B.

36. x = fraction of type A

y = fraction of type B

Constraints: $80x + 92y \leq 90$

$x + y = 1$

$x \leq 0$

$y \leq 0$

Objective function: $C = 1.13x + 1.28y$

Vertices: $\left(\frac{1}{6}, \frac{5}{6}\right)$

At $\left(\frac{1}{6}, \frac{5}{6}\right)$: $C = (1.13)\left(\frac{1}{6}\right) + (1.28)\left(\frac{5}{6}\right) = 1.255$

The minimum cost is \$1.26 and occurs with a mixture that is $\frac{1}{6}$ A and $\frac{5}{6}$ B.

38. x = number of Model A

y = number of Model B

Constraints:$2.5x + 3y \leq 4000$

$$2x + y \leq 2500$$

$$0.75x + 1.25y \leq 1500$$

$$x \leq 0$$

$$y \leq 0$$

Objective function:$P = 50x + 52y$

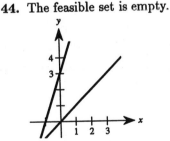

Vertices $(0, 0)$, $(0, 1200)$ $\left(\frac{4000}{7}, \frac{6000}{7}\right)$, $(1000, 500)$, $(1250, 0)$

At $(0, 0)$: $P = (50)(0) + 52(0) = 0$
At $(0, 1200)$: $P = 50(0) + 52(1200) = 62,400$
At $\left(\frac{4000}{7}, \frac{6000}{7}\right)$: $P = 50\left(\frac{4000}{7}\right) + 52\left(\frac{6000}{7}\right) \approx 73,142.86$
At $(1000, 500)$: $P = 50(1000) + 52(500) = 76,000$
At $(1250, 0)$: $P = 50(1250) + 52(0) = 62,500$

The maximum profit occurs when 1000 units of Model A and 500 units of Model B are produced.

40. Objective function: $R = 1000x + 300y$

At $(0, 0)$: $R = 1000(0) + 300(0) = 0$
At $(0, 40)$: $R = 1000(0) + 300(40) = 12,000$
At $(8, 8)$: $R = 1000(8) + 300(8) = 10,400$
At $(9, 0)$: $R = 1000(9) + 300(0) = 9000$

The revenue will be maximum if the firm does 0 audits and 40 tax returns.

42. At $(0, 0)$: $z = 0 + 0 = 0$
At $(0, 1)$: $z = 0 + 1 = 1$
At $(2, 3)$: $z = 2 + 3 = 5$

The constraints do not form a closed set of points. Therefore, $z = x + y$ is unbounded.

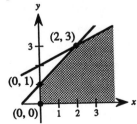

44. The feasible set is empty.

46. At $(0, 0)$: $z = 0 + 2(0) = 0$
At $(0, 2)$: $z = 0 + 2(2) = 4$
At $\left(\frac{4}{3}, \frac{4}{3}\right)$: $z = \frac{4}{3} + 2\left(\frac{4}{3}\right) = 4$
At $(2, 0)$: $z = 2 + 2(0) = 2$

The maximum value is 4 at any point along the line segment connecting $(0, 2)$ and $\left(\frac{4}{3}, \frac{4}{3}\right)$.

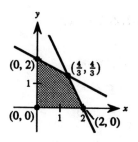

48. $z = 3x + ty$

At $(0, 0)$: $z = 3(0) + t(0) = 0$
At $(1, 0)$: $z = 3(1) + t(0) = 3$
At $(2, 1)$: $z = (3)2 + t(1) = 6 + t$
At $(0, 2)$: $z = 3(0) + t(2) = 2t$

(a) For the maximum value to be at $(2, 1)$, $z = 6 + t$ must be greater than $z = 2t$ and $z = 3$.

$6 + t > 2t$ and $6 + t > 3$

$6 > t$ $t > -3$

Thus, $-3 < t < 6$.

(b) For the maximum value to be at $(0, 2)$, $z = 2t$ must be greater than $z = 6 + t$ and $z = 3$.

$2t > 6 + t$ and $2t > 3$

$t > 6$ $t > \frac{3}{2}$

Thus, $t > 6$.

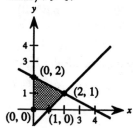

Review Exercises for Chapter 7

2. $2x = 3(y-1)$ Equation 1

$y = x$ Equation 2

Substitute y for x in Equation 1:

$2y = 3y - 3 \quad \Rightarrow \quad y = 3$

Backsubstitute $y = 3$: $x = y \quad \Rightarrow \quad x = 3$

Solution: $(3, 3)$

4. $x^2 + y^2 = 169$ Equation 1

$3x + 2y = 39$ Equation 2

Solve for x in Equation 2: $x = 13 - \frac{2}{3}y$

Substitute for x in Equation 1:

$$\left(13 - \tfrac{2}{3}y\right)^2 + y^2 = 169$$

$$169 - 4\left(\tfrac{13}{3}y\right) + \tfrac{4}{9}y^2 + y^2 = 169$$

$$\tfrac{13}{3}y\left(\tfrac{1}{3}y - 4\right) = 0 \quad \Rightarrow \quad y = 0,\ 12$$

Backsubstitute $y = 0$: $x = 13 - \frac{2}{3}(0) = 13$

Backsubstitute $y = 12$: $x = 13 - \frac{24}{3} = 5$

Solutions: $(13, 0)$, $(5, 12)$

6. $x = y + 3$ Equation 1

$x = y^2 + 1$ Equation 2

Substitute for x in Equation 1:

$y^2 + 1 = y + 3 \quad \Rightarrow \quad y^2 - y - 2 = 0$

$y = 2,\ -1$

Backsubstitute $y = 2$: $x = 2^2 + 1 = 5$

Backsubstitute $y = -1$: $x = (-1)^2 + 1 = 2$

Solutions: $(5, 2)$, $(2, -1)$

8. $40x + 30y = \ \ 24 \Rightarrow \ \ 40x + \ \ 30y = 24$

$20x - 50y = -14 \Rightarrow -40x + 100y = 28$

$130y = 52$

$y = \frac{2}{5}$

$x = \frac{3}{10}$

Solution: $\left(\frac{3}{10}, \frac{2}{5}\right)$

10. $7x + 12y = 63 \Rightarrow -7x - 12y = -63$

$2x + \ \ 3y = 15 \Rightarrow \ \ 8x + 12y = \ \ 60$

$x = -3$

$y = \ \ 7$

Solution: $(-3, 7)$

12. $1.5x + 2.5y = 8.5 \Rightarrow \ \ 3x + 5y = \ \ 17$

$6x + 10y = 24 \Rightarrow -3x - 5y = -12$

$0 = \ \ 5$

Inconsistent; no solution

14. $y = 2x^2 - 4x + 1$

$y = x^2 - 4x + 3$

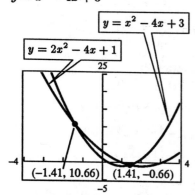

We see that there are two solutions, $(1.41, -0.66)$ and $(-1.41, 10.66)$.

16. $y = \ln(x - 1) - 3$

$y = 4 - \frac{1}{2}x$

The unique solution is $(x, y) = (9.68, -0.839)$.

18. $22,500 + 0.015x = 20,000 + 0.02x$

$$2500 = 0.005x$$

$$x = 500,000$$

You would have to sell over \$500,000 to make the second offer better.

20. $x =$ number of \$9.95 cassette tapes

$y =$ number of \$14.95 cassette tapes

$$x + y = 650 \Rightarrow y = 650 - x$$

$$9.95x + 14.95y = 7717.50$$

$$9.95x + 14.95(650 - x) = 7717.50$$

$$-5x = -2000$$

$$x = 400$$

$$y = 250$$

Solution: 400 at \$9.95 and 250 at \$14.95

22. $$2l + 2w = 480$$

$$l = 1.50w$$

$$2(1.50w) + 2w = 480$$

$$5w = 480$$

$$w = 96$$

$$l = 144$$

The dimensions are 96×144 meters.

24. Supply = Demand

$$45 + 0.0002x = 120 - 0.0001x$$

$$0.0003x = 75$$

$$x = 250,000 \text{ units}$$

$$p = \$95.00$$

Point of equilibrium: $(250,000, \quad 95)$

26.
$$x + 3y - z = 13$$
$$2x \qquad - 5z = 23$$
$$4x - y - 2z = 14$$

$$x + 3y - z = 13$$
$$- 6y - 3z = -3$$
$$- 13y + 2z = -38$$

$$x + 3y - z = 13$$
$$- 6y - 3z = -3$$
$$\tfrac{17}{2}z = -\tfrac{63}{2}$$

$$\tfrac{17}{2}z = -\tfrac{63}{2} \quad \Rightarrow \quad z = -\tfrac{63}{17}$$
$$-6y - 3\left(-\tfrac{63}{17}\right) = -3 \quad \Rightarrow \quad y = \tfrac{40}{17}$$
$$x + 3\left(\tfrac{40}{17}\right) - \left(-\tfrac{63}{17}\right) = 13 \quad \Rightarrow \quad x = \tfrac{38}{17}$$

Solution: $\left(\tfrac{38}{17},\ \tfrac{40}{17},\ -\tfrac{63}{17}\right)$

28.
$$2x \qquad + 6z = -9$$
$$3x - 2y + 11z = -16$$
$$3x - y + 7z = -11$$

$$-x + 2y - 5z = 7$$
$$3x - 2y + 11z = -16$$
$$3x - y + 7z = -11$$

$$-x + 2y - 5z = 7$$
$$4y - 4z = 5$$
$$5y - 8z = 10$$

$$-x + 2y - 5z = 7$$
$$4y - 4z = 5$$
$$-3y = 0$$

$$-3y = 0 \quad \Rightarrow \quad y = 0$$
$$4(0) - 4z = 5 \quad \Rightarrow \quad z = -\tfrac{5}{4}$$
$$-x + 2(0) - 5\left(-\tfrac{5}{4}\right) = 7 \quad \Rightarrow \quad x = -\tfrac{3}{4}$$

Solution: $\left(-\tfrac{3}{4},\ 0,\ -\tfrac{5}{4}\right)$

30.
$$\begin{aligned}
-x + 3y + 2z + 2w &= 1 \\
2x + y + z + 2w &= -1 \\
5x - 2y + z - 3w &= 0 \\
3x + 2y + 3z - 5w &= 12
\end{aligned}$$

$$\begin{aligned}
-x + 3y + 2z + 2w &= 1 \\
7y + 5z + 6w &= 1 \\
13y + 11z + 7w &= 5 \\
11y + 9z + w &= 15
\end{aligned}$$

$$\begin{aligned}
-x + 3y + 2z + 2w &= 1 \\
-y + z - 5w &= 3 \\
7y + 5z + 6w &= 1 \\
11y + 9z + w &= 15
\end{aligned}$$

$$\begin{aligned}
-x + 3y + 2z + 2w &= 1 \\
-y + z - 5w &= 3 \\
12z - 29w &= 22 \\
20z - 54w &= 48
\end{aligned}$$

$$\begin{aligned}
-x + 3y + 2z + 2w &= 1 \\
-y + z - 5w &= 3 \\
60z - 145w &= 110 \\
60z - 162w &= 144
\end{aligned}$$

$$\begin{aligned}
-x + 3y + 2z + 2w &= 1 \\
-y + z - 5w &= 3 \\
60z - 145w &= 110 \\
-17w &= 34
\end{aligned}$$

$$\begin{aligned}
-17w &= 34 &\Rightarrow&\quad w = -2 \\
60z - 145(-2) &= 110 &\Rightarrow&\quad z = -3 \\
-y - 3 - 5(-2) &= 3 &\Rightarrow&\quad y = 4 \\
-x + 3(4) + 2(-3) + 2(-2) &= 1 &\Rightarrow&\quad x = 1
\end{aligned}$$

Solution: $(1, 4, -3, -2)$

32. $\begin{bmatrix} 2 & 3 & 1 & \vdots & 10 \\ 2 & -3 & -3 & \vdots & 22 \\ 4 & -2 & 3 & \vdots & -2 \end{bmatrix} \Rightarrow \begin{bmatrix} 4 & -2 & 3 & \vdots & -2 \\ 2 & -3 & -3 & \vdots & 22 \\ 2 & 3 & 1 & \vdots & 10 \end{bmatrix}$

$\Rightarrow \begin{bmatrix} 4 & -2 & 3 & \vdots & -2 \\ 0 & 4 & 9 & \vdots & -46 \\ 0 & -8 & 1 & \vdots & -22 \end{bmatrix}$

$\Rightarrow \begin{bmatrix} 4 & -2 & 3 & \vdots & -2 \\ 0 & 4 & 9 & \vdots & -46 \\ 0 & 0 & 19 & \vdots & -114 \end{bmatrix}$

$$19z = -114 \Rightarrow z = -6$$

$$4y + 9(-6) = -46 \Rightarrow y = 2$$

$$4x - 2(2) + 3(-6) = -2 \Rightarrow x = 5$$

Answer: (5, 2, −6)

34. $\begin{bmatrix} 1 & 2 & 6 & \vdots & 1 \\ 2 & 5 & 15 & \vdots & 4 \\ 3 & 1 & 3 & \vdots & -6 \end{bmatrix} \Rightarrow \begin{bmatrix} 1 & 2 & 6 & \vdots & 1 \\ 0 & 1 & 3 & \vdots & 2 \\ 0 & -5 & -15 & \vdots & -9 \end{bmatrix} \Rightarrow \begin{bmatrix} 1 & 0 & 0 & \vdots & -3 \\ 0 & 1 & 3 & \vdots & 2 \\ 0 & 0 & 0 & \vdots & 1 \end{bmatrix}$

$$x = -3$$

$$y + 3x = 2$$

$$0 = 1$$

No solution

36. $y = ax^2 + bx + c$

At $(-5, 0)$: $\quad 0 \quad = 25a - 5b + c$

At $(1, -6)$: $\quad -6 = a + b + c$

At $(2, 14)$: $\quad 14 \quad = 4a + 2b + c$

$$a + \quad b + \quad c = \quad -6$$
$$4a + \quad 2b + \quad c = \quad 14$$
$$25a - \quad 5b + \quad c = \quad 0$$

$$a + \quad b + \quad c = \quad -6$$
$$-2b - \quad 3c = \quad 38$$
$$-30b - 24c = \quad 150$$

$$a + \quad b + \quad c = \quad -6$$
$$-2b - \quad 3c = \quad 38$$
$$21c = -420$$

$$21c = -420 \quad \Rightarrow \quad c = -20$$
$$-2b - 3(-20) = 38 \quad \Rightarrow \quad b = 11$$
$$a + 11 - 20 = -6 \quad \Rightarrow \quad a = 3$$

Solution: $y = 3x^2 + 11x - 20$

38. $x^2 + y^2 + Dx + Ey + F = 0$

At $(4, 2)$: $\quad 16 + 4 + 4D + 2E + F = 0 \Rightarrow 4D + 2E + F \quad = -20$

At $(1, 3)$: $\quad 1 + 9 + D + 3E + F \quad = 0 \Rightarrow D + 3E + F \quad = -10$

At $(-2, -6)$: $\quad 4 + 36 - 2D - 6E + F = 0 \Rightarrow -2D - 6E + F = -40$

$$D + \quad 3E + \quad F = -10$$
$$4D + \quad 2E + \quad F = -20$$
$$-2D - \quad 6E + \quad F = -40$$

$$D + \quad 3E + \quad F = -10$$
$$-10E - 3F = \quad 20$$
$$3F = -60$$

$$3F = -60 \quad \Rightarrow \quad F = -20$$
$$-10E - 3(-20) = 20 \quad \Rightarrow \quad E = 4$$
$$D + 3(4) - 20 = -10 \quad \Rightarrow \quad D = -2$$

Solution: $y = x^2 + y^2 - 2x + 4y - 20 = 0$

40. $x =$ amount invested at 7%

$y =$ amount invested at 9%

$z =$ amount invested at 11%

$$
\begin{aligned}
x + y + z &= 20{,}000 \\
0.07x + 0.09y + 0.11z &= 1780 \\
x - y &= 3000 \\
x - z &= 1000
\end{aligned}
$$

Solution: $x = \$8000, \quad y = \$5000, \quad z = \$7000$

42. $\dfrac{-x}{x^2 + 3x + 2} = \dfrac{A}{x+1} + \dfrac{B}{x+2}$

$-x = A(x+2) + B(x+1)$

Let $x = -1$: $1 = A$.

Let $x = -2$: $2 = -B \Rightarrow B = -2$.

$$\dfrac{-x}{x^2 + 3x + 2} = \dfrac{1}{x+1} - \dfrac{2}{x+2}$$

44. $\dfrac{9}{x^2 - 9} = \dfrac{A}{x-3} + \dfrac{B}{x+3}$

$9 = A(x+3) + B(x-3)$

Let $x = 3$: $9 = 6A \Rightarrow A = \frac{3}{2}$.

Let $x = -3$: $9 = -6B \Rightarrow B = -\frac{3}{2}$.

$$\dfrac{9}{x^2 - 9} = \dfrac{1}{2}\left(\dfrac{3}{x-3} - \dfrac{3}{x+3}\right)$$

46. $\dfrac{4x - 2}{3(x-1)^2} = \dfrac{A}{x-1} + \dfrac{B}{(x-1)^2}$

$\dfrac{4}{3}x - \dfrac{2}{3} = A(x-1) + B$

Let $x = 1$: $\frac{2}{3} = B$.

Let $x = 2$: $2 = A + \frac{2}{3} \Rightarrow A = \frac{4}{3}$.

$$\dfrac{4x - 2}{3(x-1)^2} = \dfrac{4}{3(x-1)} + \dfrac{2}{3(x-1)^2}$$

48. $\dfrac{4x^2}{(x-1)(x^2+1)} = \dfrac{A}{x-1} + \dfrac{Bx+C}{x^2+1}$

$4x^2 = A(x^2+1) + (Bx+C)(x-1)$

Let $x = 1$: $4 = 2A \Rightarrow A = 2$.

$4x^2 = 2(x^2+1) + (Bx+C)(x-1)$

$2x^2 - 2 = Bx^2 + (-B+C)x - C$

Equating coefficient of like powers:

$2 = B$

$-2 = -C \Rightarrow C = 2$

$$\dfrac{4x^2}{(x-1)(x^2+1)} = \dfrac{2}{x-1} + \dfrac{2x+2}{x^2+1}$$

$$= 2\left(\dfrac{1}{x-1} + \dfrac{x+1}{x^2+1}\right)$$

50. $2x + 3y \leq 24$

$2x + y \leq 16$

$x \geq 0$

$y \geq 0$

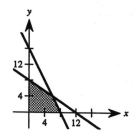

52. $2x + y \geq 16$

$x + 3y \geq 18$

$0 \leq x \leq 25$

$0 \leq y \leq 25$

54. $y \leq 6 - 2x - x^2$

$y \geq x + 6$

56. $x^2 + y^2 \leq 9$

$(x - 3)^2 + y^2 \leq 9$

58. $-x + y \leq 1$

$3x + y \leq 25$

$x + 7y \geq 15$

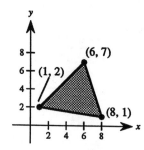

60. $x =$ number of units of Product I

$y =$ number of units of Product II

$20x + 30y \leq 24,000$

$12x + 8y \leq 12,400$

$x \geq 0$

$y \geq 0$

62.   Demand = Supply

$$130 - 0.0002x = 30 + 0.0003x$$

$$100 = 0.0005x$$

$$x = 200,000 \text{ units}$$

$$p = \$90$$

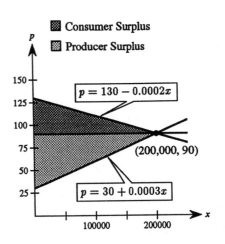

Point of equilibrium: (200,000, 90)

Consumer surplus: $\frac{1}{2}(200,000)(40) = \$4,000,000$

Producer surplus: $\frac{1}{2}(200,000)(60) = \$6,000,000$

64. $z = 10x + 7y$

At (0, 100): $z = 10(0) + 7(100) = 700$

At (25, 50): $z = 10(25) + 7(50) = 600$

At (75, 0): $z = 10(75) + 7(0) = 750$

The minimum value is 600 at (25, 50).

66. $z = 50x + 70y$

At (0, 0): $z = 50(0) + 70(0) = 0$

At (0, 75): $z = 50(0) + 70(750) = 52,500$

At (500, 500): $z = 50(500) + 70(500) = 60,000$

At (700, 0): $z = 50(700) + 70(0) = 35,000$

The maximum value is 60,000 at (500, 500).

68. x = number of haircuts

y = number of permanents

Constraints: $\frac{1}{3}x + \frac{7}{6}y \le 21$

$$x \le 0$$

$$y \le 0$$

Objective function: $R = 17x + 60y$

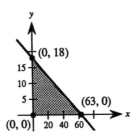

At (0, 0): $R = 17(0) + 60(0) = 0$

At (0, 18): $R = 17(0) + 60(18) = 1080$

At (63, 0): $R = 17(63) + 60(0) = 1070$

The maximum is $1080 when 18 permanents and 0 haircuts are given.

70. x = fraction of Type A

y = fraction of Type B

Constraints: $80x + 92y \leq 88$

$$x + y = 1$$

$$x \leq 0$$

$$y \leq 0$$

Objective function: $C = 1.25x + 1.55y$

At $\left(\frac{1}{3}, \frac{2}{3}\right)$: $C = 1.25\left(\frac{1}{3}\right) + 1.55\left(\frac{2}{3}\right) = 1.45$

The minimum cost is \$1.45 and occurs with a mixture of $\frac{1}{3}$A and $\frac{2}{3}$B.

CHAPTER EIGHT
Matrices and Determinants

8.1 | Matrices and Systems of Linear Equations

2. 1×4 **4.** 5×5 **6.** 1×1

8. Row-echelon form   **10.** Reduced row-echelon form

12. $\begin{bmatrix} 3 & 6 & 8 \\ 4 & -3 & 6 \end{bmatrix} \begin{array}{c} \frac{1}{3}R_1 \to \end{array} \begin{bmatrix} 1 & 2 & \frac{8}{3} \\ 4 & -3 & 6 \end{bmatrix}$

14. $\begin{bmatrix} 2 & 4 & 8 & 3 \\ 1 & -1 & -3 & 2 \\ 2 & 6 & 4 & 9 \end{bmatrix} \begin{array}{c} \frac{1}{2}R_1 \to \end{array} \begin{bmatrix} 1 & 2 & 4 & \frac{3}{2} \\ 1 & -1 & -3 & 2 \\ 2 & 6 & 4 & 9 \end{bmatrix}$

$$\begin{array}{c} \\ -R_1 + R_2 \to \\ -2R_1 + R_3 \to \end{array} \begin{bmatrix} 1 & 2 & 4 & \frac{3}{2} \\ 0 & -3 & -7 & \frac{1}{2} \\ 0 & 2 & -4 & 6 \end{bmatrix}$$

16. $\begin{bmatrix} 7 & 1 \\ 0 & 2 \\ -3 & 4 \\ 4 & 1 \end{bmatrix}$

(a) $\begin{bmatrix} 7 & 1 \\ 0 & 2 \\ -3 & 4 \\ 1 & 5 \end{bmatrix}$ (b) $\begin{bmatrix} 1 & 5 \\ 0 & 2 \\ -3 & 4 \\ 7 & 1 \end{bmatrix}$ (c) $\begin{bmatrix} 1 & 5 \\ 0 & 2 \\ 0 & 19 \\ 7 & 1 \end{bmatrix}$ (d) $\begin{bmatrix} 1 & 5 \\ 0 & 2 \\ 0 & 19 \\ 0 & -34 \end{bmatrix}$ (e) $\begin{bmatrix} 1 & 5 \\ 0 & 1 \\ 0 & 19 \\ 0 & -34 \end{bmatrix}$ (f) $\begin{bmatrix} 1 & 0 \\ 0 & 1 \\ 0 & 0 \\ 0 & 0 \end{bmatrix}$

18. $\begin{bmatrix} 1 & 2 & -1 & 3 \\ 3 & 7 & -5 & 14 \\ -2 & -1 & -3 & 8 \end{bmatrix} \Rightarrow \begin{bmatrix} 1 & 2 & -1 & 3 \\ 0 & 1 & -2 & 5 \\ 0 & 3 & -5 & 14 \end{bmatrix}$

$$\Rightarrow \begin{bmatrix} 1 & 2 & -1 & 3 \\ 0 & 1 & -2 & 5 \\ 0 & 0 & 1 & -1 \end{bmatrix}$$

20. $\begin{bmatrix} 1 & -3 & 0 & -7 \\ -3 & 10 & 1 & 23 \\ 4 & -10 & 2 & -24 \end{bmatrix} \Rightarrow \begin{bmatrix} 1 & -3 & 0 & -7 \\ 0 & 1 & 1 & 2 \\ 0 & 2 & 2 & 4 \end{bmatrix}$

$$\Rightarrow \begin{bmatrix} 1 & -3 & 0 & -7 \\ 0 & 1 & 1 & 2 \\ 0 & 0 & 0 & 0 \end{bmatrix}$$

22. $\begin{bmatrix} 1 & 3 & 2 \\ 5 & 15 & 9 \\ 2 & 6 & 10 \end{bmatrix} \Rightarrow \begin{bmatrix} 1 & 3 & 2 \\ 0 & 0 & -1 \\ 0 & 0 & 6 \end{bmatrix} \Rightarrow \begin{bmatrix} 1 & 3 & 0 \\ 0 & 0 & -1 \\ 0 & 0 & 0 \end{bmatrix} \Rightarrow \begin{bmatrix} 1 & 3 & 0 \\ 0 & 0 & 1 \\ 0 & 0 & 0 \end{bmatrix}$

24. $\begin{bmatrix} 1 & -3 \\ -1 & 8 \\ 0 & 4 \\ -2 & 10 \end{bmatrix} \Rightarrow \begin{bmatrix} 1 & -3 \\ 0 & 5 \\ 0 & 4 \\ 0 & 4 \end{bmatrix} \Rightarrow \begin{bmatrix} 1 & -3 \\ 0 & 1 \\ 0 & 4 \\ 0 & 4 \end{bmatrix} \Rightarrow \begin{bmatrix} 1 & 0 \\ 0 & 1 \\ 0 & 0 \\ 0 & 0 \end{bmatrix}$

26. $9x - 4y = 0$
    $6x + y = -4$

28. $5x + 8y + 2z = -1$
    $-2x + 15y + 5z + w = 9$
    $x + 6y - 7z = -3$

30. $x + 5y = 0$
    $y = -1$
    $x = -5y = -5(-1) = 5$

    $(5, -1)$

32. $x + 2y - 2z = -1$
    $y + z = 9$
    $z = -3$
    $y = 9 - z = 9 - (-3) = 12$
    $x = -1 - 2y + 2z$
    $ = -1 - 2(12) + 2(-3) = -31$

    $(-31, 12, -3)$

34. Row 1: $x = -2$
    Row 2: $y = 4$
    $(-2, 4)$
    No solution

36. Row 1: $x = 3$
    Row 2: $y = -1$
    Row 3: $z = 0$
    $(3, -1, 0)$

38. $2x + 6y = 16$
    $2x + 3y = 7$
    $\begin{bmatrix} 2 & 6 & \vdots & 16 \\ 2 & 3 & \vdots & 7 \end{bmatrix} \Rightarrow \begin{bmatrix} 2 & 6 & \vdots & 16 \\ 0 & -3 & \vdots & -9 \end{bmatrix}$
    $\phantom{\begin{bmatrix} 2 & 6 \end{bmatrix}} \Rightarrow \begin{bmatrix} 1 & 3 & \vdots & 8 \\ 0 & 1 & \vdots & 3 \end{bmatrix}$
    $y = 3$
    $x + 3(3) = 8 \quad \Rightarrow \quad x = -1$

    Solution: $(-1, 3)$

40. $x + 2y = 0$
    $x + y = 6$
    $3x - 2y = 8$
    $\begin{bmatrix} 1 & 2 & \vdots & 0 \\ 1 & 1 & \vdots & 6 \\ 3 & -2 & \vdots & 8 \end{bmatrix} \Rightarrow \begin{bmatrix} 1 & 2 & \vdots & 0 \\ 0 & -1 & \vdots & 6 \\ 0 & -8 & \vdots & 8 \end{bmatrix}$
    $\phantom{\begin{bmatrix} 1 & 2 \end{bmatrix}} \Rightarrow \begin{bmatrix} 1 & 2 & \vdots & 0 \\ 0 & 1 & \vdots & 6 \\ 0 & 0 & \vdots & -40 \end{bmatrix}$

    The system is inconsistent and there is no solution.

42. $2x - y = -0.1$

$3x + 2y = 1.6$

$$\begin{bmatrix} 2 & -1 & \vdots & -0.1 \\ 3 & 2 & \vdots & 1.6 \end{bmatrix} \Rightarrow \begin{bmatrix} -1 & -3 & \vdots & -1.7 \\ 3 & 2 & \vdots & 1.6 \end{bmatrix}$$

$$\Rightarrow \begin{bmatrix} -1 & -3 & \vdots & -1.7 \\ 0 & -7 & \vdots & -3.5 \end{bmatrix}$$

$$\Rightarrow \begin{bmatrix} 1 & 3 & \vdots & 1.7 \\ 0 & 1 & \vdots & 0.5 \end{bmatrix}$$

$y = 0.5$

$x + 3(0.5) = 1.7 \quad \Rightarrow \quad x = 0.2$

Solution: $(0.2, 0.5)$

44. $x - 3y = 5$

$-2x + 6y = -10$

$$\begin{bmatrix} 1 & -3 & \vdots & 5 \\ -2 & 6 & \vdots & -10 \end{bmatrix} \Rightarrow \begin{bmatrix} 1 & -3 & \vdots & 5 \\ 0 & 0 & \vdots & 0 \end{bmatrix}$$

$y = a$

$x = 3a + 5$

Solution: $(3a + 5, \; a)$

46. $2x - y + 3z = 24$

$2y - z = 14$

$7x - 5y = 6$

$$\begin{bmatrix} 2 & -1 & 3 & \vdots & 24 \\ 0 & 2 & -1 & \vdots & 14 \\ 7 & -5 & 0 & \vdots & 6 \end{bmatrix} \Rightarrow \begin{bmatrix} 1 & -2 & -9 & \vdots & -66 \\ 0 & 2 & -1 & \vdots & 14 \\ 7 & -5 & 0 & \vdots & 6 \end{bmatrix} \Rightarrow \begin{bmatrix} 1 & -2 & -9 & \vdots & -66 \\ 0 & 2 & -1 & \vdots & 14 \\ 0 & 9 & 63 & \vdots & 468 \end{bmatrix}$$

$$\Rightarrow \begin{bmatrix} 1 & -2 & -9 & \vdots & -66 \\ 0 & 8 & -4 & \vdots & 56 \\ 0 & 9 & 63 & \vdots & 468 \end{bmatrix} \Rightarrow \begin{bmatrix} 1 & -2 & -9 & \vdots & -66 \\ 0 & -1 & -67 & \vdots & -412 \\ 0 & 9 & 63 & \vdots & 468 \end{bmatrix}$$

$$\Rightarrow \begin{bmatrix} 1 & -2 & -9 & \vdots & -66 \\ 0 & -1 & -67 & \vdots & -412 \\ 0 & 0 & -540 & \vdots & -3240 \end{bmatrix} \Rightarrow \begin{bmatrix} 1 & -2 & -9 & \vdots & -66 \\ 0 & 1 & 67 & \vdots & 412 \\ 0 & 0 & 1 & \vdots & 6 \end{bmatrix}$$

$z = 6$

$y + 67(6) = 412 \quad \Rightarrow \quad y = 10$

$x - 2(10) - 9(6) = -66 \quad \Rightarrow \quad x = 8$

Solution: $(8, 10, 6)$

48.

$$2x + \quad\quad 3z = 3$$
$$4x - 3y + 7z = 5$$
$$8x - 9y + 15z = 9$$

$$\begin{bmatrix} 2 & 0 & 3 & \vdots & 3 \\ 4 & -3 & 7 & \vdots & 5 \\ 8 & -9 & 15 & \vdots & 9 \end{bmatrix} \Rightarrow \begin{bmatrix} 2 & 0 & 3 & \vdots & 3 \\ 0 & -3 & 1 & \vdots & -1 \\ 0 & -9 & 3 & \vdots & -3 \end{bmatrix}$$

$$\Rightarrow \begin{bmatrix} 2 & 0 & 3 & \vdots & 3 \\ 0 & -3 & 1 & \vdots & -1 \\ 0 & 0 & 0 & \vdots & 0 \end{bmatrix}$$

$$z = a$$
$$y = \tfrac{1}{3}a + \tfrac{1}{3}$$
$$x = -\tfrac{3}{2}a + \tfrac{3}{2}$$

$$\Rightarrow \begin{bmatrix} 1 & 0 & \tfrac{3}{2} & \vdots & \tfrac{3}{2} \\ 0 & 1 & -\tfrac{1}{3} & \vdots & \tfrac{1}{3} \\ 0 & 0 & 0 & \vdots & 0 \end{bmatrix}$$

Solution: $\left(-\tfrac{3}{2}a + \tfrac{3}{2}, \ \tfrac{1}{3}a + \tfrac{1}{3}, \ a\right)$

50.

$$4x + 12y - 7z - 20w = 22$$
$$3x + 9y - 5z - 28w = 30$$

$$\begin{bmatrix} 4 & 12 & -7 & -20 & \vdots & 22 \\ 3 & 9 & -5 & -28 & \vdots & 30 \end{bmatrix} \Rightarrow \begin{bmatrix} 1 & 3 & -2 & 8 & \vdots & -8 \\ 3 & 9 & -5 & -28 & \vdots & 30 \end{bmatrix}$$

$$\Rightarrow \begin{bmatrix} 1 & 3 & -2 & 8 & \vdots & -8 \\ 0 & 0 & 1 & -52 & \vdots & 54 \end{bmatrix}$$

$$\Rightarrow \begin{bmatrix} 1 & 3 & 0 & -96 & \vdots & 100 \\ 0 & 0 & 1 & -52 & \vdots & 54 \end{bmatrix}$$

$$w = a$$
$$z = 52a + 54$$
$$y = b$$
$$x = -3b + 96a + 100$$

Solution: $(-3b + 96a + 100, \ b, \ 52a + 54, \ a)$

52.
$$2x + 10y + 2z = 6$$
$$x + 5y + 2z = 6$$
$$x + 5y + z = 3$$
$$-3x - 15y - 3z = -9$$

$$\begin{bmatrix} 2 & 10 & 2 & \vdots & 6 \\ 1 & 5 & 2 & \vdots & 6 \\ 1 & 5 & 1 & \vdots & 3 \\ -3 & -15 & -3 & \vdots & -9 \end{bmatrix} \Rightarrow \begin{bmatrix} 1 & 5 & 1 & \vdots & 3 \\ 1 & 5 & 2 & \vdots & 6 \\ 1 & 5 & 1 & \vdots & 3 \\ -3 & -15 & -3 & \vdots & -9 \end{bmatrix}$$

$$\Rightarrow \begin{bmatrix} 1 & 5 & 1 & \vdots & 3 \\ 0 & 0 & 1 & \vdots & 3 \\ 0 & 0 & 0 & \vdots & 0 \\ 0 & 0 & 0 & \vdots & 0 \end{bmatrix} \Rightarrow \begin{bmatrix} 1 & 5 & 0 & \vdots & 0 \\ 0 & 0 & 1 & \vdots & 3 \\ 0 & 0 & 0 & \vdots & 0 \\ 0 & 0 & 0 & \vdots & 0 \end{bmatrix}$$

$$z = 3$$
$$y = a$$
$$x = -5a$$

Solution: $(-5a,\ a,\ 3)$

54.
$$x + 2y + 2z + 4w = 11$$
$$3x + 6y + 5z + 12w = 30$$

$$\begin{bmatrix} 1 & 2 & 2 & 4 & \vdots & 11 \\ 3 & 6 & 5 & 12 & \vdots & 30 \end{bmatrix} \Rightarrow \begin{bmatrix} 1 & 2 & 2 & 4 & \vdots & 11 \\ 0 & 0 & -1 & 0 & \vdots & -3 \end{bmatrix} \Rightarrow \begin{bmatrix} 1 & 2 & 0 & 4 & \vdots & 5 \\ 0 & 0 & 1 & 0 & \vdots & 3 \end{bmatrix}$$

$$w = a$$
$$z = 3$$
$$y = b$$
$$x = -2b - 4a + 5$$

Solution: $(-2b - 4a + 5,\ b,\ 3,\ a)$

56.
$$x + 2y = 0$$
$$2x + 4y = 0$$

$$\begin{bmatrix} 1 & 2 & \vdots & 0 \\ 2 & 4 & \vdots & 0 \end{bmatrix} \Rightarrow \begin{bmatrix} 1 & 2 & \vdots & 0 \\ 0 & 0 & \vdots & 0 \end{bmatrix}$$

$$y = a$$
$$x = -2a$$

Solution: $(-2a,\ a)$

58. $x + 2y + z + 3w = 0$

$x - y \quad + w = 0$

$\quad y - z + 2w = 0$

$$\begin{bmatrix} 1 & 2 & 1 & 3 & \vdots & 0 \\ 1 & -1 & 0 & 1 & \vdots & 0 \\ 0 & 1 & -1 & 2 & \vdots & 0 \end{bmatrix} \Rightarrow \begin{bmatrix} 1 & 2 & 1 & 3 & \vdots & 0 \\ 0 & -3 & -1 & -2 & \vdots & 0 \\ 0 & 1 & -1 & 2 & \vdots & 0 \end{bmatrix} \Rightarrow \begin{bmatrix} 1 & 2 & 1 & 3 & \vdots & 0 \\ 0 & 1 & -5 & 6 & \vdots & 0 \\ 0 & 1 & -1 & 2 & \vdots & 0 \end{bmatrix}$$

$$\Rightarrow \begin{bmatrix} 1 & 0 & 11 & -9 & \vdots & 0 \\ 0 & 1 & -5 & 6 & \vdots & 0 \\ 0 & 0 & 4 & -4 & \vdots & 0 \end{bmatrix} \Rightarrow \begin{bmatrix} 1 & 0 & 11 & -9 & \vdots & 0 \\ 0 & 1 & -5 & 6 & \vdots & 0 \\ 0 & 0 & 1 & -1 & \vdots & 0 \end{bmatrix}$$

$$\Rightarrow \begin{bmatrix} 1 & 0 & 0 & 2 & \vdots & 0 \\ 0 & 1 & 0 & 1 & \vdots & 0 \\ 0 & 0 & 1 & -1 & \vdots & 0 \end{bmatrix}$$

$w = a$

$z = a$

$y = -a$

$x = -2a$

Solution: $(-2a, \ -a, \ a, \ a)$

60. $x = $ amount at 9%

$y = $ amount at 10%

$z = $ amount at 12%

$x + \quad y + \quad z = 500,000$

$0.09x + 0.010y + 0.12z = \ 52,000$

$2.5x - \quad y \quad = \quad 0$

$$\begin{bmatrix} 1 & 1 & 1 & \vdots & 500,000 \\ 0.09 & 0.10 & 0.12 & \vdots & 52,000 \\ 2.5 & -1 & 0 & \vdots & 0 \end{bmatrix} \Rightarrow \begin{bmatrix} 1 & 1 & 1 & \vdots & 500,000 \\ 0 & 0.01 & 0.03 & \vdots & 7,000 \\ 0 & -3.5 & -2.5 & \vdots & -1,250,000 \end{bmatrix}$$

$$\Rightarrow \begin{bmatrix} 1 & 1 & 1 & \vdots & 500,000 \\ 0 & 1 & 3 & \vdots & 700,000 \\ 0 & -7 & -5 & \vdots & -2,500,000 \end{bmatrix}$$

$$\Rightarrow \begin{bmatrix} 1 & 0 & -2 & \vdots & -200,000 \\ 0 & 1 & 3 & \vdots & 700,000 \\ 0 & 0 & 16 & \vdots & 2,400,000 \end{bmatrix}$$

$$\Rightarrow \begin{bmatrix} 1 & 0 & -2 & \vdots & -200,000 \\ 0 & 1 & 3 & \vdots & 700,000 \\ 0 & 0 & 1 & \vdots & 150,000 \end{bmatrix}$$

$z = 150,000, \ y = 250,000, \ x = 100,000$

Solution: \$100,000 at 9%, \$250,000 at 10% \$150,000 at 12%

62.

$$I_1 - I_2 + I_3 = 0$$
$$2I_1 + 2I_2 \qquad = 7$$
$$\qquad 2I_2 + 4I_3 = 8$$

$$\begin{bmatrix} 1 & -1 & 1 & 0 \\ 2 & 2 & 0 & 7 \\ 0 & 2 & 4 & 8 \end{bmatrix} \Rightarrow \begin{bmatrix} 1 & -1 & 1 & 0 \\ 0 & 4 & -2 & 7 \\ 0 & 1 & 2 & 2 \end{bmatrix} \Rightarrow \begin{bmatrix} 1 & -1 & 1 & 0 \\ 0 & 1 & 2 & 4 \\ 0 & 0 & -10 & -9 \end{bmatrix}$$

Hence, $I_3 = \frac{9}{10}$, $I_2 = 4 - 2\left(\frac{9}{10}\right) = \frac{22}{10}$ and $I_1 = \frac{22}{10} - \frac{9}{10} = \frac{13}{10}$.

64. At $(1, 11)$: $11 = a(1)^2 + b(1) + c \quad \Rightarrow \quad a + b + c = 11$

At $(2, 10)$: $10 = a(2)^2 + b(2) + c \quad \Rightarrow \quad 4a + 2b + c = 10$

At $(3, 7)$: $7 = a(3)^2 + b(3) + c \quad \Rightarrow \quad 9a + 3b + c = 7$

$$\begin{bmatrix} 1 & 1 & 1 & \vdots & 11 \\ 4 & 2 & 1 & \vdots & 10 \\ 9 & 3 & 1 & \vdots & 7 \end{bmatrix} \Rightarrow \begin{bmatrix} 1 & 1 & 1 & \vdots & 11 \\ 0 & -2 & -3 & \vdots & -34 \\ 0 & -6 & -8 & \vdots & -92 \end{bmatrix} \Rightarrow \begin{bmatrix} 1 & 1 & 1 & \vdots & 11 \\ 0 & 1 & \frac{3}{2} & \vdots & 17 \\ 0 & 0 & 1 & \vdots & 10 \end{bmatrix}$$

$c = 10b + \frac{3}{2}(10) = 17 \quad \Rightarrow \quad b = 2$

$a + 2 + 10 = 11 \quad \Rightarrow \quad a = -1$

Solution: $y = -x^2 + 2x + 10$

66. At $(-2, 2)$: $2 = a(-2)^3 + b(-2)^2 + c(-2) + d \quad \Rightarrow \quad -8a + 4b - 2c + d = 2$

At $(-1, 17)$: $17 = a(-1)^3 + b(-1)^2 + c(-1) + d \quad \Rightarrow \quad -a + b - c + d = 17$

At $(0, 20)$: $20 = a(0)^3 + b(0)^2 + c(0) + d \quad \Rightarrow \quad d = 20$

At $(1, 23)$: $23 = a(1)^3 + b(1)^2 + c(1) + d \quad \Rightarrow \quad a + b + c + d = 23$

$$\begin{bmatrix} -8 & 4 & -2 & 1 & \vdots & 2 \\ -1 & 1 & -1 & 1 & \vdots & 17 \\ 0 & 0 & 0 & 1 & \vdots & 20 \\ 1 & 1 & 1 & 1 & \vdots & 23 \end{bmatrix} \Rightarrow \begin{bmatrix} 1 & 1 & 1 & 1 & \vdots & 23 \\ -1 & 1 & -1 & 1 & \vdots & 17 \\ -8 & 4 & -2 & 1 & \vdots & 2 \\ 0 & 0 & 0 & 1 & \vdots & 20 \end{bmatrix}$$

$$\Rightarrow \begin{bmatrix} 1 & 1 & 1 & 1 & \vdots & 23 \\ 0 & 2 & 0 & 2 & \vdots & 40 \\ 0 & 12 & 6 & 9 & \vdots & 186 \\ 0 & 0 & 0 & 1 & \vdots & 20 \end{bmatrix} \Rightarrow \begin{bmatrix} 1 & 1 & 1 & 1 & \vdots & 23 \\ 0 & 1 & 0 & 1 & \vdots & 20 \\ 0 & 0 & 6 & -3 & \vdots & -54 \\ 0 & 0 & 0 & 1 & \vdots & 20 \end{bmatrix}$$

$$\Rightarrow \begin{bmatrix} 1 & 1 & 1 & 1 & \vdots & 23 \\ 0 & 1 & 0 & 1 & \vdots & 20 \\ 0 & 0 & 1 & -\frac{1}{2} & \vdots & -9 \\ 0 & 0 & 0 & 1 & \vdots & 20 \end{bmatrix}$$

$d = 20$

$c - \frac{1}{2}(20) = -9 \quad \Rightarrow \quad c = 1$

$b + 20 = 20 \quad \Rightarrow \quad b = 0$

$a + 0 + 1 + 20 = 23 \quad \Rightarrow \quad a = 2$

Solution: $y = 2x^3 + x + 20$

8.2 Operations with Matrices

2. $x = 13,\ y = 12$

4. $\left.\begin{array}{l} x + 2 = 2x + 6 \\ \hspace{1.1cm} 2x = -8 \end{array}\right\} \Rightarrow x = -4$

$\left.\begin{array}{l} \hspace{0.6cm} 2y = 18 \\ y + 2 = 11 \end{array}\right\} \Rightarrow y = 9$

$x = -4,\ y = 9$

6. (a) $A + B = \begin{bmatrix} 1 & 2 \\ 2 & 1 \end{bmatrix} + \begin{bmatrix} -3 & -2 \\ 4 & 2 \end{bmatrix} = \begin{bmatrix} 1-3 & 2-2 \\ 2+4 & 1+2 \end{bmatrix} = \begin{bmatrix} -2 & 0 \\ 6 & 3 \end{bmatrix}$

(b) $A - B = \begin{bmatrix} 1 & 2 \\ 2 & 1 \end{bmatrix} - \begin{bmatrix} -3 & -2 \\ 4 & 2 \end{bmatrix} = \begin{bmatrix} 1+3 & 2+2 \\ 2-4 & 1-2 \end{bmatrix} = \begin{bmatrix} 4 & 4 \\ -2 & -1 \end{bmatrix}$

(c) $3A = 3\begin{bmatrix} 1 & 2 \\ 2 & 1 \end{bmatrix} = \begin{bmatrix} 3(1) & 3(2) \\ 3(2) & 3(1) \end{bmatrix} = \begin{bmatrix} 3 & 6 \\ 6 & 3 \end{bmatrix}$

(d) $3A - 2B = \begin{bmatrix} 3 & 6 \\ 6 & 3 \end{bmatrix} - 2\begin{bmatrix} -3 & -2 \\ 4 & 2 \end{bmatrix} = \begin{bmatrix} 3+6 & 6+4 \\ 6-8 & 3-4 \end{bmatrix} = \begin{bmatrix} 9 & 10 \\ -2 & -1 \end{bmatrix}$

8. (a) $A + B = \begin{bmatrix} 2 & 1 & 1 \\ -1 & -1 & 4 \end{bmatrix} + \begin{bmatrix} 2 & -3 & 4 \\ -3 & 1 & -2 \end{bmatrix} = \begin{bmatrix} 2+2 & 1-3 & 1+4 \\ -1-3 & -1+1 & 4-2 \end{bmatrix} = \begin{bmatrix} 4 & -2 & 5 \\ -4 & 0 & 2 \end{bmatrix}$

(b) $A - B = \begin{bmatrix} 2 & 1 & 1 \\ -1 & -1 & 4 \end{bmatrix} - \begin{bmatrix} 2 & -3 & 4 \\ -3 & 1 & -2 \end{bmatrix}$

$= \begin{bmatrix} 2-2 & 1-(-3) & 1-4 \\ -1-(-3) & -1-1 & 4-(-2) \end{bmatrix} = \begin{bmatrix} 0 & 4 & -3 \\ 2 & -2 & 6 \end{bmatrix}$

(c) $3A = 3\begin{bmatrix} 2 & 1 & 1 \\ -1 & -1 & 4 \end{bmatrix} = \begin{bmatrix} 3(2) & 3(1) & 3(1) \\ 3(-1) & 3(-1) & 3(4) \end{bmatrix} = \begin{bmatrix} 6 & 3 & 3 \\ -3 & -3 & 12 \end{bmatrix}$

(d) $3A - 2B = \begin{bmatrix} 6 & 3 & 3 \\ -3 & -3 & 12 \end{bmatrix} - 2\begin{bmatrix} 2 & -3 & 4 \\ -3 & 1 & -2 \end{bmatrix} = \begin{bmatrix} 6 & 3 & 3 \\ -3 & -3 & 12 \end{bmatrix} + \begin{bmatrix} -4 & 6 & -8 \\ 6 & -2 & 4 \end{bmatrix}$

$= \begin{bmatrix} 2 & 9 & -5 \\ 3 & -5 & 16 \end{bmatrix}$

10. (a) $A + B = \begin{bmatrix} 3 \\ 2 \\ -1 \end{bmatrix} + \begin{bmatrix} -4 \\ 6 \\ 2 \end{bmatrix} = \begin{bmatrix} 3-4 \\ 2+6 \\ -1+2 \end{bmatrix} = \begin{bmatrix} -1 \\ 8 \\ 1 \end{bmatrix}$

(b) $A - B = \begin{bmatrix} 3 \\ 2 \\ -1 \end{bmatrix} - \begin{bmatrix} -4 \\ 6 \\ 2 \end{bmatrix} = \begin{bmatrix} 3+4 \\ 2-6 \\ -1-2 \end{bmatrix} = \begin{bmatrix} 7 \\ -4 \\ -3 \end{bmatrix}$

(c) $3A = 3\begin{bmatrix} 3 \\ 2 \\ -1 \end{bmatrix} = \begin{bmatrix} 3(3) \\ 3(2) \\ 3(-1) \end{bmatrix} = \begin{bmatrix} 9 \\ 6 \\ -3 \end{bmatrix}$

(d) $3A - 2B = \begin{bmatrix} 9 \\ 6 \\ -3 \end{bmatrix} - 2\begin{bmatrix} -4 \\ 6 \\ 2 \end{bmatrix} = \begin{bmatrix} 9 \\ 6 \\ -3 \end{bmatrix} + \begin{bmatrix} 8 \\ -12 \\ -4 \end{bmatrix} = \begin{bmatrix} 17 \\ -6 \\ -7 \end{bmatrix}$

12. (a) $AB = \begin{bmatrix} 2 & -1 \\ 1 & 4 \end{bmatrix}\begin{bmatrix} 0 & 0 \\ 3 & -3 \end{bmatrix} = \begin{bmatrix} 2(0)+(-1)3 & 2(0)+(-1)(-3) \\ 1(0)+4(3) & 1(0)+4(-3) \end{bmatrix} = \begin{bmatrix} -3 & 3 \\ 12 & -12 \end{bmatrix}$

(b) $BA = \begin{bmatrix} 0 & 0 \\ 3 & -3 \end{bmatrix}\begin{bmatrix} 2 & -1 \\ 1 & 4 \end{bmatrix} = \begin{bmatrix} 0(2)+0(1) & 0(-1)+0(4) \\ 3(2)+(-3)(1) & 3(-1)+(-3)4 \end{bmatrix} = \begin{bmatrix} 0 & 0 \\ 3 & -15 \end{bmatrix}$

(c) $A^2 = \begin{bmatrix} 2 & -1 \\ 1 & 4 \end{bmatrix}\begin{bmatrix} 2 & -1 \\ 1 & 4 \end{bmatrix} = \begin{bmatrix} 2(2)+(-1)(1) & 2(-1)+(-1)4 \\ 1(2)+4(1) & 1(-1)+4(4) \end{bmatrix} = \begin{bmatrix} 3 & -6 \\ 6 & 15 \end{bmatrix}$

14. (a) $AB = \begin{bmatrix} 1 & -1 \\ 1 & 1 \end{bmatrix}\begin{bmatrix} 1 & 3 \\ -3 & 1 \end{bmatrix} = \begin{bmatrix} 1(1)+(-1)(-3) & 1(3)+(-1)(1) \\ 1(1)+1(-3) & 1(3)+1(1) \end{bmatrix} = \begin{bmatrix} 4 & 2 \\ -2 & 4 \end{bmatrix}$

(b) $BA = \begin{bmatrix} 1 & 3 \\ -3 & 1 \end{bmatrix}\begin{bmatrix} 1 & -1 \\ 1 & 1 \end{bmatrix} = \begin{bmatrix} 1(1)+3(1) & 1(-1)+3(1) \\ -3(1)+1(1) & -3(-1)+1(1) \end{bmatrix} = \begin{bmatrix} 4 & 2 \\ -2 & 4 \end{bmatrix}$

(c) $A^2 = \begin{bmatrix} 1 & -1 \\ 1 & -1 \end{bmatrix}\begin{bmatrix} 1 & 1 \\ 1 & 1 \end{bmatrix} = \begin{bmatrix} 1(1)+(-1)(1) & 1(-1)+(-1)(1) \\ 1(1)+(1)(1) & 1(-1)+1(1) \end{bmatrix} = \begin{bmatrix} 0 & -2 \\ 2 & 0 \end{bmatrix}$

16. (a) $AB = \begin{bmatrix} 3 & 2 & 1 \end{bmatrix}\begin{bmatrix} 2 \\ 3 \\ 0 \end{bmatrix} = [3(2)+2(3)+1(0)] = [12]$

(b) $BA = \begin{bmatrix} 2 \\ 3 \\ 0 \end{bmatrix}\begin{bmatrix} 3 & 2 & 1 \end{bmatrix} = \begin{bmatrix} 2(3) & 2(2) & 2(1) \\ 3(3) & 3(2) & 3(1) \\ 0(3) & 0(2) & 0(1) \end{bmatrix} = \begin{bmatrix} 6 & 4 & 2 \\ 9 & 6 & 3 \\ 0 & 0 & 0 \end{bmatrix}$

(c) The number of columns of A does not equal the number of rows of A; the multiplication is not possible.

18. A is 3×3 and B is 3×2 $\Rightarrow$ AB is 3×2.

$AB = \begin{bmatrix} 0 & -1 & 0 \\ 4 & 0 & 2 \\ 8 & -1 & 7 \end{bmatrix}\begin{bmatrix} 2 & 1 \\ -3 & 4 \\ 1 & 6 \end{bmatrix} = \begin{bmatrix} 4 & -4 \\ 10 & 16 \\ 29 & 46 \end{bmatrix}$

20. A is 3×3 and B is 3×3 $\Rightarrow$ AB is 3×3.

$AB = \begin{bmatrix} 1 & 0 & 0 \\ 0 & 4 & 0 \\ 0 & 0 & -2 \end{bmatrix}\begin{bmatrix} 3 & 0 & 0 \\ 0 & -1 & 0 \\ 0 & 0 & 5 \end{bmatrix} = \begin{bmatrix} 3 & 0 & 0 \\ 0 & -4 & 0 \\ 0 & 0 & -10 \end{bmatrix}$

22. A is 3×3 and B is 3×3 $\Rightarrow$ AB is 3×3.

$$AB = \begin{bmatrix} 0 & 0 & 5 \\ 0 & 0 & -3 \\ 0 & 0 & 4 \end{bmatrix} \begin{bmatrix} 6 & -11 & 4 \\ 8 & 16 & 4 \\ 0 & 0 & 0 \end{bmatrix} = \begin{bmatrix} 0 & 0 & 0 \\ 0 & 0 & 0 \\ 0 & 0 & 0 \end{bmatrix}$$

24. A is 2×5 and B is 2×2. Since the number of columns of A does not equal the number of rows of B, the multiplication is not possible.

26. $X = A - \frac{1}{2}B = \begin{bmatrix} -2 & -1 \\ 1 & 0 \\ 3 & -4 \end{bmatrix} - \frac{1}{2}\begin{bmatrix} 0 & 3 \\ 2 & 0 \\ -4 & -1 \end{bmatrix} = \begin{bmatrix} -2 & -1 \\ 1 & 0 \\ 3 & -4 \end{bmatrix} - \begin{bmatrix} 0 & \frac{3}{2} \\ 1 & 0 \\ -2 & -\frac{1}{2} \end{bmatrix} = \begin{bmatrix} -2 & -\frac{5}{2} \\ 0 & 0 \\ 5 & -\frac{7}{2} \end{bmatrix}$

28. $X = -A - 2B = -1\begin{bmatrix} -2 & -1 \\ 1 & 0 \\ 3 & -4 \end{bmatrix} - 2\begin{bmatrix} 0 & 3 \\ 2 & 0 \\ -4 & -1 \end{bmatrix} = \begin{bmatrix} 2 & 1 \\ -1 & 0 \\ -3 & 4 \end{bmatrix} + \begin{bmatrix} 0 & -6 \\ -4 & 0 \\ 8 & 2 \end{bmatrix} = \begin{bmatrix} 2 & -5 \\ -5 & 0 \\ 5 & -6 \end{bmatrix}$

30. $A = \begin{bmatrix} 2 & 3 \\ 1 & 4 \end{bmatrix}$, $X = \begin{bmatrix} x \\ y \end{bmatrix}$, $B = \begin{bmatrix} 5 \\ 10 \end{bmatrix}$

$$\begin{bmatrix} 2 & 3 & \vdots & 5 \\ 1 & 4 & \vdots & 10 \end{bmatrix} \Rightarrow \begin{bmatrix} 1 & 4 & \vdots & 10 \\ 2 & 3 & \vdots & 5 \end{bmatrix} \Rightarrow \begin{bmatrix} 1 & 4 & \vdots & 10 \\ 0 & -5 & \vdots & -15 \end{bmatrix} \Rightarrow \begin{bmatrix} 1 & 0 & \vdots & -2 \\ 0 & 1 & \vdots & 3 \end{bmatrix}$$

$x = -2, \ y = 3$

32. $A = \begin{bmatrix} 1 & 1 & -3 \\ -1 & 2 & 0 \\ 0 & -1 & 1 \end{bmatrix}$, $X = \begin{bmatrix} x \\ y \\ z \end{bmatrix}$, $B = \begin{bmatrix} -1 \\ 1 \\ 0 \end{bmatrix}$

$$\begin{bmatrix} 1 & 1 & -3 & \vdots & -1 \\ -1 & 2 & 0 & \vdots & 1 \\ 0 & -1 & 1 & \vdots & 0 \end{bmatrix} \Rightarrow \begin{bmatrix} 1 & 1 & -3 & \vdots & -1 \\ 0 & 3 & -3 & \vdots & 0 \\ 0 & -1 & 1 & \vdots & 0 \end{bmatrix}$$

$$\Rightarrow \begin{bmatrix} 1 & 1 & -3 & \vdots & -1 \\ 0 & 1 & -1 & \vdots & 0 \\ 0 & 0 & 0 & \vdots & 0 \end{bmatrix}$$

$$\Rightarrow \begin{bmatrix} 1 & 0 & -2 & \vdots & -1 \\ 0 & 1 & -1 & \vdots & 0 \\ 0 & 0 & 0 & \vdots & 0 \end{bmatrix}$$

Let $z = a$, then $y = a$, $x = 2a - 1$.

34. $AB = \begin{bmatrix} 3 & 3 \\ 4 & 4 \end{bmatrix} \begin{bmatrix} 1 & -1 \\ -1 & 1 \end{bmatrix}$

$= \begin{bmatrix} 3-3 & -3+3 \\ 4-4 & -4+4 \end{bmatrix} = \begin{bmatrix} 0 & 0 \\ 0 & 0 \end{bmatrix}$

36. $1.10\begin{bmatrix} 100 & 90 & 70 & 30 \\ 40 & 20 & 60 & 60 \end{bmatrix} = \begin{bmatrix} 110 & 99 & 77 & 33 \\ 44 & 22 & 66 & 66 \end{bmatrix}$

Thus, $AB = 0$ and neither A nor B is 0.

38. $BA = \begin{bmatrix} 20.50 & 26.50 & 29.50 \end{bmatrix} \begin{bmatrix} 5000 & 4000 \\ 6000 & 10,000 \\ 8000 & 5000 \end{bmatrix}$

$= \begin{bmatrix} \$497,500 & \$494,500 \end{bmatrix}$

The entries in the last matrix represent the cost of the product at each of the two warehouses.

40. (a) $1.6(10) + 1(8) + 0.2(5) = \$25.00$ per hour

(b) $2.5(12) + 2(9) + 0.4(6) = \$50.40$ per hour

(c) $ST = \begin{bmatrix} 1.0 & 0.5 & 0.2 \\ 1.6 & 1.0 & 0.2 \\ 2.5 & 2.0 & 0.4 \end{bmatrix} \begin{bmatrix} 12 & 10 \\ 9 & 8 \\ 6 & 5 \end{bmatrix}$

$= \begin{bmatrix} \$17.70 & \$15.00 \\ \$29.40 & \$25.00 \\ \$50.40 & \$43.00 \end{bmatrix}$

This represents the labor cost for each boat size at each plant.

42. $P^3 = \begin{bmatrix} 0.3 & 0.175 & 0.175 \\ 0.308 & 0.433 & 0.217 \\ 0.392 & 0.392 & 0.608 \end{bmatrix}$ $P^6 \approx \begin{bmatrix} 0.2125 & 0.1969 & 0.1969 \\ 0.3108 & 0.3265 & 0.2798 \\ 0.4767 & 0.4767 & 0.5233 \end{bmatrix}$

$P^4 = \begin{bmatrix} 0.25 & 0.1875 & 0.1875 \\ 0.3148 & 0.3773 & 0.2477 \\ 0.4352 & 0.4352 & 0.5648 \end{bmatrix}$ $P^7 \approx \begin{bmatrix} 0.2063 & 0.1984 & 0.1984 \\ 0.3077 & 0.3156 & 0.2876 \\ 0.4860 & 0.4860 & 0.5140 \end{bmatrix}$

$P^5 \approx \begin{bmatrix} 0.225 & 0.1938 & 0.1938 \\ 0.3139 & 0.3451 & 0.2674 \\ 0.4611 & 0.4611 & 0.5389 \end{bmatrix}$ $P^8 \approx \begin{bmatrix} 0.2031 & 0.1992 & 0.1992 \\ 0.3053 & 0.3092 & 0.2924 \\ 0.4916 & 0.4916 & 0.5084 \end{bmatrix}$

8.3 | Inverse Matrices and Systems

2. $AB = \begin{bmatrix} 1 & -1 \\ -1 & 2 \end{bmatrix} \begin{bmatrix} 2 & 1 \\ 1 & 1 \end{bmatrix} = \begin{bmatrix} 2-1 & 1-1 \\ -2+2 & -1+2 \end{bmatrix} = \begin{bmatrix} 1 & 0 \\ 0 & 1 \end{bmatrix}$

$BA = \begin{bmatrix} 2 & 1 \\ 1 & 1 \end{bmatrix} \begin{bmatrix} 1 & -1 \\ -1 & 2 \end{bmatrix} = \begin{bmatrix} 2-1 & -2+2 \\ 1-1 & -1+2 \end{bmatrix} = \begin{bmatrix} 1 & 0 \\ 0 & 1 \end{bmatrix}$

4. $AB = \begin{bmatrix} 1 & -1 \\ 2 & 3 \end{bmatrix} \begin{bmatrix} \frac{3}{5} & \frac{1}{5} \\ -\frac{2}{5} & \frac{1}{5} \end{bmatrix} = \begin{bmatrix} \frac{3}{5}+\frac{2}{5} & \frac{1}{5}-\frac{1}{5} \\ \frac{6}{5}-\frac{6}{5} & \frac{2}{5}+\frac{3}{5} \end{bmatrix} = \begin{bmatrix} 1 & 0 \\ 0 & 1 \end{bmatrix}$

$BA = \begin{bmatrix} \frac{3}{5} & \frac{1}{5} \\ -\frac{2}{5} & \frac{1}{5} \end{bmatrix} \begin{bmatrix} 1 & -1 \\ 2 & 3 \end{bmatrix} = \begin{bmatrix} \frac{3}{5}+\frac{2}{5} & -\frac{3}{5}+\frac{3}{5} \\ -\frac{2}{5}+\frac{2}{5} & \frac{2}{5}+\frac{3}{5} \end{bmatrix} = \begin{bmatrix} 1 & 0 \\ 0 & 1 \end{bmatrix}$

6. $AB = \begin{bmatrix} 2 & -17 & 11 \\ -1 & 11 & -7 \\ 0 & 3 & -2 \end{bmatrix} \begin{bmatrix} 1 & 1 & 2 \\ 2 & 4 & -3 \\ 3 & 6 & -5 \end{bmatrix}$

$= \begin{bmatrix} 2-34+33 & 2-68+66 & 4+51-55 \\ -1+22-21 & -1+44-42 & -2-33+35 \\ 6-6 & 12-12 & -9+10 \end{bmatrix} = \begin{bmatrix} 1 & 0 & 0 \\ 0 & 1 & 0 \\ 0 & 0 & 1 \end{bmatrix}$

$BA = \begin{bmatrix} 1 & 1 & 2 \\ 2 & 4 & -3 \\ 3 & 6 & -5 \end{bmatrix} \begin{bmatrix} 2 & -17 & 11 \\ -1 & 11 & -7 \\ 0 & 3 & -2 \end{bmatrix} = \begin{bmatrix} 2-1 & -17+11+6 & 11-7-4 \\ 4-4 & -34+44-9 & 22-28+6 \\ 6-6 & -51+66-15 & 33-42+10 \end{bmatrix} = I_3$

8. $AB = \frac{1}{3} \begin{bmatrix} -1 & 1 & 0 & -1 \\ 1 & -1 & 1 & 0 \\ -1 & 1 & 2 & 0 \\ 0 & -1 & 1 & 1 \end{bmatrix} \begin{bmatrix} -3 & 1 & 1 & -3 \\ -3 & -1 & 2 & -3 \\ 0 & 1 & 1 & 0 \\ -3 & -2 & 1 & 0 \end{bmatrix}$

$= \frac{1}{3} \begin{bmatrix} 3-3+0+3 & -1-1+0+2 & -1+2+0-1 & 3-3+0+0 \\ -3+3+0+0 & 1+1+1+0 & 1-2+1+0 & -3+3+0+0 \\ 3-3+0+0 & -1-1+2+0 & -1+2+2+0 & 3-3+0+0 \\ 0+3+0-3 & 0+1+1-2 & 0-2+1+1 & 0+3+0+0 \end{bmatrix}$

$= \frac{1}{3} \begin{bmatrix} 3 & 0 & 0 & 0 \\ 0 & 3 & 0 & 0 \\ 0 & 0 & 3 & 0 \\ 0 & 0 & 0 & 3 \end{bmatrix} = I_4$

$BA = \frac{1}{3} \begin{bmatrix} -3 & 1 & 1 & -3 \\ -3 & -1 & 2 & -3 \\ 0 & 1 & 1 & 0 \\ -3 & -2 & 1 & 0 \end{bmatrix} \begin{bmatrix} -1 & 1 & 0 & -1 \\ 1 & -1 & 1 & 0 \\ -1 & 1 & 2 & 0 \\ 0 & -1 & 1 & 1 \end{bmatrix}$

$= \frac{1}{3} \begin{bmatrix} 3+1-1+0 & -3-1+1+3 & 0+1+2-3 & 3+0+0-3 \\ 3-1-2+0 & -3+1+2+3 & 0-1+4-3 & 3+0+0-3 \\ 0+1-1+0 & 0-1+1+0 & 0+1+2+0 & 0+0+0+0 \\ 3-2-1+0 & -3+2+1+0 & 0-2+2+0 & 3+0+0+0 \end{bmatrix} = \frac{1}{3} \begin{bmatrix} 3 & 0 & 0 & 0 \\ 0 & 3 & 0 & 0 \\ 0 & 0 & 3 & 0 \\ 0 & 0 & 0 & 3 \end{bmatrix} = I_4$

10. $A^{-1} = \frac{1}{ad-bc} \begin{bmatrix} d & -b \\ -c & a \end{bmatrix} = \frac{1}{7-6} \begin{bmatrix} 7 & -2 \\ -3 & 1 \end{bmatrix} = \begin{bmatrix} 7 & -2 \\ -3 & 1 \end{bmatrix}$

12. $A^{-1} = \frac{1}{ad-bc} \begin{bmatrix} d & -b \\ -c & a \end{bmatrix} = \frac{1}{133-132} \begin{bmatrix} -19 & -33 \\ -4 & -7 \end{bmatrix} = \begin{bmatrix} -19 & -33 \\ -4 & -7 \end{bmatrix}$

14. $A^{-1} = \frac{1}{ad-bc} \begin{bmatrix} d & -b \\ -c & a \end{bmatrix} = \frac{1}{0+1} \begin{bmatrix} 0 & -1 \\ 1 & 11 \end{bmatrix} = \begin{bmatrix} 0 & -1 \\ 1 & 11 \end{bmatrix}$

16. $A^{-1} = \frac{1}{ad-bc} \begin{bmatrix} d & -b \\ -c & a \end{bmatrix} = \frac{1}{8-3} \begin{bmatrix} 4 & -3 \\ -1 & 2 \end{bmatrix} = \frac{1}{5} \begin{bmatrix} 4 & -3 \\ -1 & 2 \end{bmatrix}$

18. $A = \begin{bmatrix} -2 & 5 \\ 6 & -15 \\ 0 & 1 \end{bmatrix}$

A has no inverse because it is not square.

20. Using a graphing utility,

$A^{-1} = \begin{bmatrix} -13 & 6 & 4 \\ 12 & -5 & -3 \\ -5 & 2 & 1 \end{bmatrix}.$

22. Using a graphing utility,

$$A^{-1} = \begin{bmatrix} -10 & -4 & 27 \\ 2 & 1 & -5 \\ -13 & -5 & 35 \end{bmatrix}.$$

24. Using a graphing utility,

$$A^{-1} = \frac{1}{2} \begin{bmatrix} 2 & -2 & 0 \\ 14 & -17 & 2 \\ -16 & 20 & -2 \end{bmatrix}.$$

26. Using a graphing utility,

$$A^{-1} = \frac{1}{30} \begin{bmatrix} 15 & 0 & 0 \\ 0 & 10 & 0 \\ 0 & 0 & 6 \end{bmatrix}.$$

28. Using a graphing utility, we see that the inverse of A does not exist.

30. Using a graphing utility,

$$A^{-1} = \begin{bmatrix} 1 & 0 & 1 & 0 \\ 0 & 1 & 0 & 1 \\ 2 & 0 & 1 & 0 \\ 0 & 1 & 0 & 2 \end{bmatrix}.$$

32. Using a graphing utility,

$$A^{-1} = \frac{1}{10} \begin{bmatrix} 10 & -15 & -40 & 26 \\ 0 & 5 & 10 & -8 \\ 0 & 0 & -5 & 1 \\ 0 & 0 & 0 & 2 \end{bmatrix}.$$

34. Using a graphing utility,

$$A^{-1} = \begin{bmatrix} 27 & -10 & 4 & -29 \\ -16 & 5 & -2 & 18 \\ -17 & 4 & -2 & 20 \\ -7 & 2 & -1 & 8 \end{bmatrix}.$$

36. Using a graphing utility, the solution is $\left(\frac{1}{2}, \frac{1}{3}\right)$.

38. Using a graphing utility, we see that the system is inconsistent.

40. Using a graphing utility, the solution is $(-0.0769, 0.6154) = \left(-\frac{1}{13}, \frac{8}{13}\right)$.

42. Using a graphing utility, the solution is $(5, 8, -2)$.

44. Using a graphing utility, we see that the coefficient matrix is not invertible. However, using row reduction we find there are an infinite number of solutions: $(-1 + 2a, \ 2 - 3a, \ a)$.

46. Using a graphing utility, the solution is $(7, -1, -3, 2)$.

For Exercises 48 and 50 we have:

$$A = \begin{bmatrix} 1 & 1 & 1 \\ 0.065 & 0.07 & 0.09 \\ 0 & 2 & -1 \end{bmatrix}$$

Using a graphing utility, we obtain

$$A^{-1} = \frac{1}{11} \begin{bmatrix} 50 & -600 & -4 \\ -13 & 200 & 5 \\ -26 & 400 & -1 \end{bmatrix}.$$

48. $B = \begin{bmatrix} 45{,}000 \\ 3{,}750 \\ 0 \end{bmatrix}$

$$X = A^{-1}B = \frac{1}{11} \begin{bmatrix} 50 & -600 & -4 \\ -13 & 200 & 5 \\ -26 & 400 & -1 \end{bmatrix} \begin{bmatrix} 45{,}000 \\ 3{,}750 \\ 0 \end{bmatrix} = \begin{bmatrix} 0 \\ 15{,}000 \\ 30{,}000 \end{bmatrix}$$

Answer: 0 in AAA-rated bonds, \$15,000 in A-rated bonds, \$30,000 in B-rated bonds

50. $B = \begin{bmatrix} 500,000 \\ 38,000 \\ 0 \end{bmatrix}$

$X = A^{-1}B = \begin{bmatrix} 50 & -600 & -4 \\ -13 & 200 & 5 \\ -26 & 400 & -1 \end{bmatrix} \begin{bmatrix} 500,000 \\ 38,000 \\ 0 \end{bmatrix} = \begin{bmatrix} 200,000 \\ 100,000 \\ 200,000 \end{bmatrix}$

Answer: $200,000 in AAA-rated bonds, $100,000 in A-rated bonds, $200,000 in B-rated bonds

For Exercise 52 we have:

$A = \begin{bmatrix} 2 & 0 & 4 \\ 0 & 1 & 4 \\ 1 & 1 & -1 \end{bmatrix}$

Using a graphing utility, we obtain

$A^{-1} = \frac{1}{14}\begin{bmatrix} 5 & -4 & 4 \\ -4 & 6 & 8 \\ 1 & 2 & -2 \end{bmatrix}.$

52. $B = \begin{bmatrix} 10 \\ 10 \\ 0 \end{bmatrix}$

$X = A^{-1}B = \frac{1}{14}\begin{bmatrix} 5 & -4 & 4 \\ -4 & 6 & 8 \\ 1 & 2 & -2 \end{bmatrix} \begin{bmatrix} 10 \\ 10 \\ 0 \end{bmatrix} = \begin{bmatrix} \frac{5}{7} \\ \frac{10}{7} \\ \frac{15}{7} \end{bmatrix}$

Answer:

$I_1 = \frac{5}{7}$ amps, $I_2 = \frac{10}{7}$ amps, $I_3 = \frac{15}{7}$ amps

8.4 The Determinant of a Square Matrix

2. -8

4. $\begin{vmatrix} -3 & 1 \\ 5 & 2 \end{vmatrix} = -3(2) - 1(5) = -6 - 5 = -11$

6. $\begin{vmatrix} 2 & -2 \\ 4 & 3 \end{vmatrix} = 2(3) - (-2)(4) = 6 + 8 = 14$

8. $\begin{vmatrix} 4 & -3 \\ 0 & 0 \end{vmatrix} = 4(0) - (-3)(0) = 0$

10. $\begin{vmatrix} 2 & -3 \\ -6 & 9 \end{vmatrix} = 2(9) - (-3)(-6) = 0$

12. $\begin{vmatrix} -2 & 2 & 3 \\ 1 & -1 & 0 \\ 0 & 1 & 4 \end{vmatrix} = 8 + 0 + 3 - 0 - 0 - 8 = 3$

14. $\begin{vmatrix} 0.1 & 0.2 & 0.3 \\ -0.3 & 0.2 & 0.2 \\ 0.5 & 0.4 & 0.4 \end{vmatrix} = 0.008 + 0.020 + (-0.036) - 0.030 - 0.008 - (-0.024) = -0.022$

16. $\begin{vmatrix} 2 & 3 & 1 \\ 0 & 5 & -2 \\ 0 & 0 & -2 \end{vmatrix} = 2(5)(-2) = -20$

(Upper Triangular Matrix)

18. $\begin{vmatrix} 1 & 1 & 2 \\ 3 & 1 & 0 \\ -2 & 0 & 3 \end{vmatrix} = 3 + 0 + 0 - (-4) - 0 - 9$
$= -2$

20. $\begin{vmatrix} 1 & 0 & 0 \\ -4 & -1 & 0 \\ 5 & 1 & 5 \end{vmatrix} = (1)(-1)(5) = -5$

(Lower Triangular Matrix)

22. $\begin{vmatrix} -2 & 0 & 0 & 0 & 0 \\ 0 & 0 & -1 & 0 & 0 \\ 0 & 0 & 0 & 2 & 0 \\ 0 & 0 & 0 & 0 & -4 \end{vmatrix} = (-2)(3)(-1)(2)(-4) = -48$

(Diagonal Matrix)

24. $\begin{vmatrix} 3-\lambda & 2 \\ 4 & 1-\lambda \end{vmatrix} = (3-\lambda)(1-\lambda) - 8 = \lambda^2 - 4\lambda - 5$

26. (a) $M_{11} = 2$ (b) $C_{11} = M_{11} = 2$

$M_{12} = -3$ $C_{12} = -M_{12} = 3$

$M_{21} = 0$ $C_{21} = -M_{21} = 0$

$M_{22} = 11$ $C_{22} = M_{22} = 11$

28. (a) $M_{11} = \begin{vmatrix} -6 & 0 \\ 7 & -6 \end{vmatrix} = 36$

$M_{12} = \begin{vmatrix} 7 & 0 \\ 6 & -6 \end{vmatrix} = -42$

$M_{13} = \begin{vmatrix} 7 & -6 \\ 6 & 7 \end{vmatrix} = 85$

$M_{21} = \begin{vmatrix} 9 & 4 \\ 7 & -6 \end{vmatrix} = -82$

$M_{22} = \begin{vmatrix} -2 & 4 \\ 6 & -6 \end{vmatrix} = -12$

$M_{23} = \begin{vmatrix} -2 & 9 \\ 6 & 7 \end{vmatrix} = -68$

$M_{31} = \begin{vmatrix} 9 & 4 \\ -6 & 0 \end{vmatrix} = 24$

$M_{32} = \begin{vmatrix} -2 & 4 \\ 7 & 0 \end{vmatrix} = -28$

$M_{33} = \begin{vmatrix} -2 & 9 \\ 7 & -6 \end{vmatrix} = -51$

(b) $C_{12} = (-1)^3 M_{12} = 42$

$C_{13} = (-1)^4 M_{13} = 85$

$C_{21} = (-1)^3 M_{21} = 82$

$C_{22} = (-1)^4 M_{22} = -12$

$C_{23} = (-1)^5 M_{23} = 68$

$C_{31} = (-1)^4 M_{31} = 24$

$C_{32} = (-1)^5 M_{32} = 28$

$C_{33} = (-1)^6 M_{33} = -51$

30. (a) $\begin{vmatrix} -3 & 4 & 2 \\ 6 & 3 & 1 \\ 4 & -7 & -8 \end{vmatrix} = 6C_{21} + 3C_{22} + 1C_{23}$

$\qquad\qquad\qquad\qquad = 6(18) + 3(16) + 1(-5)$

$\qquad\qquad\qquad\qquad = 151$

(b) $\begin{vmatrix} -3 & 4 & 2 \\ 6 & 3 & 1 \\ 4 & -7 & -8 \end{vmatrix} = 2C_{13} + 1C_{23} - 8C_{33}$

$\qquad\qquad\qquad\qquad = 2(-54) + 1(-5) - 8(-33)$

$\qquad\qquad\qquad\qquad = 151$

32. (a) $\begin{vmatrix} 10 & -5 & 5 \\ 30 & 0 & 10 \\ 0 & 10 & 1 \end{vmatrix} = 0C_{31} + 10C_{32} + 1C_{33}$

$\qquad\qquad\qquad\qquad = 0 + 10(50) + 1(150)$

$\qquad\qquad\qquad\qquad = 650$

(b) $\begin{vmatrix} 10 & -5 & 5 \\ 30 & 0 & 10 \\ 0 & 10 & 1 \end{vmatrix} = 10C_{11} + 30C_{21} + 0C_{31}$

$\qquad\qquad\qquad\qquad = 10(-100) + 30(55) + 0$

$\qquad\qquad\qquad\qquad = 650$

34. (a) $\begin{vmatrix} 10 & 8 & 3 & -7 \\ 4 & 0 & 5 & -6 \\ 0 & 3 & 2 & 7 \\ 1 & 0 & -3 & 2 \end{vmatrix} = 0C_{31} + 3C_{32} + 2C_{33} + 7C_{34}$

$\qquad\qquad\qquad\qquad\qquad = 0 + 3(3) + 2(-112) + 7(-136)$

$\qquad\qquad\qquad\qquad\qquad = -1167$

(b) $\begin{vmatrix} 10 & 8 & 3 & -7 \\ 4 & 0 & 5 & -6 \\ 0 & 3 & 2 & 7 \\ 1 & 0 & -3 & 2 \end{vmatrix} = 10C_{11} + 4C_{21} + 0C_{31} + 1C_{41}$

$\qquad\qquad\qquad\qquad\qquad = 10(24) + 4(-245) + 0 + 1(-427)$

$\qquad\qquad\qquad\qquad\qquad = -1167$

36. Expansion along the third row:

$\begin{vmatrix} 2 & -1 & 3 \\ 1 & 4 & 4 \\ 1 & 0 & 2 \end{vmatrix} = 1 \begin{vmatrix} -1 & 3 \\ 4 & 4 \end{vmatrix} + 0 + 2 \begin{vmatrix} 2 & -1 \\ 1 & 4 \end{vmatrix}$

$\qquad\qquad\qquad = -16 + 2(9) = 2$

38. Expansion along the first row:

$\begin{vmatrix} -3 & 0 & 0 \\ 7 & 11 & 0 \\ 1 & 2 & 2 \end{vmatrix} = -3 \begin{vmatrix} 11 & 0 \\ 2 & 2 \end{vmatrix} + 0 + 0$

$\qquad\qquad\qquad = -3(22) = -66$

40. Expansion along the third column:

$$\begin{vmatrix} 2 & 6 & 6 & 2 \\ 2 & 7 & 3 & 6 \\ 1 & 5 & 0 & 1 \\ 3 & 7 & 0 & 7 \end{vmatrix} = 6 \begin{vmatrix} 2 & 7 & 6 \\ 1 & 5 & 1 \\ 3 & 7 & 7 \end{vmatrix} - 3 \begin{vmatrix} 2 & 6 & 2 \\ 1 & 5 & 1 \\ 3 & 7 & 7 \end{vmatrix} + 0 + 0$$

$$= 6\left[2(28) - 1(7) + 3(-23)\right] - 3\left[2(28) - 1(28) + 3(-4)\right]$$

$$= 6\left[56 - 7 - 69\right] - 3\left[56 - 28 - 12\right] = 6(-20) - 3(16) = -168$$

42. Expansion along the third row:

$$\begin{vmatrix} 1 & 4 & 3 & 2 \\ -5 & 6 & 2 & 1 \\ 0 & 0 & 0 & 0 \\ 3 & -2 & 1 & 5 \end{vmatrix} = 0 + 0 + 0 + 0 = 0$$

44. Expansion along the first column:

$$\begin{vmatrix} 5 & 2 & 0 & 0 & -2 \\ 0 & 1 & 4 & 3 & 2 \\ 0 & 0 & 2 & 6 & 3 \\ 0 & 0 & 3 & 4 & 1 \\ 0 & 0 & 0 & 0 & 2 \end{vmatrix} = 5 \begin{vmatrix} 1 & 4 & 3 & 2 \\ 0 & 2 & 6 & 3 \\ 0 & 3 & 4 & 1 \\ 0 & 0 & 0 & 2 \end{vmatrix}$$

$$= 5(1) \begin{vmatrix} 2 & 6 & 3 \\ 3 & 4 & 1 \\ 0 & 0 & 2 \end{vmatrix} \quad \text{Expansion along the first column}$$

$$= 5\left[0 + 0 + 2(-10)\right] = 5(-20) = -100$$

46. $$\begin{vmatrix} 1 & 7 & -3 \\ 1 & 3 & 1 \\ 4 & 8 & 1 \end{vmatrix} = \begin{vmatrix} 1 & 7 & 3 \\ 0 & -4 & 4 \\ 0 & -20 & 13 \end{vmatrix}$$

$$= 1(-1)^2 \begin{vmatrix} -4 & 4 \\ -20 & 13 \end{vmatrix} = 28$$

48. $$\begin{vmatrix} 4 & 3 & -2 \\ 5 & 4 & 1 \\ -2 & 3 & 4 \end{vmatrix} = \begin{vmatrix} 14 & 11 & 0 \\ 5 & 4 & 1 \\ -22 & -13 & 0 \end{vmatrix}$$

$$= (-1)^5 \begin{vmatrix} 14 & 11 \\ -22 & -13 \end{vmatrix} = -60$$

50. $$\begin{vmatrix} 5 & -8 & 0 \\ 9 & 7 & 4 \\ -8 & 7 & 1 \end{vmatrix} = \begin{vmatrix} 5 & -8 & 0 \\ 41 & -21 & 0 \\ -8 & 7 & 1 \end{vmatrix}$$

$$= (-1)^6 \begin{vmatrix} 5 & -8 \\ 41 & -21 \end{vmatrix} = 223$$

52. $$\begin{vmatrix} 2 & 0 & 1 \\ 4 & -4 & 0 \\ -1 & 5 & 2 \end{vmatrix} = \begin{vmatrix} 2 & 0 & 1 \\ 4 & -4 & 0 \\ -5 & 5 & 0 \end{vmatrix} = 0$$

(Row 3 is a multiple of Row 2.)

54. $\begin{vmatrix} 3 & 0 & 0 \\ -2 & 5 & 0 \\ 12 & 5 & 7 \end{vmatrix} = 3 \cdot 5 \cdot 7 = 105$

56. $\begin{vmatrix} 4 & -7 & 9 & 1 \\ 6 & 2 & 7 & 0 \\ 3 & 6 & -3 & 3 \\ 0 & 7 & 4 & -1 \end{vmatrix} = \begin{vmatrix} 4 & 0 & 13 & 0 \\ 6 & 2 & 7 & 0 \\ 3 & 27 & 9 & 0 \\ 0 & 7 & 4 & -1 \end{vmatrix}$

$= -1(-1)^8 \begin{vmatrix} 4 & 0 & 13 \\ 6 & 2 & 7 \\ 3 & 27 & 9 \end{vmatrix}$

$= -1(3) \begin{vmatrix} 4 & 0 & 13 \\ 6 & 2 & 7 \\ 1 & 9 & 3 \end{vmatrix}$

$= -3 \begin{vmatrix} 0 & -36 & 1 \\ 0 & -52 & -11 \\ 1 & 9 & 3 \end{vmatrix}$

$= -3(-1)^4 \begin{vmatrix} -36 & 1 \\ -52 & -11 \end{vmatrix}$

$= -3(396 + 52) = -1344$

58. $\begin{vmatrix} 1 & -1 & 8 & 4 \\ 2 & 6 & 0 & -4 \\ 2 & 0 & 2 & 6 \\ 0 & 2 & 8 & 0 \end{vmatrix} = \begin{vmatrix} 1 & -1 & 12 & 4 \\ 2 & 6 & -24 & -4 \\ 2 & 0 & 2 & 6 \\ 0 & 2 & 0 & 0 \end{vmatrix}$

$= 2(-1)^6 \begin{vmatrix} 1 & 12 & 4 \\ 2 & -24 & -4 \\ 2 & 2 & 6 \end{vmatrix}$

$= 2 \cdot 2^2 \begin{vmatrix} 1 & 6 & 2 \\ 2 & -12 & -2 \\ 2 & 1 & 3 \end{vmatrix}$

$= 8 \begin{vmatrix} 1 & 6 & 2 \\ 0 & -24 & -6 \\ 0 & -11 & -1 \end{vmatrix}$

$= 8(1)(-1)^2(-6) \begin{vmatrix} 4 & 1 \\ -11 & -1 \end{vmatrix}$

$= -48(-4 + 11) = -336$

60. $\begin{vmatrix} 4 & 2 & -1 & 0 & 3 \\ 0 & 1 & 1 & 2 & -3 \\ 0 & 0 & -2 & 8 & 12 \\ 0 & 0 & 0 & 5 & 13 \\ 0 & 0 & 0 & 0 & 3 \end{vmatrix} = 4(1)(-2)(5)(3)$

$= -120$

62. $\begin{vmatrix} x-2 & -1 \\ -3 & x \end{vmatrix} = 0$

$$(x-2)x - (-3)(-1) = 0$$
$$x^2 - 2x - 3 = 0$$
$$(x+1)(x-3) = 0$$
$$x = -1 \text{ or } x = 3$$

64. $\begin{vmatrix} 3x^2 & -3y^2 \\ 1 & 1 \end{vmatrix} = 3x^2 + 3y^2$

66. $\begin{vmatrix} e^{-x} & xe^{-x} \\ -e^{-x} & (1-x)e^{-x} \end{vmatrix} = (1-x)e^{-2x} + xe^{-2x} = e^{-2x}$

68. $\begin{vmatrix} x & x\ln x \\ 1 & 1+\ln x \end{vmatrix} = x(1+\ln x) - x\ln x = x$

70. (a) $\begin{vmatrix} -2 & 1 \\ 4 & -2 \end{vmatrix} = 4 - 4 = 0$ (b) $\begin{vmatrix} 1 & 2 \\ 0 & -1 \end{vmatrix} = -1 - 0 = -1$

(c) $\begin{bmatrix} -2 & 1 \\ 4 & -2 \end{bmatrix}\begin{bmatrix} 1 & 2 \\ 0 & -1 \end{bmatrix} = \begin{bmatrix} -2+0 & -4-1 \\ 4+0 & 8+2 \end{bmatrix}$ (d) $\begin{vmatrix} -2 & -5 \\ 4 & 10 \end{vmatrix} = -20 + 20 = 0$

$$= \begin{bmatrix} -2 & -5 \\ 4 & 10 \end{bmatrix}$$

72. (a) $\begin{vmatrix} 2 & 0 & 1 \\ 1 & -1 & 2 \\ 3 & 1 & 0 \end{vmatrix} = \begin{vmatrix} 2 & 0 & 1 \\ -3 & -1 & 0 \\ 3 & 1 & 0 \end{vmatrix} = 0$

(b) $\begin{vmatrix} 2 & -1 & 4 \\ 0 & 1 & 3 \\ 3 & -2 & 1 \end{vmatrix} = 2\begin{vmatrix} 1 & 3 \\ -2 & 1 \end{vmatrix} - 0 + 3\begin{vmatrix} -1 & 4 \\ 1 & 3 \end{vmatrix} = 2(7) + 3(-7) = -7$

(c) $\begin{bmatrix} 2 & 0 & 1 \\ 1 & -1 & 2 \\ 3 & 1 & 0 \end{bmatrix}\begin{bmatrix} 2 & -1 & 4 \\ 0 & 1 & 3 \\ 3 & -2 & 1 \end{bmatrix} = \begin{bmatrix} 4+0+3 & -2+0-2 & 8+0+1 \\ 2+0+6 & -1-1-4 & 4-3+2 \\ 6+0+0 & -3+1+0 & 12+3+0 \end{bmatrix} = \begin{bmatrix} 7 & -4 & 9 \\ 8 & -6 & 3 \\ 6 & -2 & 15 \end{bmatrix}$

(d) $\begin{vmatrix} 7 & -4 & 9 \\ 8 & -6 & 3 \\ 6 & -2 & 15 \end{vmatrix} = (-2)(3)\begin{vmatrix} 7 & 2 & 3 \\ 8 & 3 & 1 \\ 6 & 1 & 5 \end{vmatrix} = -6\begin{vmatrix} -17 & -7 & 0 \\ 8 & 3 & 1 \\ -34 & -14 & 0 \end{vmatrix} = 0$

74. The determinant of A is -11. To illustrate the property, let B be the matrix obtained from A by adding (-7) times row 1 to row 3:

$$B = \begin{vmatrix} 1 & 10 & -6 \\ 2 & -3 & 4 \\ 0 & -64 & 45 \end{vmatrix}$$

Once again, the determinant of B is -11.

8.5 Applications of Determinants and Matrices

2. $D = \begin{vmatrix} 2 & -1 \\ 3 & 2 \end{vmatrix} = 7$

$x = \dfrac{D_x}{D} = \dfrac{\begin{vmatrix} -10 & -1 \\ -1 & 2 \end{vmatrix}}{7}$

$\qquad = \dfrac{-21}{7} = -3$

$y = \dfrac{D_y}{D} = \dfrac{\begin{vmatrix} 2 & -10 \\ 3 & -1 \end{vmatrix}}{7}$

$\qquad = \dfrac{28}{7} = 4$

Answer: $(-3, 4)$

4. $D = \begin{vmatrix} 18 & 12 \\ 30 & 24 \end{vmatrix} = 72$

$x = \dfrac{D_x}{D} = \dfrac{\begin{vmatrix} 13 & 12 \\ 23 & 24 \end{vmatrix}}{72} = \dfrac{36}{72} = \dfrac{1}{2}$

$y = \dfrac{D_y}{D} = \dfrac{\begin{vmatrix} 18 & 13 \\ 30 & 23 \end{vmatrix}}{72} = \dfrac{24}{72} = \dfrac{1}{3}$

Answer: $(\frac{1}{2}, \frac{1}{3})$

6. $D = \begin{vmatrix} 13 & -6 \\ 26 & -12 \end{vmatrix} = 0$

Since the determinant of the coefficient matrix is zero, Cramer's Rule does not apply.

8. $D = \begin{vmatrix} -0.4 & 0.8 \\ 0.2 & 0.3 \end{vmatrix} = -0.28$

$x = \dfrac{D_x}{D} = \dfrac{\begin{vmatrix} 1.6 & 0.8 \\ 2.2 & 0.3 \end{vmatrix}}{-0.28} = \dfrac{-1.18}{-0.28} = \dfrac{32}{7}$

$y = \dfrac{D_y}{D} = \dfrac{\begin{vmatrix} -0.4 & 1.6 \\ 0.2 & 2.2 \end{vmatrix}}{-0.28} = \dfrac{-1.20}{-0.28} = \dfrac{30}{7}$

Answer: $(\frac{32}{7}, \frac{30}{7})$

10. $D = \begin{vmatrix} 3 & 2 \\ 2 & 10 \end{vmatrix} = 26$

$x = \dfrac{D_x}{D} = \dfrac{\begin{vmatrix} 1 & 2 \\ 6 & 10 \end{vmatrix}}{26} = \dfrac{-2}{26} = -\dfrac{1}{13}$

$y = \dfrac{D_y}{D} = \dfrac{\begin{vmatrix} 3 & 1 \\ 2 & 6 \end{vmatrix}}{26} = \dfrac{16}{26} = \dfrac{8}{13}$

Answer: $(-\frac{1}{13}, \frac{8}{13})$

12. $D = \begin{vmatrix} 4 & -2 & 3 \\ 2 & 2 & 5 \\ 8 & -5 & -2 \end{vmatrix} = -82$

$x = \dfrac{D_x}{D} = \dfrac{\begin{vmatrix} -2 & -2 & 3 \\ 16 & 2 & 5 \\ 4 & -5 & -2 \end{vmatrix}}{-82} = \dfrac{-410}{-82} = 5$

14. $D = \begin{vmatrix} 14 & -21 & -7 \\ -4 & 2 & -2 \\ 56 & -21 & 7 \end{vmatrix} = 1568$

$x = \dfrac{D_x}{D} = \dfrac{\begin{vmatrix} 10 & -21 & -7 \\ 4 & 2 & -2 \\ 5 & -21 & 7 \end{vmatrix}}{1568} = \dfrac{1176}{1568} = \dfrac{3}{4}$

16. $D = \begin{vmatrix} 2 & 3 & 5 \\ 3 & 5 & 9 \\ 5 & 9 & 17 \end{vmatrix} = 0$

Since the determinant of the coefficient matrix is zero, Cramer's Rule does not apply.

18. $D = \begin{vmatrix} 3 & 2 & 5 \\ 4 & -3 & -4 \\ -8 & 2 & 3 \end{vmatrix} = -43$

$x = \dfrac{D_x}{D} = \dfrac{\begin{vmatrix} 4 & 2 & 5 \\ 1 & -3 & -4 \\ 0 & 2 & 3 \end{vmatrix}}{-43} = \dfrac{0}{-43} = 0$

20. $D = \begin{vmatrix} 2 & 5 & 0 & 1 \\ 1 & 4 & 2 & -2 \\ 2 & -2 & 5 & 1 \\ 1 & 0 & 0 & -3 \end{vmatrix} = \begin{vmatrix} 2 & 5 & 0 & 7 \\ 1 & 4 & 2 & 1 \\ 2 & -2 & 5 & 7 \\ 1 & 0 & 0 & 0 \end{vmatrix} = - \begin{vmatrix} 5 & 0 & 7 \\ 4 & 2 & 1 \\ -2 & 5 & 7 \end{vmatrix} = -213$

$x = \dfrac{D_x}{D} = \dfrac{\begin{vmatrix} 11 & 5 & 0 & 1 \\ -7 & 4 & 2 & -2 \\ 3 & -2 & 5 & 1 \\ 1 & 0 & 0 & -3 \end{vmatrix}}{-213} = \dfrac{\begin{vmatrix} 11 & 5 & 0 & 34 \\ -7 & 4 & 2 & -233 & -2 & 5 & 10 \\ 1 & 0 & 0 & 0 \end{vmatrix}}{-213}$

$= \dfrac{-|5 \quad 0 \quad 344 \quad 2 \quad -23-2 \quad 5 \quad 10|}{-213} = \dfrac{-1491}{-213} = 7$

22. $D = \begin{vmatrix} 4 & 0 & 8 \\ 0 & 2 & 8 \\ 1 & 1 & -1 \end{vmatrix} = -56$

$I_1 = \dfrac{\begin{vmatrix} 2 & 0 & 8 \\ 6 & 2 & 8 \\ 0 & 1 & -1 \end{vmatrix}}{-56} = \dfrac{28}{-56} = -\dfrac{1}{2}$

$I_2 = \dfrac{\begin{vmatrix} 4 & 2 & 8 \\ 0 & 6 & 8 \\ 1 & 0 & -1 \end{vmatrix}}{-56} = \dfrac{-56}{-56} = 1$

$I_3 = \dfrac{\begin{vmatrix} 4 & 0 & 2 \\ 0 & 2 & 6 \\ 1 & 1 & 0 \end{vmatrix}}{-56} = \dfrac{-28}{-56} = \dfrac{1}{2}$

24. $D = \begin{vmatrix} 1 & -2 & 0 \\ 1 & 0 & -3 \\ 0 & 1 & 2 \end{vmatrix} = 7$

$a = \dfrac{\begin{vmatrix} 1 & -2 & 0 \\ 1 & 0 & 192 \\ 0 & 1 & 64 \end{vmatrix}}{7} = \dfrac{-64}{7}$ ft/sec^2

Answer: $I_1 = -\frac{1}{2}$ amp, $I_2 = 1$ amp, $I_3 = \frac{1}{2}$ amp

26. $A = \pm\frac{1}{2}\begin{vmatrix} 4 & 5 & 1 \\ 5 & -2 & 1 \\ 0 & 0 & 1 \end{vmatrix} = \pm\frac{1}{2}(-33)$
$= \frac{33}{2}$ square units

28. $A = \pm\frac{1}{2}\begin{vmatrix} 1 & 6 & 1 \\ 3 & -1 & 1 \\ -2 & 1 & 1 \end{vmatrix} = \pm\frac{1}{2}(-31)$
$= \frac{31}{2}$ square units

30. $A = \pm\frac{1}{2}\begin{vmatrix} -4 & -5 & 1 \\ 6 & -1 & 1 \\ 6 & 10 & 1 \end{vmatrix} = \pm\frac{1}{2}(110)$
$= 55$ square units

32. $A = \pm\frac{1}{2}\begin{vmatrix} 0 & -2 & 1 \\ -1 & 4 & 1 \\ 3 & 5 & 1 \end{vmatrix} = \pm\frac{1}{2}(-25)$
$= \frac{25}{2}$ square units

34. $A = \pm\frac{1}{2}\begin{vmatrix} -2 & 4 & 1 \\ 1 & 5 & 1 \\ 3 & -2 & 1 \end{vmatrix} = \pm\frac{1}{2}(-23)$

$= \frac{23}{2}$ square units

36. First vertex: $(0,\ 0)$
Second vertex: $(65,\ 50)$
Third vertex: $(-20,\ 80)$

$A = \pm\frac{1}{2}\begin{vmatrix} 0 & 0 & 1 \\ 65 & 50 & 1 \\ -20 & 80 & 1 \end{vmatrix} = \pm\frac{1}{2}(6200)$

$= 3100$ square feet

38. $\begin{vmatrix} -3 & -5 & 1 \\ 6 & 1 & 1 \\ 10 & 2 & 1 \end{vmatrix} = -15 \neq 0$

The points are not collinear.

40. $\begin{vmatrix} 0 & 1 & 1 \\ 4 & -2 & 1 \\ -8 & 7 & 1 \end{vmatrix} = 0$

The points are collinear.

42. $\begin{vmatrix} 2 & 3 & 1 \\ 3 & 3.5 & 1 \\ -1 & 2 & 1 \end{vmatrix} = 0.5 \neq 0$

The points are not collinear.

44. $\begin{vmatrix} x & y & 1 \\ 0 & 0 & 1 \\ -2 & 2 & 1 \end{vmatrix} = 0$

$x\begin{vmatrix} 0 & 1 \\ 2 & 1 \end{vmatrix} - y\begin{vmatrix} 0 & 1 \\ -2 & 1 \end{vmatrix} + 1\begin{vmatrix} 0 & 0 \\ -2 & 2 \end{vmatrix} = 0$

$-2x - 2y = 0$

$x + y = 0$

46. $\begin{vmatrix} x & y & 1 \\ 10 & 7 & 1 \\ -2 & -7 & 1 \end{vmatrix} = 0$

$x\begin{vmatrix} 7 & 1 \\ -7 & 1 \end{vmatrix} - y\begin{vmatrix} 10 & 1 \\ -2 & 1 \end{vmatrix} + 1\begin{vmatrix} 10 & 7 \\ -2 & -7 \end{vmatrix} = 0$

$14x - 12y - 56 = 0$

$7x - 6y - 28 = 0$

48. $\begin{vmatrix} x & y & 1 \\ \frac{2}{3} & 4 & 1 \\ 6 & 12 & 1 \end{vmatrix} = 0$

$x\begin{vmatrix} 4 & 1 \\ 12 & 1 \end{vmatrix} - y\begin{vmatrix} \frac{2}{3} & 1 \\ 6 & 1 \end{vmatrix} + 1\begin{vmatrix} \frac{2}{3} & 4 \\ 6 & 12 \end{vmatrix} = 0$

$-8x + \frac{16}{3}y - 16 = 0$

$3x - 2y + 6 = 0$

50. B E A M — M E — U P — S C O T T Y —
[2 5 1] [13 0 13] [5 0 21] [16 0 19] [3 15 20] [20 25 0]

$[2 \quad 5 \quad 1]A = [16 \quad 35 \quad 42]$

$[13 \quad 0 \quad 13]A = [0 \quad -26 \quad -65]$

$[5 \quad 0 \quad 21]A = [-16 \quad -74 \quad -137]$

$[16 \quad 0 \quad 19]A = [-3 \quad -44 \quad -101]$

$[3 \quad 15 \quad 20]A = [28 \quad 31 \quad 1]$

$[20 \quad 25 \quad 0]A = [95 \quad 215 \quad 265]$

Cryptogram: 16 35 42 0 −26 −65 −16 −74 −137 −3 −44 −101 28 31 1 95 215 265

52. O P E R A T I O N — O V E R L O A D
[15 16 5] [18 1 20] [9 15 14] [0 15 22] [5 18 12] [15 1 4]

$[15 \quad 16 \quad 5]A = [58 \quad 122 \quad 139]$

$[18 \quad 1 \quad 20]A = [1 \quad -37 \quad -95]$

$[9 \quad 15 \quad 14]A = [40 \quad 67 \quad 55]$

$[0 \quad 15 \quad 22]A = [23 \quad 17 \quad -19]$

$[5 \quad 18 \quad 12]A = [47 \quad 88 \quad 88]$

$[15 \quad 1 \quad 4]A = [14 \quad 21 \quad 11]$

Cryptogram: 58 122 139 1 −37 −95 40 67 55 23 17 −19 47 88 88 14 21 11

54. From Exercise 53, $A^{-1} = \begin{bmatrix} -13 & 6 & 4 \\ 12 & -5 & -3 \\ -5 & 2 & 1 \end{bmatrix}$

$[13 \quad -9 \quad -59]A^{-1} = [18 \quad 5 \quad 20]$

$[61 \quad 112 \quad 106]A^{-1} = [21 \quad 18 \quad 14]$

$[-17 \quad -73 \quad -131]A^{-1} = [0 \quad 1 \quad 20]$

$[11 \quad -24 \quad 29]A^{-1} = [0 \quad 4 \quad 1]$

$[65 \quad 144 \quad 172]A^{-1} = [23 \quad 14 \quad 0]$

18 5 20 21 18 14 0 1 20 0 4 1 23 14 0
R E T U R N — A T — D A W N —

8.R Review

2.
$$\begin{bmatrix} 2 & -5 & \vdots & 2 \\ 3 & -7 & \vdots & 1 \end{bmatrix} \Rightarrow \begin{bmatrix} 1 & -2 & \vdots & -1 \\ 3 & -7 & \vdots & 1 \end{bmatrix}$$

$$\Rightarrow \begin{bmatrix} 1 & -2 & \vdots & -1 \\ 0 & -1 & \vdots & 4 \end{bmatrix}$$

$$\Rightarrow \begin{bmatrix} 1 & 0 & \vdots & -9 \\ 0 & 1 & \vdots & -4 \end{bmatrix}$$

$x = -9$

$y = -4$

Answer: $(-9, -4)$

4.
$$\begin{bmatrix} 2 & 1 & \vdots & 0.3 \\ 3 & -1 & \vdots & -1.3 \end{bmatrix} \Rightarrow \begin{bmatrix} 1 & -2 & \vdots & -1.6 \\ 3 & -1 & \vdots & -1.3 \end{bmatrix} \Rightarrow \begin{bmatrix} 1 & -2 & \vdots & -1.6 \\ 0 & 5 & \vdots & 3.5 \end{bmatrix} \Rightarrow \begin{bmatrix} 1 & 0 & \vdots & -0.2 \\ 0 & 1 & \vdots & 0.7 \end{bmatrix}$$

$x = -0.2$

$y = 0.7$

Answer: $(-0.2,\ 0.7)$

6.
$$\begin{bmatrix} 4 & 4 & 4 & \vdots & 5 \\ 4 & -2 & -8 & \vdots & 1 \\ 5 & 3 & 8 & \vdots & 6 \end{bmatrix} \Rightarrow \begin{bmatrix} 1 & -1 & 4 & \vdots & 1 \\ 4 & -2 & -8 & \vdots & 1 \\ 5 & 3 & 8 & \vdots & 6 \end{bmatrix} \Rightarrow \begin{bmatrix} 1 & -1 & 4 & \vdots & 1 \\ 0 & -6 & -12 & \vdots & -4 \\ 0 & 8 & -12 & \vdots & 1 \end{bmatrix}$$

$$\Rightarrow \begin{bmatrix} 1 & -1 & 4 & \vdots & 1 \\ 0 & 1 & 2 & \vdots & \frac{2}{3} \\ 0 & 8 & -12 & \vdots & 1 \end{bmatrix} \Rightarrow \begin{bmatrix} 1 & -1 & 4 & \vdots & 1 \\ 0 & 1 & 2 & \vdots & \frac{2}{3} \\ 0 & 0 & -28 & \vdots & -\frac{13}{3} \end{bmatrix}$$

$$-28z = -\frac{13}{3} \Rightarrow z = \frac{13}{84}$$

$$y + 2\left(\frac{13}{84}\right) = \frac{2}{3} \Rightarrow y = \frac{5}{14}$$

$$x - \frac{5}{14} + 4\left(\frac{13}{84}\right) = 1 \Rightarrow x = \frac{31}{42}$$

Answer: $\left(\frac{31}{42},\ \frac{5}{14},\ \frac{13}{84}\right)$

8. $\begin{bmatrix} 3 & 21 & -29 & \vdots & -1 \\ 2 & 15 & -21 & \vdots & 0 \end{bmatrix} \Rightarrow \begin{bmatrix} 1 & 6 & -8 & \vdots & -1 \\ 2 & 15 & -21 & \vdots & 0 \end{bmatrix}$

$\Rightarrow \begin{bmatrix} 1 & 6 & -8 & \vdots & -1 \\ 0 & 3 & -5 & \vdots & 2 \end{bmatrix} \Rightarrow \begin{bmatrix} 1 & 0 & 2 & \vdots & -5 \\ 0 & 3 & -5 & \vdots & 2 \end{bmatrix}$

Let $z = a$.

$3y - 5a = 2 \Rightarrow y = \frac{5}{3}a + \frac{2}{3}$

$x + 2a = -5 \Rightarrow x = -2a - 5$

Answer: $(-2a - 5, \frac{5}{3}a + \frac{2}{3}, a)$

10. $-2 \begin{bmatrix} 1 & 2 \\ 5 & -4 \\ 6 & 0 \end{bmatrix} + 8 \begin{bmatrix} 7 & 1 \\ 1 & 2 \\ 1 & 4 \end{bmatrix} = \begin{bmatrix} -2 & -4 \\ -10 & 8 \\ -12 & 0 \end{bmatrix} + \begin{bmatrix} 56 & 8 \\ 8 & 16 \\ 8 & 32 \end{bmatrix} = \begin{bmatrix} 54 & 4 \\ -2 & 24 \\ -4 & 32 \end{bmatrix}$

12. $\begin{bmatrix} 1 & 5 & 6 \\ 2 & -4 & 0 \end{bmatrix} \begin{bmatrix} 6 & -2 & 8 \\ 4 & 0 & 0 \end{bmatrix}$ is undefined.

14. $\begin{bmatrix} 4 \\ 6 \end{bmatrix} \begin{bmatrix} 6 & -2 \end{bmatrix} = \begin{bmatrix} 24 & -8 \\ 36 & -12 \end{bmatrix}$

16. $\begin{bmatrix} 2 & 1 \\ 6 & 0 \end{bmatrix} \left(\begin{bmatrix} 4 & 2 \\ -3 & 1 \end{bmatrix} + \begin{bmatrix} -2 & 4 \\ 0 & 4 \end{bmatrix} \right) = \begin{bmatrix} 2 & 1 \\ 6 & 0 \end{bmatrix} \begin{bmatrix} 2 & 6 \\ -3 & 5 \end{bmatrix} = \begin{bmatrix} 1 & 17 \\ 12 & 36 \end{bmatrix}$

18. $X = \frac{1}{6} \left(4 \begin{bmatrix} -4 & 0 \\ 1 & -5 \\ -3 & 2 \end{bmatrix} + 3 \begin{bmatrix} 1 & 2 \\ -2 & 1 \\ 4 & 4 \end{bmatrix} \right)$

$= \frac{1}{6} \begin{bmatrix} -13 & 6 \\ -2 & -17 \\ 0 & 20 \end{bmatrix}$

20. $X = \frac{1}{3} \left(2 \begin{bmatrix} -4 & 0 \\ 1 & -5 \\ -3 & 2 \end{bmatrix} - 5 \begin{bmatrix} 1 & 2 \\ -2 & 1 \\ 4 & 4 \end{bmatrix} \right)$

$= \frac{1}{3} \begin{bmatrix} -13 & -10 \\ 12 & -15 \\ -26 & -16 \end{bmatrix}$

22. $\begin{bmatrix} 2 & 3 & 1 \\ 2 & -3 & -3 \\ 4 & -2 & 3 \end{bmatrix} \begin{bmatrix} x \\ y \\ z \end{bmatrix} = \begin{bmatrix} 10 \\ 22 \\ -2 \end{bmatrix}$

24. $\begin{bmatrix} 3 & -10 & \vdots & 1 & 0 \\ 4 & 2 & \vdots & 0 & 1 \end{bmatrix} \Rightarrow \begin{bmatrix} 1 & 12 & \vdots & -1 & 1 \\ 4 & 2 & \vdots & 0 & 1 \end{bmatrix}$

$\Rightarrow \begin{bmatrix} 1 & 12 & \vdots & -1 & 1 \\ 0 & -46 & \vdots & 4 & -3 \end{bmatrix}$

$\Rightarrow \begin{bmatrix} 1 & 12 & \vdots & -1 & 1 \\ 0 & 1 & \vdots & -\frac{2}{23} & \frac{3}{46} \end{bmatrix} \Rightarrow \begin{bmatrix} 1 & 0 & \vdots & \frac{1}{23} & \frac{5}{23} \\ 0 & 1 & \vdots & -\frac{2}{23} & \frac{3}{46} \end{bmatrix}$

26. $\begin{bmatrix} 1 & 4 & 6 & \vdots & 1 & 0 & 0 \\ 2 & -3 & 1 & \vdots & 0 & 1 & 0 \\ -1 & 18 & 16 & \vdots & 0 & 0 & 1 \end{bmatrix} \Rightarrow \begin{bmatrix} 1 & 4 & 6 & \vdots & 1 & 0 & 0 \\ 0 & -11 & -11 & \vdots & -2 & 1 & 0 \\ 0 & 22 & 22 & \vdots & 1 & 0 & 1 \end{bmatrix}$

$\Rightarrow \begin{bmatrix} 1 & 4 & 6 & \vdots & 1 & 0 & 0 \\ 0 & -11 & -11 & \vdots & -2 & 1 & 0 \\ 0 & 0 & 0 & \vdots & -3 & 2 & 1 \end{bmatrix}$

The inverse of A does not exist.

28. $\begin{vmatrix} 8 & 5 \\ 2 & -4 \end{vmatrix} = -32 - 10 = -42$

30. $\begin{vmatrix} -5 & 6 & 0 & 0 \\ 0 & 1 & -1 & 2 \\ -3 & 4 & -5 & 1 \\ 1 & 6 & 0 & 3 \end{vmatrix} = -5 \begin{vmatrix} 1 & -1 & 2 \\ 4 & -5 & 1 \\ 6 & 0 & 3 \end{vmatrix} - 6 \begin{vmatrix} 0 & -1 & 2 \\ -3 & -5 & 1 \\ 1 & 0 & 3 \end{vmatrix}$

$= -5[6(9) - 0 + 3(-1)] - 6[0 - (-1)(-10) + 2(5)]$

$= -5[51] - 6[0] = -255$

32. (a) $\begin{bmatrix} x \\ y \end{bmatrix} = A^{-1}B = \begin{bmatrix} \frac{1}{10} & -\frac{3}{20} \\ \frac{3}{10} & \frac{1}{20} \end{bmatrix} \begin{bmatrix} 23 \\ -18 \end{bmatrix} = \begin{bmatrix} 5 \\ 6 \end{bmatrix}$

(b) $D = \begin{vmatrix} 1 & 3 \\ -6 & 2 \end{vmatrix} = 20$

$x = \dfrac{D_x}{D} = \dfrac{\begin{vmatrix} 23 & 3 \\ -18 & 2 \end{vmatrix}}{20} = \dfrac{100}{20} = 5$

$y = \dfrac{D_y}{D} = \dfrac{\begin{vmatrix} 1 & 23 \\ -6 & -18 \end{vmatrix}}{20} = \dfrac{120}{20} = 6$

Answer: (5, 6)

34. (a) $\begin{bmatrix} x \\ y \\ z \end{bmatrix} = \begin{bmatrix} -18 & -7 & 5 \\ -3 & -1 & 1 \\ -5 & -2 & 1 \end{bmatrix} \begin{bmatrix} 8 \\ -19 \\ 3 \end{bmatrix} = \begin{bmatrix} 4 \\ -2 \\ 1 \end{bmatrix}$

(b) $D = \begin{vmatrix} 1 & -3 & -2 \\ -2 & 7 & 3 \\ 1 & -1 & -3 \end{vmatrix} = 1$

$x = \dfrac{D_x}{D} = \dfrac{\begin{vmatrix} 8 & -3 & -2 \\ -19 & 7 & 3 \\ 3 & -1 & -3 \end{vmatrix}}{1} = \dfrac{4}{1} = 4$

$y = \dfrac{D_y}{D} = \dfrac{\begin{vmatrix} 1 & 8 & -2 \\ -2 & -19 & 3 \\ 1 & 3 & -3 \end{vmatrix}}{1} = \dfrac{-2}{1} = -2$

$z = \dfrac{D_z}{D} = \dfrac{\begin{vmatrix} 1 & -3 & -2 \\ -2 & 7 & 3 \\ 1 & -1 & -3 \end{vmatrix}}{1} = \dfrac{1}{1} = 1$

Answer: $(4, -2, 1)$

36. (a) $\begin{bmatrix} x \\ y \\ z \end{bmatrix} = \begin{bmatrix} \frac{5}{2} & -2 & -2 \\ -1 & 1 & 1 \\ \frac{7}{4} & -\frac{3}{2} & -1 \end{bmatrix} \begin{bmatrix} -12 \\ -14 \\ -6 \end{bmatrix} = \begin{bmatrix} 10 \\ -8 \\ 6 \end{bmatrix}$

(b) $D = \begin{vmatrix} 2 & 4 & 0 \\ 3 & 4 & -2 \\ -1 & 1 & 2 \end{vmatrix} = 4$

$x = \dfrac{D_x}{D} = \dfrac{\begin{vmatrix} -12 & 4 & 0 \\ -14 & 4 & -2 \\ -6 & 1 & 2 \end{vmatrix}}{4} = \dfrac{40}{4} = 10$

$y = \dfrac{D_y}{D} = \dfrac{\begin{vmatrix} 2 & -12 & 0 \\ 3 & -14 & -2 \\ -1 & -6 & 2 \end{vmatrix}}{4} = \dfrac{-32}{4} = -8$

$z = \dfrac{D_z}{D} = \dfrac{\begin{vmatrix} 2 & 4 & -12 \\ 3 & 4 & -14 \\ -1 & 1 & -6 \end{vmatrix}}{4} = \dfrac{24}{4} = 6$

Answer: $(10, -8, 6)$

38. (a) $\begin{bmatrix} -1 & 1 & 1 & \vdots & 1 & 0 & 0 \\ 4 & -3 & 1 & \vdots & 0 & 1 & 0 \\ 2 & -1 & 3 & \vdots & 0 & 0 & 1 \end{bmatrix} \Rightarrow \begin{bmatrix} 1 & -1 & -1 & \vdots & -1 & 0 & 0 \\ 0 & 1 & 5 & \vdots & 4 & 1 & 0 \\ 0 & 1 & 5 & \vdots & 2 & 0 & 1 \end{bmatrix}$

$\Rightarrow \begin{bmatrix} 0 & -1 & -1 & \vdots & -1 & 0 & 0 \\ 0 & 1 & 5 & \vdots & 4 & 1 & 0 \\ 0 & 0 & 0 & \vdots & -2 & -1 & 1 \end{bmatrix}$

(b) $D = \begin{vmatrix} -1 & 1 & 1 \\ 4 & -3 & 1 \\ 2 & -1 & 3 \end{vmatrix} = 0$

Answer: A has no inverse. The system is inconsistent.

40. Area $= \pm\frac{1}{2} \begin{vmatrix} -4 & 0 & 1 \\ 4 & 0 & 1 \\ 0 & 6 & 1 \end{vmatrix} = \pm\frac{1}{2}(48)$
$= 24$ square units

42. Area $= \pm\frac{1}{2} \begin{vmatrix} \frac{3}{2} & 1 & 1 \\ 4 & -\frac{1}{2} & 1 \\ 4 & 2 & 1 \end{vmatrix} = \pm\frac{1}{2}\left(\frac{25}{4}\right)$
$= \frac{25}{8}$

44.
$$\begin{vmatrix} x & y & 1 \\ 2 & 5 & 1 \\ 6 & -1 & 1 \end{vmatrix} = 0$$

$$x \begin{vmatrix} 5 & 1 \\ -1 & 1 \end{vmatrix} - y \begin{vmatrix} 2 & 1 \\ 6 & 1 \end{vmatrix} + 1 \begin{vmatrix} 2 & 5 \\ 6 & -1 \end{vmatrix} = 0$$

$$6x + 4y - 32 = 0$$

$$3x + 2y - 16 = 0$$

46.
$$\begin{vmatrix} x & y & 1 \\ -0.8 & 0.2 & 1 \\ 0.7 & 3.2 & 1 \end{vmatrix} = 0$$

$$x \begin{vmatrix} 0.2 & 1 \\ 3.2 & 1 \end{vmatrix} - y \begin{vmatrix} -0.8 & 1 \\ 0.7 & 1 \end{vmatrix} + 1 \begin{vmatrix} -0.8 & 0.2 \\ 0.7 & 3.2 \end{vmatrix} = 0$$

$$-3.0x + 1.5y - 2.7 = 0$$

$$10x - 5y + 9 = 0$$

48. Let $x =$ the number of gallons of 75% solution, and $y =$ the number of gallons of 50% solution. Then,

$$x + y = 100$$

$$0.75x + 0.50y = 60.$$

By Cramer's Rule we have

$$x = \frac{D_x}{D} = \frac{\begin{vmatrix} 100 & 1 \\ 60 & 0.5 \end{vmatrix}}{\begin{vmatrix} 1 & 1 \\ 0.75 & 0.50 \end{vmatrix}} = \frac{-10}{-0.25} = 40.$$

Using back-substitution in the first equation yields $y = 60$. To obtain the desired mixture, use 40 gallons of 75% solution and 60 gallons of 50% solution.

50.
$$5c \quad + 10a = \quad 9.1$$
$$10b \quad\quad = \quad 8.0$$
$$10c \quad + 34a = \quad 19.8$$

$$-10c \quad - 20a = -18.2$$
$$10b \quad\quad = \quad 8.0$$
$$10c \quad + 34a = \quad 19.8$$

$$-10c \quad - 20a = -18.2$$
$$10b \quad\quad = \quad 8.0$$
$$14a = \quad 1.6$$

$$14a = 1.6 \quad \Rightarrow \quad a \approx 0.114$$

$$10b = 8.0 \quad \Rightarrow \quad b = 0.8$$

$$-10c - 20(0.114) = -18.2 \quad \Rightarrow \quad c \approx 1.591$$

Least squares regression parabola:
$$y = 0.114x^2 + 0.8x + 1.591$$

52. Let $x =$ the number of units produced, and $y =$ the number of units sold. Then,

$$x - y = 0$$

$$-3.75x + 5.25y = 25{,}000.$$

By Cramer's Rule we have

$$y = \frac{\begin{vmatrix} 1 & 0 \\ -3.75 & 25{,}000 \end{vmatrix}}{\begin{vmatrix} 1 & -1 \\ -3.75 & 5.25 \end{vmatrix}} = \frac{25{,}000}{1.50} \approx 16{,}667.$$

Approximately 16,667 units must be sold before the business breaks even.

54.
$$\begin{vmatrix} a_{11} & a_{12} & a_{13} \\ a_{21} & a_{22} & a_{23} \\ a_{31}+c_1 & a_{32}+c_2 & a_{33}+c_3 \end{vmatrix} = (a_{31}+c_1)\begin{vmatrix} a_{12} & a_{13} \\ a_{22} & a_{23} \end{vmatrix} - (a_{32}+c_2)\begin{vmatrix} a_{11} & a_{13} \\ a_{21} & a_{23} \end{vmatrix} + (a_{33}+c_3)\begin{vmatrix} a_{11} & a_{12} \\ a_{21} & a_{22} \end{vmatrix}$$

$$= \left(a_{31}\begin{vmatrix} a_{12} & a_{13} \\ a_{22} & a_{23} \end{vmatrix} - a_{32}\begin{vmatrix} a_{11} & a_{13} \\ a_{21} & a_{23} \end{vmatrix} + a_{33}\begin{vmatrix} a_{11} & a_{12} \\ a_{21} & a_{22} \end{vmatrix}\right)$$

$$+ \left(c_1\begin{vmatrix} a_{12} & a_{13} \\ a_{22} & a_{23} \end{vmatrix} - c_2\begin{vmatrix} a_{11} & a_{13} \\ a_{21} & a_{23} \end{vmatrix} + c_3\begin{vmatrix} a_{11} & a_{12} \\ a_{21} & a_{22} \end{vmatrix}\right)$$

$$= \begin{vmatrix} a_{11} & a_{12} & a_{13} \\ a_{21} & a_{22} & a_{23} \\ a_{31} & a_{32} & a_{33} \end{vmatrix} + \begin{vmatrix} a_{11} & a_{12} & a_{13} \\ a_{21} & a_{22} & a_{23} \\ c_1 & c_2 & c_3 \end{vmatrix}$$

Cumulative Test, Chapters 6—8

1. $\dfrac{1+\cos\beta}{\sin\beta} + \dfrac{\sin\beta}{1+\cos\beta} = \dfrac{(1+\cos\beta)^2 + (\sin\beta)^2}{\sin\beta\,(1+\cos\beta)} = \dfrac{1 + 2\cos\beta + \cos^2\beta + \sin^2\beta}{\sin\beta\,(1+\cos\beta)}$

$$= \dfrac{2 + 2\cos\beta}{\sin\beta\,(1+\cos\beta)} = \dfrac{2}{\sin\beta} = 2\csc\beta$$

2. $2\cos^2 x - \cos x = 0$

$\cos x(2\cos x - 1) = 0$

$\cos x = 0 \Rightarrow x = \dfrac{\pi}{2}, \dfrac{3\pi}{2}$

$2\cos x - 1 = 0 \Rightarrow \cos x = \dfrac{1}{2} \Rightarrow x = \dfrac{\pi}{3}, \dfrac{5\pi}{3}$

Solutions: $\dfrac{\pi}{3}, \dfrac{\pi}{2}, \dfrac{3\pi}{2}, \dfrac{5\pi}{3}$

3. $\sin x = \dfrac{2}{3} \Rightarrow$

$\cos^2 x = 1 - \sin^2 x$

$= 1 - \dfrac{4}{9} = \dfrac{5}{9} \Rightarrow$

$\cos x = \dfrac{\sqrt{5}}{3}$ (first quadrant)

$\sin(2x) = 2\sin x \cdot \cos x$

$$= 2\left(\dfrac{2}{3}\right)\left(\dfrac{\sqrt{5}}{3}\right) = \dfrac{4\sqrt{5}}{9}$$

4. $\cos 105° = \cos(135° - 30°) = \cos 135° \cos 30° + \sin 135° \sin 30°$

$$= -\dfrac{\sqrt{2}}{2}\left(\dfrac{\sqrt{3}}{2}\right) + \dfrac{\sqrt{2}}{2}\left(\dfrac{1}{2}\right) = -\dfrac{\sqrt{6}}{4} + \dfrac{\sqrt{2}}{4} = \dfrac{\sqrt{3}}{4}(1 - \sqrt{3})$$

5. (a) $\dfrac{44}{\sin 45°} = \dfrac{24}{\sin B} \Rightarrow$

$\sin B = \dfrac{24}{44}\sin 45°$

$= \dfrac{6}{11} \cdot \dfrac{\sqrt{2}}{2} \Rightarrow$

$B \approx 22.69° \Rightarrow$

$C \approx 180° - 22.69° - 45°$

$= 112.31°$

Finally, $\dfrac{c}{\sin 112.31°} = \dfrac{44}{\sin 45°} \Rightarrow c = 57.57.$

(b) $a^2 = b^2 + c^2 - 2bc \cdot \cos A$

$= 64 + 256 - 256\cos 30°$

$= 98.297 \Rightarrow$

$a = 9.91$

Now, $\dfrac{8}{\sin B} = \dfrac{9.91}{\sin 30°} \Rightarrow$

$\sin B = 0.4036 \Rightarrow$

$B \approx 23.8°$

and $C = 180° - A° - B° \approx 126.2°.$

6. $a^2 = 45^2 + 60^2 - 2(45)(60)\cos(45°)$

$= 1806.62$

$a = 42.5$ feet

7. $z = 2(1 - i)$

$r = \sqrt{2^2 + 2^2} = \sqrt{8} = 2\sqrt{2}$

$\theta = \dfrac{7\pi}{4}$ or $315°$

$z = 2\sqrt{2}(\cos 315° + i\sin 315°)$

$z^3 = (2\sqrt{2})^3(\cos(3 \cdot 315) + i\sin(3 \cdot 315))$

$= 16\sqrt{2}(\cos 225° + i\sin 225°)$

Note: $3 \cdot 315° = 945° = 225°$

8. $y = 3 - x^2$

$2(y - 2) = x - 1 \Rightarrow$

$2(3 = x^2 - 2) = x - 1$

$2 - 2x^2 = x - 1$

$0 = 2x^2 + x - 3$

$0 = (2x + 3)(x - 1)$

$x = 1, -\dfrac{3}{2}$

The solutions are $(1, 2)$ and $\left(-\dfrac{3}{2}, \dfrac{3}{4}\right)$.

9. $x + 3y = -1$

$2x + 4y = 0$

Using a graphing utility, we find that the solution is $(2, -1)$.

10. $x + 3y - 2z = -7$

$-2x + y - z = -5$

$4x + y + z = 3$

$$\begin{bmatrix} 1 & 3 & -2 & \vdots & -7 \\ -2 & 1 & -1 & \vdots & -5 \\ 4 & 1 & 1 & \vdots & 3 \end{bmatrix}$$

$$\begin{bmatrix} 1 & 3 & -2 & \vdots & -7 \\ 0 & 7 & -5 & \vdots & -19 \\ 0 & -11 & 9 & \vdots & 31 \end{bmatrix}$$

$$\begin{bmatrix} 1 & 3 & -2 & \vdots & -7 \\ 0 & 1 & -\frac{5}{7} & \vdots & -\frac{19}{7} \\ 0 & 0 & \frac{8}{7} & \vdots & \frac{8}{7} \end{bmatrix}$$

$$\begin{bmatrix} 1 & 0 & 0 & \vdots & 1 \\ 0 & 1 & 0 & \vdots & -2 \\ 0 & 0 & 1 & \vdots & 1 \end{bmatrix}$$

Solution: $(1, -2, 1)$

11. $ax - 8y = 9$

$3x + 4y = 0$

Two times Equation 2 added to Equation 1 gives

$6x + ax = 9$

$(6 + a)x = 9$

$x = \dfrac{9}{6 + a}.$

Hence, $a = -6$ would render the system inconsistent.

12. $3x + 4y \geq 16$

   $3x - 4y \leq 8$

   $y \leq 4$

The vertices are found by solving pairs of equalities:

$\left.\begin{array}{l} 3x + 4y = 16 \\ 3x - 4y = 8 \end{array}\right\} \Rightarrow \quad 6x = 24 \Rightarrow x = 4(4, 1)$

$\left.\begin{array}{l} 3x + 4y = 16 \\ y = 4 \end{array}\right\} (0, 4)$

$\left.\begin{array}{l} 3x - 4y = 8 \\ y = 4 \end{array}\right\} 3x - 16 = 8 \Rightarrow x = 8(8, 4)$

13. $a_1 = 54$ and $r = -\frac{1}{3} \quad \Rightarrow \quad a_2 = -\frac{1}{3}(54) = -18$

   $a_3 = -\frac{1}{3}(-18) = 6$

   $a_4 = -\frac{1}{3}(6) = -2$

   $a_5 = -\frac{1}{3}(-2) = \frac{2}{3}$

14. $2 \begin{bmatrix} 6 & -1 \\ 2 & 4 \\ -3 & 5 \end{bmatrix} - \begin{bmatrix} 1 & 4 \\ -1 & 5 \\ 1 & 10 \end{bmatrix}$

   $= \begin{bmatrix} 11 & -6 \\ 5 & 3 \\ -7 & 0 \end{bmatrix}$

15. $AB = \begin{bmatrix} 4 & -3 \\ 2 & 1 \\ 5 & 0 \end{bmatrix} \begin{bmatrix} 3 & -2 \\ 1 & -3 \end{bmatrix}$

   $= \begin{bmatrix} 9 & 1 \\ 7 & -7 \\ 15 & -10 \end{bmatrix}$

16. Using a graphing utility, the inverse of A is

   $A^{-1} = \begin{bmatrix} -175 & 37 & -13 \\ 95 & -20 & 7 \\ 14 & -3 & 1 \end{bmatrix}$

17. $\begin{vmatrix} 1 & 1 & 1 \\ 2 & -1 & -2 \\ 1 & -2 & -1 \end{vmatrix} = 1(1 - 4) - 1(-2 + 2) + 1(-4 + 1)$

   $= -3 - 3 = -6$

18. Area $= \pm\frac{1}{2} \begin{vmatrix} x_1 & y_1 & 1 \\ x_2 & y_2 & 1 \\ x_3 & y_3 & 1 \end{vmatrix}$

   $= \pm\frac{1}{2} \begin{vmatrix} 0 & 0 & 1 \\ 6 & 2 & 1 \\ 8 & 10 & 1 \end{vmatrix}$

   $= \pm\frac{1}{2}(44) = 22$

CHAPTER NINE
Sequences, Probability, and Statistics

9.1 | Sequences and Summation Notation

2. $a_n = 4n - 3$

$a_1 = 4(1) - 3 = 1$

$a_2 = 4(2) - 3 = 5$

$a_3 = 4(3) - 3 = 9$

$a_4 = 4(4) - 3 = 13$

$a_5 = 4(5) - 3 = 17$

4. $a_n = \left(\frac{1}{2}\right)^n$

$a_1 = \left(\frac{1}{2}\right)^1 = \frac{1}{2}$

$a_2 = \left(\frac{1}{2}\right)^2 = \frac{1}{4}$

$a_3 = \left(\frac{1}{2}\right)^3 = \frac{1}{8}$

$a_4 = \left(\frac{1}{2}\right)^4 = \frac{1}{16}$

$a_5 = \left(\frac{1}{2}\right)^5 = \frac{1}{32}$

6. $a_n = \left(-\frac{1}{2}\right)^n$

$a_1 = \left(-\frac{1}{2}\right)^1 = -\frac{1}{2}$

$a_2 = \left(-\frac{1}{2}\right)^2 = \frac{1}{4}$

$a_3 = \left(-\frac{1}{2}\right)^3 = -\frac{1}{8}$

$a_4 = \left(-\frac{1}{2}\right)^4 = \frac{1}{16}$

$a_5 = \left(-\frac{1}{2}\right)^5 = -\frac{1}{32}$

8. $a_n = \frac{n}{n+1}$

$a_1 = \frac{1}{1+1} = \frac{1}{2}$

$a_2 = \frac{2}{2+1} = \frac{2}{3}$

$a_3 = \frac{3}{3+1} = \frac{3}{4}$

$a_4 = \frac{4}{4+1} = \frac{4}{5}$

$a_5 = \frac{5}{5+1} = \frac{5}{6}$

10. $a_n = \frac{3^n}{4^n}$

$a_1 = \frac{3^1}{4^1} = \frac{3}{4}$

$a_2 = \frac{3^2}{4^2} = \frac{9}{16}$

$a_3 = \frac{3^3}{4^3} = \frac{27}{64}$

$a_4 = \frac{3^4}{4^4} = \frac{81}{256}$

$a_5 = \frac{3^5}{4^5} = \frac{243}{1024}$

12. $a_n = \frac{3n^2 - n + 4}{2n^2 + 1}$

$a_1 = \frac{3(1)^2 - 1 + 4}{2(1)^2 + 1} = 2$

$a_2 = \frac{3(2)^2 - 2 + 4}{2(2)^2 + 1} = \frac{14}{9}$

$a_3 = \frac{3(3)^2 - 3 + 4}{2(3)^2 + 1} = \frac{28}{19}$

$a_4 = \frac{3(4)^2 - 4 + 4}{2(4)^2 + 1} = \frac{16}{11}$

$a_5 = \frac{3(5)^2 - 5 + 4}{2(5)^2 + 1} = \frac{74}{51}$

14. $a_n = \frac{n!}{n}$

$a_1 = \frac{1!}{1} = 1$

$a_2 = \frac{2!}{2} = 1$

$a_3 = \frac{3!}{3} = 2$

$a_4 = \frac{4!}{4} = 6$

$a_5 = \frac{5!}{5} = 24$

16. $a_n = (-1)^n \left(\frac{n}{n+1}\right)$

$a_1 = (-1)^1 \frac{1}{1+1} = -\frac{1}{2}$

$a_2 = (-1)^2 \frac{2}{2+1} = \frac{2}{3}$

$a_3 = (-1)^3 \frac{3}{3+1} = -\frac{3}{4}$

$a_4 = (-1)^4 \frac{4}{4+1} = \frac{4}{5}$

$a_5 = (-1)^5 \frac{5}{5+1} = -\frac{5}{6}$

18. $a_1 = 4$ and $a_{k+1} = \left(\frac{k+1}{2}\right) a_k$

$a_1 = 4$

$a_2 = \left(\frac{1+1}{2}\right) 4 = 4$

$a_3 = \left(\frac{2+1}{2}\right) 4 = 6$

$a_4 = \left(\frac{3+1}{2}\right) 6 = 12$

$a_5 = \left(\frac{4+1}{2}\right) 12 = 30$

20. $\frac{25!}{23!} = \frac{25 \cdot 24 \cdot 23!}{23!} = 600$

22. $\frac{(n+2)!}{n!} = \frac{(n+2)(n+1)n!}{n!} = (n+2)(n+1)$

24. $\dfrac{(2n+2)!}{(2n)!} = \dfrac{(2n+2)(2n+1)(2n)!}{(2n!)}$

$= (2n+2)(2n+1)$

26. $a_1 = 3 = 4(1) - 1$

$a_2 = 7 = 4(2) - 1$

$a_3 = 11 = 4(3) - 1$

$a_4 = 15 = 4(4) - 1$

$a_5 = 19 = 4(5) - 1$

$a_n = 4n - 1$

28. $a_1 = 1 = \dfrac{1}{1^2}$

$a_2 = \dfrac{1}{4} = \dfrac{1}{2^2}$

$a_3 = \dfrac{1}{9} = \dfrac{1}{3^2}$

$a_4 = \dfrac{1}{16} = \dfrac{1}{4^2}$

$a_5 = \dfrac{1}{25} = \dfrac{1}{5^2}$

$a_n = \dfrac{1}{n^2}$

30. $a_1 = \dfrac{1}{3} = \dfrac{2^{1-1}}{3^1}$

$a_2 = \dfrac{2}{9} = \dfrac{2^{2-1}}{3^2}$

$a_3 = \dfrac{4}{27} = \dfrac{2^{3-1}}{3^3}$

$a_4 = \dfrac{8}{81} = \dfrac{2^{4-1}}{3^4}$

$a_n = \dfrac{2^{n-1}}{3^n}$

32. $a_1 = 1 + \dfrac{1}{2} = 1 + \dfrac{2^1 - 1}{2^1}$

$a_2 = 1 + \dfrac{3}{4} = 1 + \dfrac{2^2 - 1}{2^2}$

$a_3 = 1 + \dfrac{7}{8} = 1 + \dfrac{2^3 - 1}{2^2}$

$a_4 = 1 + \dfrac{15}{16} = 1 + \dfrac{2^4 - 1}{2^4}$

$a_5 = 1 + \dfrac{31}{32} = 1 + \dfrac{2^5 - 1}{2^5}$

$a_n = 1 + \dfrac{2^n - 1}{2^n}$

34. $a_1 = 2 = (-1)^{1+1} 2(1)$

$a_2 = -4 = (-1)^{2+1} 2(2)$

$a_3 = 6 = (-1)^{3+1} 2(3)$

$a_4 = -8 = (-1)^{4+1} 2(4)$

$a_5 = 10 = (-1)^{5+1} 2(5)$

$a_n = (-1)^{n+1} 2n$

36. $a_1 = 1 = \dfrac{2^{1-1}}{(1-1)!}$

$a_2 = 2 = \dfrac{2^{2-1}}{(2-1)!}$

$a_3 = \dfrac{2^2}{2} = \dfrac{2^{3-1}}{(3-1)!}$

$a_4 = \dfrac{2^3}{6} = \dfrac{2^{4-1}}{(4-1)!}$

$a_5 = \dfrac{2^4}{24} = \dfrac{2^{5-1}}{(5-1)!}$

$a_6 = \dfrac{2^5}{120} = \dfrac{2^{6-1}}{(6-1)!}$

$a_n = \dfrac{2^{n-1}}{(n-1)!}$

38. $\displaystyle\sum_{i=1}^{6} (3i - 1) = (3 \cdot 1 - 1) + (3 \cdot 2 - 1) + (3 \cdot 3 - 1) + (3 \cdot 4 - 1) + (3 \cdot 5 - 1) + (3 \cdot 6 - 1) = 57$

40. $\displaystyle\sum_{k=1}^{5} 6 = 6 + 6 + 6 + 6 + 6 = 30$

42. $\displaystyle\sum_{i=0}^{5} 3i^2 = 3\displaystyle\sum_{i=0}^{5} i^2 = 3(0^2 + 1^2 + 2^2 + 3^2 + 4^2 + 5^2) = 165$

44. $\displaystyle\sum_{j=3}^{5}\frac{1}{j} = \frac{1}{3} + \frac{1}{4} + \frac{1}{5} = \frac{47}{60}$

46. $\displaystyle\sum_{k=2}^{5}(k+1)(k-3) = (2+1)(2-3) + (3+1)(3-3) + (4+1)(4-3) + (5+1)(5-3) = 14$

48. $\displaystyle\sum_{j=0}^{4}(-2)^j = (-2)^0 + (-2)^1 + (-2)^2 + (-2)^3 + (-2)^4 = 11$

50. $\displaystyle\sum_{k=0}^{4}\frac{(-1)^k}{k!} = \frac{(-1)^0}{0!} + \frac{(-1)^1}{1!} + \frac{(-1)^2}{2!} + \frac{(-1)^3}{3!} + \frac{(-1)^4}{4!} = \frac{3}{8}$

52. $\displaystyle\frac{5}{1+1} + \frac{5}{1+2} + \frac{5}{1+3} + \cdots + \frac{5}{1+15} = \sum_{i=1}^{15}\frac{5}{1+i}$

54. $\displaystyle\left[1 - \left(\frac{1}{6}\right)^2\right] + \left[1 - \left(\frac{2}{6}\right)^2\right] + \cdots + \left[1 - \left(\frac{6}{6}\right)^2\right] = \sum_{k=1}^{6}\left[1 - \left(\frac{k}{6}\right)^2\right]$

56. $\displaystyle 1 - \frac{1}{2} + \frac{1}{4} - \frac{1}{8} + \cdots - \frac{1}{128} = \frac{1}{2^0} - \frac{1}{2^1} + \frac{1}{2^2} - \frac{1}{2^3} + \cdots - \frac{1}{2^7} = \sum_{n=0}^{7}\left(-\frac{1}{2}\right)^n$

58. $\displaystyle\frac{1}{1\cdot3} + \frac{1}{2\cdot4} + \frac{1}{3\cdot5} + \cdots + \frac{1}{10\cdot12} = \sum_{k=1}^{10}\frac{1}{k(k+2)}$

60. $\displaystyle\frac{1}{2} + \frac{2}{4} + \frac{6}{8} + \frac{24}{16} + \frac{120}{32} + \frac{720}{64} = \sum_{k=1}^{6}\frac{k!}{2^k}$

62. (a) $A_1 = 100(101)[(1.01)^1 - 1] = \$101.00$

$A_2 = 100(101)[(1.01)^2 - 1] = \$203.01$

$A_3 = 100(101)[(1.01)^3 - 1] \approx \$306.04$

$A_4 = 100(101)[(1.01)^4 - 1] \approx \$410.10$

$A_5 = 100(101)[(1.01)^5 - 1] \approx \$515.20$

$A_6 = 100(101)[(1.01)^6 - 1] \approx \$621.35$

(b) $A_{60} = 100(101)[(1.01)^{60} - 1] \approx \$8248.64$

(c) $A_{240} = 100(101)[(1.01)^{240} - 1] \approx \$99,914.79$

64.
$$a_0 = 0.1\sqrt{82 + 9 \cdot 0^2} \approx 0.91$$
$$a_1 = 0.1\sqrt{82 + 9 \cdot 1^2} \approx 0.95$$
$$a_2 = 0.1\sqrt{82 + 9 \cdot 2^2} \approx 1.09$$
$$a_3 = 0.1\sqrt{82 + 9 \cdot 3^2} \approx 1.28$$
$$a_4 = 0.1\sqrt{82 + 9 \cdot 4^2} \approx 1.50$$
$$a_5 = 0.1\sqrt{82 + 9 \cdot 5^2} \approx 1.75$$
$$a_6 = 0.1\sqrt{82 + 9 \cdot 6^2} \approx 2.01$$
$$a_7 = 0.1\sqrt{82 + 9 \cdot 7^2} \approx 2.29$$
$$a_8 = 0.1\sqrt{82 + 9 \cdot 8^2} \approx 2.57$$
$$a_9 = 0.1\sqrt{82 + 9 \cdot 9^2} \approx 2.85$$
$$a_{10} = 0.1\sqrt{82 + 9 \cdot 10^2} \approx 3.13$$

66.
$$\sum_{n=0}^{9}(48.217n^2 + 228.1n + 311.28) = 48.71\sum_{n=0}^{9}n^2 + 228.1\sum_{n=0}^{9}n + 311.28\sum_{n=0}^{9}1$$
$$= 48.217(0^2 + 1^2 + 2^2 + 3^2 + 4^2 + 5^2 + 6^2 + 7^2 + 8^2 + 9^2)$$
$$+ 228.1(0 + 1 + 2 + 3 + 4 + 5 + 6 + 7 + 8 + 9)$$
$$+ 311.28(1 + 1 + 1 + 1 + 1 + 1 + 1 + 1 + 1 + 1)$$
$$\approx 27{,}119.1 \text{ million}$$

68.
$$\sum_{i=1}^{n}(x_i - \overline{x})^2 = \sum_{i=1}^{n}(x_i^2 - 2x_i\overline{x} + \overline{x}^2) = \sum_{i=1}^{n}x_i^2 - 2\overline{x}\sum_{i=1}^{n}x_i + n\overline{x}^2$$
$$= \sum_{i=1}^{n}x_i^2 - 2 \cdot \frac{1}{n}\sum_{i=1}^{n}x_i\sum_{i=1}^{n}x_i + n \cdot \frac{1}{n}\sum_{i=1}^{n}x_i \cdot \frac{1}{n}\sum_{i=1}^{n}x_i$$
$$= \sum_{i=1}^{n}x_i^2 + \sum_{i=1}^{n}x_i\sum_{i=1}^{n}x_i\left(-\frac{2}{n} + \frac{1}{n}\right) = \sum_{i=1}^{n}x_i^2 - \frac{1}{n}\left(\sum_{i=1}^{n}x_i\right)^2$$

9.2 Arithmetic Sequences

2. 10, 8, 6, 4, 2, ...
Arithmetic sequence, $d = -2$

4. 3, $\frac{5}{2}$, 2, $\frac{3}{2}$, 1, ...
Arithmetic sequence, $d = -\frac{1}{2}$

6. $-12, -8, -4, 0, 4, \ldots$
Arithmetic sequence, $d = 4$

8. $\ln 1, \ln 2, \ln 3, \ln 4, \ln 5, \ldots$
Not an arithmetic sequence

10. $1^2, 2^2, 3^2, 4^2, 5^2, \ldots$
Not an arithmetic sequence

12. $a_n = (2^n)n$
2, 8, 24, 64, 160
Not an arithmetic sequence

14. $a_n = 1 + (n-1)4$
1, 5, 9, 13, 17
Arithmetic sequence, $d = 4$

16. $a_n = 2^{n-1}$
1, 2, 4, 8, 16
Not an arithmetic sequence

18. $a_n = (-1)^n$
$-1, 1, -1, 1, -1$
Not an arithmetic sequence

20. $a_1 = 15$, $d = 4$
$a_n = dn + c$
$a_1 = 15 = 4(1) + c \Rightarrow c = 11$
$a_n = 4n + 11$

22. $a_1 = 0$, $d = -\frac{2}{3}$
$a_n = dn + c$
$a_1 = 0 = -\frac{2}{3}(1) + c \Rightarrow c = \frac{2}{3}$
$a_n = -\frac{2}{3}n + \frac{2}{3}$

24. $a_1 = -y$, $d = 5y$
$a_n = dn + c$
$a_1 = -y = 5y(1) + c \Rightarrow c = 6y$
$a_n = 5yn - 6y$

26. 10, 5, 0, -5, -10, $\ldots \Rightarrow d = -5$
$a_n = dn + c$
$a_1 = 10 = -5(1) + c \Rightarrow c = 15$
$a_n = -5n + 15$

28. $a_1 = -4$, $a_5 = 16$
$a_n = dn + c$
$a_1 = -4 = d(1) + c$, $a_5 = 16 = d(5) + c$
$d + c = -4$
$5d + c = 16$

Solving this system of equations yields $d = 5$ and $c = -9$.
$a_n = 5n - 9$

30. $a_5 = 190$, $a_{10} = 115$

$a_n = dn + c$

$a_5 = 190 = d(5) + c$, $a_{10} = 115 = d(10) + c$

$5d + c = 190$

$10d + c = 115$

Solving this system of equalities yields $d = -15$ and $c = 265$.

$a_n = -15n + 265$

32. $a_1 = 5$, $d = -\frac{3}{4}$

$a_1 = 5$

$a_2 = 5 - \frac{3}{4} = \frac{17}{4}$

$a_3 = \frac{17}{4} - \frac{3}{4} = \frac{7}{2}$

$a_4 = \frac{7}{2} - \frac{3}{4} = \frac{11}{4}$

$a_5 = \frac{11}{4} - \frac{3}{4} = 2$

34. $a_1 = 16.5$, $d = 0.25$

$a_1 = 16.5$

$a_2 = 16.5 + 0.25 = 16.75$

$a_3 = 16.75 + 0.25 = 17.00$

$a_4 = 17.00 + 0.25 = 17.25$

$a_5 = 17.25 + 0.25 = 17.50$

36. $a_1 = 6$, $a_{k+1} = a_k + 12 \Rightarrow d = 12$

$a_1 = 6$

$a_2 = 6 + 12 = 18$

$a_3 = 18 + 12 = 30$

$a_4 = 30 + 12 = 42$

$a_5 = 42 + 12 = 54$

38. $a_4 = 16$, $a_{10} = 46$

$a_4 = 16 = d(4) + c$, $a_{10} = 46 = d(10) + c$

$4d + c = 16$

$10d + c = 46$

Solving this system yields $d = 5$ and $c = -4$.

$a_n = 5n - 4$

$a_1 = 1$

$a_2 = 6$

$a_3 = 11$

$a_4 = 16$

$a_5 = 21$

40. $a_3 = 19$, $a_{15} = -1.7$

$a_3 = 19 = d(3) + c$, $a_{15} = -1.7 = d(15) + c$

$3d + c = 19$

$15d + c = -1.7$

Solving this system yields $d = -1.725$ and $c = 24.175$.

$a_n = -1.725n + 24.175$

$a_1 = 22.45$

$a_2 = 20.725$

$a_3 = 19$

$a_4 = 17.275$

$a_5 = 15.55$

42. 2, 8, 14, 20, ..., $n = 25$

$a_n = 6n - 4$

$a_1 = 2$ and $a_{25} = 146$

$S_{25} = \frac{25}{2}(2 + 146) = 1850$

44. 0.5, 0.9, 1.3, 1.7, ..., $n = 10$

$a_n = 0.4n + 0.1$

$a_1 = 0.5$ and $a_{10} = 4.1$

$S_{10} = \frac{10}{2}(0.5 + 4.1) = 23$

46. 1.50, 1.45, 1.40, 1.35, ..., $n = 20$

$a_n = -0.05n + 1.55$

$a_1 = 1.50$ and $a_{20} = 0.55$

$S_{20} = \frac{20}{2}(1.50 + 0.55) = 20.5$

48. $a_1 = 15$, $a_{100} = 307$, $n = 100$

$S_{100} = \frac{100}{2}(15 + 307) = 16,100$

50. $a_n = 2n$

$a_1 = 2$, $a_{100} = 200$

$S_{100} = \frac{100}{2}(2 + 200) = 10{,}100$

52. $a_n = 7n$

$a_{51} = 357$, $a_{100} = 700$

$S = \frac{50}{2}(357 + 700) = 26{,}425$

54. $\displaystyle\sum_{n=51}^{100} n - \sum_{n=1}^{50} n = \frac{50}{2}(51 + 100) - \frac{50}{2}(1 + 50) = 3775 - 1275 = 2500$

56. $a_n = 1000 - n$

$a_1 = 999$, $a_{250} = 750$

$S_{250} = \frac{250}{2}(999 + 750) = 218{,}625$

58. $a_n = \dfrac{n + 4}{2}$

$a_1 = \dfrac{5}{2}$, $a_{100} = 52$

$S_{100} = \dfrac{100}{2}\left(\dfrac{5}{2} + 52\right) = 2725$

60. $\displaystyle\sum_{n=0}^{100} \frac{8 - 3n}{16} = \frac{1}{2} + \sum_{n=1}^{100} \frac{8 - 3n}{16} = \frac{1}{2} + \frac{100}{2}\left(\frac{5}{16} - \frac{292}{16}\right) = -896.375$

62. 24, m_1, m_2, m_3, 56

$a_5 = 56 = 24 + 4d$

$d = 8$

$m_1 = 24 + 8 = 32$

$m_2 = 32 + 8 = 40$

$m_3 = 40 + 8 = 48$

64. 2, m_1, m_2, m_3, m_4, 5

$a_6 = 5 = 2 + 5d$

$d = \frac{3}{5}$

$m_1 = 2 + \frac{3}{5} = \frac{13}{5}$

$m_2 = \frac{13}{5} + \frac{3}{5} = \frac{16}{5}$

$m_3 = \frac{16}{5} + \frac{3}{5} = \frac{19}{5}$

$m_4 = \frac{19}{5} + \frac{3}{5} = \frac{22}{5}$

66. $a_1 = -10$, $a_{61} = 50$, $n = 61$

$S_{61} = \frac{61}{2}(-10 + 50) = 1220$

68. (a) $a_n = a_1 + (n - 1)d$

$\qquad = 32{,}800 + (n - 1)1750$

$\qquad = 1750n + 31{,}050$

$a_6 = 1750(6) + 31{,}050 = \$41{,}550$

(b) $S_6 = \frac{6}{2}(32{,}800 + 41{,}550) = \$223{,}050$

70. $a_n = 12 + 3n$

$a_1 = 15$, $a_{36} = 120$

$S_{36} = \frac{36}{2}(15 + 120)$

$\qquad = 2430$ seats

72. $a_1 = 4.9$, $a_2 = 14.7$,

$a_3 = 24.5$, $a_4 = 34.3 \Rightarrow d = 9.8$

$a_1 = 4.9 = 9.8(1) + c \Rightarrow c = -4.9$

$a_n = 9.8n - 4.9$

$a_{10} = 9.8(10) - 4.9 = 93.1$

$S_{10} = \frac{10}{2}(4.9 + 93.1) = 490$ meters

$\boxed{9.3}$ Geometric Sequences

2. 3, 12, 48, 192, ...
Geometric sequence, $r = 4$

4. 1, -2, 4, -8, ...
Geometric sequence, $r = -2$

6. 5, 1, 0.2, 0.04, ...
Geometric sequence, $r = 0.2$

8. 9, -6, 4, $-\frac{8}{3}$, ...
Geometric sequence, $r = -\frac{2}{3}$

10. $\frac{1}{5}$, $\frac{2}{3}$, $\frac{3}{9}$, $\frac{4}{11}$, ...
Not a geometric sequence

12. $a_1 = 6$, $r = 2$
$a_1 = 6$
$a_2 = 6(2)^1 = 12$
$a_3 = 6(2)^2 = 24$
$a_4 = 6(2)^3 = 48$
$a_5 = 6(2)^4 = 96$

14. $a_1 = 1$, $r = \frac{1}{3}$
$a_1 = 1$
$a_2 = 1\left(\frac{1}{3}\right)^1 = \frac{1}{3}$
$a_3 = 1\left(\frac{1}{3}\right)^2 = \frac{1}{9}$
$a_4 = 1\left(\frac{1}{3}\right)^3 = \frac{1}{27}$
$a_5 = 1\left(\frac{1}{3}\right)^4 = \frac{1}{81}$

16. $a_1 = 6$, $r = -\frac{1}{4}$
$a_1 = 6$
$a_2 = 6\left(-\frac{1}{4}\right)^1 = -\frac{3}{2}$
$a_3 = 6\left(-\frac{1}{4}\right)^2 = \frac{3}{8}$
$a_4 = 6\left(-\frac{1}{4}\right)^3 = -\frac{3}{32}$
$a_5 = 6\left(-\frac{1}{4}\right)^4 = \frac{3}{128}$

18. $a_1 = 2$, $r = \sqrt{3}$
$a_1 = 2$
$a_2 = 2(\sqrt{3})^1 = 2\sqrt{3}$
$a_3 = 2(\sqrt{3})^2 = 6$
$a_4 = 2(\sqrt{3})^3 = 6\sqrt{3}$
$a_5 = 2(\sqrt{3})^4 = 18$

20. $a_1 = 5$, $r = 2x$
$a_1 = 5$
$a_2 = 5(2x)^1 = 10x$
$a_3 = 5(2x)^2 = 20x^2$
$a_4 = 5(2x)^3 = 40x^3$
$a_5 = 5(2x)^4 = 80x^4$

22. $a_1 = 5$, $r = \frac{3}{2}$, $n = 8$
$a_8 = 5\left(\frac{3}{2}\right)^7 = \frac{10,935}{128}$

24. $a_1 = 8$, $r = \sqrt{5}$, $n = 9$
$a_9 = 8(\sqrt{5})^8 = 5000$

26. $a_1 = 1$, $r = -\frac{x}{3}$, $n = 7$
$a_7 = 1\left(-\frac{x}{3}\right)^6 = \frac{x^6}{729}$

28. $a_1 = 1000$, $r = 1.005$, $n = 60$
$a_6 = 1000(1.005)^{59}$

30. $a_2 = 3$, $a_5 = \frac{3}{64}$, $n = 1$
$a_2 r^3 = a_5$
$3r^3 = \frac{3}{64}$
$r^3 = \frac{1}{64}$
$r = \frac{1}{4}$
$a_2 = a_1 r$
$3 = a_1 \frac{1}{4}$
$a_1 = 12$

32. $a_3 = \frac{16}{3}$, $a_5 = \frac{64}{27}$, $n = 7$
$a_3 r^2 = a_5$
$\frac{16}{3} r^2 = \frac{64}{27}$
$r^2 = \frac{4}{9}$
$r = \pm\frac{2}{3}$
$a_7 = a_5 r^2 = \frac{64}{27}\left(\pm\frac{2}{3}\right)^2 = \frac{256}{243}$

34. $A = P\left(1 + \dfrac{r}{n}\right)^{nt} = 2500\left(1 + \dfrac{0.12}{n}\right)^{n(20)}$

     (a) $n = 1,$ $A = 2500(1 + 0.12)^{20}$ $\approx \$24{,}115.73$

     (b) $n = 2,$ $A = 2500\left(1 + \dfrac{0.12}{2}\right)^{2(20)}$ $\approx \$25{,}714.29$

     (c) $n = 4,$ $A = 2500\left(1 + \dfrac{0.12}{4}\right)^{4(20)}$ $\approx \$26{,}602.23$

     (d) $n = 12,$ $A = 2500\left(1 + \dfrac{0.12}{12}\right)^{12(20)}$ $\approx \$27{,}231.38$

     (e) $n = 365,$ $A = 2500\left(1 + \dfrac{0.12}{365}\right)^{365(20)}$ $\approx \$27{,}547.07$

36. $P =$ population after n years

     $P_0 =$ initial population $= 250{,}000$

     $r =$ rate of increase $= 1.3\%$

     $n =$ number of years $= 30$

     $P = P_0(1 + r)^n = 250{,}000(1.013)^{30} \approx 368{,}318$

38. $\displaystyle\sum_{n=1}^{9}(-2)^{n-1} = \dfrac{1(1 - (-2)^9)}{1 - (-2)} = 171$

40. $\displaystyle\sum_{i=1}^{6}32\left(\dfrac{1}{4}\right)^{i-1} = \dfrac{32\left(1 - \left(\frac{1}{4}\right)^6\right)}{1 - \frac{1}{4}} = \dfrac{1365}{32}$

42. $\displaystyle\sum_{i=1}^{10}5\left(-\dfrac{1}{3}\right)^{i-1} = \dfrac{5\left(1 - \left(-\frac{1}{3}\right)^{10}\right)}{1 - \left(-\frac{1}{3}\right)}$

$= \dfrac{15\left(1 - \left(-\frac{1}{3}\right)^{10}\right)}{4} \approx 3.75$

44. $\displaystyle\sum_{n=0}^{15}2\left(\dfrac{4}{3}\right)^{n} = \dfrac{2\left(1 - \left(\frac{4}{3}\right)^{16}\right)}{1 - \frac{4}{3}}$

$= -6\left(1 - \left(\dfrac{4}{3}\right)^{16}\right) \approx 592.65$

46. $\displaystyle\sum_{n=0}^{6}500(1.04)^{n} = \dfrac{500(1 - 1.04^7)}{1 - 1.04} \approx 3949.15$

48. $a = \displaystyle\sum_{n=1}^{60}50\left(1 + \dfrac{0.12}{12}\right)^{n} = 50(1.01) \cdot \dfrac{(1 - (1.01)^{60})}{(1 - 1.01)} \approx \$4124.32$

50. Let $N = 12t$ be the total number of deposits.

$$A = Pe^{r/12} + Pe^{2r/12} + \cdots + Pe^{Nr/12}$$

$$= \sum_{n=1}^{N} Pe^{(r/12) \cdot n} = Pe^{r/12} \frac{\left(1 - \left(e^{r/12}\right)^N\right)}{\left(1 - e^{r/12}\right)}$$

$$= Pe^{r/12} \frac{\left(1 - \left(e^{r/12}\right)^{12t}\right)}{1 - e^{r/12}} = \frac{Pe^{r/12}(e^{rt} - 1)}{(e^{r/12} - 1)}$$

52. (a) $A = 75 \left[\left(1 + \dfrac{0.09}{12}\right)^{12(25)} - 1\right] \left(1 + \dfrac{12}{0.09}\right) \approx \$84{,}714.78$

(b) $A = \dfrac{75e^{0.09/12}(e^{0.09(25)} - 1)}{e^{0.09/12} - 1} \approx \$85{,}196.05$

54. (a) $A = 20 \left[\left(1 + \dfrac{0.06}{12}\right)^{12(50)} - 1\right] \left(1 + \dfrac{12}{0.06}\right) \approx \$76{,}122.54$

(b) $A = \dfrac{20e^{0.06/12}\left(e^{0.06(50)} - 1\right)}{e^{0.06/12} - 1} \approx \$76{,}533.16$

56. $P = W \left(\dfrac{12}{r}\right) \left[1 - \left(1 + \dfrac{r}{12}\right)^{-12t}\right] = 2000 \left(\dfrac{12}{0.09}\right) \left[1 - \left(1 + \dfrac{0.09}{12}\right)^{-12(20)}\right] = 222{,}289.91$

58. (a) $\displaystyle\sum_{n=1}^{29} 0.01(2)^{n-1} = 0.01\frac{(1 - 2)^{29}}{(1 - 2)} = \$5{,}368{,}709.11$

(b) $\displaystyle\sum_{n=1}^{30} 0.01(2)^{n-1} = 0.01\frac{(1 - 2^{30})}{(1 - 2)} = \$10{,}737{,}418.23$

(c) $\displaystyle\sum_{n=1}^{31} 0.01(2)^{n-1} = 0.01\frac{(1 - 2^{31})}{(1 - 2)} = \$21{,}474{,}836.47$

60. $A = \displaystyle\sum_{n=1}^{6} 16^2 \left(\frac{1}{9}\right)\left(\frac{8}{9}\right)^{n-1} - \frac{16^2}{9}\frac{\left(1 - \left(\frac{8}{9}\right)^6\right)}{\left(1 - \frac{8}{9}\right)} \approx 129.72 \text{ square inches}$

62. $\displaystyle\sum_{n=0}^{\infty} 2\left(\frac{2}{3}\right)^n = 2 + \frac{4}{3} + \frac{8}{9} + \frac{16}{27} + \cdots = \frac{1}{1 - \frac{2}{3}} = 6$

64. $\displaystyle\sum_{n=0}^{\infty} 2\left(-\frac{2}{3}\right)^n = 2 - \frac{4}{3} + \frac{8}{9} - \frac{16}{27} + \cdots = \frac{2}{1 - \left(-\frac{2}{3}\right)} = \frac{6}{5}$

66. $\displaystyle\sum_{n=0}^{\infty} \left(\frac{1}{10}\right)^n = 1 + 0.1 + 0.01 + 0.001 + \cdots = \frac{1}{1 - \frac{1}{10}} = \frac{10}{9}$

68. $3 - 1 + \dfrac{1}{3} - \dfrac{1}{9} + \cdots = \displaystyle\sum_{n=0}^{\infty} 3\left(-\frac{1}{3}\right)^n = \frac{3}{1 - \left(-\frac{1}{3}\right)} = \frac{9}{4}$

70. $2 + \sqrt{2} + 1 + \dfrac{1}{\sqrt{2}} + \cdots \displaystyle\sum_{n=0}^{\infty} 2\left(\frac{1}{\sqrt{2}}\right)^n = \frac{2}{1 - \dfrac{1}{\sqrt{2}}} = 4 + 2\sqrt{2}$

72. The horizontal asymptote is the sum of the series.

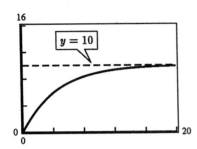

74. $t = 1 + 2\sum_{n=1}^{\infty}(0.9)^n$

$$= 1 + 2\left(\frac{0.9}{1-0.9}\right)$$

$$= 19 \text{ seconds}$$

$\boxed{9.4}$ Mathematical Induction

2. $P_{k+1} = \dfrac{1}{[(k+1)+1][(k+1)+3]}$

$\qquad = \dfrac{1}{(k+2)(k+4)}$

4. $P_{k+1} = \dfrac{k+1}{2}[3(k+1)-1] = \dfrac{k+1}{2}(3k+2)$

6. 1. When $n = 1$, $S_1 = 3 = 1(2 \cdot 1 + 1)$.

   2. Assume that

$$S_k = 3 + 7 + 11 + 15 + \cdots + (4k-1) = k(2k+1).$$

   Then,

$$S_{k+1} = 3 + 7 + 11 + 15 + \cdots + (4k-1) + [4(k+1)-1]$$

$$= S_k + [4(k+1)-1]$$

$$= k(2k+1) + (4k+3)$$

$$= 2k^2 + 5k + 3$$

$$= (k+1)(2k+3)$$

$$= (k+1)[2(k+1)+1].$$

We conclude by mathematical induction that the formula is valid for all positive integer values of n.

8. 1. When $n = 1$,

$$S_1 = 1 = \frac{1}{2}(3 \cdot 1 - 1).$$

2. Assume that

$$S_k = 1 + 4 + 7 + 10 + \cdots + (3k - 2) = \frac{k}{2}(3k - 1).$$

Then,

$$\begin{aligned} S_{k+1} &= 1 + 4 + 7 + 10 + \cdots + (3k - 2) + (3(k+1) - 2) \\ &= S_k + (3(k+1) - 2) \\ &= \frac{k}{2}(3k - 1) + (3k + 1) \\ &= \frac{3k^2 - k + 6k + 2}{2} \\ &= \frac{3k^2 + 5k + 2}{2} \\ &= \frac{(k+1)(3k+2)}{2} \\ &= \frac{k+1}{2}[3(k+1) - 1]. \end{aligned}$$

Therefore, we conclude that this formula holds for all positive integer values of n.

10. 1. When $n = 1$, $S_1 = 2 = 3^1 - 1$.

2. Assume that

$$S_k = 2(1 + 3 + 3^2 + 3^3 + \cdots + 3^{k+1}) = 3^k - 1.$$

Then,

$$\begin{aligned} S_{k+1} &= 2(1 + 3 + 3^2 + 3^3 + \cdots + 3^{k-1}) + 2 \cdot 3^{k+1-1} \\ &= S_k + 2 \cdot 3^k \\ &= 3^k - 1 + 2 \cdot 3^k \\ &= 3 \cdot 3^k - 1 \\ &= 3^{k+1} - 1. \end{aligned}$$

Therefore, we conclude that this formula holds for all positive integer values of n.

12. 1. When $n = 1$,

$$S_1 = 1 = \frac{1(1 + 1)(2 \cdot 1 + 1)}{6}.$$

2. Assume that

$$S_k = 1^2 + 2^2 + 3^2 + 4^2 + \cdots + k^2 = \frac{k(k + 1)(2k + 1)}{6}.$$

Then,

$$S_{k+1} = 1^2 + 2^2 + 3^2 + 4^2 + \cdots + k^2 + (k + 1)^2$$

$$= S_k + (k + 1)^2$$

$$= \frac{k(k + 1)(2k + 1)}{6} + (k + 1)^2$$

$$= \frac{k(k + 1)(2k + 1) + 6(k + 1)^2}{6}$$

$$= \frac{(k + 1)[2k^2 + k + 6k + 6]}{6}$$

$$= \frac{(k + 1)(k + 2)(2k + 3)}{6}.$$

Therefore, we conclude that this formula holds for all positive integer values of n.

14. 1. When $n = 1$, $S_1 = 2 = 1 + 1$.

2. Assume that

$$S_k = \left(1 + \frac{1}{1}\right)\left(1 + \frac{1}{2}\right)\left(1 + \frac{1}{3}\right)\cdots\left(1 + \frac{1}{k}\right) = k + 1$$

Then,

$$S_{k+1} = \left(1 + \frac{1}{1}\right)\left(1 + \frac{1}{2}\right)\left(1 + \frac{1}{3}\right)\cdots\left(1 + \frac{1}{k}\right)\left(1 + \frac{1}{k + 1}\right)$$

$$= (S_k)\left(1 + \frac{1}{k + 1}\right)$$

$$= (k + 1)\left(1 + \frac{1}{k + 1}\right)$$

$$= k + 1 + 1$$

$$= k + 2.$$

Therefore, we conclude that this formula holds for all positive integer values of n.

16. 1. When $n = 1$,

$$S_1 = 1^4 = \frac{1(1+1)(2 \cdot 1 + 1)(3 \cdot 1^2 + 3 \cdot 1 - 1)}{30}.$$

2. Assume that

$$S_k = \sum_{i=1}^{k} i^4 = \frac{k(k+1)(2k+1)(3k^2 + 3k - 1)}{30}.$$

Then,

$$S_{k+1} = S_k + (k+1)^4$$

$$= \frac{k(k+1)(2k+1)(3k^2 + 3k - 1)}{30} + (k+1)^4$$

$$= \frac{k(k+1)(2k+1)(3k^2 + 3k - 1) + 30(k+1)^4}{30}$$

$$= \frac{(k+1)[k(2k+1)(3k^2 + 3k - 1) + 30(k+1)^3]}{30}$$

$$= \frac{(k+1)(6k^4 + 39k^3 + 91k^2 + 89k + 30)}{30}$$

$$= \frac{(k+1)(k+2)(2k+3)(3k^2 + 9k + 5)}{30}$$

$$= \frac{(k+1)(k+2)(2(k+1)+1)(3(k+1)^2 + 3(k+1) - 1)}{30}.$$

Therefore, we conclude that this formula holds for all positive integer values of n.

18. 1. When $n = 1$,

$$S_1 = \frac{1}{2} = \frac{1}{2 \cdot 1 + 1}.$$

2. Assume that

$$S_k = \sum_{i=0}^{k} \frac{1}{(2i-1)(2i+1)} = \frac{k}{2k+1}.$$

Then,

$$S_{k+1} = S_k + \frac{1}{(2(k+1)-1)(2(k+1)+1)}$$

$$= \frac{k}{2k+1} + \frac{1}{(2k+1)(2k+3)}$$

$$= \frac{k(2k+3)+1}{(2k+1)(2k+3)}$$

$$= \frac{2k^2 + 3k + 1}{(2k+1)(2k+3)}$$

$$= \frac{(2k+1)(k+1)}{(2k+1)(2k+3)}$$

$$= \frac{k+1}{2(k+1)+1}.$$

Therefore, we conclude that this formula holds for all positive integer values of n.

20. $\displaystyle\sum_{n=1}^{50} = \frac{50(50+1)}{2} = 1275$

22. $\displaystyle\sum_{n=1}^{10} n^2 = \frac{10(10+1)(2 \cdot 10 + 1)}{6} = 385$

24. $\displaystyle\sum_{n=1}^{8} n^3 = \frac{8^2(8+1)^2}{4} = 1296$

26. $\displaystyle\sum_{n=1}^{4} n^5 = \frac{4^2(4+1)^2(2 \cdot 4^2 + 2 \cdot 4 - 1)}{12} = 1300$

28. $\displaystyle\sum_{n=1}^{10} (n^3 - n^2) = \sum_{n=1}^{10} n^3 - \sum_{n=1}^{10} n^2 = \frac{10^2(10+1)^2}{4} - \frac{10(10+1)(2 \cdot 10 + 1)}{6} = 3025 - 385 = 2640$

30. 25, 22, 19, 16, ... is an arithmetic sequence with $a_1 = 25$, common difference $d = -3$, and $a_n = a_1 + (n-1)d = 25 + (n-1)(-3) = 28 - 3n$. Hence, the sum of the first n terms is

$$S = \frac{n}{2}(a_1 + a_n) = \frac{n}{2}(25 + (28 - 3n)) = \frac{n}{2}(53 - 3n).$$

32. $3, -\frac{9}{2}, \frac{27}{4}, -\frac{81}{8}, \ldots$ is a geometric sequence with $a_1 = 3$ and common ratio $r = -\left(\frac{3}{2}\right)$. Hence, the sum of the first n terms is

$$S = a_1\left(\frac{1-r^n}{1-r}\right) = 3\left(\frac{1-(-3/2)^n}{1-(-3/2)}\right) = \frac{6}{5}\left(1 - \left(\frac{-3}{2}\right)^n\right).$$

34. $\dfrac{1}{2\cdot 3}, \dfrac{1}{3\cdot 4}, \dfrac{1}{4\cdot 5}, \cdots, \dfrac{1}{(n+1)(n+2)}, \cdots$. The sum of the first n terms is

$$\sum_{i=1}^{n} \frac{1}{(i+1)(i+2)} = \sum_{i=1}^{n}\left(\frac{1}{i+1}-\frac{1}{i+2}\right) = \left(\frac{1}{2}-\frac{1}{3}\right) + \left(\frac{1}{3}-\frac{1}{4}\right) + \cdots + \left(\frac{1}{n+1}-\frac{1}{n+2}\right)$$

$$= \frac{1}{2} - \frac{1}{n+2}$$

$$= \frac{n}{2(n+2)}.$$

36. 1. When $n = 2$,

$$\frac{1}{\sqrt{1}} + \frac{1}{\sqrt{2}} = \frac{\sqrt{2}+1}{\sqrt{2}}\cdot\frac{\sqrt{2}}{\sqrt{2}} = \left(\frac{\sqrt{2}+1}{2}\right)\sqrt{2} > \sqrt{2}.$$

2. Assume that

$$\frac{1}{\sqrt{1}} + \frac{1}{\sqrt{2}} + \frac{1}{\sqrt{3}} + \cdots + \frac{1}{\sqrt{k}} > \sqrt{k},\; k > 2.$$

Then,

$$\frac{1}{\sqrt{1}} + \frac{1}{\sqrt{2}} + \frac{1}{\sqrt{3}} + \cdots + \frac{1}{\sqrt{k}} + \frac{1}{\sqrt{k+1}} > \sqrt{k} + \frac{1}{\sqrt{k+1}} = \left(\sqrt{k} + \frac{1}{\sqrt{k+1}}\right)\left(\frac{\sqrt{k+1}}{\sqrt{k+1}}\right)$$

$$= \left(\frac{\sqrt{k^2+k}+1}{k+1}\right)\sqrt{k+1} > \sqrt{k+1}.$$

Therefore,

$$\frac{1}{\sqrt{1}} + \frac{1}{\sqrt{2}} + \frac{1}{\sqrt{3}} + \cdots \frac{1}{\sqrt{n}} > \sqrt{n},\; n \geq 2.$$

38. 1. When $n = 1$,

$$\left(\frac{x}{y}\right)^2 < \left(\frac{x}{y}\right) \text{ and } (0 < x < y).$$

2. Assume that

$$\left(\frac{x}{y}\right)^{k+1} < \left(\frac{x}{y}\right)^{k}$$

$$\left(\frac{x}{y}\right)^{k+1} < \left(\frac{x}{y}\right)^{k} \Rightarrow \left(\frac{x}{y}\right)\left(\frac{x}{y}\right)^{k+1} < \left(\frac{x}{y}\right)\left(\frac{x}{y}\right)^{k} \Rightarrow \left(\frac{x}{y}\right)^{k+2} < \left(\frac{x}{y}\right)^{k+1}$$

Therefore,

$$\left(\frac{x}{y}\right)^{n+1} < \left(\frac{x}{y}\right)^{n} \text{ for all integers } n \geq 1.$$

40. 1. When $n = 1$, $\left(\dfrac{a}{b}\right)^1 = \dfrac{a^1}{b^1}$.

 2. Assume that $\left(\dfrac{a}{b}\right)^k = \dfrac{a^k}{b^k}$.

  Then, $\left(\dfrac{a}{b}\right)^{k+1} = \left(\dfrac{a}{b}\right)^k \left(\dfrac{a}{b}\right) = \dfrac{a^k}{b^k} \cdot \dfrac{a}{b} = \dfrac{a^{k+1}}{b^{k+1}}$.

  Thus, $\left(\dfrac{a}{b}\right)^n = \dfrac{a^n}{b^n}$.

42. 1. When $n = 1$, $\ln x_1 = \ln x_1$.

 2. Assume that
$$\ln(x_1 x_2 x_3 \ldots x_k) = \ln x_1 + \ln x_2 + \ln x_3 + \cdots + \ln x_k.$$
  Then, $\ln(x_1 x_2 x_3 \ldots x_k x_{k+1}) = \ln[(x_1 x_2 x_3 \ldots x_k) x_{k+1}]$
$$= \ln(x_1 x_2 x_3 \ldots x_k) + \ln x_{k+1}$$
$$= \ln x_1 + \ln x_2 + \ln x_3 + \cdots + \ln x_k + \ln x_{k+1}.$$

  Thus, $\ln(x_1 x_2 x_3 \ldots x_n) = \ln x_1 + \ln x_2 + \ln x_3 + \cdots + \ln x_n$.

44. 1. When $n = 1$, $a + bi$ and $a - bi$ are complex conjugates by definition.

 2. Assume that $(a + bi)^k$ and $(a - bi)^k$ are complex conjugates.
  That is, if $(a + bi)^k = c + di$, then $(a - bi)^k = c - di$.

  Then, $(a + bi)^{k+1} = (a + bi)^k(a + bi) = (c + di)(a + bi)$
$$= (ac - bd) + i(bc + ad)$$
  and $(a - bi)^{k+1} = (a - bi)^k(a - bi) = (c - di)(a - bi)$
$$= (ac - bd) - i(bc + ad).$$

  This implies that $(a + bi)^{k+1}$ and $(a - bi)^{k+1}$ are complex conjugates.
  Therefore, $(a + bi)^n$ and $(a - bi)^n$ are complex conjugates for $n \geq 1$.

46. 1. When $n = 1$, $\left(2^{2(1)-1} + 3^{2(1)-1}\right) = 2 + 3 = 5$ and 5 is a factor.

2. Assume that 5 is a factor of $\left(2^{2k-1} + 3^{2k-1}\right)$.

$$\text{Then, } \left(2^{2(k+1)-1} + 3^{2(k+1)-1}\right) = \left(2^{2k+2-1} + 3^{2k+2-1}\right)$$
$$= \left(2^{2k-1}2^2 + 3^{2k-1}3^2\right)$$
$$= \left(4 \cdot 2^{2k-1} + 9 \cdot 3^{2k-1}\right)$$
$$= \left(2^{2k-1} + 3^{2k-1}\right) + \left(2^{2k-1} + 3^{2k-1}\right)$$
$$\quad + \left(2^{2k-1} + 3^{2k-1}\right) + \left(2^{2k-1} + 3^{2k-1}\right) + 5 \cdot 3^{2k-1}.$$

Since 5 is a factor of each set of parenthesis and 5 is a factor of $5 \cdot 3^{2k-1}$, then 5 is a factor of the whole sum. Thus, 5 is a factor of $\left(2^{2n-1} + 3^{2n-1}\right)$ for every positive integer n.

48. $a_1 = 10$, $a_n = 4a_{n-1}$

$a_2 = 4a_1 = 4(10) = 40$

$a_3 = 4a_2 = 4(40) = 160$

$a_4 = 4a_3 = 4(160) = 640$

$a_5 = 4a_4 = 4(640) = 2560$

50. $a_0 = 0$, $a_1 = 2$, $a_n = a_{n-1} + 2a_{n-2}$

$a_2 = a_1 + 2a_0 = 2 + 2(0) = 2$

$a_3 = a_2 + 2a_1 = 2 + 2(2) = 6$

$a_4 = a_3 + 2a_2 = 6 + 2(2) = 10$

52. $f(1) = 2$, $a_n = n - a_{n-1}$

	2		0		3		1		4
		−2		3		−2		3	
			5		−5		5		

Neither linear nor quadratic

54. $f(2) = -3$, $a_n = -2a_{n-1}$

	−3		6		−12		24		−48
		9		−18		36		−72	
			−27		54		−108		

Neither linear nor quadratic

56. $a_0 = 2$, $a_n = [a_{n-1}]^2$

	2		4		16		256		65,536
		2		12		240		65,280	
			10		228		65,040		

Neither linear nor quadratic

58. $f(1) = 0$, $a_n = a_{n-1} + 2_n$

$$\begin{array}{ccccccccc} 0 & & 4 & & 10 & & 18 & & 28 \\ & 4 & & 6 & & 8 & & 10 & \\ & & 2 & & 2 & & 2 & & \end{array}$$

Quadratic

60. $a_0 = 0$, $a_n = a_{n-1} - 1$

$$\begin{array}{ccccccccc} 0 & & -1 & & -2 & & -3 & & -4 \\ & -1 & & -1 & & -1 & & -1 & \\ & & 0 & & 0 & & 0 & & \end{array}$$

Linear

62. $a_0 = 7$, $a_1 = 6$, $a_3 = 10$

$f(n) = an^2 + bn + c$

$f(0) = a(0)^2 + b(0) + c = 7$

$f(1) = a(1)^2 + b(1) + c = 6$

$f(3) = a(3)^2 + b(3) + c = 10$

The system of three equations is

$$\begin{aligned} c &= 7 \\ a + b + c &= 6 \\ 9a + 3b + c &= 10. \end{aligned}$$

The solution is

$a = 1$, $b = -2$, $c = 7$

$\Rightarrow f(n) = n^2 - 2n + 7.$

64. $a_0 = 3$, $a_2 = 0$, $a_6 = 36$

$f(n) = an^2 + bn + c$

$f(0) = a(0)^2 + b(0) + c = 3$

$f(2) = a(2)^2 + b(2) + c = 0$

$f(6) = a(6)^2 + b(6) + c = 36$

The system of three equations is

$$\begin{aligned} c &= 3 \\ 4a + 2b + c &= 0 \\ 36a + 6b + c &= 36. \end{aligned}$$

The solution is

$a = \frac{7}{4}$, $b = -5$, $c = 3$

$\Rightarrow f(n) = \frac{7}{4}n^2 - 5n + 3.$

9.5 The Binomial Theorem

2. $_8C_6 = \dfrac{8!}{6!2!} = \dfrac{8 \cdot 7}{2 \cdot 1} = 28$

4. $_{20}C_{20} = \dfrac{20!}{20!0!} = 1$

6. $_{12}C_5 = \dfrac{12!}{5!7!} = \dfrac{12 \cdot 11 \cdot 10 \cdot 9 \cdot 8}{5 \cdot 4 \cdot 3 \cdot 2 \cdot 1} = 792$

8. $_{10}C_4 = \dfrac{10!}{6!4!} = \dfrac{10 \cdot 9 \cdot 8 \cdot 7}{4 \cdot 3 \cdot 2 \cdot 1} = 210$

10. $_{10}C_6 = \dfrac{10!}{6!4!} = 210$

12. $(x+1)^6 = {}_6C_0 x^6 + {}_6C_1 x^5(1) + {}_6C_2 x^4(1)^2 + {}_6C_3 x^3(1)^3 + {}_6C_4 x^2(1)^4 + {}_6C_5 x(1)^5 + {}_6C_6(1)^6$
$\quad\quad\quad\quad = x^6 + 6x^5 + 15x^4 + 20x^3 + 15x^2 + 6x + 1$

14. $(a+3)^4 = {}_4C_0 a^4 + {}_4C_1 a^3(3) + {}_4C_2 a^2(3)^2 + {}_4C_3 a(3)^3 + {}_4C_4(3)^4$
$\quad\quad\quad\quad = a^4 + 12a^3 + 54a^2 + 108a + 81$

16. $(y-2)^5 = {}_5C_0 y^5 - {}_5C_1 y^4(2) + {}_5C_2 y^3(2)^2 - {}_5C_3 y^2(2)^3 + {}_5C_4 y(2)^4 - {}_5C_5(2)^5$
$\quad\quad\quad\quad = y^5 - 10y^4 + 40y^3 - 80y^2 + 80y - 32$

18. $(x+y)^6 = {}_6C_0 x^6 + {}_6C_1 x^5 y + {}_6C_2 x^4 y^2 + {}_6C_3 x^3 y^3 + {}_6C_4 x^2 y^4 + {}_6C_5 x y^5 + {}_6C_6 y^6$
$\quad\quad\quad\quad = x^6 + 6x^5 y + 15x^4 y^2 + 20x^3 y^3 + 15x^2 y^4 + 6x y^5 + y^6$

20. $(x+2y)^4 = {}_4C_0 x^4 + {}_4C_1 x^3(2y) + {}_4C_2 x^2(2y)^2 + {}_4C_3 x(2y)^3 + {}_4C_4(2y)^4$
$\quad\quad\quad\quad = x^4 + 4x^3(2y) + 6x^2(4y^2) + 4x(8y^3) + 16y^4$
$\quad\quad\quad\quad = x^4 + 8x^3 y + 24x^2 y^2 + 32x y^3 + 16y^4$

22. $(2x-y)^5 = {}_5C_0(2x)^5 - {}_5C_1(2x)^4 y + {}_5C_2(2x)^3 y^2 - {}_5C_3(2x)^2 y^3 + {}_5C_4(2x)y^4 - {}_5C_5 y^5$
$\quad\quad\quad\quad = 32x^5 - 5(16x^4)y + 10(8x^3)y^2 - 10(4x^2)y^3 + 5(2x)y^4 - y^5$
$\quad\quad\quad\quad = 32x^5 - 80x^4 y + 80x^3 y^2 - 40x^2 y^3 + 10x y^4 - y^5$

24. $(5-3y)^3 = 5^3 - 3(5)^2 3y + 3(5)(3y)^2 - (3y)^3$
$\quad\quad\quad\quad = 125 - 225y + 135y^2 - 27y^3$

26. $(x^2+y^2)^6 = {}_6C_0(x^2)^6 + {}_6C_1(x^2)^5(y^2) + {}_6C_2(x^2)^4(y^2)^2 + {}_6C_3(x^2)^3(y^2)^3 + {}_6C_4(x^2)^2(y^2)^4$
$\quad\quad\quad\quad\quad\quad + {}_6C_5(x^2)(y^2)^5 + {}_6C_6(y^2)^6$
$\quad\quad\quad\quad = x^{12} + 6x^{10}y^2 + 15x^8 y^4 + 20x^6 y^6 + 15x^4 y^8 + 6x^2 y^{10} + y^{12}$

28. $\left(\dfrac{1}{x} + 2y\right)^6 = {}_6C_0\left(\dfrac{1}{x}\right)^6 + {}_6C_1\left(\dfrac{1}{x}\right)^5(2y) + {}_6C_2\left(\dfrac{1}{x}\right)^4(2y)^2 + {}_6C_3\left(\dfrac{1}{x}\right)^3(2y)^3$

$$+ {}_6C_4\left(\dfrac{1}{x}\right)^2(2y)^4 + {}_6C_5\left(\dfrac{1}{x}\right)(2y)^5 + {}_6C_6(2y)^6$$

$$= 1\left(\dfrac{1}{x}\right)^6 + 6(2)\left(\dfrac{1}{x}\right)^5 y + 15(4)\left(\dfrac{1}{x}\right)^4 y^2 + 20(8)\left(\dfrac{1}{x}\right)^3 y^3 + 15(16)\left(\dfrac{1}{x}\right)^2 y^4$$

$$+ 6(32)\left(\dfrac{1}{x}\right)y^5 + 1(64)y^6$$

$$= \dfrac{1}{x^6} + \dfrac{12y}{x^5} + \dfrac{60y^2}{x^4} + \dfrac{160y^3}{x^3} + \dfrac{240y^4}{x^2} + \dfrac{192y^5}{x} + 64y^6$$

30. $3(x+1)^5 - 4(x+1)^3 = 3[{}_5C_0x^5 + {}_5C_1x^4(1) + {}_5C_2x^3(1)^2 + {}_5C_3x^2(1)^3 + {}_5C_4x(1)^4 + {}_5C_5(1)^5]$

$$- 4[{}_3C_0x^3 + {}_3C_1x^2(1) + {}_3C_2x(1)^2 + {}_3C_3(1)^3]$$

$$= 3[(1)x^5 + 5x^4 + 10x^3 + 10x^2 + 5x + 1] - 4[(1)x^3 + 3x^2 + 3x + 1]$$

$$= 3x^5 + 15x^4 + 26x^3 + 18x^2 + 3x - 1$$

32. $(2-i)^5 = {}_5C_02^5 - {}_5C_12^4i + {}_5C_22^3i^2 - {}_5C_32^2i^3 + {}_5C_42i^4 - {}_5C_5i^5$

$$= 32 - 80i - 80 + 40i + 10 - i$$

$$= -38 - 41i$$

34. $(5 + \sqrt{-9})^3 = (5 + 3i)^3$

$$= 5^3 + 3 \cdot 5^2(3i) + 3 \cdot 5(3i)^2 + (3i)^3$$

$$= 125 + 225i - 135 - 27i$$

$$= -10 + 198i$$

36. $(5 - \sqrt{3}\,i)^4 = 5^4 - 4 \cdot 5^3(\sqrt{3}\,i) + 6 \cdot 5^2(\sqrt{3}\,i)^2 - 4 \cdot 5(\sqrt{3}\,i)^3 + (\sqrt{3}\,i)^4$

$$= 625 - 500\sqrt{3}\,i - 450 + 60\sqrt{3}\,i + 9$$

$$= 184 - 440\sqrt{3}\,i$$

38. 5th row of Pascal's Triangle: 1   5   10   10   5   1

$$(x + 2y)^5 = (1)x^5 + 5x^4 2y + 10x^3(2y)^2 + 10x^2(2y)^3 + 5x(2y)^4 + (2y)^5$$

$$= x^5 + 10x^4y + 40x^3y^2 + 80x^2y^3 + 80xy^4 + 32y^5$$

40. 5th Row of Pascal's Triangle: 1   5   10   10   5   1

$$(3y + 2)^5 = (3y)^5 + 5(3y)^4(2) + 10(3y)^3(2)^2 + 10(3y)^2(2)^3 + 5(3y)(2)^4 + (2)^5$$

$$= 243y^5 + 810y^4 + 1080y^3 + 720y^2 + 240y + 32$$

42. ${}_{12}C_8(x^2)^4(3)^8 = \dfrac{12!}{(12-8)!8!} \cdot 3^8 x^8 = 3{,}247{,}695x^8$

$$a = 3{,}247{,}695$$

44. $_{10}C_8(4x)^2(-y)^8 = \dfrac{10!}{(10-8)!8!} \cdot 16x^2y^8 = 720x^2y^8$

$a = 720$

46. $_8C_2(2x)^6(-3y)^2 = \dfrac{8!}{(8-6)!2!}(64x^6)(9y^2) = 16{,}128x^6y^2$

$a = 16{,}128$

48. $_{12}C_9(z^2)^3(-1)^9 = \dfrac{12}{(12-9)!9!}z^6(-1) = -220z^6$

$a = -220$

50. $_{10}C_3\left(\dfrac{1}{4}\right)^3\left(\dfrac{3}{4}\right)^7 = \dfrac{10 \cdot 9 \cdot 8}{3 \cdot 2} \cdot \dfrac{3^7}{4^{10}} \approx 0.250$

52. $_8C_4\left(\dfrac{1}{2}\right)^4\left(\dfrac{1}{2}\right)^4 = \dfrac{8 \cdot 7 \cdot 6 \cdot 5}{4 \cdot 3 \cdot 2}\left(\dfrac{1}{2}\right)^8 \approx 0.273$

54. $(2.005)^{10} = (2 + 0.005)^{10} = 2^{10} + 10(2)^9(0.005) + 45(2)^8(0.005)^2 + 120(2)^7(0.005)^3 + 210(2)^6(0.005)^4$

$\qquad + \; 252(2)^5(0.005)^5 + 210(2)^4(0.005)^6 + 120(2)^3(0.005)^7 + 45(2)^8(0.005)^2$

$\qquad + \; 10(2)(0.005)^9 + (0.005)^{10}$

$\quad = 1024 + 25.6 + 0.288 + 0.00192 + 0.0000084 + \ldots$

$\quad \approx 1049.890$

56. $(1.98)^9 = (2 - 0.02)^9 = 2^9 - 9(2)^8(0.02) + 36(2)^7(0.02)^2 - 84(2)^6(0.02)^3 + 126(2)^5(0.02)^4$

$\qquad - \; 126(2)^4(0.02)^5 + 84(2)^3(0.02)^6 - 36(2)^2(0.02)^7 + 9(2)(0.02)^8 - (0.02)^9$

$\quad = 512 - 46.08 + 1.8432 - 0.043008 + 0.00064512$

$\quad \approx 467{,}721$

58. $f(x) = 2x^2 - 4x + 1$

$\quad g(x) = f(x + 3)$

$\qquad = 2(x + 3)^2 - 4(x + 3) + 1$

$\qquad = 2(x^2 + 6x + 9) - 4x - 12 + 1$

$\qquad = 2x^2 + 8x + 7$

g is f shifted 3 units to the left.

60. $f(x) = -x^4 + 4x^2 - 1$

$$g(x) = f(x - 3)$$
$$= -(x - 3)^4 + 4(x - 3)^2 - 1$$
$$= -(x^4 - 12x^3 + 54x^2 - 108x + 81) + 4(x^2 - 6x + 9) - 1$$
$$= -x^4 + 12x^3 - 50x^2 + 84x - 46$$

g is f shifted 3 units to the right.

62. $f(t) = 215t^2 - 470t + 6700$

$$g(t) = f(t + 4)$$
$$= 215(t + 4)^2 - 470(t + 4) + 6700$$
$$= 215(t^2 + 8t + 16) - 470t - 1880 + 6700$$
$$= 215t^2 + 1250t + 8260$$

64. $0 = (1 - 1)^n = {}_nC_0 - {}_nC_1 + {}_nC_2 - {}_nC_3 + \cdots + (\pm {}_nC_n) = 0$

66. We use the fact that $(1 + x)^{2n} = (1 + x)^n(1 + x)^n$. The coefficient of x^n on the left-hand side is ${}_{2n}C_n$. We then want to find the coefficient of the x^n term on the right-hand side.

$$(1 + x)^n(1 + x)^n = \left(\sum_{j=0}^{n} {}_nC_j x^j \right) \left(\sum_{j=0}^{n} {}_nC_j x^j \right)$$

The contributions to the coefficient of the x^n term are of the form

$${}_nC_k \, x^k \, {}_nC_{n-k} x^{n-k} = {}_nC_k \, {}_nC_{n-k} \, x^n, \quad k = 0, 1, 2, 3, \ldots, n$$

So the coefficient of the x^n term is

$$\sum_{k=0}^{n} {}_nC_k \, {}_nC_{n-k}.$$

But ${}_nC_{n-k} = {}_nC_k$. Therefore, the coefficient is

$$\sum_{k=0}^{n} ({}_nC_k)^2.$$

That gives us ${}_{2n}C_n = ({}_nC_0)^2 + ({}_nC_1)^2 + \cdots + ({}_nC_n)^2$.

68. f and p are identical by the binomial expansion:

$$(1 - x)^3 = 1 - 3x + 3x^2 - x^3$$

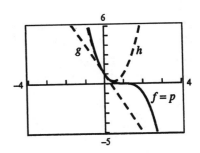

9.6 Counting Principles, Permutations, and Combinations

2.

First Number	Second Number
1	7
2	6
3	5
5	3
6	2
7	1

A total of eight may be obtained 6 different ways.

4. $3 \cdot 2 \cdot 4 = 24$ ways to choose a system

6. $2 \cdot 3 \cdot 5 = 30$ possible schedules

8. $24 \cdot 24 \cdot 10 \cdot 10 \cdot 10 \cdot 10 = 5{,}760{,}000$ distinct license plate numbers

10. $4^{10} = 1{,}048{,}576$ different ways

12. (a) $9 \cdot 10 \cdot 10 \cdot 10 = 9000$ ways
(b) $9 \cdot 9 \cdot 8 \cdot 7 = 4536$ ways
(c) $4 \cdot 10 \cdot 10 \cdot 10 = 4000$ ways (The first digit must be greater than 0 and less than 5.)
(d) $9 \cdot 10 \cdot 10 \cdot 5 = 4500$ ways (The last digit must be 0, 2, 4, 6, or 8.)

14. $50 \cdot 50 \cdot 50 = 125{,}000$ different lock combinations

16. (a) $5! = 5 \cdot 4 \cdot 3 \cdot 2 \cdot 1 = 120$ different ways
(b) $2!3! = 2 \cdot 1 \cdot 3 \cdot 2 \cdot 1 = 12$ different ways
(c) $3!2! = 3 \cdot 2 \cdot 1 \cdot 2 \cdot 1 = 12$ different ways

18. $_5P_5 = \dfrac{5!}{(5-5)!} = \dfrac{5!}{0!} = 120$

20. $_{20}P_2 = \dfrac{20!}{(20-2)!} = \dfrac{20!}{18!} = 380$

22. $_{100}P_1 = \dfrac{100!}{(100-1)!} = \dfrac{100!}{99!} = 100$

24. $_{10}P_2 = \dfrac{10!}{(10-2)!} = \dfrac{10!}{8!} = 90$

26. $_7P_4 = \dfrac{7!}{(7-4)!} = \dfrac{7!}{3!} = 840$

28. ABCD ACBD DBCA DCBA

30. $6! = 720$ ways

32. $4! = 24$ different orders

34. $\dfrac{8!}{3!5!} = 56$

36. $\dfrac{9!}{2!3!2!2!} = 7560$

38. $\dfrac{11!}{1!4!4!2!} = 34{,}650$

40. ABC, ABD, ABE, ABF, ACD, ACE, ACF, ADE, ADF, AEF, BCD, BCE, BCF, BDE, BDF, BEF, CDE, CDF, CEF, DEF

42. $_{12}C_{10} = \dfrac{12!}{(12-10)!10!} = \dfrac{12!}{2!10!} = 66$

44. $_{50}C_6 = \dfrac{50!}{(50-6)!6!} = \dfrac{50!}{44!6!} = 15{,}890{,}700$

46. $_{80}C_5 = \dfrac{80!}{(80-5)!5!} = \dfrac{80!}{75!5!} = 24{,}040{,}016$

48. There are 9 good units and 3 defective units.

(a) $_9C_4 = \dfrac{9!}{(9-4)!4!} = \dfrac{9!}{5!4!} = 126$

(b) $_9C_2 \, _3C_2 = \dfrac{9!}{(9-2)!2!} \cdot \dfrac{3!}{(3-2)!2!} = 108$

(c) $_9C_2 \, _3C_2 + {_9C_3} \, _3C_1 + {_9C_4} \, _3C_0 = \dfrac{9!}{(9-2)!2!} \cdot \dfrac{3!}{(3-2)!2!} + \dfrac{9!}{(9-3)!3!} \cdot \dfrac{3!}{(3-1)!1!}$

$$+ \dfrac{9!}{(9-4)!4!} \cdot \dfrac{3!}{(3-0)!0!}$$

$$= \dfrac{9!3!}{7!2!2!} + \dfrac{9!3!}{6!3!2!} + \dfrac{9!3!}{5!4!3!} = 486$$

50. Select type of card for three-of-a-kind: $_{13}C_1$

Select 3 of 4 cards for three-of-a-kind: $_4C_3$

Select type of card for pair: $_{12}C_1$

Select 2 of 4 cards for pair: $_4C_2$

$$_{13}C_1 \, _4C_3 \, _{12}C_1 \, _4C_2 = \dfrac{13!}{(13-1)!1!} \cdot \dfrac{4!}{(4-3)!3!} \cdot \dfrac{12!}{(12-1)!1!} \cdot \dfrac{4!}{(4-2)!2!}$$

$$= \dfrac{13!}{12!} \cdot \dfrac{4!}{3!} \cdot \dfrac{12!}{11!} \cdot \dfrac{4!}{2!2!} = 3744$$

52. (a) $_3C_2 = \dfrac{3!}{(3-2)!2!} = 3$

(b) $_8C_2 = \dfrac{8!}{(8-2)!2!} = 28$

(c) $_{12}C_2 = \dfrac{12!}{(12-2)!2!} = 66$

(d) $_{20}C_2 = \dfrac{20!}{(20-2)!2!} = 190$

54. $_6C_2 - 6 = \dfrac{6!}{(6-2)!2!} - 6 = 15 - 6 = 9$

56. $_{10}C_2 - 10 = \dfrac{10!}{(10-2)!2!} - 10 = 45 - 10 = 35$

58. $_nP_5 = 18_{n-2}P_4$

$$n(n-1)(n-2)(n-3)(n-4) = 18(n-2)(n-3)(n-4)(n-5)$$

$$n(n-1)(n-2)(n-3)(n-4) - 18(n-2)(n-3)(n-4)(n-5) = 0$$

$$(n-2)(n-3)(n-4)[n(n-1) - 18(n-5)] = 0$$

$$(n-2)(n-3)(n-4)[n^2 - n - 18n + 90] = 0$$

$$(n-2)(n-3)(n-4)(n^2 - 19n + 90) = 0$$

$$(n-2)(n-3)(n-4)(n-9)(n-10) = 0$$

Since $n \geq 6$, $n = 9$ or $n = 10$.

60. $_nP_1 = n = \dfrac{n(n-1)(n-2)\cdots 3 \cdot 2 \cdot 1}{(n-1)(n-2)\cdots 3 \cdot 2 \cdot 1} = \dfrac{n!}{(n-1)!1!} = \ _nC_1$

62. $_nC_n = \dfrac{n!}{(n-n)!n!} = \dfrac{n!}{0!n!} = \dfrac{n}{(n-0)!0!} = \ _nC_0$

9.7 | Probability

2. $\{2, 3, 4, 5, 6, 7, 8, 9, 10, 11, 12\}$

4. $\{$(red, red), (red, blue), (red, black), (blue, blue), (blue, black)$\}$

6. $\{SSS, SSF, SFS, SFF, FSS, FSF, FFS, FFF\}$

8. $n(E) = \{HHH, HHT, HTH, HTT\}$

$P(E) = \dfrac{n(E)}{n(S)} = \dfrac{4}{8} = \dfrac{1}{2}$

10. $n(E) = \{HHH, HHT, HTH, THH\}$

$P(E) = \dfrac{n(E)}{n(S)} = \dfrac{4}{8} = \dfrac{1}{2}$

12. The probability of not getting a face card is the complement of getting a face card. From Exercise 11, the probability of getting a face card is $P(E) = \dfrac{12}{52} = \dfrac{3}{13}$. Hence,

$$P(E') = 1 - P(E) = 1 - \tfrac{3}{13} = \tfrac{10}{13}.$$

14. There are six possible cards (A, 2, 3, 4, 5, 6) in each of four suits.

$$P(E) = \tfrac{24}{52} = \tfrac{6}{13}$$

16. $n(S) = 6 \cdot 6 = 36$

$E' = \{(5,6),\ (6,6),\ (6,5)\}$

$n(E) = n(S) - n(E') = 36 - 3 = 33$

$P(E) = \tfrac{33}{66} = \tfrac{11}{12}$

18. $n(S) = 6 \cdot 6 = 36$

$E = \{(1,1),\ (1,2),\ (2,1),\ (6,6)\}$

$P(E) = \tfrac{4}{36} = \tfrac{1}{9}$

20. $n(S) = 6 \cdot 6 = 36$

$E = \{(1,1),\ (1,2),\ (1,4),\ (1,6),\ (2,1),\ (2,3),\ (2,5),\ (3,2),\ (3,4),\ (3,6),$

$\quad (4,1),\ (4,3),\ (4,5),\ (5,2),\ (5,4),\ (5,6),\ (6,1),\ (6,3),\ (6,5)\}$

$P(E) = \tfrac{19}{36}$

22. $n(S) = {}_6C_2 = 15$

$n(E) = {}_2C_2 = 1$

$P(E) = \tfrac{1}{15}$

24. $n(S) = {}_6C_2 = 15$

$n(E) = {}_1C_1\,{}_2C_1 + {}_1C_1\,{}_3C_1 + {}_2C_1\,{}_3C_1$

$\qquad = 2 + 3 + 6 = 11$

$P(E) = \tfrac{11}{15}$

26. $1 - p = 1 - 0.36 = 0.64$

28. $1 - p = 1 - 0.84 = 0.16$

30. $n(S) = 72$

(a) $n(E) = 18 + 12 = 30$

$P(E) = \frac{30}{72} = \frac{5}{12}$

(b) $n(E) = 72 - 30 = 42$

$P(E) = \frac{42}{72} = \frac{7}{12}$

(c) $n(E) = 28 - 18 = 10$

$P(E) = \frac{10}{72} = \frac{5}{36}$

34. $n(S) = {}_8C_5 = 56$

(a) $n(E) = {}_6C_5 = 6$

$P(E) = \frac{6}{56} = \frac{3}{28}$

(b) $n(E) = {}_6C_4\, {}_2C_1 = 30$

$P(E) = \frac{30}{56} = \frac{15}{28}$

(c) $n(E) = {}_6C_4\, {}_2C_1 + {}_6C_5$

$= 30 + 6 = 36$

$P(E) = \frac{36}{56} = \frac{9}{14}$

38. (a) $n(S) = 4! = 24$

$n(E) = 1$

$P(E) = \frac{1}{24}$

(b) $n(S) = 3! = 6$

$n(E) = 1$

$p(E) = \frac{1}{6}$

42. $n(S) = {}_{20}C_5 = 15,504$

(a) $n(E) = {}_{16}C_5 = 4368$

$P(E) = \frac{4368}{15,504} = \frac{91}{323} \approx 0.282$

(b) $n(E) = {}_{16}C_4\, {}_4C_1 = 7280$

$P(E) = \frac{7280}{15,504} = \frac{455}{969} \approx 0.470$

(c) $n(E) = {}_{16}C_4\, {}_4C_1 + {}_{16}C_3\, {}_4C_2 + {}_{16}C_2\, {}_4C_3 + {}_{16}C_1\, {}_4C_4 = 11,136$

$P(E) = \frac{11,136}{15,504} = \frac{232}{323} \approx 0.718$

32. $1 - (0.37 + 0.44) = 0.19$

36. $n(S) = 5! = 120$

5 correct: 1 way
4 correct: not possible
3 correct: 10
2 correct: 20
1 correct: 45
0 correct: 44

(a) $\dfrac{45}{120} = \dfrac{3}{8}$

(b) $\dfrac{1 + 10 + 20 + 45}{120} = \dfrac{76}{120} = \dfrac{19}{30}$

40. $n(S) = {}_{52}C_5 = 2,598,960$

$n(E) = {}_{13}C_1\, {}_4C_3\, {}_{12}C_1\, {}_4C_2 = 3744$

$p(E) = \dfrac{3744}{2,598,960} = \dfrac{6}{4165}$

44. (a) $P(EE) = \frac{20}{40} \cdot \frac{20}{40} = \frac{1}{4}$

(b) $P(EO \text{ or } OE) = 2\left(\frac{20}{40}\right)\left(\frac{20}{40}\right) = \frac{1}{2}$

(c) $P(N_1 \le 30, N_2 \le 30) = \left(\frac{30}{40}\right)\left(\frac{30}{40}\right) = \frac{9}{16}$

(d) $P(N_1 N_1) = \frac{40}{40} \cdot \frac{1}{40} = \frac{1}{40}$

46. (a) $P(AA) = (0.9)^2 = 0.81$

(b) $P(NN) = (0.1)^2 = 0.01$

(c) $P(A) = 1 - P(NN) = 1 - 0.01 = 0.99$

48. (a) $P(SSSS) = \left(\frac{1}{3}\right)^4 = \frac{1}{81}$

(b) $P(NNNN) = \left(\frac{2}{3}\right)^4 = \frac{16}{81}$

(c) $P \text{ (at least one contract)} = 1 - P(NNNN) = 1 - \frac{16}{81} = \frac{65}{81}$

50. (a) $P(GGGGGG) = \left(\frac{1}{2}\right)\left(\frac{1}{2}\right)\left(\frac{1}{2}\right)\left(\frac{1}{2}\right)\left(\frac{1}{2}\right)\left(\frac{1}{2}\right) = \frac{1}{64}$

(b) $P(BBBBBB) + P(GGGGGG) = \frac{1}{64} + \frac{1}{64} = \frac{1}{32}$

(c) $1 - P(BBBBBB) = 1 - \frac{1}{64} = \frac{63}{64}$

52. $P(FFF) = (0.78)(0.78)(0.78) = 0.474552$

9.8 Exploring Data: Measures of Central Tendency

2. mean $= \dfrac{30 + 32 + 32 + 33 + 34 + 37 + 39}{7} = \dfrac{237}{2} \approx 33.86$

median $= 33$

mode $= 32$

4. mean $= \dfrac{20 + 32 + 32 + 33 + 34 + 37 + 39}{7} = \dfrac{227}{7} \approx 32.43$

median$=33$

mode $= 32$

6. mean $= \dfrac{30 + 32 + 32 + 34 + 37 + 39}{6} = \dfrac{204}{6} = 34$

median $= \dfrac{32 + 34}{2} = 33$

mode $= 32$

8. (a) mean $= \dfrac{11+13+13+14+15+18+20}{7} = \dfrac{104}{7} \approx 14.86$

   median $= 14$

   mode $= 13$

   The mean, median and mode are each increased by 6.

   (b) The mean, median and mode will each be increased by k. If $a_1, \ldots a_n$ are the original measurements, then the new mean is

   $$\frac{(a_1+k)+\ldots+(a_n+k)}{n} = \frac{a_1+\ldots+a_n}{n} + \frac{nk}{n}.$$

   $$= \text{(original mean)} + k$$

10. mean $= \dfrac{150+260+320+320+410+460}{6} = \dfrac{1920}{6} = 320$

    median $= 320$
    mode $= 320$

12. (a) average $= \dfrac{0(14)+1(26)+2(7)+3(2)+4(1)}{50} = \dfrac{50}{50} = 1$

    (b) batting average $= \dfrac{50}{200} = .250$

14. (a) mean $= \dfrac{19.5(0.9)+29.5(3.5)+39.5(11.8)+49.5(30.9)+59.5(81.4)+69.5(165.8)}{294.3}$

    $= \dfrac{18482.85}{294.3} \approx 62.8$ years

    (b) mean $= \dfrac{19.5(16.7)+29.5(16.6)+39.5(11.9)+49.5(7.5)+59.5(7.6)+69.5(8.8)}{69.1}$

    $= \dfrac{2720.45}{69.1} \approx 39.4$ years

16. mean $= \dfrac{8+9.5+10+10.5+10.5+10.5+12}{7} = \dfrac{71}{7} \approx 10.14$

    median $= 10.5$
    mode $= 10.5$

    The median and mode give the most representative description.

9.9 Exploring Data: Measures of Dispersion

2. $\bar{x} = \dfrac{2+3+6+9+15}{5} = \dfrac{35}{5} = 7$

$v = \dfrac{(2-7)^2 + (3-7)^2 + (6-7)^2 + (9-7)^2 + (15-7)^2}{5} = \dfrac{110}{5} = 22$

$\sigma = \sqrt{v} = \sqrt{22} \approx 4.69$

4. $\bar{x} = 2$

$v = 0$ (see the discussion preceeding Example 1)

$\sigma = 0$

6. $\bar{x} = \dfrac{1+1+1+5+5+5}{6} = \dfrac{18}{6} = 3$

$v = \dfrac{(1-3)^2 + (1-3)^2 + (1-3) + (5-3)^2 + (5-3)^2 + (5-3)^2}{6} = \dfrac{24}{6} = 4$

$\sigma = \sqrt{v} = \sqrt{4} = 2$

8. $\bar{x} = \dfrac{1.5. + 0.4 + 2.1 + 0.7 + 0.8}{5} = \dfrac{5.5}{5} = 1.1$

$v = \dfrac{(1.5-1.1)^2 + (0.4-1.1)^2 + (2.1-1.1)^2 + (0.7-1.1)^2 + (0.8-1.1)^2}{5} = \dfrac{1.9}{5} = 0.38$

$\sigma = \sqrt{v} = \sqrt{0.38} \approx 0.616$

10. The mean is $\bar{x} = \dfrac{192}{8} = 24$. Then,

$\sigma = \sqrt{\dfrac{10^2 + 25^2 + 50^2 + 26^2 + 15^2 + 33^2 + 29^2 + 4^2}{8} - 24^2}$

$= \sqrt{\dfrac{6072}{8} - 576} = \sqrt{183} \approx 13.53$

12. The mean is $\bar{x} = \dfrac{38.6}{5} = 7.72$. Then,

$\sigma = \sqrt{\dfrac{6^2 + 9.1^2 + 4.4^2 + 8.7^2 + 10.4^2}{5} - 7.72^2}$

$= \sqrt{\dfrac{322.02}{5} - 7.72^2} = \sqrt{4.8056} \approx 2.19$

14. The mean is $\bar{x} = \dfrac{33.2}{5} = 6.64$. Then,

$\sigma = \sqrt{\dfrac{9.0^2 + 7.5^2 + 3.3^2 + 7.4^2 + 6.0^2}{5} - 6.64^2}$

$= \sqrt{\dfrac{238.9}{5} - 6.64^2} = \sqrt{3.6904} \approx 1.92$

16. All the numbers must be equal.

18. (a) $\bar{x} = \dfrac{3(11) + 2(13) + 15 + 2(17) + 3(19)}{11} = \dfrac{165}{11} = 15$

$\sigma = \sqrt{\dfrac{3(11)^2 + 2(13)^2 + 15^2 + 2(17)^2 + 3(19)^2}{11} - 15^2}$

$= \sqrt{\dfrac{2587}{11} - 15^2} = \sqrt{10.\overline{18}} \approx 3.19$

(b) $\bar{x} = \dfrac{2(11) + 3(13) + 15 + 3(17) + 2(19)}{11} = \dfrac{165}{11} = 15$

$\sigma = \sqrt{\dfrac{2(11)^2 + 3(13)^2 + 15^2 + 3(17)^2 + 2(19)^2}{11} - 15^2}$

$= \sqrt{\dfrac{2563}{11} - 15^2} = \sqrt{8} \approx 2.83$

(c) $\bar{x} = \dfrac{2(21) + 3(23) + 25 + 3(27) + 2(29)}{11} = \dfrac{275}{11} = 25$

$\sigma = \sqrt{\dfrac{2(21)^2 + 3(23)^2 + 25^2 + 3(27)^2 + 2(29)^2}{11} - 25^2}$

$= \sqrt{\dfrac{6963}{11} - 25^2} = \sqrt{8} \approx 2.83$

(d) $\bar{x} = \dfrac{3(1) + 2(3) + 1(5) + 2(7) + 3(9)}{11} = \dfrac{55}{11} = 5$

$\sigma = \sqrt{\dfrac{3(1)^2 + 2(3)^2 + 5^2 + 2(7)^2 + 3(9)^2}{11} - 5^2}$

$= \sqrt{\dfrac{387}{11} - 5^2} = \sqrt{10.\overline{18}} \approx 3.19$

20. A: $\bar{x} = \dfrac{4(1) + 4(2) + 4(3) + 4(4)}{16} = \dfrac{40}{16} = 2.5$

$v = \sigma^2 = \dfrac{4(1)^2 + 4(2)^2 + 4(3)^2 + 4(4)^2}{16} - (2.5)^2 = 1.25$

B: $\bar{x} = \dfrac{8(1) + 1(2) + 2(3) + 4(4)}{15} = \dfrac{32}{15}$

$v = \sigma^2 = \dfrac{8(1)^2 + 1(2)^2 + 2(3)^2 + 4(4)^2}{15} - \left(\dfrac{32}{15}\right)^2 \approx 1.716$

C: $\bar{x} = \dfrac{6(2) + 8(3) + 2(4)}{16} = \dfrac{44}{16} = \dfrac{11}{4}$

$v = \sigma^2 = \dfrac{6(2)^2 + 8(3)^2 + 2(4)^2}{16} - \left(\dfrac{11}{4}\right)^2 \approx 0.4375$

D: $\bar{x} = \dfrac{2(1) + 6(2) + 6(3) + 2(4)}{16} = \dfrac{40}{16} = 2.5$

$v = \sigma^2 = \dfrac{2(1)^2 + 6(2)^2 + 6(3)^2 + 2(4)^2}{16} - (2.5)^2 = 0.75$

Here the variances are ordered from smallest to largest as follows

$C < D < A < B$

22. $\bar{x} \approx 40.7177$

$v \approx 46.74$

$\sigma \approx 6.837$

$\dfrac{26}{30} \approx 86.7\%$ lie between 2 standard deviations. (By Chebyshev's Theorem, at least 75% of the data lies between 2 standard deviations.)

24. 8, 10, 11, 11, 11, 14, 14, 14, 16, 17, 20

median = 14

26. 19, 20, 21, 22, 24, 25, 25, 27, 28, 28, 28, 29

Review Exercises for Chapter 9

2. $2(1^2) + 2(2^2) + 2(3^2) + \cdots + 2(9)^2 = \sum\limits_{k=1}^{9} 2k^2$

4. $1 - \dfrac{1}{3} + \dfrac{1}{9} - \dfrac{1}{27} + \cdots = \sum\limits_{k=0}^{\infty} \left(-\dfrac{1}{3}\right)^k$

6. $\sum\limits_{k=2}^{5} 4k = 4(2 + 3 + 4 + 5) = 4(14) = 56$

8. Arithmetic, $a_1 = 17$, $a_8 = -4$, $n = 8$

$$\sum\limits_{j=1}^{8} (20 - 3j) = \tfrac{8}{2}(17 + (-4)) = 52$$

10. Geometric, $a_0 = 1$, $r = 3$, $n = 5$

$$\sum\limits_{i=0}^{4} 3^i = 1\left(\dfrac{1 - 3^5}{1 - 3}\right) = 121$$

12. $\sum\limits_{i=0}^{\infty} \left(\dfrac{1}{3}\right)^i = \dfrac{1}{1 - r} = \dfrac{1}{1 - (1/3)} = \dfrac{3}{2}$

(sum of infinite geometric sequence)

14. Geometric, $a_0 = 1.3$, $r = \dfrac{1}{10}$

$$\sum\limits_{k=0}^{\infty} 1.3 \left(\dfrac{1}{10}\right)^k = \dfrac{1.3}{1 - (1/10)} = \dfrac{13}{9}$$

16. Arithmetic, $a_1 = 1$, $a_{25} = 19$, $n = 25$

$$\sum\limits_{k=1}^{25} \left(\dfrac{3k + 1}{4}\right) = \dfrac{25}{2}(1 + 19) = 250$$

18. $\sum\limits_{n=1}^{100} = \left(\dfrac{1}{1} - \dfrac{1}{2}\right) + \left(\dfrac{1}{2} - \dfrac{1}{3}\right) + \left(\dfrac{1}{3} - \dfrac{1}{4}\right) + \cdots + \left(\dfrac{1}{100} - \dfrac{1}{101}\right) = \dfrac{1}{1} - \dfrac{1}{101} = \dfrac{100}{101}$

20. $a_1 = 8$

$a_2 = 8 + (-2) = 6$

$a_3 = 6 + (-2) = 4$

$a_4 = 4 + (-2) = 2$

$a_5 = 2 + (-2) = 0$

22. $a_2 = 14 = d(2) + c$

$a_6 = 22 = d(6) + c$

$2d + c = 14$

$6d + c = 22$

Solving this system yields $d = 2$ and $c = 10$.

Thus, $a_n = 2n + 10$

$a_1 = 12$

$a_2 = 14$

$a_3 = 16$

$a_4 = 18$

$a_5 = 20.$

24. $a_1 = 10 = d(1) + c$

$a_3 = 28 = d(3) + c$

$d + c = 10$

$3d + c = 28$

Solving this system of equations yields $d = 9$ and $c = 1$.

Thus, $a_n = 9n + 1$

$a_{20} = 181, \ n = 20.$

$$\sum_{n=1}^{20} (9n + 1) = \tfrac{20}{2}(10 + 181) = 1910$$

26. $\displaystyle\sum_{i=20}^{80} i = \tfrac{61}{2}(20 + 80)$

$= 3050$

28. $a_1 = 2$

$a_2 = 2 \cdot 2 = 4$

$a_3 = 4 \cdot 2 = 8$

$a_4 = 8 \cdot 2 = 16$

$a_5 = 16 \cdot 2 = 32$

30. $a_3 = 12 = 2r^2$

$r^2 = 6$

$r = \pm\sqrt{6}$

$a_1 = 2$	$a_1 = 2$
$a_2 = 2\sqrt{6}$	$a_2 = 2(-\sqrt{6}) = -2\sqrt{6}$
$a_3 = 2(\sqrt{6})^2 = 12$	$a_3 = 2(-\sqrt{6})^2 = 12$
$a_4 = 2(\sqrt{6})^3 = 12\sqrt{6}$	$a_4 = 2(-\sqrt{6})^3 = -12\sqrt{6}$
$a_5 = 2(\sqrt{6})^4 = 72$	$a_5 = 2(-\sqrt{6})^4 = 72$

or

32. $a_n = 100(1.05)^{n-1}$

$$\sum_{n=1}^{20} 100(1.05)^{n-1} = 100\left(\frac{1 - 1.05^{20}}{1 - 1.05}\right)$$

≈ 3306.595

34. $\displaystyle\sum_{n=1}^{40} 32{,}000(1.055)^{n-1} = 32{,}000\left(\frac{1 - 1.055^{40}}{1 - 1.055}\right)$

$\approx \$4{,}371{,}379.65$

36. $A = \displaystyle\sum_{i=1}^{120} 100 \left(1 + \frac{0.065}{12}\right)^i = \sum_{i=1}^{120} 100 \left(\frac{12.065}{12}\right)^i$

$= 100 \left(\frac{12.065}{12}\right) \left[\dfrac{1 - \left(\frac{12.065}{12}\right)^{120}}{1 - \left(\frac{12.065}{12}\right)}\right]$

$\approx \$16{,}931.53$

38. 1. When $n = 1$,

$1 = \dfrac{1}{4}(1 + 3) = 1.$

2. Assume that

$1 + \dfrac{3}{2} + 2 + \dfrac{5}{2} + \cdots + \dfrac{1}{2}(k + 1) = \dfrac{k}{4}(k + 3).$

Then,

$\begin{aligned}
1 + \dfrac{3}{2} + 2 + \dfrac{5}{2} + \cdots + \dfrac{1}{2}(k + 1) + \dfrac{1}{2}(k + 2) &= \dfrac{k}{4}(k + 3) + \dfrac{1}{2}(k + 2) \\
&= \dfrac{k(k + 3) + 2(k + 2)}{4} \\
&= \dfrac{k^2 + 5k + 4}{4} \\
&= \dfrac{(k + 1)(k + 4)}{4} \\
&= \dfrac{k + 1}{4}\left[(k + 1) + 3\right].
\end{aligned}$

Thus, the formula holds for all positive integers n.

40. 1. When $n = 1$,

$$a + 0 \cdot d = a = \frac{1}{2}[2a + (1-1)d] = a.$$

2. Assume that

$$\sum_{k=0}^{i-1} (a + kd) = \frac{i}{2}[2a + (i-1)d].$$

Then,

$$\sum_{k=0}^{i+1-1} (a + kd) = \frac{i}{2}[2a + (i-1)d] + [a + id]$$

$$= \frac{2ia + i(i-1)d + 2a + 2id}{2}$$

$$= \frac{2a(i+1) + id(i+1)}{2}$$

$$= \left(\frac{i+1}{2}\right)[2a + id].$$

Thus, the formula holds for all positive integers n.

42. $_{10}C_7 = \dfrac{10!}{(10-7)!7!} = 120$

44. $_{12}C_3 = \dfrac{12 \cdot 11 \cdot 10}{3 \cdot 2 \cdot 1} = 220$

46. $(a - 3b)^5 = a^5 - 5a^4(3b) + 10a^3(3b)^2 - 10a^2(3b)^3 + 5a(3b)^4 - (3b)^5$

$\qquad = a - 15a^4b + 90a^3b^2 - 270a^2b^3 + 405ab^4 - 243b^5$

48. $(3x + y^2)^7$

$\qquad = (3x)^7 + 7(3x)^6y^2 + 21(3x)^5(y^2)^2 + 35(3x)^4(y^2)^3 + 35(3x)^3(y^2)^4$

$\qquad\quad + 21(3x)^2(y^2)^5 + 7(3x)(y^2)^6 + (y^2)^7$

$\qquad = 2187x^7 + 5103x^6y^2 + 5103x^5y^4 + 2835x^4y^6 + 945x^3y^8 + 189x^2y^{10} + 21xy^{12} + y^{14}$

50. $(4 - 5i)^3 = 4^3 - 3(4)^2(5i) + 3(4)(5i)^2 - (5i)^3$ 　　**52.** $2 \cdot 2 \cdot 2 = 8$

$\qquad = 64 - 240i - 300 + 125i$

$\qquad = -236 - 115i$

54. (a) $_5C_2 = 10$ 　　**56.** $n(S) = 5! = 120$

　　(b) $_{10}C_2 = 45$ 　　　　　$n(E) = 1$

　　　　　　　　　　　　　　　$P(E) = \frac{1}{120}$

58. $\left(\frac{6}{6}\right)\left(\frac{5}{6}\right)\left(\frac{4}{6}\right)\left(\frac{3}{6}\right)\left(\frac{2}{6}\right)\left(\frac{1}{6}\right) = \frac{5}{324}$ 　　**60.** $P(NNN) = (0.8)(0.8)(0.8)$

　　　　　　　　　　　　　　　　　　　　　　　　　　　　　$= 0.512$

62. (a) With 10 people, the probability that no two have the same birthday is

$$\frac{365 \cdot 364 \cdot 363 \cdot \cdots \cdot 356}{(365)^{10}} \approx 0.883$$

So, the probability that two will have the same birthday is

$$1 - 0.883 = 0.117.$$

(b) When the group is 23 or larger, the probability is at least 50% since

$$1 - \frac{365 \cdot 364 \cdot 363 \cdots 343}{(365)^{23}} \approx 0.507.$$

64. (a) $\bar{x} = \dfrac{9 + 3(10) + 11 + 13 + 3(14) + 2(15)}{11} = \dfrac{135}{11} \approx 12.27$

$$\sigma = \sqrt{\frac{9^2 + 3(10)^2 + 11^2 + 13^2 + 3(14)^2 + 2(15)^2}{11} - \left(\frac{135}{11}\right)^2}$$

$$= \sqrt{\frac{1709}{11} - \left(\frac{135}{11}\right)^2} \approx \sqrt{4.7438} \approx 2.18$$

(b) $\bar{x} = \dfrac{28 + 3(29) + 4(30) + 3(31) + 32}{12} = \dfrac{360}{12} = 30$

$$\sigma = \sqrt{\frac{28^2 + 3(29)^2 + 4(30)^2 + 3(31)^2 + 32^2}{12} - 30^2}$$

$$= \sqrt{\frac{10814}{12} - 30^2} \approx \sqrt{1.1667} \approx 1.08$$

(c) $\bar{x} = \dfrac{58 + 3(59) + 4(60) + 3(61) + 62}{12} = \dfrac{720}{12} = 60$

$$\sigma = \sqrt{\frac{58^2 + 3(59)^2 + 4(60)^2 + 3(61)^2 + 62^2}{12} - 60^2}$$

$$= \sqrt{\frac{43214}{12} - 60^2} \approx \sqrt{1.1667} \approx 1.08$$

(d) $\bar{x} = \dfrac{2(1) + 3(2) + 3 + 5 + 3(6) + 2(7)}{12} = \dfrac{48}{12} = 4$

$$\sigma = \sqrt{\frac{2(1)^2 + 3(2)^2 + 3^2 + 5^2 + 3(6)^2 + 2(7)^2}{12} - 4^2}$$

$$= \sqrt{\frac{254}{12} - 4^2} \approx \sqrt{5.1667} \approx 2.27$$

66. 5, 8, 10, 11, 12, 15, 18, 18, 20

$$\text{mean} = 13$$

$$\text{median} = 12$$

$$\text{variance} = 22.89$$

$$\text{standard deviation} = 4.78$$

68. 228.1, 298.6, 309.1, 335.2, 356.6, 393.8

$$\text{mean} = 320.23$$

$$\text{median} = 322.15$$

$$\text{variance} = 2673.2$$

$$\text{standard deviation} = 51.70$$

70.

35   46 51 56   64

CHAPTER TEN
Conics, Parametric Equations, and Polar Coordinates

10.1 Introduction to Conics: Parabolas

2. $x^2 = -2y$
Vertex: $(0, 0)$
$p = -\frac{1}{2} < 0$
Opens downward
Matches graph (f)

4. $y^2 = -12x$
Vertex: $(0, 0)$
$p = -3 < 0$
Opens to the left
Matches graph (c)

6. $(x + 3)^2 = -2(y - 2)$
Vertex: $(-3, 2)$
$p = -\frac{1}{2} < 0$
Opens downward
Matches graph (b)

8. $y = 2x^2$

$x^2 = \frac{1}{2}y = 4\left(\frac{1}{8}\right)y$

Vertex: $(0, 0)$
Focus: $\left(0, \frac{1}{8}\right)$
Directrix: $y = -\frac{1}{8}$

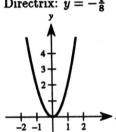

10. $y^2 = 3x = 4\left(\frac{3}{4}\right)x$

Vertex: $(0, 0)$
Focus: $\left(\frac{3}{4}, 0\right)$
Directrix: $x = -\frac{3}{4}$

12. $x + y^2 = 0$

$y^2 = -x = 4\left(-\frac{1}{4}\right)x$

Vertex: $(0, 0)$
Focus: $\left(-\frac{1}{4}, 0\right)$
Directrix: $x = \frac{1}{4}$

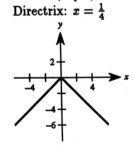

14. $(x + 3) + (y - 2)^2 = 0$

$(y - 2)^2 = 4\left(-\frac{1}{4}\right)(x + 3)$

Vertex: $(-3, 2)$
Focus: $(-3.25, 2)$
Directrix: $x = -2.75$

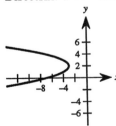

16. $\left(x + \frac{1}{2}\right)^2 - 4(y - 3) = 0$

$$\left(x + \frac{1}{2}\right)^2 = 4(1)(y - 3)$$

Vertex: $\left(-\frac{1}{2}, 3\right)$
Focus: $\left(-\frac{1}{2}, 4\right)$
Directrix: $y = 2$

18. $4x - y^2 - 2y - 33 = 0$
$y^2 + 2y + 1 = 4x - 33 + 1$
$(y + 1)^2 = 4(1)(x - 8)$
Vertex: $(8, -1)$
Focus: $(9, -1)$
Directrix: $x = 7$

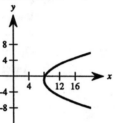

20. $y^2 + 6y + 8x + 25 = 0$
$y^2 + 6y + 9 = -8x - 25 + 9$
$(y + 3)^2 = 4(-2)(x + 2)$
Vertex: $(-2, -3)$
Focus: $(-4, -3)$
Directrix: $x = 0$

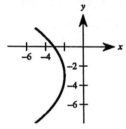

22. $x^2 + 4x + 4y - 4 = 0$

$$x^2 + 4x + 4 = -4y + 4 + 4$$

$$(x + 2)^2 = 4(-1)(y - 2)$$

Vertex: $(-2, 2)$
Focus: $(-2, 1)$
Directrix: $y = 3$

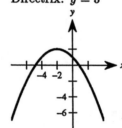

24.
$$x^2 - 2x + 8y + 9 = 0$$
$$x^2 - 21x + 1 = -8y - 9 + 1$$
$$(x - 1)^2 = 4(-2)(y + 1)$$

Vertex: $(1, -1)$
Focus: $(1, -3)$
Directrix: $x = 1$

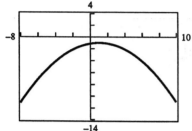

26. $y^2 - 4x - 4 = 0$
$y^2 = 4x + 4 = 4(1)(x + 1)$
Vertex: $(-1, 0)$
Focus: $(0, 0)$
Directrix: $x = -2$

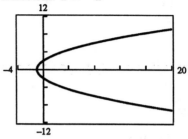

28. Vertex: $(0, 0) \Rightarrow h = 0, \quad k = 0$
Focus: $(-2, 0) \Rightarrow p = -2$
$$(y - k)^2 = 4p(x - h)$$
$$y^2 = 4(-2)x$$
$$y^2 = -8x$$

30. Vertex: $(3, 2) \Rightarrow h = 3, \quad k = 2$
Focus: $(1, 2) \Rightarrow p = -2$
$$(y - k)^2 = 4p(x - h)$$
$$(y - 2)^2 = 4(-2)(x - 3)$$
$$(y - 2)^2 = -8(x - 3)$$
$$y^2 - 4y + 8x - 20 = 0$$

32. Vertex: $(0, 0)$
Directrix: $y = 4 \Rightarrow p = -4$
$$(x - h)^2 = 4p(y - k)$$
$$(x - 0)^2 = 4(-4)(y - 0)$$
$$y = -\tfrac{1}{16}x^2$$

34. $(4, -2)$: $(-2 - k)^2 = 4p(4 - h) \Rightarrow (k + 2)^2 = 16p - 4ph$

  $(0, \; 0)$: $(0 - k)^2 = 4p(0 - h) \Rightarrow k^2 = -4ph$

  $(3, -3)$: $(-3 - k)^2 = 4p(3 - h) \Rightarrow (k + 3)^2 = 12p - 4ph$

By subtraction:

$$(k + 2)^2 - k^2 = 16p - 4ph - (-4ph) \Rightarrow 4k + 4 = 16p$$

$$(k + 3)^2 - k^2 = 12p - 4ph - (-4ph) \Rightarrow 6k + 9 = 12p$$

By elimination:

$$12k + 12 = 48p$$
$$\underline{-12k - 18 = -24p}$$
$$-6 = 24p \Rightarrow p = -\tfrac{1}{4}$$

$$12k + 12 = 48 \left(-\frac{1}{4}\right) \Rightarrow k = -2$$

$$(-2)^2 = -4 \left(-\frac{1}{4}\right) h \Rightarrow h = 4$$

$$(y - (-2))^2 = 4 \left(-\frac{1}{4}\right)(x - 4)$$

$$y^2 + 4y + 4 = -x + 4$$

$$y^2 + 4y + x = 0$$

36. The x-intercepts occur at $(0, 0)$ and $(4, 0)$ and the parabola opens downward.

$$y = -(x - 0)(x - 4)$$

$$y = -x^2 + 4x$$

$$x^2 - 4x + y = 0$$

38. Superimpose the coordinate plane over the parabola so that its vertex is at $(0, 0)$. The points $(-6, 4)$ and $(6, 4)$ are also on the parabola.

$$(x - 0)^2 = 4p(y - 0)$$

$$x^2 = 4py$$

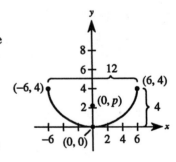

At $(6, 4)$: $6^2 = 4p(4) \Rightarrow p = \frac{9}{4}$. So the pipe is $\frac{9}{4}$ or $2\frac{1}{4}$ feet from the vertex.

40. $(x-0)^2 = 4p(y-0)$

$x^2 = 4py$

At $\left(50, \frac{1}{4}\right)$:

$50^2 = 4p\left(\frac{1}{4}\right) \Rightarrow p = 2500.$

$x^2 = 4(2500)y$

$y = \frac{1}{10,000}x^2$

42. Vertex: $(0, 0)$

$(y-0)^2 = 4p(x-0)$

$y^2 = 4px$

At $(1000, 800)$:

$800^2 = 4p(1000) \Rightarrow p = 160.$

$y^2 = 4(160)x$

$x = \frac{1}{640}y^2$

44. (a) $y = -\dfrac{16}{32^2}x^2 + 100$

$y = -\dfrac{1}{64}x^2 + 100$

(b) $0 = -\dfrac{1}{64}x^2 + 100$

$x^2 = 6400$

$x = 80$ feet

46. $y = ax^2 \Rightarrow p = \dfrac{1}{4a}$

$d_1 = \sqrt{x_0{}^2 + [ax_0{}^2 - (1/4a)]^2}$

$\quad = \sqrt{x_0{}^2 + [(4a^2x_0{}^2 - 1)/4a]^2} = \sqrt{\dfrac{16a^2x_0{}^2 + 16a^4x_0{}^4 - 8a^2x_0{}^2 + 1}{16a^2}} = \dfrac{4a^2x_0{}^2 + 1}{4a}$

$d_2 = \dfrac{1}{4a} - b$

$d_1 = d_2 \Rightarrow \dfrac{1}{4a} - b = \dfrac{4a^2x_0{}^2 + 1}{4a}$

$b = -ax_0{}^2$

Slope of tangent line: $m = \dfrac{2ax_0{}^2}{x_0} = 2ax_0$

Equation of tangent line: $y - ax_0{}^2 = 2ax_0(x - x_0)$

$\qquad\qquad\qquad\qquad y = 2ax_0x - ax_0{}^2$

x-intercept of tangent line: $2ax_0x - ax_0{}^2 = 0$

$\qquad\qquad\qquad\qquad 2ax_0x = ax_0{}^2$

$\qquad\qquad\qquad\qquad x = \dfrac{x_0}{2}$

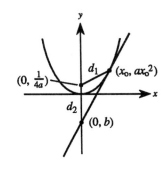

48.
$$y = \frac{1}{2}x^2$$
$$2y = x^2$$
$$4\left(\frac{1}{2}\right)y = x^2$$
$$p = \frac{1}{2}$$

Focus: $\left(0, \frac{1}{2}\right)$

$$d_1 = \frac{1}{2} - b$$
$$d_2 = \sqrt{(-3-0)^2 + \left(\frac{9}{2} - \frac{1}{2}\right)^2} = 5$$
$$\frac{1}{2} - b = 5$$
$$b = -\frac{9}{2}$$
$$m = \frac{-(9/2) - (9/2)}{0 + 3} = -3$$

Tangent line: $y = -3x - \frac{9}{2}$

x-intercept: $\left(-\frac{3}{2}, 0\right)$

50.
$$y = -2x^2$$
$$-\frac{1}{2}y = x^2$$
$$4\left(-\frac{1}{8}\right)y = x^2$$
$$p = -\frac{1}{8}$$

Focus: $\left(0, -\frac{1}{8}\right)$

$$d_1 = \frac{1}{8} + b$$
$$d_2 = \sqrt{(3-0)^2 + \left(-18 - \left(-\frac{1}{8}\right)\right)^2} = \frac{145}{8}$$
$$\frac{1}{8} + b = \frac{145}{8}$$
$$b = \frac{144}{8} = 18$$
$$m = \frac{-18 - 18}{3 - 0} = -12$$

Tangent line: $y = -12x + 18$

x-intercept: $\left(\frac{3}{2}, 0\right)$

10.2 Ellipses

2. $\frac{x^2}{9} + \frac{y^2}{1} = 1$
Center: $(0, 0)$
$a = 3$, $b = 1$
Horizontal major axis
Matches graph (a)

4. $\frac{x^2}{9} + \frac{y^2}{9} = 1$
Center: $(0, 0)$
$a = b = 3$
Circle
Matches graph (b)

6. $\frac{(x+2)^2}{4} + \frac{(y+2)^2}{25} = 1$
Center: $(-2, -2)$
$a = 5$, $b = 2$
Vertical major axis
Matches graph (d)

8. $\dfrac{x^2}{144} + \dfrac{y^2}{169} = 1$

$a^2 = 169, \ b^2 = 144, \ c^2 = 25$

Center: $(0, 0)$

Foci: $(0, \pm 5)$

Vertices: $(0, \pm 13), \ e = \frac{5}{13}$

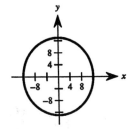

10. $\dfrac{x^2}{169} + \dfrac{y^2}{144} = 1$

$a^2 = 169, \ b^2 = 144, \ c^2 = 25$

Center: $(0, 0)$

Foci: $(\pm 5, 0)$

Vertices: $(\pm 13, 0)$

$e = 5/13$

12. $\dfrac{x^2}{28} + \dfrac{y^2}{64} = 1$

$a^2 = 64, \ b^2 = 28, \ c^2 = 36$

Center: $(0, 0)$

Foci: $(0, \pm 6)$

Vertices: $(0, \pm 8)$

$e = 3/4$

14. $5x^2 + 3y^2 = 15$

$\dfrac{x^2}{3} + \dfrac{y^2}{5} = 1$

$a^2 = 5, \ b^2 = 3, \ c^2 = 2$

Center: $(0, 0)$

Foci: $\left(0, \pm\sqrt{2}\right)$

Vertices: $\left(0 \pm \sqrt{5}\right)$

$e = \sqrt{2/5}$

16. $5x^2 + 7y^2 = 70$

$\dfrac{x^2}{14} + \dfrac{y^2}{10} = 1$

$a^2 = 14, \ b^2 = 10, \ c^2 = 4$

Center: $(0, 0)$

Foci: $(\pm 2, 0)$

Vertices: $\left(\pm\sqrt{14}, 0\right)$

$e = 2/\sqrt{14} = \sqrt{14}/7$

18. $\dfrac{(x + 2)^2}{1} + \dfrac{(y + 4)^2}{1/4} = 1$

$a^2 = 1, \ b^2 = \dfrac{1}{4}, \ c^2 = \dfrac{3}{4}$

Center: $(-2, -4)$

Foci: $\left(-2 \pm \dfrac{\sqrt{3}}{2}, -4\right)$

Vertices: $(-1, -4), \ (-3, -4), \ e = \dfrac{\sqrt{3}}{2}$

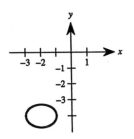

20. $\qquad 9x^2 + 4y^2 - 36x + 8y + 31 = 0$

$$9(x^2 - 4x + 4) + 4(y^2 + 2y + 1) = -31 + 36 + 4$$

$$\frac{(x-2)^2}{1} + \frac{(y+1)^2}{9/4} = 1$$

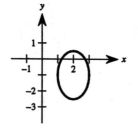

$a^2 = \dfrac{9}{4}, \quad b^2 = 1, \quad c^2 = \dfrac{5}{4}$

Center: $(2, -1)$

Foci: $\left(2, \ -1 \pm \dfrac{\sqrt{5}}{2}\right)$

Vertices: $\left(2, \ \dfrac{1}{2}\right), \ \left(2, \ -\dfrac{5}{2}\right), \quad e = \dfrac{\sqrt{5}}{3}$

22. $\qquad 9x^2 + 25y^2 - 36x - 50y + 61 = 0$

$$9(x^2 - 4x + 4) + 25(y^2 - 2y + 1) = -61 + 36 + 25$$

$$9(x-2)^2 + 25(y-1)^2 = 0$$

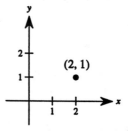

Degenerate ellipse with center $(2, 1)$ as the only point

24. $16x^2 + 25y^2 = 1$

$$\frac{x^2}{1/16} + \frac{y^2}{1/25} = 1$$

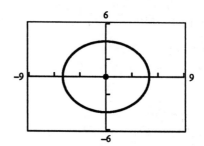

$a^2 = 1/16, \quad b^2 = 1/25, \quad c^2 = 9/400$

Center: $(0, 0)$

Foci: $(\pm 3/20, 0)$

Vertices: $(\pm 1/4, 0)$

$e = (3/20)/(1/4) = 3/5$

26. $36x^2 + 9y^2 + 48x - 36y + 43 = 0$

$36\left(x^2 + \frac{4}{3}x + \frac{4}{9}\right) + 9(y^2 - 4y + 4) = -43 + 16 + 36$

$$\frac{[x + (2/3)]^2}{1/4} + \frac{(y-2)^2}{1} = 1$$

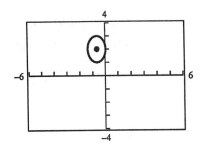

$a^2 = 1,\ \ b^2 = \frac{1}{4},\ \ c^2 = \frac{3}{4}$

Center: $\left(-\frac{2}{3}, 2\right)$

Foci: $\left(-\frac{2}{3}, 2 \pm \frac{\sqrt{3}}{2}\right)$

Vertices: $\left(-\frac{2}{3}, 3\right), \left(-\frac{2}{3}, 1\right)$

$e = \dfrac{\sqrt{3}}{2}$

28. $2x^2 + y^2 + 4.8x - 6.4y + 3.12 = 0$

$2(x^2 + 2.4x + 1.44) + (y^2 - 6.4y + 10.24) = -3.12 + 2.88 + 10.24$

$$2(x + 1.2)^2 + (y - 3.2)^2 = 10$$

$$\frac{(x + 1.2)^2}{\left(\sqrt{5}\right)^2} + \frac{(y - 3.2)^2}{\left(\sqrt{10}\right)^2} = 1$$

center: $(-1.2, 3.2)$

$a = \sqrt{10},\ \ b = \sqrt{5}\ \ \Rightarrow c^2 = a^2 - b^2 = 5$

$c = \sqrt{5}$

foci: $\left(-1.2,\ 3.2 \pm \sqrt{5}\right)$

vertices: $\left(-1.2,\ 3.2 \pm \sqrt{10}\right)$

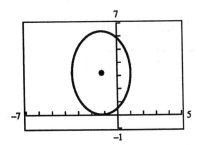

30. Vertices: $(0, \pm 8) \Rightarrow a = 8$

Foci: $(0, \pm 4) \Rightarrow c = 4$

$b^2 = a^2 - c^2 = 64 - 16 = 48$

Center: $(0, 0) = (h, k)$

$$\frac{(y - k)^2}{a^2} + \frac{(x - h)^2}{b^2} = 1$$

$$\frac{y^2}{64} + \frac{x^2}{48} = 1$$

32. Vertices: $(0, 2), (4, 2) \Rightarrow a = 2$

Minor axis of length $2 \Rightarrow b = 1$

Center: $(2, 2) = (h, k)$

$$\frac{(x - h)^2}{a^2} + \frac{(y - k)^2}{b^2} = 1$$

$$\frac{(x - 2)^2}{4} + (y - 2)^2 = 1$$

34. Vertices: $(0, \pm 5) \Rightarrow a = 5$

Solution point: $(4, 2)$

Center: $(0, 0) = (h, k)$

$$\frac{(y-k)^2}{a^2} + \frac{(x-h)^2}{b^2} = 1$$

$$\frac{y^2}{25} + \frac{x^2}{b^2} = 1 \Rightarrow b^2 = \frac{25x^2}{25 - y^2} = \frac{25(4^2)}{25 - 2^2} = \frac{400}{21}$$

$$\frac{y^2}{25} + \frac{x^2}{400/21} = 1$$

36. Vertices: $(\pm 5, 0) \Rightarrow a = 5$

Eccentricity: $\frac{3}{5} \Rightarrow c = \frac{3}{5}a = 3$

$b^2 = a^2 - c^2 = 25 - 9 = 16$

Center: $(0, 0) = (h, k)$

$$\frac{(x-h)^2}{a^2} + \frac{(y-k)^2}{b^2} = 1$$

$$\frac{x^2}{25} + \frac{y^2}{16} = 1$$

38. $a = 50$, $b = 30$

$$\frac{x^2}{50^2} + \frac{y^2}{30^2} = 1$$

$$\frac{45^2}{50^2} + \frac{y^2}{30^2} = 1$$

$$\frac{2025}{2500} + \frac{y^2}{900} = 1$$

$$y^2 = \frac{475}{2500}(900)$$

$$y^2 = 171$$

$$y = \pm\sqrt{19}$$

The height five feet from the edge of the tunnel is $3\sqrt{19} \approx 13$ feet.

40. For $\dfrac{x^2}{a^2} + \dfrac{y^2}{b^2} = 1$, we have $c^2 = a^2 - b^2$.

When $x = c$,

$$\frac{c^2}{a^2} + \frac{y^2}{b^2} = 1 \Rightarrow y^2 = b^2\left(1 - \frac{a^2 - b^2}{a^2}\right)$$

$$\Rightarrow y^2 = \frac{b^4}{a^2}$$

$$\Rightarrow 2y = \frac{2b^2}{a}.$$

42. $\dfrac{x^2}{9} + \dfrac{y^2}{16} = 1$

$a = 4$, $b = 3$, $c = \sqrt{7}$

Points on the ellipse: $(\pm 3, 0)$, $(0, \pm 4)$

Length of latus recta: $\dfrac{2b^2}{a} = \dfrac{2(3)^2}{4} = \dfrac{9}{2}$

Additional points: $\left(\pm\dfrac{9}{4}, -\sqrt{7}\right)$, $\left(\pm\dfrac{9}{4}, \sqrt{7}\right)$

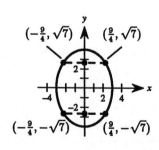

44. $5x^2 + 3y^2 = 15$

$\dfrac{x^3}{3} + \dfrac{y^2}{5} = 1$

$a = \sqrt{5}, \ b = \sqrt{3}, \ c = \sqrt{2}$

Points on the ellipse: $(\pm\sqrt{3}, \ 0), \ (0, \ \pm\sqrt{5})$

Length of latus recta: $\dfrac{2b^2}{a} = \dfrac{2 \cdot 3}{\sqrt{5}} = \dfrac{6\sqrt{5}}{5}$

Additional points: $\left(\pm\dfrac{3\sqrt{5}}{5}, \ -\sqrt{2}\right), \ \left(\pm\dfrac{3\sqrt{5}}{5}, \ \sqrt{2}\right)$

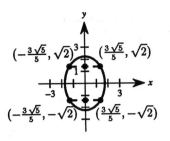

46. $a = 3.666 \times 10^9, \ e = 0.248$

$e = \dfrac{c}{a}$

$0.248 = \dfrac{c}{3.666 \times 10^9}$

$c \approx 909{,}168{,}000$

Least distance:

$a - c = 3.666 \times 10^9 - 909{,}168{,}000 = 2{,}756{,}832{,}000$

Greatest distance:

$a + c = 3.666 \times 10^9 + 909{,}168{,}000 = 4{,}575{,}168{,}000$

48. $e = \dfrac{c}{a}$

$A + P = 2a, \quad a = \dfrac{A + P}{2}$

$c = a - P = \dfrac{A + P}{2} - P = \dfrac{A - P}{2}$

$e = \dfrac{c}{a} = \dfrac{(A - P)/2}{(A + P)/2} = \dfrac{A - P}{A + P}$

50. $A = 122{,}000 + 4000 = 126{,}000 \text{ mi}, \ P = 4119 \text{ mi}$

$e = \dfrac{A - P}{A + P} = 0.937$

10.3 Hyperbolas

2. $\dfrac{y^2}{9} - \dfrac{x^2}{4} = 1$

Center: $(0, 0)$

$a = 3, \ b = 2$

Vertical transverse axis

Matches graph (a)

4. $\dfrac{y^2}{16} - \dfrac{x^2}{1} = 1$

Center: $(0, 0)$

$a = 4, \ b = 1$

Vertical transverse axis

Matches graph (c)

6. $\dfrac{(x+1)^2}{16} - \dfrac{(y-3)^2}{9} = 1$

Center: $(-1, \ 3)$

$a = 4, \ b = 3$

Horizontal transverse axis

Matches graph (b)

8. $\dfrac{x^2}{9} - \dfrac{y^2}{16} = 1$

$a = 3, \ b = 4, \ c = 5$

Center: $(0, 0)$

Vertices: $(\pm\,3, \, 0)$

Foci: $(\pm\,5, \, 0)$

Asymptotes: $y = \pm\dfrac{4}{3}x$

10. $\dfrac{y^2}{9} - \dfrac{x^2}{1} = 1$

$a = 3, \ b = 1, \ c = \sqrt{10}$

Center: $(0, 0)$

Vertices: $(0, \pm\,3)$

Foci: $(0, \pm\sqrt{10})$

Asymptotes: $y = \pm 3x$

12. $\dfrac{x^2}{36} - \dfrac{y^2}{4} = 1$

$a = 6, \ b = 2, \ c = 2\sqrt{10}$

Center: $(0, 0)$

Vertices: $(\pm 6, 0)$

Foci: $(\pm 2\sqrt{10}, 0)$

Asymptotes: $y = \pm\dfrac{1}{3}x$

14. $7x^2 - 3y^2 = 21$

$$\frac{x^2}{3} - \frac{y^2}{7} = 1$$

$a = \sqrt{3}$, $b = \sqrt{7}$, $c = \sqrt{10}$
Center: $(0, 0)$
Vertices: $(\pm\sqrt{3}, 0)$
Foci: $(\pm\sqrt{10}, 0)$

Asymptotes: $y = \pm\frac{\sqrt{7}}{\sqrt{3}}x$

16. $\frac{(x+1)^2}{144} - \frac{(y-4)^2}{25} = 1$

$a = 12$, $b = 5$, $c = 13$
Center: $(-1, 4)$
Vertices: $(-13, 4)$, $(11, 4)$
Foci: $(-14, 4)$, $(12, 4)$
Asymptotes: $y = 4 \pm \frac{5}{12}(x + 1)$

18. $\frac{(y-1)^2}{1/4} - \frac{(x+3)^2}{1/9} = 1$

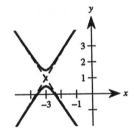

$a = \frac{1}{2}$, $b = \frac{1}{3}$, $c = \frac{\sqrt{13}}{6}$
Center: $(-3, 1)$
Vertices: $\left(-3, \frac{1}{2}\right)$, $\left(-3, \frac{3}{2}\right)$
Foci: $\left(-3, 1 \pm \frac{1}{6}\sqrt{13}\right)$
Asymptotes: $y = 1 \pm \frac{3}{2}(x + 3)$

20. $x^2 - 9y^2 + 36y - 72 = 0$

$$x^2 - 9(y^2 - 4y + 4) = 72 - 36$$

$$\frac{x^2}{36} - \frac{(y-2)^2}{4} = 1$$

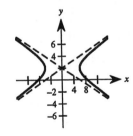

$a = 6$, $b = 2$, $c = 2\sqrt{10}$
Center: $(0, 2)$
Vertices: $(-6, 2)$, $(6, 2)$
Foci: $(\pm\sqrt{10}, 2)$
Asymptotes: $y = 2 \pm 13x$

22.
$$16y^2 - x^2 + 2x + 64y + 63 = 0$$
$$16(y^2 + 4y + 4) - (x^2 - 2x + 1) = -63 + 64 - 1$$
$$16(y + 2)^2 - (x - 1)^2 = 0$$
$$y + 2 = \pm \tfrac{1}{4}(x - 1)$$

Degenerate hyperbola is two intersecting lines

24. $3y^2 - 5x^2 = 15$

$$\frac{y^2}{5} - \frac{x^2}{3} = 1$$

$a = \sqrt{5}, \; b = \sqrt{3}, \; c = 2\sqrt{2}$
Center: $(0, 0)$
Vertices: $(0, \pm\sqrt{5})$
Foci: $(0, \pm 2\sqrt{2})$

Asmyptotes: $y = \pm \dfrac{sqrt5}{\sqrt{3}}x$

26.
$$9x^2 - y^2 + 54x + 10y + 55 = 0$$
$$9(x^2 + 6x + 9) - (y^2 - 10y + 25) = -55 + 81 - 25$$
$$\frac{(x + 3)^2}{1/9} - \frac{(y - 5)^2}{1} = 1$$

$a = \dfrac{1}{3}, \quad b = 1, \quad c = \dfrac{\sqrt{10}}{3}$
Center: $(-3, 5)$
Vertices: $\left(-3 \pm \dfrac{1}{3}, \; 5\right)$
Foci: $\left(-3 \pm \dfrac{\sqrt{10}}{3}, \; 5\right)$
Asymptotes: $y = 5 \pm 3(x + 3)$

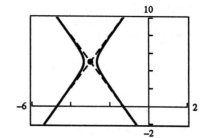

28.
$$3y^2 - x^2 + 6x - 12y = 0$$
$$3(y^2 - 4y + 4) - (x^2 - 6x + 9) = 12 - 9$$
$$3(y-2)^2 - (x-3)^2 = 3$$
$$\frac{(y-2)^2}{1^2} - \frac{(x-3)^2}{(\sqrt{3})^2} = 1$$

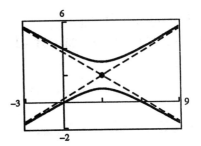

Center: $(3, 2)$

$a = 1,\ b = \sqrt{3},\ c^2 = b^2 + a^2 = 4,\ c = 2$

Foci: $(3, 2 \pm 2) = (3, 0)$ and $(3, 4)$

Vertices: $(3, 2 \pm \sqrt{3})$

Asymptotes: $y - 2 = \dfrac{\pm 1}{\sqrt{3}}(x - 3)$
$$y = 2 \pm \frac{1}{\sqrt{3}}(x - 3)$$

30. Vertices: $(0, \pm 3) \ \Rightarrow \ a = 3$

Asymptotes: $y = \pm 3x \ \Rightarrow \ \dfrac{a}{b} = 3,\ b = 1$

Center: $(0, 0) = (h, k)$
$$\frac{(y-k)^2}{a^2} - \frac{(x-h)^2}{b^2} = 1$$
$$\frac{y^2}{9} - x^2 = 1$$

32. Vertices: $(0, \pm 3) \ \Rightarrow \ a = 3$

Solution point: $(-2, 5)$

Center: $(0, 0) = (h, k)$
$$\frac{(y-k)^2}{a^2} - \frac{(x-h)^2}{b^2} = 1$$
$$\frac{y^2}{9} - \frac{x^2}{b^2} = 1 \ \Rightarrow \ b^2 = \frac{9x^2}{y^2 - 9} = \frac{9(-2)^2}{5^2 - 9} = \frac{36}{16} = \frac{9}{4}$$
$$\frac{y^2}{9} - \frac{x^2}{9/4} = 1$$

34. Vertices: $(4, 1),\ (4, 9) \ \Rightarrow \ a = 4$

Foci: $(4, 0),\ (4, 10) \ \Rightarrow \ c = 5$

$b^2 = c^2 - a^2 = 25 - 16 = 9$

Center: $(4, 5) = (h, k)$
$$\frac{(y-k)^2}{a^2} - \frac{(x-h)^2}{b^2} = 1$$
$$\frac{(y-5)^2}{16} - \frac{(x-4)^2}{9} = 1$$

36. Vertices: $(-2, 1)$, $(2, 1)$ $\Rightarrow$ $a = 2$

Solution point: $(4, 3)$

Center: $(0, 1) = (h, k)$

$$\frac{(x-h)^2}{a^2} - \frac{(y-k)^2}{b^2} = 1$$

$$\frac{x^2}{4} - \frac{(y-1)^2}{b^2} = 1 \quad \Rightarrow \quad b^2 = \frac{4(y-1)^2}{x^2-4} = \frac{4(2^2)}{16-4} = \frac{16}{12} = \frac{4}{3}$$

$$\frac{x^2}{4} - \frac{(y-1)^2}{4/3} = 1$$

38. Vertices: $(3, 0)$, $(3, 4)$ $\Rightarrow$ $a = 2$

Asymptotes: $y = \frac{2}{3}x$, $y = 4 - \frac{2}{3}x$

$\frac{a}{b} = \frac{2}{3}$ $\Rightarrow$ $b = 3$

Center: $(3, 2) = (h, k)$

$$\frac{(y-k)^2}{a^2} - \frac{(x-h)^2}{b^2} = 1$$

$$\frac{(y-2)^2}{4} - \frac{(x-3)^2}{9} = 1$$

40. Foci: $(\pm 150, 0) \Rightarrow c = 150$

Center: $(0, 0) = (h, k)$

$$\frac{d_2}{186{,}000} - \frac{d_1}{186{,}000} = 0.001 \Rightarrow 2a = 186, \quad a = 93$$

$$b^2 = c^2 - a^2 = 150^2 - 93^2 = 13{,}851$$

$$\frac{x^2}{93^2} - \frac{y^2}{13{,}851} = 1$$

$$x^2 = 93^2\left(1 + \frac{75^2}{13{,}851}\right) \approx 12{,}161$$

$$x \approx 110.3 \text{ miles}$$

42.

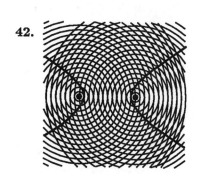

44. $x^2 + 4y^2 - 6x + 16y + 21 = 0$

$A = 1, \ C = 4$

$AC = 1(4) = 4 > 0 \Rightarrow$ Ellipse

46. $y^2 - 4y - 4x = 0$

   $A = 0,\ C = 1$

   $AC = 0(1) = 0 \Rightarrow$ Parabola

48. $4y^2 - 2x^2 - 4y - 8x - 15 = 0$

   $A = -2,\ C = 4$

   $AC = (-2)(4)$

   $\qquad = -8 < 0 \Rightarrow$ Hyperbola

50. $4x^2 + 4y^2 - 16y + 15 = 0$

   $A = 4,\ C = 4$

   $A = C \Rightarrow$ Circle

10.4 Rotation and General Second-Degree Equations

2. $xy - 4 = 0$

   $A = 0,\ B = 1,\ C = 0$

   $\cot 2\theta = \dfrac{A - C}{B} = 0 \Rightarrow 2\theta = \dfrac{\pi}{2} \Rightarrow \theta = \dfrac{\pi}{4}$

$$x = x' \cos \frac{\pi}{4} - y' \sin \frac{\pi}{4} \qquad\qquad y = x' \sin \frac{\pi}{4} + y' \cos \frac{\pi}{4}$$

$$\quad = x'\left(\frac{\sqrt{2}}{2}\right) - y'\left(\frac{\sqrt{2}}{2}\right) \qquad\qquad = x'\left(\frac{\sqrt{2}}{2}\right) + y'\left(\frac{\sqrt{2}}{2}\right)$$

$$\quad = \frac{x' - y'}{\sqrt{2}} \qquad\qquad\qquad\qquad\quad = \frac{x' + y'}{\sqrt{2}}$$

$$xy - 4 = 0$$

$$\left(\frac{x' - y'}{\sqrt{2}}\right)\left(\frac{x' + y'}{\sqrt{2}}\right) - 4 = 0$$

$$\frac{(x')^2 - (y')^2}{2} = 4$$

$$\frac{(x')^2}{8} - \frac{(y')^2}{8} = 1$$

4. $xy + x - 2y + 3 = 0$

$A = 0, \ B = 1, \ C = 0$

$\cot 2\theta = \dfrac{A - C}{B} = 0 \Rightarrow 2\theta = \dfrac{\pi}{2} \Rightarrow \theta = \dfrac{\pi}{4}$

$x = x' \cos \dfrac{\pi}{4} - y' \sin \dfrac{\pi}{4}$

$\qquad y = x' \sin \dfrac{\pi}{4} + y' \cos \dfrac{\pi}{4}$

$= x' \left(\dfrac{\sqrt{2}}{2}\right) - y' \left(\dfrac{\sqrt{2}}{2}\right) \qquad = x' \left(\dfrac{\sqrt{2}}{2}\right) + y' \left(\dfrac{\sqrt{2}}{2}\right)$

$= \dfrac{x' - y'}{\sqrt{2}} \qquad\qquad\qquad = \dfrac{x' + y'}{\sqrt{2}}$

$$xy + x - 2y + 3 = 0$$

$$\left(\frac{x' - y'}{\sqrt{2}}\right)\left(\frac{x' + y'}{\sqrt{2}}\right) + \left(\frac{x' - y'}{\sqrt{2}}\right) - 2\left(\frac{x' + y'}{\sqrt{2}}\right) + 3 = 0$$

$$\frac{(x')^2}{2} - \frac{(y')^2}{2} + \frac{x'}{\sqrt{2}} - \frac{y'}{\sqrt{2}} - \frac{2x'}{\sqrt{2}} - \frac{2y'}{\sqrt{2}} + 3 = 0$$

$$\left[(x')^2 - \sqrt{2}\,x' + \left(\frac{\sqrt{2}}{2}\right)^2\right] - \left[(y')^2 + 3\sqrt{2}\,y' + \left(\frac{3\sqrt{2}}{2}\right)^2\right] = -6 + \left(\frac{\sqrt{2}}{2}\right)^2 - \left(\frac{3\sqrt{2}}{2}\right)^2$$

$$\left(x' - \frac{\sqrt{2}}{2}\right)^2 - \left(y' + \frac{3\sqrt{2}}{2}\right)^2 = -10$$

$$\frac{\left(y' + \frac{3\sqrt{2}}{2}\right)^2}{10} - \frac{\left(x' - \frac{\sqrt{2}}{2}\right)^2}{10} = 1$$

6. $13x^2 + 6\sqrt{3}xy + 7y^2 - 16 = 0$

$A = 13, \ B = 6\sqrt{3}, \ C = 7$

$\cot 2\theta = \dfrac{A-C}{B} = \dfrac{1}{\sqrt{3}} \Rightarrow 2\theta = \dfrac{\pi}{3} \Rightarrow \theta = \dfrac{\pi}{6}$

$x = x' \cos \dfrac{\pi}{6} - y' \sin \dfrac{\pi}{6} \qquad\qquad y = x' \sin \dfrac{\pi}{6} + y' \cos \dfrac{\pi}{6}$

$\qquad = x'\left(\dfrac{\sqrt{3}}{2}\right) - y'\left(\dfrac{1}{2}\right) \qquad\qquad = x'\left(\dfrac{1}{2}\right) + y'\left(\dfrac{\sqrt{3}}{2}\right)$

$\qquad = \dfrac{\sqrt{3}x' - y'}{2} \qquad\qquad\qquad\qquad = \dfrac{x' + \sqrt{3}y'}{2}$

$13x^2 + 6\sqrt{3}xy + 7y^2 - 16 = 0$

$13\left(\dfrac{\sqrt{3}x' - y'}{2}\right)^2 + 6\sqrt{3}\left(\dfrac{\sqrt{3}x' - y'}{2}\right)\left(\dfrac{x' + \sqrt{3}y'}{2}\right) + 7\left(\dfrac{x' + \sqrt{3}y'}{2}\right)^2 - 16 = 0$

$\dfrac{39(x')^2}{4} - \dfrac{13\sqrt{3}x'y'}{2} + \dfrac{13(y')^2}{4} + \dfrac{18(x')^2}{4} + \dfrac{18\sqrt{3}x'y'}{4} - \dfrac{6\sqrt{3}x'y'}{4}$

$\qquad\qquad - \dfrac{18(y')^2}{4} + \dfrac{7(x')^2}{4} + \dfrac{7\sqrt{3}x'y'}{2} + \dfrac{21(y')^2}{4} - 16 = 0$

$16(x')^2 + 4(y')^2 = 16$

$\dfrac{(x')^2}{1} + \dfrac{(y')^2}{4} = 1$

8. $2x^2 - 3xy - 2y^2 + 10 = 0$

$A = 2, \ B = -3, \ C = -2$

$\cot 2\theta = \dfrac{A - C}{B} = -\dfrac{4}{3} \Rightarrow \theta \approx 71.57°$

$\cos 2\theta = -\dfrac{4}{5}$

$\sin \theta = \sqrt{\dfrac{1 - \cos 2\theta}{2}} = \sqrt{\dfrac{1 - (-4/5)}{2}} = \dfrac{3}{\sqrt{10}}$

$\cos \theta = \sqrt{\dfrac{1 + \cos 2\theta}{2}} = \sqrt{\dfrac{1 + (-4/5)}{2}} = \dfrac{1}{\sqrt{10}}$

$$
\begin{aligned}
x &= x' \cos \theta - y' \sin \theta & y &= x' \sin \theta + y' \cos \theta \\
&= x'\left(\dfrac{1}{\sqrt{10}}\right) - y'\left(\dfrac{3}{\sqrt{10}}\right) & &= x'\left(\dfrac{3}{\sqrt{10}}\right) + y'\left(\dfrac{1}{\sqrt{10}}\right) \\
&= \dfrac{x' - 3y'}{\sqrt{10}} & &= \dfrac{3x' + y'}{\sqrt{10}}
\end{aligned}
$$

$$2x^2 - 3xy - 2y^2 + 10 = 0$$

$$2\left(\dfrac{x' - 3y'}{\sqrt{10}}\right)^2 - 3\left(\dfrac{x' - 3y'}{\sqrt{10}}\right)\left(\dfrac{3x' + y'}{\sqrt{10}}\right) - 2\left(\dfrac{3x' + y'}{\sqrt{10}}\right)^2 + 10 = 0$$

$$\dfrac{(x')^2}{5} - \dfrac{6x'y'}{5} + \dfrac{9(y')^2}{5} - \dfrac{9(x')^2}{10} + \dfrac{24x'y'}{10} + \dfrac{9(y')^2}{10} - \dfrac{9(x')^2}{5} - \dfrac{6x'y'}{5} - \dfrac{(y')^2}{5} + 10 = 0$$

$$-\dfrac{5}{2}(x')^2 + \dfrac{5}{2}(y')^2 = -10$$

$$\dfrac{(x')^2}{4} - \dfrac{(y')^2}{4} = 1$$

10. $16x^2 - 24xy + 9y^2 - 60x - 80y + 100 = 0$

$A = 16, \ B = -24, \ C = 9$

$\cot 2\theta = \dfrac{A-C}{B} = -\dfrac{7}{24} \Rightarrow \theta \approx 53.13°$

$\cos 2\theta = -\dfrac{7}{25}$

$\sin \theta = \sqrt{\dfrac{1 - \cos 2\theta}{2}} = \sqrt{\dfrac{1 - (-7/25)}{2}} = \dfrac{4}{5}$

$\cos \theta = \sqrt{\dfrac{1 + \cos 2\theta}{2}} = \sqrt{\dfrac{1 - (-7/25)}{2}} = \dfrac{3}{5}$

$x = x' \cos \theta - y' \sin \theta$

$\qquad = x'\left(\dfrac{3}{5}\right) - y'\left(\dfrac{4}{5}\right) = \dfrac{3x' - 4y'}{5}$

$y = x' \sin \theta + y' \cos \theta$

$\qquad = x'\left(\dfrac{4}{3}\right) + y'\left(\dfrac{3}{5}\right) = \dfrac{4x' + 3y'}{5}$

$16x^2 - 24xy + 9y^2 - 60x - 80y + 100 = 0$

$16\left(\dfrac{3x'-4y'}{5}\right)^2 - 24\left(\dfrac{3x'-4y'}{5}\right)\left(\dfrac{4x'+3y'}{5}\right) + 9\left(\dfrac{4x'+3y'}{5}\right)^2 - 60\left(\dfrac{3x'-4y'}{5}\right)$

$\qquad\qquad\qquad - 80\left(\dfrac{4x'+3y'}{5}\right) + 100 = 0$

$\dfrac{144(x')^2}{25} - \dfrac{384x'y'}{25} + \dfrac{256(y')^2}{25} - \dfrac{288(x')^2}{25} + \dfrac{168x'y'}{25} + \dfrac{288(y')^2}{25} + \dfrac{144(x')^2}{25} + \dfrac{216x'y'}{25}$

$\qquad\qquad + \dfrac{81(y')^2}{25} - 36x' + 48y' - 64x' - 48y' + 100 = 0$

$25(y')^2 - 100x' + 100 = 0$

$(y')^2 = 4(x' - 1)$

12. $9x^2 + 24xy + 16y^2 + 80x - 60y = 0$

$A = 9, \ B = 24, \ C = 16$

$\cot 2\theta = \dfrac{A-C}{B} = -\dfrac{7}{24} \Rightarrow \theta \approx 53.13°$

$\cos 2\theta = -\dfrac{7}{25}$

$\sin \theta = \sqrt{\dfrac{1 - \cos 2\theta}{2}} = \sqrt{\dfrac{1 - (-7/25)}{2}} = \dfrac{4}{5}$

$\cos \theta = \sqrt{\dfrac{1 + \cos 2\theta}{2}} = \sqrt{\dfrac{1 + (-7/25)}{2}} = \dfrac{3}{5}$

14. $x^2 - 4xy + 2y^2 = 6$

$\cot 2\theta = \dfrac{A - C}{B} = \dfrac{1 - 2}{-4} = \dfrac{1}{4} \Rightarrow \tan 2\theta = 4$

$2\theta = 1.3258$

$\theta = 0.6629 \approx 37.98°$

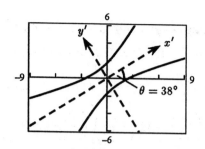

16. $40x^2 + 36xy + 25y^2 - 52 = 0$

$A = 40, \ B = 36, \ C = 25, \ D = 0, \ E = 0, \ F = -52$

$\cot 2\theta = \dfrac{5}{12}, \ \cos 2\theta = \dfrac{5}{13}, \ \theta \approx 33.69°, \ \sin\theta = \dfrac{\sqrt{1 - (5/13)}}{\sqrt{2}} = \dfrac{2}{\sqrt{13}}, \ \cos\theta = \dfrac{\sqrt{1 + (5/13)}}{\sqrt{2}} = \dfrac{3}{\sqrt{13}}$

$A' = 40\left(\dfrac{9}{13}\right) + 36\left(\dfrac{2}{\sqrt{13}}\right)\left(\dfrac{3}{\sqrt{13}}\right) + 25\left(\dfrac{4}{13}\right) = 52$

$C' = 40\left(\dfrac{4}{13}\right) - 36\left(\dfrac{2}{\sqrt{13}}\right)\left(\dfrac{3}{\sqrt{13}}\right) + 25\left(\dfrac{9}{13}\right) = 13$

$D' = E' = 0$

$E' = -52$

$52(x')^2 + 13(y')^2 - 52 = 0$ or $\dfrac{(x')^2}{1} + \dfrac{(y')^2}{4} = 1$

18. $4x^2 - 12xy + 9y^2 + (4\sqrt{13} - 12)x - (6\sqrt{13} + 8)y - 91 = 0$

$A = 4, \ B = -12, \ C = 9, \ D = 4\sqrt{13} - 12, \ E = -(6\sqrt{13} + 8), F = -91$

$\cot 2\theta = \dfrac{5}{12}, \ \cos 2\theta = \dfrac{5}{13}, \ \theta \approx 33.69°, \ \sin\theta = \dfrac{\sqrt{1 - (5/13)}}{\sqrt{2}} = \dfrac{2}{\sqrt{13}}, \ \cos\theta = \dfrac{\sqrt{1 + (5/13)}}{\sqrt{2}} = \dfrac{3}{\sqrt{13}}$

$A' = 4\left(\dfrac{9}{13}\right) - 12\left(\dfrac{2}{\sqrt{13}}\right)\left(\dfrac{3}{\sqrt{13}}\right) + 9\left(\dfrac{4}{13}\right) = 0$

$C' = 4\left(\dfrac{4}{13}\right) + 12\left(\dfrac{2}{\sqrt{13}}\right)\left(\dfrac{3}{\sqrt{13}}\right) + 9\left(\dfrac{9}{13}\right) = 13$

$D' = (4\sqrt{13} - 12)\left(\dfrac{3}{\sqrt{13}}\right) - (6\sqrt{13} + 8)\left(\dfrac{2}{\sqrt{13}}\right) = -4\sqrt{13}$

$E' = -(4\sqrt{13} - 12)\left(\dfrac{2}{\sqrt{13}}\right) - (6\sqrt{13} + 8)\left(\dfrac{3}{\sqrt{13}}\right) = -26$

$F' = -91 \quad 13(y')^2 - 4\sqrt{13}x' - 26y' - 91 = 0 \text{ or } (y' - 1)^2 = \dfrac{3}{\sqrt{13}}(x' + 2\sqrt{13})$

20. $B^2 - 4AC = (-4)^2 - 4(1)(-2) = 24$
Hyperbola

22. $B^2 - 4AC = (4)^2 - 4(2)(5) = -24$
Ellipse or circle

24. $B^2 - 4AC = (-6)^2 - 4(1)(-5) = 56$
Hyperbola

26. $B^2 - 4AC = (-60)^2 - 4(36)(25) = 0$
Parabola

28. $-x^2 - y^2 - 8x + 20y - 7 = 0$

$x^2 + 9y^2 + 8x + 4y + 7 = 0$

Adding the two equations yields

$8y^2 + 24y = 0$

$\quad y^2 + 3y = 0 \Rightarrow y = 0, -3$

When $y = 0$: $\quad x^2 + 8x + 7 = 0$

$\quad\quad\quad\quad (x + 7)(x + 1) = 0 \Rightarrow x = -1, -7$

When $y = -3$: $x^2 + 81 + 8x - 12 + 7 = 0$

$\quad\quad\quad\quad\quad\quad x^2 + 8x + 76 = 0 \;$ no solution

Solutions: $(-1, 0)$ and $(7, 0)$

30. $x^2 - 4y^2 - 20x - 64y - 172 = 0$

$16x^2 + 4y^2 - 320x + 64y + 1600 = 0$

Adding the two equations yields

$17x^2 - 340x + 1428 = 0$

$\quad x^2 - 20x + 84 = 0$

$\quad (x - 6)(x - 14) = 0 \Rightarrow x = 6, 14$

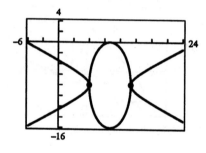

When $x = 6$: $16(36) + 4y^2 - 320(6) + 64y + 1600 = 0$

$4y^2 + 64y + 256 = 0$

$\quad y^2 + 16y + 64 = 0$

$\quad\quad (y + 8)^2 = 0 \Rightarrow y = -8$

When $x = 14$: $16(14^2) + 4y^2 - 320(14) + 64y + 1600 = 0$

$4y^2 + 64y + 256 = 0$

$\quad y^2 + 16y + 64 = 0$

$\quad\quad (y + 8)^2 = 0 \Rightarrow y = -8$

Solutions: $(6, -8)$, $(14, -8)$

32. $x^2 + 4y^2 = 2x - 8y + 1 = 0$

$$-x^2 + 2x - 4y - 1 = 0$$

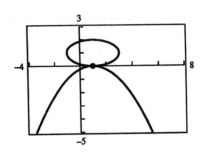

Adding the two equations yields

$4y^2 - 12y = 0$

$y^2 - 3y = 0 \Rightarrow y = 0, 3$

When $y = 0$: $x^2 - 2x + 1 = 0$

$(x - 1)^2 = 0 \Rightarrow x = 1$

When $y = 3$: $x^2 + 4(9) - 2x - 8(3) + 1 = 0$

$x^2 - 2x + 13 = 0$ no solution

Solutions: $(1, 0)$

34. $16x^2 - y^2 + 16y - 128 = 0$

$y^2 - 48x - 16y - 32 = 0$

Adding the two equations yields

$16x^2 - 48x - 160 = 0$

$x^2 - 3x - 10 = 0$

$(x - 5)(x + 2) = 0 \Rightarrow x = 5, -2$

When $x = 5$: $y^2 - 48(5) - 16y - 32 = 0$

$y^2 - 16y - 272 = 0$ no solution

When $x = -2$: $y^2 - 48(-2) - 16y - 32 = 0$

$y^2 - 16y + 64 = 0$

$(y - 8)^2 = 0 \Rightarrow y = 8$

Solution: $(-2, 8)$

36. $4x^2 + 9y^2 - 36y = 0$

$\qquad x^2 + 9y - 27 = 0$

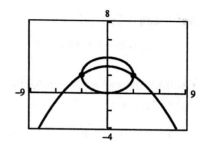

From the second equation, $x^2 = 27 - 9y$. Substituting into the first equation,

$\qquad 4(27 - 9y) + 9y^2 - 36y = 0$

$\qquad\qquad 9y^2 - 72y + 108 = 0$

$\qquad\qquad\quad y^2 - 8y + 12 = 0$

$\qquad\quad (y - 2)(y - 6) = 0 \Rightarrow y = 2, 6$

When $y = 2$: $x^2 = 27 - 9(2) = 9 \Rightarrow x = \pm 3$

When $y = 6$: $x^2 = 27 - 9(6) = -9$ no solution

Solutions: $(3, 2), (-3, 2)$

38. $x^2 + 2y^2 - 4x + 6y - 5 = 0$

$\qquad x^2 - 4x - y + 4 = 0$

From the first equation, $x^2 - 4x = 5 - 6y - 2y^2$. Substituting into the second equation,

$\qquad 5 - 6y - 2y^2 - y + 4 = 0$

$\qquad\qquad -2y^2 - 7y + 9 = 0$

$\qquad\qquad\quad 2y^2 + 7y - 9 = 0$

$(2y + 9)(y - 1) \Rightarrow y = 1, -\dfrac{9}{2}$

When $y = 1$, $x^2 - 4x = 5 - 6 - 2 = -3$

$\qquad\qquad\quad x^2 - 4x + 3 = 0$

$\qquad (x - 3)(x - 1) = 0 \Rightarrow x = 1, 3$

When $y = -\dfrac{9}{2}$, $\qquad x^2 - 4x = 5 - 6\left(-\dfrac{9}{2}\right) - 2\left(-\dfrac{9}{2}\right)^2$

$\qquad\qquad x^2 - 4x + 8.4 = 0$ no solution

Solutions: $(1, 1), (3, 1)$

40. $5x^2 - 2xy + 5y^2 - 12 = 0$

$$x + y - 1 = 0$$

From the second equation, $y = 1 - x$. Substituting into the first equation,

$5x^2 - 2x(1 - x) + 5(1 - x)^2 - 12 = 0$

$$12x^2 - 12x - 7 = 0 \Rightarrow x = 1.41287, -0.41287$$

When $x = 1.41287$, $y = 1 - 1.41287 = -0.41287$
When $x = 0.41287$, $y = 1.41287$.
Solutions: $(1.41287, -0.14287)$ and $(-0.14287, 1.41287)$

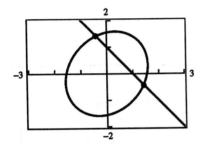

10.5 Plane Curves and Parametric Equations

2. $x = t$, $y = \dfrac{1}{2}t$

$$y = \dfrac{1}{2}x$$

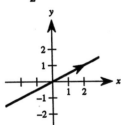

4. $x = 3 - 2t$, $y = 2 + 3t$

$$y = 2 + 3\left(\dfrac{3 - x}{2}\right)$$

$$3x + 2y = 13$$

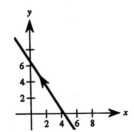

6. $x = t,\ y = t^3$

$y = x^3$

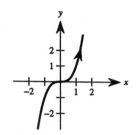

8. $x = \sqrt{t}$

$y = 1 - t$

$y = 1 - x^2,\ x \geq 0$

10. $x = t - 1,\ y = \dfrac{t}{t-1}$

$y = \dfrac{x+1}{x+1-1}$

$y = \dfrac{x+1}{x}$

12. $x = 4\sin 2\theta \Rightarrow \left(\dfrac{x}{4}\right)^2 = \sin^2 2\theta$

$y = 2\cos 2\theta \Rightarrow \left(\dfrac{y}{2}\right)^2 = \cos^2 2\theta$

$\left(\dfrac{x}{4}\right)^2 + \left(\dfrac{y}{2}\right)^2 = 1$

$\dfrac{x^2}{16} + \dfrac{y^2}{4} = 1$

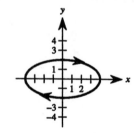

14. $x = \sec\theta$

$y = \cos\theta$

$xy = 1$

$y = \dfrac{1}{x}, \quad -1 \le y \le 1$

16. $x = 4\sec\theta \Rightarrow \left(\dfrac{x}{4}\right)^2 = \sec^2\theta$

$y = 3\tan\theta \Rightarrow \left(\dfrac{y}{3}\right)^2 = \tan^2\theta$

$1 + \left(\dfrac{y}{3}\right)^2 = \left(\dfrac{x}{4}\right)^2$

$\dfrac{x^2}{16} - \dfrac{y^2}{9} = 1$

18. $x = e^{2t}$

$y = e^{t} \Rightarrow y^2 = e^{2t}$

$y^2 = x, \quad y > 0$

20. $x = \ln t \Rightarrow e^x = t$

$y = t^2$

$y = (e^x)^2$

$y = e^{2x}$

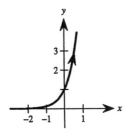

22. By eliminating the parameters in (a)–(d), we get $y = x^2 - 1$. They differ from each other in restricted domain and in orientation.

    (a) Domain: $-\infty < x < \infty$
        Orientation: Left to right

    (b) Domain: $0 \le x < \infty$
        Orientation: Left to right

    (c) Domain: $-1 \le x \le 1$
        Orientation: Oscillates

    (d) Domain: $0 < x < \infty$
        Orientation: Left to right

24. $x = h + r\cos\theta, \quad y = k + r\sin\theta$

$\cos\theta = \dfrac{x-h}{r}, \quad \sin\theta = \dfrac{y-k}{r}$

$\cos^2\theta + \sin^2\theta = \dfrac{(x-h)^2}{r^2} + \dfrac{(y-k)^2}{r^2} = 1$

$(x-h)^2 + (y-k)^2 = r^2$

26. $x = h + a\sec\theta, \quad y = k + b\tan\theta$

$\dfrac{x-h}{a} = \sec\theta, \quad \dfrac{y-k}{b} = \tan\theta$

$\dfrac{(x-h)^2}{a^2} - \dfrac{(y-k)^2}{b^2} = 1$

28. From Exercise 23:

$x = 1 + 4t$

$y = 4 - 6t$

Solution not unique

30. From Exercise 24:

$x = -3 + 3\cos\theta$

$y = 1 + 3\sin\theta$

Solution not unique

32. From Exercise 25:

$x = 4 + 5\cos\theta$

$y = 2 + 4\sin\theta$

$b = 4$

Center: $(4, 2)$
Solution not unique

34. From Exercise 26:

$x = 2\sqrt{6}\tan\theta$

$y = \sec\theta$

$b = 2\sqrt{6}$

Center: $(0, 0)$

Solution not unique. The transverse axis is vertical, therefore, x and y are interchanged.

36. $y = x^2$

<u>Examples</u>

$x = t, \quad y = t^2$

$x = t^2, \quad y = t^4$

$x = \sin t, \quad y = \sin^2 t$

38. $x = \theta + \sin\theta, \quad y = 1 - \cos\theta$

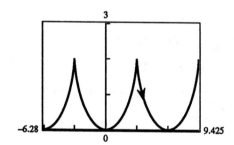

40. $x = 2\theta - \sin\theta$, $y = 2 - \cos\theta$

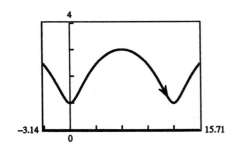

42. $x = \dfrac{3t}{1+t^3}$, $y = \dfrac{3t^2}{1+t^3}$,
undefined when $t = -1$.

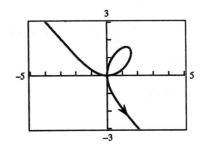

44. $x = \cos^3\theta$, $y = 2\sin^3\theta$

θ	0	$\pi/2$	π	$3\pi/2$	2π
x	1	0	-1	0	1
y	0	2	0	-2	0

Matches graph (c).

46. $x = \cot\theta$, $y = 4\sin\theta\cos\theta$, undefined when $\theta = 0$.

θ	$\pi/6$	$\pi/4$	$\pi/3$	$2\pi/3$	$3\pi/4$	$5\pi/6$
x	1.73	1	0.58	-0.58	-1	-1.73
y	1.73	2	1.73	-1.73	-2	-1.73

Matches graph (a).

48. $2\theta = \text{arc}TS = \text{arc}TP = \phi$

$$y = \overline{OA} - \overline{OC}$$

$$\overline{OA} = 3\sin\theta$$

$$\overline{OC} = \sin(180° - \theta - \phi)$$

$$= \sin(180° - 3\theta)$$

$$= \sin 180° \cos 3\theta - \sin 3\theta \cos 180°$$

$$= \sin 3\theta$$

Therefore, $y = 3\sin\theta - \sin 3\theta$.
Similarly, $x = 3\cos\theta - \cos 3\theta$.

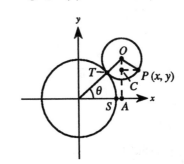

10.6 Polar Coordinates

2. Polar coordinates: $\left(4, \dfrac{3\pi}{2}\right)$

$x = 4\cos\left(\dfrac{3\pi}{2}\right) = 0, \quad y = 4\sin\left(\dfrac{3\pi}{2}\right) = -4$

Rectangular coordinates: $(0, -4)$

4. Polar coordinates: $(0, -\pi)$

$x = 0\cos(-\pi) = 0, \quad y = 0\sin(-\pi) = 0$

Rectangular coordinates: $(0, 0)$

6. Polar coordinates: $\left(-1, \dfrac{-3\pi}{4}\right)$

$x = -1\cos\left(\dfrac{-3\pi}{4}\right) = \dfrac{\sqrt{2}}{2},$

$y = -1\sin\left(\dfrac{-3\pi}{4}\right) = \dfrac{\sqrt{2}}{2}$

Rectangular coordinates: $\left(\dfrac{\sqrt{2}}{2}, \dfrac{\sqrt{2}}{2}\right)$

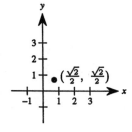

8. Polar coordinates: $\left(\dfrac{3}{2}, \dfrac{5\pi}{2}\right)$

$x = \dfrac{3}{2}\cos\left(\dfrac{5\pi}{2}\right) = 0, \quad y = \dfrac{3}{2}\sin\left(\dfrac{5\pi}{2}\right) = \dfrac{3}{2}$

Rectangular coordinates: $\left(0, \dfrac{3}{2}\right)$

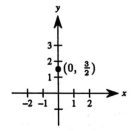

10. Polar coordinates: $(-3, -1.57)$

$x = -3\cos(-1.57) \approx -0.0024$

$y = -3\sin(-1.57) \approx 3.000$

Rectangular coordinates: $(-0.0024, 3)$

12. Rectangular coordinates: $(0, -5)$

$r = 5$, $\tan\theta$ undefined, $\theta = \dfrac{\pi}{2}$

Polar coordinates: $\left(5, \dfrac{3\pi}{2}\right)$, $\left(-5, \dfrac{\pi}{2}\right)$

14. Rectangular coordinates: $(-3, -3)$

$r = 3\sqrt{2}$, $\tan\theta = 1$, $\theta = \dfrac{\pi}{4}$

Polar coordinates:

$$\left(3\sqrt{2}, \dfrac{5\pi}{4}\right), \left(-3\sqrt{2}, \dfrac{\pi}{4}\right)$$

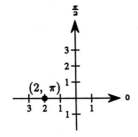

16. Rectangular coordinates: $(3, -1)$

$r = \sqrt{9+1} = \sqrt{10}$, $\tan\theta = -\dfrac{1}{3}$, $\theta \approx -0.322$

Polar coordinates: $(-\sqrt{10}, 2.820)$, $(\sqrt{10}, 5.961)$

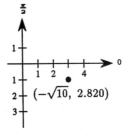

18. Rectangular coordinates: $(-2, 0)$

$r = 2$, $\tan\theta = 0$, $\theta = 0$

Polar coordinates: $(2, \pi)$, $(-2, 0)$

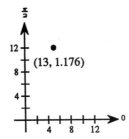

20. Rectangular coordinates: $(5, 12)$

$r = \sqrt{25+144} = 13$, $\tan\theta = \dfrac{12}{5}$, $\theta \approx 1.176$

Polar coordinates: $(13, 1.176)$, $(-13, 4.318)$

22. $x^2 + y^2 = a^2$

$\quad\quad r = a$

24. $x^2 + y^2 - 2ay = 0$

$\quad\quad r^2 - 2ar\sin\theta = 0$

$\quad\quad r(r - 2a\sin\theta) = 0$

$\quad\quad\quad r = 2a\sin\theta$

26. $\quad\quad y = b$

$\quad\quad r\sin\theta = b$

$\quad\quad\quad r = b\csc\theta$

28. $\quad\quad x = a$

$\quad r\cos\theta = a$

$\quad\quad r = a\sec\theta$

30. $\quad\quad 4x + 7y - 2 = 0$

$\quad 4r\cos\theta + 7r\sin\theta - 2 = 0$

$\quad\quad r(4\cos\theta + 7\sin\theta) = 2$

$$r = \frac{2}{4\cos\theta + 7\sin\theta}$$

32. $\quad\quad y = x$

$\quad r\cos\theta = r\sin\theta$

$\quad\quad 1 = \tan\theta$

$\quad\quad \theta = \dfrac{\pi}{4}$

34. $\quad\quad y^2 - 8x - 16 = 0$

$\quad\quad r^2\sin^2\theta - 8r\cos\theta = 16$

$\quad r^2 - r^2\cos^2\theta - 8r\cos\theta - 16 = 0$

$\quad\quad r^2\cos^2\theta + 8r\cos\theta + 16 = r^2$

$\quad\quad\quad (r\cos\theta + 4)^2 = r^2$

$\quad\quad\quad\quad r = \pm(r\cos\theta + 4)$

$$r = \frac{4}{1 - \cos\theta}$$

$$\text{or } r = \frac{-4}{1 + \cos\theta}$$

36. $\quad\quad r = 4\cos\theta$

$\quad\quad r^2 = 4r\cos\theta$

$\quad x^2 + y^2 = 4x$

$x^2 + y^2 - 4x = 0$

38. $\quad\quad r = 4$

$\quad\quad r^2 = 16$

$\quad x^2 + y^2 = 16$

40. $\quad\quad r^2 = \sin 2\theta$

$\quad\quad\quad = 2\sin\theta\cos\theta$

$\quad\quad r^2 = 2\left(\dfrac{y}{r}\right)\left(\dfrac{x}{r}\right)$

$\quad\quad\quad = \dfrac{2xy}{r^2}$

$\quad\quad r^4 = 2xy$

$\quad (x^2 + y^2)^2 = 2xy$

42. $\quad\quad r = \dfrac{1}{1 - \cos\theta}$

$\quad\quad r - r\cos\theta = 1$

$\quad \sqrt{x^2 + y^2} - x = 1$

$\quad\quad x^2 + y^2 = 1 + 2x + x^2$

$\quad\quad\quad y^2 = 2x + 1$

44. $\quad r = \dfrac{6}{2\cos\theta - 3\sin\theta}$

$\quad r = \dfrac{6}{2(x/r) - 3(y/r)}$

$\quad r = \dfrac{6r}{2x - 3y}$

$\quad 1 = \dfrac{6}{2x - 3y}$

$\quad\quad 2x - 3y = 6$

46.

$$r = 8$$

$$r^2 = 64$$

$$x^2 + y^2 = 64$$

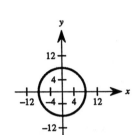

48.

$$\theta = \frac{5\pi}{6}$$

$$\tan \theta = \tan \frac{5\pi}{6}$$

$$\frac{y}{x} = -\frac{1}{\sqrt{3}}$$

$$\sqrt{3}y = -x$$

$$x + \sqrt{3}y = 0$$

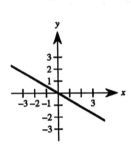

50.

$$r = 2 \csc \theta$$

$$r \sin \theta = 2$$

$$y = 2$$

$$y - 2 = 0$$

52. Let $(1, \pi/2)$ and $(-2, \pi)$ be two points in the polar coordinate system. From Exercise 51, the distance between them is

$$\sqrt{r_1{}^2 + r_2{}^2 - 2r_1r_2 \cos(\theta_1 - \theta_2)} = \sqrt{1 + 4 - 2(1)(-2) \cos[(\pi/2) - \pi]}$$

$$= \sqrt{5 + 4\cos(-\pi/2)}$$

$$= \sqrt{5} \, .$$

Now let $(-1, 3\pi/2)$ and $(2, 0)$ be different polar coordinate representations of the same two points. The distance formula from Exercise 51 gives

$$\sqrt{r_1{}^2 + r_2{}^2 - 2r_1r_2 \cos(\theta_1 - \theta_2)} = \sqrt{1 + 4 - (-1)(2) \cos[(3\pi/2) - 0]}$$

$$= \sqrt{5 + 2\cos(3\pi/2)}$$

$$= \sqrt{5} \, .$$

Even though the polar representations are different, the distance formula gives the same result.

54. $r = \cos \theta + 3 \sin \theta = 2(\frac{1}{2} \cos \theta + \frac{3}{2} \sin \theta)$

Using Exercise 53, we see that this is the equation of a circle of radius

$$\sqrt{h^2 + k^2} = \sqrt{\left(\frac{1}{2}\right)^2 + \left(\frac{3}{2}\right)^2} = \sqrt{\frac{1}{4} + \frac{9}{4}} = \frac{1}{2}\sqrt{10}$$

and center

$$(h, \ k) = \left(\frac{1}{2}, \frac{3}{2}\right) :$$

$$\left(x - \frac{1}{2}\right)^2 + \left(y - \frac{3}{2}\right)^2 = \frac{10}{4}.$$

10.7 Graphs of Polar Equations

2. $r = 16\cos 3\theta$

$\theta = \dfrac{\pi}{2}$: $\quad -r = 16\cos(3(-\theta))$

$\qquad\qquad -r = 16\cos(-3\theta)$

$\qquad\qquad -r = 16\cos 3\theta$

$\qquad\qquad$ Not an equivalent equation

Polar axis: $\quad r = 16\cos(3(-\theta))$

$\qquad\qquad r = 16\cos(-3\theta)$

$\qquad\qquad r = 16\cos 3\theta$

$\qquad\qquad$ Equivalent equation

Pole: $\quad -r = 16\cos 3\theta$

$\qquad\qquad$ Not an equivalent equation

Answer: Symmetric with respect to polar axis

4. $r = 6\sin\theta$

$\theta = \dfrac{\pi}{2}$: $\quad -r = 6\sin(-\theta)$

$\qquad\qquad -r = -6\sin\theta$

$\qquad\qquad r = 6\sin\theta$

$\qquad\qquad$ Equivalent equation

Polar axis: $\quad r = 6\sin(-\theta)$

$\qquad\qquad r = -6\sin\theta$

$\qquad\qquad$ Not an equivalent equation

Pole: $\quad -r = 6\sin\theta$

$\qquad\qquad$ Not an equivalent equation

Answer: Symmetric with respect to $\theta = \pi/2$

6. $r^2 = 25\sin 2\theta$

$\theta = \dfrac{\pi}{2}$: $\quad (-r)^2 = 25\sin(2(-\theta))$

$\qquad\qquad r^2 = -25\sin 2\theta$

$\qquad\qquad$ Not an equivalent equation

Polar axis: $\quad r^2 = 25\sin(2(-\theta))$

$\qquad\qquad r^2 = -25\sin 2\theta$

$\qquad\qquad$ Not an equivalent equation

Pole: $\quad (-r)^2 = 25\sin 2\theta$

$\qquad\qquad r^2 = 25\sin 2\theta$

$\qquad\qquad$ Equivalent equation

Answer: Symmetric with respect to pole

8. $|r| = |-2\cos\theta| = 2|\cos\theta| \le 2$

$\qquad |\cos\theta| = 1$

$\qquad \cos\theta = \pm 1$

$\qquad\quad \theta = 0,\ \pi$

Maximum: $|r| = 2$ when $\theta = 0,\ \pi$

10. $|r| = |6 + 12\cos\theta| \le |6| + |12\cos\theta|$

$\qquad = 6 + 12|\cos\theta| \le 18$

$\qquad\quad \cos\theta = 1$

$\qquad\qquad \theta = 0$

Maximum: $|r| = 18$ when $\theta = 0$

12. Circle: $r = 2$

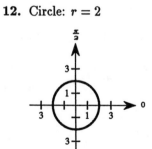

14. $\theta = -\dfrac{\pi}{4}$

$\tan\theta = \tan\left(-\dfrac{\pi}{4}\right)$

$\dfrac{y}{x} = -1$

$y = -x \Rightarrow$ Line

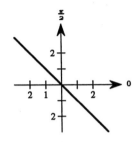

16. $r = 3 - 3\cos\theta$

Symmetric with respect to polar axis

$\dfrac{a}{b} = \dfrac{3}{3} = 1 \Rightarrow$ Cardioid

$|r| = 6$ when $\theta = \pi$.
$r = 0$ when $\theta = 0$.

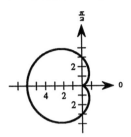

18. $r = 3 - 2\cos\theta$

Symmetric with respect to polar axis

$\dfrac{a}{b} = \dfrac{3}{2} > 1 \Rightarrow$ Dimpled limaçon

$|r| = 5$ when $\theta = \pi$.

20. $r = 2 + 4\sin\theta$

Symmetric with respect to $\theta = \dfrac{\pi}{2}$

$\dfrac{a}{b} = \dfrac{2}{4} < 1 \Rightarrow$ Limaçon with inner loop

$|r| = 6$ when $\theta = \dfrac{\pi}{2}$.

$r = 0$ when $\theta = \dfrac{7\pi}{6},\ \dfrac{11\pi}{6}$.

22. $r = 2\cos 3\theta$

Symmetric with respect to polar axis

Rose curve ($n = 3$) with 3 petals

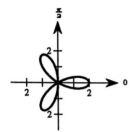

24. $r = 2\sec\theta$

$$r = \frac{2}{\cos\theta}$$

$r\cos\theta = 2$

$x = 2 \Rightarrow$ Line

26. $$r = \frac{6}{2\sin\theta - 3\cos\theta}$$

$r(2\sin\theta - 3\cos\theta) = 6$

$2y - 3x = 6$

$y = \dfrac{3}{2}x + 3 \Rightarrow$ Line

28. $r^2 = 4\sin\theta$

Symmetric with respect to the pole

Lemniscate

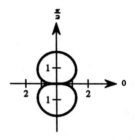

30. $r = \theta$

Symmetric with respect to $\theta = \dfrac{\pi}{2}$

Spiral

32. $r = \dfrac{\theta}{2}$

$-20 \le \theta \le 20$
$-10 \le x \le 10$
$-10 \le y \le 10$

34. $r = \cos 2\theta$

36. $r = 3\cos 2\theta \sec \theta = \dfrac{3\cos 2\theta}{\cos \theta}$

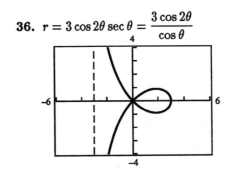

38. $r = 2\cos(3\theta - 2)$

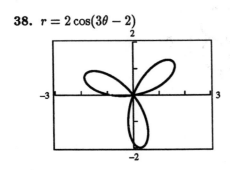

40. $r = 2 + \csc \theta = 2 + \dfrac{1}{\sin \theta}$

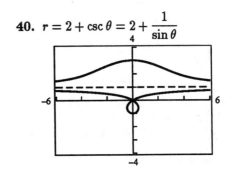

42. $r = 2\cos 2\theta \cdot \sec \theta = \dfrac{2\cos 2\theta}{\cos \theta}$

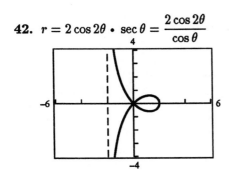

44. $r = 2(1 - 2\sin \theta)$

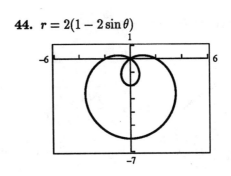

46. $r = 4 + 3\cos\theta$

48. $r = 35\sin\left(\dfrac{5\theta}{2}\right)$

50. $r^2 = 3\cos 4\theta$

$r = \pm\sqrt{3\cos 4\theta}$

52.

$$r = 2 + \csc\theta = 2 + \frac{1}{\sin\theta}$$

$$r\sin\theta = 2\sin\theta + 1$$

$$r(r\sin\theta) = 2r\sin\theta + r$$

$$(\pm\sqrt{x^2 + y^2})(y) = 2y + (\pm\sqrt{x^2 + y^2})$$

$$(\pm\sqrt{x^2 + y^2})(y - 1) = 2y$$

$$(\pm\sqrt{x^2 + y^2}) = \frac{2y}{y - 1}$$

$$x^2 + y^2 = \frac{4y^2}{(y - 1)^2}$$

$$x^2 = \frac{y^2(3 + 2y - y^2)}{(y - 1)^2}$$

$$x = \pm\sqrt{\frac{y^2(3 + 2y - y^2)}{(y - 1)^2}}$$

$$= \pm\left|\frac{y}{y - 1}\right|\sqrt{3 + 2y - y^2}$$

The graph has an asymptote at $y = 1$.

54. Use the result of Exercise 33.

(a) Rotation: $\phi = \dfrac{\pi}{2}$

Original graph: $r = f(\sin \theta)$

Rotated graph: $r = f\left(\sin\left(\theta - \dfrac{\pi}{2}\right)\right) = f(-\cos \theta)$

(b) Rotation: $\phi = \pi$

Original graph: $r = f(\sin \theta)$

Rotated graph: $r = f(\sin(\theta - \pi)) = f(-\sin \theta)$

(c) Rotation: $\phi = \dfrac{3\pi}{2}$

Original graph: $r = f(\sin \theta)$

Rotated graph: $r = f\left(\sin\left(\theta - \dfrac{3\pi}{2}\right)\right) = f(\cos \theta)$

56. (a) $r = 2\sin\left[2\left(\theta - \dfrac{\pi}{6}\right)\right]$

$\qquad = 2\sin\left(2\theta - \dfrac{\pi}{3}\right)$

$\qquad = \sin 2\theta - \sqrt{3}\cos 2\theta$

(b) $r = 2\sin\left[2\left(\theta - \dfrac{\pi}{2}\right)\right]$

$\qquad = 2\sin(2\theta - \pi)$

$\qquad = -2\sin 2\theta$

(c) $r = 2\sin\left[2\left(\theta - \dfrac{2\pi}{3}\right)\right]$

$\qquad = 2\sin\left(2\theta - \dfrac{4\pi}{3}\right)$

$\qquad = \sqrt{3}\cos 2\theta - \sin 2\theta$

(d) $r = 2\sin[2(\theta - \pi)]$

$\qquad = 2\sin(2\theta - 2\pi)$

$\qquad = 2\sin 2\theta$

58. (a) $r = 3\sec \theta$

(b) $r = 3\sec\left(\theta - \dfrac{\pi}{4}\right)$

(c) $r = 3\sec\left(\theta + \dfrac{\pi}{3}\right)$

(d) $r = 3\sec\left(\theta - \dfrac{\pi}{2}\right)$

10.8 | Polar Equations of Conics

2. $r = \dfrac{2}{2 - \cos\theta} = \dfrac{1}{1 - (1/2)\cos\theta} \Rightarrow e = \dfrac{1}{2}$

Ellipse with horizontal major axis and vertices
(2, 0) and (2/3, π)

Matches graph (f)

4. $r = \dfrac{2}{1 + \sin\theta} \Rightarrow e = 1$

Parabola with vertical axis and vertex (1, π/2)

Matches graph (e)

6. $r = \dfrac{2}{2 + 3\cos\theta} \Rightarrow e = \dfrac{3}{2}$

Hyperbola with horizontal transverse axis and vertices (2/5, 0) and (−2, π)

Matches graph (d)

8. $r = \dfrac{4}{1 + \sin\theta}$

$e = 1$ so the graph is a parabola.

Vertex: (2, π/2)

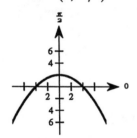

10. $r = \dfrac{6}{1 + \cos\theta}$

$e = 1$, the graph is a parabola.

Vertex: (3, 0)

12. $r = \dfrac{3}{3 + \sin\theta} = \dfrac{1}{1 + \frac{1}{3}\sin\theta}$

$e = \frac{1}{3} < 1$, the graph is an ellipse.

Vertices: $\left(\dfrac{3}{4}, \dfrac{\pi}{2}\right)$, $\left(\dfrac{3}{2}, \dfrac{3\pi}{2}\right)$

14. $r = \dfrac{6}{3 - 2\cos\theta} = \dfrac{2}{1 - \frac{2}{3}\cos\theta}$

$e = \frac{2}{3} < 1$, the graph is an ellipse.

Vertices: $(6, 0)$, $\left(\dfrac{6}{5}, \pi\right)$

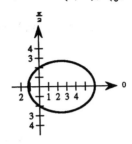

16. $r = \dfrac{5}{-1 + 2\cos\theta} = \dfrac{-5}{1 - 2\cos\theta}$

$e = 2 > 1$, the graph is a hyperbola.

Vertices: $(5, 0)$, $\left(-\dfrac{5}{3}, \pi\right)$

18. $r = \dfrac{3}{2 + 6\sin\theta} = \dfrac{\frac{3}{2}}{1 + 3\sin\theta}$

$e = 3 > 1$, the graph is a hyperbola.

Vertices: $\left(\dfrac{3}{8}, \dfrac{\pi}{2}\right)$, $\left(-\dfrac{3}{4}, \dfrac{3\pi}{2}\right)$

20. $e = 1$, $y = -2$, $p = 2$

Horizontal directrix below the pole

$$r = \dfrac{1(2)}{1 - 1\sin\theta} = \dfrac{2}{1 - \sin\theta}$$

22. $e = \frac{3}{4}$, $y = -2$, $p = 2$

Horizontal directrix below the pole

$$r = \dfrac{\frac{3}{4}(2)}{1 - \frac{3}{4}\sin\theta} = \dfrac{6}{4 - 3\sin\theta}$$

24. $e = \frac{3}{2}$, $x = -1$, $p = 1$

Vertical directrix to the left of the pole

$$r = \dfrac{\frac{3}{2}(1)}{1 - \frac{3}{2}\cos\theta} = \dfrac{3}{2 - 3\cos\theta}$$

26. Vertex: $(4, 0) \Rightarrow e = 1$, $p = 8$

Vertical directrix to the right of the pole

$$r = \dfrac{1(8)}{1 + 1\cos\theta} = \dfrac{8}{1 + \cos\theta}$$

28. Vertex: $\left(10, \frac{\pi}{2}\right) \Rightarrow e = 1$, $p = 20$

Horizontal directrix above the pole

$$r = \frac{1(20)}{1 + 1 \sin\theta} = \frac{20}{1 + \sin\theta}$$

30. Center: $\left(1, \frac{3\pi}{2}\right)$; $c = 1$, $a = 3$, $e = \frac{1}{3}$

Horizontal directrix above the axis

$$r = \frac{\frac{1}{3}p}{1 + \frac{1}{3}\sin\theta} = \frac{p}{3 + \sin\theta}$$

$$2 = \frac{p}{3 + \sin(\pi/2)}$$

$$p = 8$$

$$r = \frac{8}{3 + \sin\theta}$$

32. Center: $(6, 0)$; $c = 6$, $a = 4$, $e = \frac{3}{2}$

Vertical directrix to the right of the pole

$$r = \frac{\frac{3}{2}p}{1 + \frac{3}{2}\cos\theta} = \frac{3p}{2 + 3\cos\theta}$$

$$2 = \frac{3p}{2 + 3\cos 0}$$

$$p = \frac{10}{3}$$

$$r = \frac{3\left(\frac{10}{3}\right)}{2 + 3\cos\theta} = \frac{10}{2 + 3\cos\theta}$$

34. Center: $\left(\frac{5}{2}, \frac{\pi}{2}\right)$; $c = \frac{5}{2}$, $a = \frac{3}{2}$, $e = \frac{5}{3}$

Horizontal directrix above the pole

$$r = \frac{\frac{5}{3}p}{1 + \frac{5}{3}\sin\theta} = \frac{5p}{3 + 5\sin\theta}$$

$$1 = \frac{5p}{3 + 5\sin(-3\pi/2)}$$

$$p = \frac{8}{5}$$

$$r = \frac{5\left(\frac{8}{5}\right)}{3 + 5\sin\theta} = \frac{8}{3 + 5\sin\theta}$$

36.

$$\frac{x^2}{a^2} - \frac{y^2}{b^2} = 1$$

$$\frac{r^2\cos^2\theta}{a^2} - \frac{r^2\sin^2\theta}{b^2} = 1$$

$$\frac{r^2\cos^2\theta}{a^2} - \frac{r^2(1 - \cos^2\theta)}{b^2} = 1$$

$$r^2b^2\cos^2\theta - r^2a^2 + r^2a^2\cos^2\theta = a^2b^2$$

$$r^2(b^2 + a^2)\cos^2\theta - r^2a^2 = a^2b^2$$

$$a^2 + b^2 = c^2$$

$$r^2c^2\cos^2\theta - r^2a^2 = a^2b^2$$

$$r^2\left(\frac{c}{a}\right)^2\cos^2\theta - r^2 = b^2, \quad e = \frac{c}{a}$$

$$r^2e^2\cos^2\theta - r^2 = b^2$$

$$r^2(e^2\cos^2\theta - 1) = b^2$$

$$r^2 = \frac{b^2}{e^2\cos^2\theta - 1}$$

$$= \frac{-b^2}{1 - e^2\cos^2\theta}$$

38. $\dfrac{x^2}{25} + \dfrac{y^2}{16} = 1$

$$a = 5, \quad b = 4, \quad c = 3, \quad e = \frac{3}{5}$$

$$r^2 = \frac{400}{25 - 9\cos^2\theta}$$

40. $\dfrac{x^2}{36} - \dfrac{y^2}{4} = 1$

$a = 6, \quad b = 2, \quad c = 2\sqrt{10}, \quad e = \dfrac{\sqrt{10}}{3}$

$r^2 = \dfrac{-4}{1 - (10/9)\cos^2\theta} = \dfrac{-36}{9 - 10\cos^2\theta}$

42. Ellipse

One focus: $(4, 0)$

Vertices: $(5, 0), (5, \pi)$

$a = 5, \quad c = 4, \quad b = 3, \quad e = \dfrac{4}{5}$

$r^2 = \dfrac{9}{1 - (16/25)\cos^2\theta} = \dfrac{225}{25 - 16\cos^2\theta}$

44. $r = \dfrac{4}{1 + \sin[\theta - (\pi/3)]}$

46. Minimum distance occurs when $\theta = \pi$.

$r = \dfrac{(1 - e^2)a}{1 - e\cos\pi} = \dfrac{(1 - e)(1 + e)a}{1 + e} = a(1 - e)$

Maximum distance occurs when $\theta = 0$.

$r = \dfrac{(1 - e^2)a}{1 - e\cos 0} = \dfrac{(1 - e)(1 + e)a}{1 - e} = a(1 + e)$

48. $r = \dfrac{[1 - (0.2481)^2](3.666 \times 10^9)}{1 - 0.2481\cos\theta}$

$\approx \dfrac{3.4403 \times 10^9}{1 - 0.2481\cos\theta}$

Perihelion distance:

$r = 3.666 \times 10^9(1 - 0.2481) \approx 2.7565 \times 10^9$

Aphelion distance:

$r = 3.666 \times 10^9(1 + 0.2481) \approx 4.5755 \times 10^9$

50. Assume the earth's radius is 4000 miles.

$e = \dfrac{c}{a} = \dfrac{126,000 - 4119}{126,000 + 4119} \approx 0.937$

$r = \dfrac{ep}{1 - e\cos\theta}$

$(4119, \pi)\colon 4119 = \dfrac{0.937}{1 - 0.937(-1)} \Rightarrow p \approx 8516$

$r \approx \dfrac{7977.2}{1 - 0.937\cos\theta}$

When $\theta = 60°\colon r = \dfrac{7977.2}{1 - 0.937(1/2)} \approx 15,009.$

Distance: $15,000 - 4000 = 11,008$ miles

Review Exercises for Chapter 10

2. $x^2 = -8y$
Parabola
Vertex: $(0,\ 0)$
Vertical axis

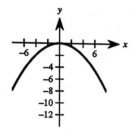

4. $y^2 - 12y - 8x + 20 = 0$

$$(y-6)^2 = 8(x+2)$$

Parabola
Vertex: $(-2,\ 6)$
Focus: $(0,\ 6)$

6. $16x^2 + 16y^2 - 16x + 24y - 3 = 0$

$$16\left(x - \tfrac{1}{2}\right)^2 + 16\left(y + \tfrac{3}{4}\right)^2 = 16$$

$$\left(x - \tfrac{1}{2}\right)^2 + \left(y + \tfrac{3}{4}\right)^2 = 1$$

Circle
Center: $\left(\tfrac{1}{2},\ -\tfrac{3}{4}\right)$
Radius: 1

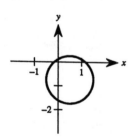

8. $2x^2 + 6y^2 = 18$

$$\frac{x^2}{9} + \frac{y^2}{3} = 0$$

Ellipse
Center: $(0,\ 0)$
Vertices: $(-3,\ 0),\ (3,\ 0)$

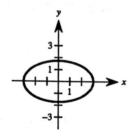

10. $4x^2 + y^2 - 16x + 15 = 0$

$$4(x-2)^2 + y^2 = 1$$

Ellipse
Center: $(2, 0)$
Vertices: $(2, -1)$, $(2, 1)$

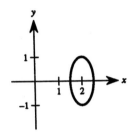

12. $x^2 - 9y^2 + 10x + 18y + 7 = 0$

$$(x+5)^2 - 9(y-1)^2 = 9$$

$$\frac{(x+5)^2}{9} - (y-1)^2 = 1$$

Hyperbola
Horizontal transverse axis
Center: $(-5, 1)$
Vertices: $(-8, 1)$, $(-2, 1)$
Asymptotes: $y = \pm\frac{1}{3}(x+5) + 1$

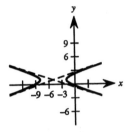

14. $40x^2 + 36xy + 25y^2 - 52 = 0$

$25y^2 + 36xy + (40x^2 - 52) = 0$

$$y = \frac{-36x \pm \sqrt{(36x)^2 - 4(25)(40x^2 - 52)}}{50}$$

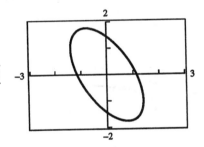

16. Horizontal axis, $p = -2$

$(y-k)^2 = 4p(x-h)$

$(y-0)^2 = 4(-2)(x-2)$

$y^2 = -8(x-2)$

18. Vertical axis, $p = 2$

$(x-h)^2 = 4p(y-k)$

$(x-2)^2 = 4(2)(y-2)$

$(x-2)^2 = 8(y-2)$

20. Vertical major axis
Center: $(2, 2)$
$a = 2$, $c = 1$,
$b = \sqrt{4-1} = \sqrt{3}$

$$\frac{(x-h)^2}{b^2} + \frac{(y-k)^2}{a^2} = 1$$

$$\frac{(x-2)^2}{4} + \frac{(y-2)^2}{3} = 1$$

22. Horizontal major axis
Center: $(2, 1)$
$a = 2$, $b = 1$

$$\frac{(x-h)^2}{a^2} + \frac{(y-k)^2}{b^2} = 1$$

$$\frac{(x-2)^2}{4} + \frac{(y-1)^2}{1} = 1$$

24. Horizontal transverse axis
Center: $(0, 2)$
$a = 2$, $c = 4$,
$b = \sqrt{16-4} = \sqrt{12}$

$$\frac{(x-h)^2}{a^2} - \frac{(y-k)^2}{b^2} = 1$$

$$\frac{x^2}{4} - \frac{(y-2)^2}{12} = 1$$

26. Vertical transverse axis

Center: $(3, 0) \Rightarrow C = 2$

$\dfrac{a}{b} = 2 \Rightarrow a = 2b$

$a^2 + b^2 = c^2$

$(2b)^2 + b^2 = 4$

$b^2 = \dfrac{4}{5}$

$a^2 = \dfrac{16}{5}$

$\dfrac{(y-k)^2}{a^2} - \dfrac{(x-h)^2}{b^2} = 1$

$\dfrac{5y^2}{16} - \dfrac{5(x-3)^2}{4} = 1$

28. $a = 5, \quad b = 4, \quad c = \sqrt{a^2 - b^2} = 3$

The foci should be placed 3 feet on either side of center and 10 feet high.

30. $x = t^2, \quad y = \sqrt{t}$

$t = y^2$

$x = (y^2)^2 = y^4, \quad y \geq 0$

32. $x = t + 4, \quad y = t^2$

$t = x - 4$

$y = (x - 4)^2$

34. $x = \dfrac{1}{t}, \quad y = 2t + 3$

$t = \dfrac{1}{x}$

$y = \dfrac{2}{x} + 3$

$y = \dfrac{2 + 3x}{x}$

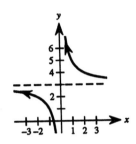

36. $x = 3 + 3\cos\theta, \quad y = 2 + 5\sin\theta$

$\cos\theta = \dfrac{x - 3}{3}, \quad \sin\theta = \dfrac{y - 2}{5}$

$\dfrac{(x - 3)^2}{9} + \dfrac{(y - 2)^2}{25} = 1$

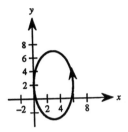

38. $x = \sec\theta, \quad y = \tan\theta$

$y^2 + 1 = x^2$

40. $x = 2\theta - \sin\theta, \quad y = 2 - \cos\theta$

$\cos\theta = 2 - y$

$x = 2\arccos(2 - y) - \sqrt{1 - (2 - y)^2}$

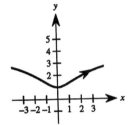

42. $\theta = \dfrac{\pi}{12}$

Line

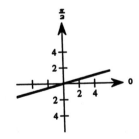

44. $r = 2\theta$

Symmetric with respect to $\theta = \pi/2$

Spiral

46. $r = 3 - 4\cos\theta$

Symmetric with respect to polar axis

$\dfrac{a}{b} = \dfrac{3}{4} < 0 \Rightarrow$ Limaçon with inner loop

48. $r = \cos 5\theta$

Symmetric with respect to polar axis

Rose curve ($n = 5$) with 5 petals

50. $r^2 = \cos 2\theta$

Symmetric with respect to polar axis

Lemniscate

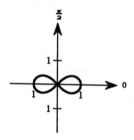

52. $r = 3\csc\theta$

$r\sin\theta = 3$

$y = 3$

Line

54. $r = \dfrac{4}{5 - 3\cos\theta}$

$r = \dfrac{4/5}{1 - (3/5)\cos\theta}, \quad e = \dfrac{3}{5}$

Ellipse symmetric with polar axis and having vertices at $(2, 0)$ and $(1/2, \pi)$

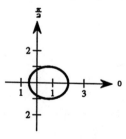

56. $r = \dfrac{1}{1 + 2\sin\theta}, \quad e = 2$

Hyperbola symmetric with $\theta = \pi/2$ and having vertices at $(1/3, \pi/2)$ and $(-1, 3\pi/2)$

58.
$$r = 4\sec\left(\theta - \frac{\pi}{3}\right)$$
$$r\cos\left(\theta - \frac{\pi}{3}\right) = 4$$
$$r\left(\cos\theta\cos\frac{\pi}{3} + \sin\theta\sin\frac{\pi}{3}\right) = 4$$
$$\frac{1}{2}x + \frac{\sqrt{3}}{2}y = 4$$
$$x + \sqrt{3}\,y = 8$$

60.
$$r = \frac{1}{2 - \cos\theta}$$
$$2r - r\cos\theta = 1$$
$$2r = 1 + r\cos\theta$$
$$4r^2 = (1 + r\cos\theta)^2$$
$$4(x^2 + y^2) = (1 + x)^2$$
$$4x^2 + 4y^2 = 1 + 2x + x^2$$
$$3x^2 + 4y^2 - 2x - 1 = 0$$

62.
$$r = 10$$
$$r^2 = 100$$
$$x^2 + y^2 = 100$$

64. $x^2 + y^2 - 4x = 0$
$$r^2 - 4r\cos\theta = 0$$
$$r = 4\cos\theta$$

66.
$$m = \sqrt{3}$$
$$\tan\theta = \sqrt{3}$$
$$\theta = \frac{\pi}{3}$$

68. Parabola: $r = \dfrac{ep}{1 + e\sin\theta}, \quad e = 1$

Vertex: $\left(2, \dfrac{\pi}{2}\right)$

Focus: $(0,\ 0) \Rightarrow p = 4$
$$r = \frac{4}{1 + \sin\theta}$$

70. Hyperbola: $r = \dfrac{ep}{1 + e\cos\theta}$

Vertices: $(1,\ 0),\ (7,\ 0) \Rightarrow a = 3$

One focus: $(0,\ 0) \Rightarrow c = 4$

$e = \dfrac{c}{a} = \dfrac{4}{3}, \quad p = \dfrac{7}{4}$

$r = \dfrac{(4/3)(7/4)}{1 + (4/3)\cos\theta} = \dfrac{7/3}{1 + (4/3)\cos\theta} = \dfrac{7}{3 + 4\cos\theta}$

72. $\dfrac{y^2}{16} - \dfrac{x^2}{9} = 1$

$x = 3\tan\theta$

$y = 4\sec\theta$

This solution is not unique.

74. $y = \overline{QB} - \overline{QA}$

$\overline{QP} = \text{arc}QC = r\theta$

$\overline{QA} = r\theta \sin(90° - \theta)$

$\quad\quad = r\theta \cos\theta$

$\overline{QB} = r\sin\theta$

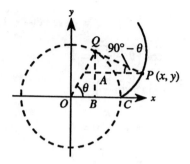

Therefore,

$y = r\sin\theta - r\theta\cos\theta$

$\quad = r(\sin\theta - \theta\cos\theta).$

Similarly, $x = \overline{OB} + \overline{AP}.$
Therefore,

$x = r\cos\theta + r\theta\sin\theta$

$\quad = r(\cos\theta + \theta\sin\theta).$

CHAPTER ELEVEN
Vectors in the Plane and in Space

2. $3\mathbf{v}$

4. $\mathbf{u} + 2\mathbf{v}$

6. $\mathbf{v} - \frac{1}{2}\mathbf{u}$

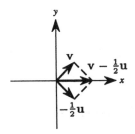

8. Initial point: $(0, 0)$
Terminal point: $(-4, -2)$
$\mathbf{v} = \langle -4 - 0, \ -2 - 0 \rangle = \langle -4, -2 \rangle$
$\|\mathbf{v}\| = \sqrt{(-4)^2 + (-2)^2} = 2\sqrt{5}$

10. Initial point: $(-1, -2)$
Terminal point: $(3, 4)$
$\mathbf{v} = \langle 3 - (-1), \ 4 - (-2) \rangle = \langle 4, 6 \rangle$
$\|\mathbf{v}\| = \sqrt{4^2 + 6^2} = 2\sqrt{13}$

12. Initial point: $(-4, -2)$
Terminal point: $(3, -2)$
$\mathbf{v} = \langle 3 - (-4), \ -2 - (-2) \rangle = \langle 7, 0 \rangle$
$\|\mathbf{v}\| = \sqrt{7^2 + 0^2} = 7$

14. Initial point: $(1, 11)$
Terminal point: $(9, 3)$
$\mathbf{v} = \langle 9 - 1, \ 3 - 11 \rangle = \langle 8, -8 \rangle$
$\|\mathbf{v}\| = \sqrt{8^2 + (-8)^2} = 8\sqrt{2}$

16. Initial point: $(-3, 11)$
Terminal point: $(9, 40)$
$\mathbf{v} = \langle 9 - (-3), \ 40 - 11 \rangle = \langle 12, 29 \rangle$
$\|\mathbf{v}\| = \sqrt{12^2 + 29^2} = \sqrt{985}$

18. $\mathbf{u} = \langle 2, 3 \rangle$, $\mathbf{v} = \langle 4, 0 \rangle$
(a) $\mathbf{u} + \mathbf{v} = \langle 6, 3 \rangle$
(b) $\mathbf{u} - \mathbf{v} = \langle -2, 3 \rangle$
(c) $2\mathbf{u} - 3\mathbf{v} = \langle -8, 6 \rangle$

20. $\mathbf{u} = \langle 0, 1 \rangle$, $\mathbf{v} = \langle 0, -1 \rangle$
(a) $\mathbf{u} + \mathbf{v} = \langle 0, 0 \rangle$
(b) $\mathbf{u} - \mathbf{v} = \langle 0, 2 \rangle$
(c) $2\mathbf{u} - 3\mathbf{v} = \langle 0, 5 \rangle$

22. $\mathbf{u} = \langle 0, 0 \rangle$, $\mathbf{v} = \langle 2, 1 \rangle$
(a) $\mathbf{u} + \mathbf{v} = \langle 2, 1 \rangle$
(b) $\mathbf{u} - \mathbf{v} = \langle -2, -1 \rangle$
(c) $2\mathbf{u} - 3\mathbf{v} = \langle -6, -3 \rangle$

24. $\mathbf{u} = 2\mathbf{i} - \mathbf{j}$, $\mathbf{v} = -\mathbf{i} + \mathbf{j}$

   (a) $\mathbf{u} + \mathbf{v} = \mathbf{i}$

   (b) $\mathbf{u} - \mathbf{v} = 3\mathbf{i} - 2\mathbf{j}$

   (c) $2\mathbf{u} - 3\mathbf{v} = 7\mathbf{i} - 5\mathbf{j}$

26. $\mathbf{u} = 3\mathbf{j}$, $\mathbf{v} = 2\mathbf{i}$

   (a) $\mathbf{u} + \mathbf{v} = 2\mathbf{i} + 3\mathbf{j}$

   (b) $\mathbf{u} - \mathbf{v} = -2\mathbf{i} + 3\mathbf{j}$

   (c) $2\mathbf{u} - 3\mathbf{v} = -6\mathbf{i} + 6\mathbf{j}$

28. $\|\mathbf{v}\| = 8$, $\theta = 135°$

30. $\|\mathbf{v}\| = \sqrt{(-2)^2 + (5)^2} = \sqrt{29}$

   $\tan\theta = \dfrac{5}{-2}$

   Since $\mathbf{v}$ lies in Quadrant II, $\theta = 111.8°$.

32. $\mathbf{v} = \langle \cos 45°,\ \sin 45° \rangle$

   $= \left\langle \dfrac{\sqrt{2}}{2},\ \dfrac{\sqrt{2}}{2} \right\rangle$

34. $\mathbf{v} = \left\langle \dfrac{5}{2}\cos 45°,\ \dfrac{5}{2}\sin 45° \right\rangle$

   $= \left\langle \dfrac{5\sqrt{2}}{4},\ \dfrac{5\sqrt{2}}{4} \right\rangle$

36. $\mathbf{v} = \langle 8\cos 90°,\ 8\sin 90° \rangle$

   $= \langle 0,\ 8 \rangle$

38. $\mathbf{v} = 3\left(\dfrac{1}{\sqrt{3^2 + 4^2}}\right)(3\mathbf{i} + 4\mathbf{j})$

   $= \dfrac{3}{5}(3\mathbf{i} + 4\mathbf{j})$

   $= \dfrac{9}{5}\mathbf{i} + \dfrac{12}{5}\mathbf{j}$

40. $\mathbf{v} = \mathbf{u} + \mathbf{w}$

$= (2\mathbf{i} - \mathbf{j}) + (\mathbf{i} + 2\mathbf{j})$

$= 3\mathbf{i} + \mathbf{j}$

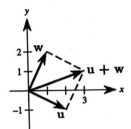

42. $\mathbf{v} = -\mathbf{u} + \mathbf{w}$

$= -(2\mathbf{i} - \mathbf{j}) + (\mathbf{i} + 2\mathbf{j})$

$= -\mathbf{i} + 3\mathbf{j}$

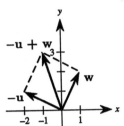

44. $\mathbf{v} = \mathbf{u} - 2\mathbf{w}$

$= (2\mathbf{i} - \mathbf{j}) - 2(\mathbf{i} + 2\mathbf{j})$

$= 2\mathbf{i} - \mathbf{j} - 2\mathbf{i} - 4\mathbf{j}$

$= -5\mathbf{j}$

46. $\mathbf{u} = \langle 2\cos 30°, \ 2\sin 30° \rangle = \langle \sqrt{3}, \ 1 \rangle$

$\mathbf{v} = \langle 2\cos 90°, \ 2\sin 90° \rangle = \langle 0, \ 2 \rangle$

$\mathbf{u} + \mathbf{v} = \langle \sqrt{3}, \ 3 \rangle$

48. $\mathbf{u} = \langle 35\cos 25°, \ 35\sin 25° \rangle \approx \langle 31.72, \ 14.79 \rangle$

$\mathbf{v} = \langle 50\cos 120°, \ 50\sin 120° \rangle = \langle -25, \ 25\sqrt{3} \rangle$

$\mathbf{u} + \mathbf{v} \approx \langle 6.72, \ 58.09 \rangle$

50. $\mathbf{u} = \dfrac{1}{\|\mathbf{v}\|}$

$\mathbf{v} = \dfrac{1}{\sqrt{2}}(\mathbf{i} + \mathbf{j})v = \dfrac{\sqrt{2}}{2}\mathbf{i} + \dfrac{\sqrt{2}}{2}\mathbf{j}$

52. $\mathbf{u} = \dfrac{1}{\|\mathbf{v}\|}$

$\mathbf{v} = \dfrac{1}{\sqrt{5}}(\mathbf{i} - 2\mathbf{j}) = \dfrac{\sqrt{5}}{5}\mathbf{i} - \dfrac{2\sqrt{5}}{5}\mathbf{j}$

54. $\mathbf{v} = 3\mathbf{i} + \mathbf{j}$

$\mathbf{w} = 2\mathbf{i} - \mathbf{j}$

$\mathbf{u} = \mathbf{v} - \mathbf{w} = \mathbf{i} + 2\mathbf{j}$

$\cos\theta = \dfrac{\|\mathbf{v}\|^2 + \|\mathbf{w}\|^2 - \|\mathbf{v} - \mathbf{w}\|^2}{2\|\mathbf{v}\|\,\|\mathbf{w}\|} = \dfrac{10 + 5 - 5}{2\sqrt{10}\sqrt{5}} = \dfrac{\sqrt{2}}{2}$

$\theta = 45°$

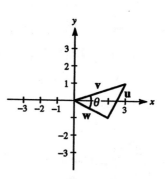

56. $\mathbf{v} = \mathbf{i} + 2\mathbf{j}$

$\mathbf{w} = 2\mathbf{i} - \mathbf{j}$

$\mathbf{u} = \mathbf{v} - \mathbf{w} = -\mathbf{i} + 3\mathbf{j}$

$\cos\theta = \dfrac{\|\mathbf{v}\|^2 + \|\mathbf{w}\|^2 - \|\mathbf{v} - \mathbf{w}\|^2}{2\|\mathbf{v}\|\,\|\mathbf{w}\|} = \dfrac{5 + 5 - 10}{2\sqrt{5}\sqrt{5}} = 0$

$\theta = 90°$

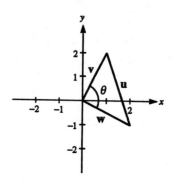

58. Force One: $\mathbf{u} = 3000\mathbf{i}$

Force Two: $\mathbf{v} = 1000\cos\theta\,\mathbf{i} + 1000\sin\theta\,\mathbf{j}$

Resultant Force: $\mathbf{u} + \mathbf{v} = (3000 + 1000\cos\theta)\mathbf{i} + 1000\sin\theta\,\mathbf{j}$

$\|\mathbf{u} + \mathbf{v}\| = \sqrt{(3000 + 1000\cos\theta)^2 + (1000\sin\theta)^2} = 3750$

$9{,}000{,}000 + 6{,}000{,}000\cos\theta + 1{,}000{,}000 = 14{,}062{,}500$

$6{,}000{,}000\cos\theta = 4{,}062{,}500$

$\cos\theta = \dfrac{4{,}062{,}500}{6{,}000{,}000} \approx 0.6771$

$\theta \approx 47.4°$

60. $\mathbf{u} = (500\cos 30°)\mathbf{i} + (500\sin 30°)\mathbf{j} \approx 433.01\mathbf{i} + 250.00\mathbf{j}$

$\mathbf{v} = (200\cos(-45°))\mathbf{i} + (200\sin(-45°))\mathbf{j} \approx 141.42\mathbf{i} - 144.42\mathbf{j}$

$\mathbf{u} + \mathbf{v} \approx 574.43\mathbf{i} + 108.58\mathbf{j}$

$\|\mathbf{u} + \mathbf{v}\| \approx \sqrt{574.43^2 + 108.58^2} \approx 584.6$ pounds

$\tan\theta = \dfrac{108.58}{574.43} \Rightarrow \theta \approx 10.7°$

62.
$$\mathbf{u} = (70\cos 30°)\mathbf{i} - (70\sin 30°)\mathbf{j} \approx 60.62\mathbf{i} - 35\mathbf{j}$$
$$\mathbf{v} = (40\cos 45°)\mathbf{i} + (40\sin 45°)\mathbf{j} \approx 28.28\mathbf{i} + 28.28\mathbf{j}$$
$$\mathbf{w} = (60\cos 135°)\mathbf{i} + (60\sin 135°)\mathbf{j} \approx -42.43\mathbf{i} + 42.43\mathbf{j}$$
$$\mathbf{u} + \mathbf{v} + \mathbf{w} \approx 46.47\mathbf{i} + 35.71\mathbf{j}$$
$$\|\mathbf{u} + \mathbf{v} + \mathbf{w}\| \approx 58.61$$
$$\tan\theta \approx \frac{35.71}{46.47} \approx 0.7684$$
$$\theta \approx 37.5°$$

64. Horizontal component of velocity: $1200\cos 6° \approx 1193.4$ ft/sec

Vertical component of velocity: $1200\sin 6° \approx 125.4$ ft/sec

66. Rope $\overrightarrow{AC}$: $\mathbf{u} = 10\mathbf{i} - 24\mathbf{j}$

The vector lies in Quadrant IV and its reference angle is $\arctan\left(\frac{12}{5}\right)$.
$$\mathbf{u} = \|\mathbf{u}\|\left[\cos\left(\arctan\tfrac{12}{5}\right)\mathbf{i} - \sin\left(\arctan\tfrac{12}{5}\right)\mathbf{j}\right]$$

Rope $\overrightarrow{BC}$: $\mathbf{v} = -20\mathbf{i} - 24\mathbf{j}$

The vector lies in Quadrant III and its reference angle is $\arctan\left(\frac{6}{5}\right)$.
$$\mathbf{v} = \|\mathbf{v}\|\left[-\cos\left(\arctan\tfrac{6}{5}\right)\mathbf{i} - \sin\left(\arctan\tfrac{6}{5}\right)\mathbf{j}\right]$$

Resultant: $\mathbf{u} + \mathbf{v} = -4000\mathbf{j}$
$$\|\mathbf{u}\|\cos\left(\arctan\tfrac{12}{5}\right) - \|\mathbf{v}\|\cos\left(\arctan\tfrac{6}{5}\right) = 0$$
$$-\|\mathbf{u}\|\sin\left(\arctan\tfrac{12}{5}\right) - \|\mathbf{v}\|\sin\left(\arctan\tfrac{6}{5}\right) = 0$$

Solving this system of equations yields:
$$T_{AC} = \|\mathbf{u}\| \approx 2888.9 \text{ pounds}$$
$$T_{BC} = \|\mathbf{v}\| \approx 1735.6 \text{ pounds}$$

68. Rope 1: $\mathbf{u} = \|\mathbf{u}\|(\cos 60°\mathbf{i} - \sin 60°\mathbf{j})$

Rope 2: $\mathbf{v} = \|\mathbf{u}\|(-\cos 60°\mathbf{i} - \sin 60°\mathbf{j})$

Resultant: $\mathbf{u} + \mathbf{v} = -100\mathbf{j}$
$$-\|\mathbf{u}\|\sin 60° - \|\mathbf{u}\|\sin 60° = -100$$
$$\|\mathbf{u}\| \approx 57.7$$

Therefore, the tension on each rope is $\|\mathbf{u}\| \approx 57.7$ pounds.

70. Plane: $\mathbf{u} = (580\cos 148°)\mathbf{i} + (580\sin 148°)\mathbf{j} \approx -491.9\mathbf{i} + 307.4\mathbf{j}$

Wind: $\mathbf{v} = (60\cos 45°)\mathbf{i} + (60\sin 45°)\mathbf{j} \approx 42.43\mathbf{i} + 42.43\mathbf{j}$

$$\mathbf{u} + \mathbf{v} \approx -449.47\mathbf{i} + 349.83\mathbf{j}$$

$$\|\mathbf{u} + \mathbf{v}\| \approx 569.5$$

$$\tan\theta \approx -\frac{349.83}{449.47} \approx -0.7783$$

$$\theta \approx 142.1°$$

The ground speed is 569.5 mph and the direction is N 52.1° W.

72. Horizontal force: $\mathbf{u} = \|\mathbf{u}\|\mathbf{i}$

Weight: $\mathbf{w} = -\mathbf{j}$

Rope: $t = \|t\|(\cos 120°\mathbf{i} + \sin 120°\mathbf{j})$

$$\mathbf{u} + \mathbf{w} + t = \mathbf{O} \Rightarrow \|\mathbf{u}\| + \|t\|\cos 120° = 0$$

$$-1 + \|t\|\sin 120° = 0$$

$$\|t\| = \frac{2}{\sqrt{3}} \approx 1.1547 \text{ lb}$$

$$\|\mathbf{u}\| = \left(-\frac{2}{\sqrt{3}}\right)\left(-\frac{1}{2}\right) = \frac{1}{\sqrt{3}} \approx 0.5774 \text{ lb}$$

11.2   The Dot Product of Two Vectors

2. $\mathbf{u} \cdot \mathbf{v} = \langle 4, 10 \rangle \cdot \langle -3, 1 \rangle = 4(10) + (-3)(1) = 40 - 3 = 37$

4. $\mathbf{u} \cdot \mathbf{v} = \mathbf{j} \cdot \mathbf{j} = 1$

6. $\mathbf{u} \cdot \mathbf{v} = (2\mathbf{i} + 5\mathbf{j}) \cdot (9\mathbf{i} - 3\mathbf{j}) = 18 - 15 = 3$

8. $\|\mathbf{u}\|^2 = \mathbf{u} \cdot \mathbf{u} = \langle 2, 2 \rangle \cdot \langle 2, 2 \rangle = 4 + 4 = 8$

10. $\mathbf{u} \cdot (2\mathbf{v}) = \langle 2, 2 \rangle \cdot \langle -6, 8 \rangle = -12 + 16 = 4$ scalar

12. $\mathbf{u} \cdot \mathbf{v} = \|\mathbf{u}\|\,\|\mathbf{v}\|\cos\theta = (100)(250)\cos\frac{pi}{6} = 25{,}000\left(\frac{\sqrt{3}}{2}\right)$

$$= 12{,}500\sqrt{3}$$

14. $\langle \mathbf{u} \cdot \mathbf{v} \rangle = \langle 1245, 2600 \rangle \cdot \langle 12.81, 8.93 \rangle$

$\qquad = 15948.45 + 23218$

$\qquad = 39166.45$

This represents the total revenue if all the units are sold at the new price.

16. $\cos \theta = \dfrac{\mathbf{u} \cdot \mathbf{v}}{\parallel \mathbf{u} \parallel \parallel \mathbf{v} \parallel} = \dfrac{\langle 4.4 \rangle \cdot \langle 2.0 \rangle}{\parallel \langle 4, 4 \rangle \parallel \parallel \langle 2, 0 \rangle \parallel} = \dfrac{8}{4\sqrt{2} \cdot 2} = \dfrac{1}{\sqrt{2}} \Rightarrow angle = 45°$

18. $\cos \theta = \dfrac{\mathbf{u} \cdot \mathbf{v}}{\parallel \mathbf{u} \parallel \parallel \mathbf{v} \parallel} = \dfrac{\langle -2, 4 \rangle \cdot \langle 1, 4 \rangle}{\parallel \langle -2, 5 \rangle \parallel \parallel \langle 1, 4 \rangle \parallel} = \dfrac{18}{\sqrt{29}\sqrt{17}} \Rightarrow \theta \approx 35.84°$

20. $\cos \theta = \dfrac{\mathbf{u} \cdot \mathbf{v}}{\parallel \mathbf{u} \parallel \parallel \mathbf{v} \parallel} = \dfrac{(\mathbf{i} - 6\mathbf{j}) \cdot (2\mathbf{i} - 12\mathbf{j})}{\sqrt{37}\sqrt{148}} = \dfrac{74}{\sqrt{37}\sqrt{148}} = \dfrac{74}{\sqrt{37} \cdot \sqrt[2]{37}} = 1 \Rightarrow \theta = 0°$

(vectors are parallel)

22. $\cos \theta = \dfrac{\mathbf{u} \cdot \mathbf{v}}{\parallel \mathbf{u} \parallel \parallel \mathbf{v} \parallel} = \dfrac{\left(\frac{\sqrt{2}}{2}\mathbf{i} + \frac{\sqrt{2}}{2}\mathbf{j}\right) \cdot (0\mathbf{i} + \mathbf{j})}{(1)(1)} = \dfrac{\sqrt{2}}{2} \Rightarrow \theta = 45°$

Or, notice that the vector $\mathbf{v}$ is obtained from $\mathbf{u}$ by a rotation of $\dfrac{\pi}{2} - \dfrac{\pi}{4} = \dfrac{\pi}{4} = 45°$

24. $\theta \approx 53.13°$

26. $\theta \approx 93.18$

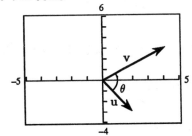

28. $\cos \theta = \dfrac{\mathbf{u} \cdot \mathbf{v}}{\parallel \mathbf{u} \parallel \parallel \mathbf{v} \parallel} = \dfrac{24 + 294}{\sqrt{212}\sqrt{477}} = 1 \Rightarrow parallel$

30. $\cos \theta = \dfrac{\mathbf{u} \cdot \mathbf{v}}{\parallel \mathbf{u} \parallel \parallel \mathbf{v} \parallel} = \dfrac{-75 + 540}{\sqrt{2250}\sqrt{169}} \approx \dfrac{465}{616.64}$ neither

32. $\cos \theta = \dfrac{\mathbf{u} \cdot \mathbf{v}}{\parallel \mathbf{u} \parallel \parallel \mathbf{v} \parallel} = \dfrac{\cos 20° \sin 70° + \sin 20° \cos 70°}{(1)(1)}$

$\qquad = \sin(70° + 20°) = \sin 90° = 1$ parallel

34. $\operatorname{proj}_{\mathbf{v}}\mathbf{u} = \left(\dfrac{\mathbf{u} \cdot \mathbf{v}}{\parallel \mathbf{v} \parallel^2}\right)\mathbf{v} = \left(\dfrac{36}{104}\right)\langle 10, 2 \rangle = \left\langle \dfrac{45}{13}, \dfrac{9}{13} \right\rangle$

$\qquad \mathbf{u} - \operatorname{proj}_{\mathbf{v}}\mathbf{u} = \langle 4, -2 \rangle - \left\langle \dfrac{45}{13}, \dfrac{9}{13} \right\rangle = \left\langle \dfrac{7}{13}, \dfrac{-35}{13} \right\rangle$

36. $\text{proj}_{\mathbf{u}} = \left(\dfrac{\mathbf{u} \cdot \mathbf{v}}{\| \mathbf{v} \|^2}\right) \mathbf{v} = \dfrac{45}{229}\langle 2, 15\rangle = \left\langle \dfrac{90}{229}, \dfrac{675}{229}\right\rangle$

$\mathbf{u} - \text{proj}_{\mathbf{v}}\mathbf{u} = \langle 0, 3\rangle - \left\langle \dfrac{90}{229}, \dfrac{675}{229}\right\rangle = \left\langle -\dfrac{90}{229}, \dfrac{12}{229}\right\rangle$

38. $\text{proj}_{\mathbf{v}}\mathbf{u} = \left(\dfrac{\mathbf{u} \cdot \mathbf{v}}{\| \mathbf{v} \|^2}\right) \mathbf{v} = \dfrac{4}{2}\langle -1, 1\rangle = \langle -2, 2\rangle$

$\mathbf{u} - \text{proj}_{\mathbf{v}}\mathbf{u} = \langle -5, -1\rangle - \langle -2, 2\rangle = \langle -3, -3\rangle$

40. $W = \overrightarrow{F} \cdot \overrightarrow{PQ} = \langle 45\cos 30°, 45\sin 30°\rangle \cdot \langle 20, 0\rangle$

$= 779.42 \ ft \cdot lbs$

42. $o \cdot \mathbf{v} = \langle 0, 0\rangle \cdot \langle \mathbf{v_1 v_2}\rangle = \langle o\mathbf{v_1}, o\mathbf{v_2}\rangle = \langle 0, 0\rangle = \mathbf{o}$

44. $c(\mathbf{u} \cdot \mathbf{v}) = c(\langle u_1, u_2\rangle \cdot \langle v_1, v_2\rangle)$

$= c(u_1 v_1 + u_2 v_2)$

$= (cu_1)v_1 + (cu_2)v_2$

$= \langle cu_1, cu_2\rangle \cdot \langle v_1, v_2\rangle = c\mathbf{u} \cdot \mathbf{v}$

$= u_1(cv_1) + u_2(cv_2)$

$= \langle u_1, u_2\rangle \cdot \langle cv_1, cv_2\rangle = \mathbf{u} \cdot c\mathbf{v}$

11.3 The Three-Dimensional Coordinate System

2. (3, −1, 4)

4.

6. $AB = \sqrt{(5-3)^2 + (-1-1)^2 + (1-2)^2} = \sqrt{4+4+1} = 3$

$BC = \sqrt{(1-5)^2 + (3+1)^2 + (1-1)^2} = \sqrt{16+16} = 4\sqrt{2}$

$AC = \sqrt{(1-3)^2 + (3-1)^2 + (1-2)^2} = \sqrt{4+4+1} = 3$

isosceles

8. $AB = \sqrt{(0-6)^2 + (3-0)^2 + (0-0)^2} = \sqrt{36+9} = \sqrt{45}$

$BC = \sqrt{(0-0)^2 + (0-3)^2 + (-2-0)^2} = \sqrt{9+4} = \sqrt{13}$

$AC = \sqrt{(0-6)^2 + (0-0)^2 + (-2-0)^2} = \sqrt{36+4} = \sqrt{40}$

neither

10. $\text{midpoint} = \left(\dfrac{2+5}{2}, \dfrac{-2+6}{2}, \dfrac{-8+18}{2} \right) = \left(\dfrac{7}{2}, 2, 5 \right)$

12. $\text{midpoint} = \left(\dfrac{-3-6}{2}, \dfrac{5+4}{2}, \dfrac{7+10}{2} \right) = \left(\dfrac{9}{2}, \dfrac{9}{2}, \dfrac{17}{2} \right)$

14. $(x-1)^2 + (y+2)^2 + (z-3)^2 = 5^2 = 25$

16. $\text{radius} = \text{diameter} /2 = 3 : (x-0)^2 + (y-5)^2 + (z+9)^2 = 3^2 = 9$

18. $\text{center} = \left(\dfrac{2-1}{2}, \dfrac{-2+4}{2}, \dfrac{2+6}{2} \right) = \left(\dfrac{1}{2}, 1, 3 \right)$

$\text{radius} = \sqrt{\left(2 - \dfrac{1}{2} \right)^2 + (-2-1)^2 + (2-3)^2} = \sqrt{\dfrac{9}{4} + 9 + 1} = \dfrac{7}{2}$

$\text{sphere} = \left(x - \dfrac{1}{2} \right)^2 + (y-1)^2 + (z-3)^2 = \dfrac{49}{4}$

20. $(x^2 - 6x + 9) + (y^2 + 4y + 4) + z^2 = -9 + 9 + 4$

$(x-3)^2 + (y+2)^2 + z^2 = 4 \text{ center: } (3, -2, 0)$

$$\text{radius: } 2$$

22. $x^2 + (y^2 - 8y + 16) + (z^2 - 6z + 9) = -13 + 16 + 9$

$x^2 + (y-4)^2 + (z-3)^2 = 12 \text{ center: } (0, 4, 3)$

$$\text{radius: } \sqrt{12} = 2\sqrt{3}$$

24. $x^2 + y^2 + z^2 - x - 3y - 2z = -\dfrac{5}{2}$

$\left(x^2 - x + \dfrac{1}{4} \right) + \left(y^2 - 3y + \dfrac{9}{4} \right) + (z^2 - 2z + 1) = -\dfrac{5}{2} + \dfrac{1}{4} + \dfrac{9}{4} + 1$

$\left(x - \dfrac{1}{2} \right)^2 + \left(y - \dfrac{3}{2} \right)^2 + (z-1)^2 = 1 \text{ center: } \left(\dfrac{1}{2}, \dfrac{3}{2}, 1 \right)$

$$\text{radius: } 1$$

26. (a)

(b)

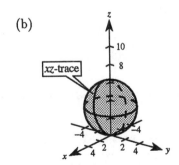

28. $z^2 = -x^2 - y^2 + 4y + 4 \Rightarrow$
$$\begin{cases} z_1 = \sqrt{-x^2 - y^2 + 4y + 4} \\ z_2 = -\sqrt{-x^2 - y^2 + 4y + 4} \end{cases}$$

30. $x^2 + y^2 + 6y + (z^2 - 8z + 16) = -21 + 16$

$x^2 + y^2 + 6y + (z - 4)^2 = -5$

$z_1 = 4 + \sqrt{-5 - x^2 - y^2 - 6y}$

$z_2 = 4 - \sqrt{-5 - x^2 - y^2 - 6y}$

11.4 Vectors in Space

2. $\mathbf{v} = \langle 2 - (-1), 1 - 4, 3 - 2 \rangle = \langle 3, -3, 1 \rangle$

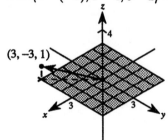

4. $\mathbf{v} = \langle 1 - 1, 4 - 4, 0 - 4 \rangle = \langle 0, 0, -4 \rangle$

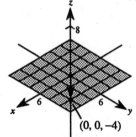

6. $\mathbf{v} = \langle 4 - (-4), 0 - 5, 0 - 5 \rangle = \langle 8, -5, -5 \rangle$

any vector of form $c\langle 8, -5, -5 \rangle, c > 0$, is parallel to $\mathbf{v}$.

any vector of form $c\langle 8, -5, -5 \rangle, c < 0$, is opposite direction.

8. $\mathbf{z} = 3\mathbf{u} - \mathbf{v} + \mathbf{w} = 3\langle -1, 3, 2 \rangle, -\langle 1, -2, -2 \rangle + \langle 5, 0, -5 \rangle = \langle 1, 11, 3 \rangle$

10. $\mathbf{z} = -7\mathbf{u} + \mathbf{v} - \dfrac{1}{5}\mathbf{w} = 7\langle -1, 3, 2 \rangle + \langle 1, -2, -2 \rangle - \dfrac{1}{5}\langle 1, -2, -2 \rangle$
$$= \langle 7\tfrac{4}{5}, -22\tfrac{3}{5}, -15\tfrac{3}{5} \rangle$$

12. $\mathbf{u} + \mathbf{v} - 2\mathbf{w} + \mathbf{z} = \mathbf{o} \Rightarrow \mathbf{z} = 2\mathbf{w} - \mathbf{u} - \mathbf{v}$

$$= 2\langle 5, 0, -5 \rangle - \langle -1, 3, 2 \rangle - \langle 1, -2, -2 \rangle$$

$$= \langle 10, -1, -10 \rangle$$

14. $\| \mathbf{v} \| = \sqrt{4^2 + (-3)^2 + (-7)^2} = \sqrt{74}$

16. $\dfrac{\mathbf{u}}{\| \mathbf{u} \|} = \dfrac{\langle -3, 5, 10 \rangle}{\sqrt{134}} = \dfrac{1}{\sqrt{134}}(-3\mathbf{i} + 5\mathbf{j} + 10\mathbf{k})$

18. $\mathbf{u} \cdot \mathbf{v} = 4(2) + 4(-5) + (-1)(-8) = -4$

20. $\mathbf{u} \cdot \mathbf{v} = 2(9) + 5(-3) + (-3)(1) = 0$

22. $\cos\theta = \dfrac{\mathbf{u} \cdot \mathbf{v}}{\| \mathbf{u} \| \, \| \mathbf{v} \|} = \dfrac{6}{\sqrt{21}\sqrt{21}} = 2 \Rightarrow \theta \approx 73.40°$

24. $\cos\theta = \dfrac{\mathbf{u} \cdot \mathbf{v}}{\| \mathbf{u} \| \, \| \mathbf{v} \|} = \dfrac{20}{\sqrt{41}\sqrt{29}} \approx 54.55°$

26. Since $\mathbf{u} \cdot \mathbf{v} = 0$, the vectors are orthogonal.

28. $\cos\theta = \dfrac{\mathbf{u} \cdot \mathbf{v}}{\| \mathbf{u} \| \, \| \mathbf{v} \|} = \dfrac{-56}{\sqrt{686}\sqrt{6}} =$ Neither

30. $\mathbf{w}_1 = proj_{\mathbf{v}}\mathbf{u} = \left(\dfrac{\mathbf{u} \cdot \mathbf{v}}{\| \mathbf{v} \|^2} \right)\mathbf{v} = \dfrac{6}{3}\langle -1, 1, 1 \rangle = \langle -2, 2, 2 \rangle$

$\mathbf{w}_z = \mathbf{u} - \mathbf{w}_1 = \langle -5 - (-2), -1 - 2, 2 - 2 \rangle = \langle -3, -3, 0 \rangle$

32. $\mathbf{v} = \langle -1 - 0, 5 - 4, 6 - 4 \rangle = \langle -1, 1, 2 \rangle$

$\mathbf{u} = \langle -2 - (-1), 6 - 5, 7 - 6 \rangle = \langle -1, 1, 1 \rangle$

Since $\mathbf{u}$ and $\mathbf{v}$ are not parallel, the points are not collinear.

34. $\mathbf{v} = \langle -4 - (-2), 8 - 7, 1 - 4 \rangle = \langle -2, 1, -3 \rangle$

$\mathbf{u} = \langle 0 - (-4), 6 - 8, 7 - 1 \rangle = \langle 4, -2, 6 \rangle$

Since $\mathbf{u} = -2\mathbf{v}$, the points are collinear.

36. $\mathbf{v} = \left\langle 4, \dfrac{3}{2}, -\dfrac{1}{4} \right\rangle = \left\langle q_1 - (-2), q_2 - 1, q_3 - \left(-\dfrac{3}{2} \right) \right\rangle \Rightarrow$

$\left. \begin{array}{l} 4 = q_1 + 2 \\ \dfrac{3}{2} = q_2 - 1 \\ -\dfrac{1}{4} = q_3 + \dfrac{3}{2} \end{array} \right\} \Rightarrow \left. \begin{array}{l} q_1 = 2 \\ q_2 = \dfrac{5}{2} \\ q_3 = -\dfrac{7}{4} \end{array} \right\} \Rightarrow$ Terminal point is $\langle 2, \dfrac{5}{2}, -\dfrac{7}{4} \rangle$

38. $\mathbf{v} = \langle q_1, q_2, q_3 \rangle$. Since $\mathbf{v}$ lies in the xz-plane, $q_z = 0$. Since $\mathbf{v}$ makes an angle of $60°$, $q_1 = c \cdot \sqrt{3}/2$, and $q_3 = c\frac{1}{2}$, for some constant c. Finally $\| \mathbf{v} \| = 10$ implies that $\sqrt{c^2 \frac{3}{4} + c^2 \frac{2}{4}} = 10 \Rightarrow c = 10$. Thus, $\mathbf{v} = \langle 10\sqrt{3}/2, 0, 5 \rangle$.

40. $\overrightarrow{PQ_1} = \langle 0, -24, -12\sqrt{21} \rangle$

$\overrightarrow{PQ_2} = \langle 12\sqrt{3}, 12, -12\sqrt{21} \rangle$

$\overrightarrow{PQ_3} = \langle -12\sqrt{3}, 12, -12\sqrt{21} \rangle (0, -24, 0)$

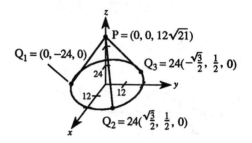

Let $\mathbf{F}_1, \mathbf{F}_2$ and $\mathbf{F}_3$ be the tension on each wire. Since $\| \mathbf{F}_1 \| = \| \mathbf{F}_2 \| = \| \mathbf{F}_3 \|$, there exists a constant c such that

$\mathbf{F}_1 = C\langle 0, -24, -12\sqrt{21} \rangle$

$\mathbf{F}_2 = C\langle 12\sqrt{3}, 12, -12\sqrt{21} \rangle$

$\mathbf{F}_3 = C\langle -12\sqrt{3}, 12, -12\sqrt{21} \rangle$

The total force is $-30\mathbf{k} = \mathbf{F}_1 + \mathbf{F}_2 + \mathbf{F}_3 \Rightarrow$ the vertical (**k**) component satisfies $-10 = -12\sqrt{21}c \Rightarrow c = \dfrac{5}{6\sqrt{21}}$

Hence,

$\mathbf{F}_1 = \left\langle 0, \dfrac{-20}{\sqrt{21}}, -10 \right\rangle$

$\mathbf{F}_2 = \left\langle \dfrac{10}{\sqrt{7}}, \dfrac{10}{\sqrt{21}}, -10 \right\rangle$

$\mathbf{F}_3 = \left\langle \dfrac{-10}{\sqrt{7}}, \dfrac{10}{\sqrt{21}}, -10 \right\rangle$

42. $W = \mathbf{F} \cdot \overrightarrow{PQ} = \langle -6, 2, 6 \rangle \cdot \langle -4 - 2, 4 - 4, 10 - 0 \rangle$

$= 36 + 60 = 96$ work units

11.5 The Cross Product of Two Vectors

2. $\mathbf{j} \times \mathbf{k} = \begin{vmatrix} \mathbf{i} & \mathbf{j} & \mathbf{k} \\ 0 & 1 & 0 \\ 0 & 0 & 1 \end{vmatrix} = \mathbf{i}$

4. $\mathbf{k} \times \mathbf{i} = \begin{vmatrix} \mathbf{i} & \mathbf{j} & \mathbf{k} \\ 0 & 0 & 1 \\ 1 & 0 & 0 \end{vmatrix} = \mathbf{j}$

6. $\mathbf{u} \times \mathbf{v} = \begin{vmatrix} \mathbf{i} & \mathbf{j} & \mathbf{k} \\ -3 & 2 & 3 \\ 0 & 1 & 0 \end{vmatrix} = -3\mathbf{i} - 3\mathbf{k} = \langle -3, 0, -3 \rangle$

8. $\mathbf{u} \times \mathbf{v} = \begin{vmatrix} \mathbf{i} & \mathbf{j} & \mathbf{k} \\ -5 & 5 & 11 \\ 2 & 2 & 3 \end{vmatrix} = \langle -7, 37, -20 \rangle$

10. $\mathbf{u} \times \mathbf{v} = \begin{vmatrix} \mathbf{i} & \mathbf{j} & \mathbf{k} \\ 4 & -2 & 6 \\ -1 & 5 & 7 \end{vmatrix} = \langle -44, -34, 18 \rangle$

12. $\mathbf{u} \times \mathbf{v} = \begin{vmatrix} \mathbf{i} & \mathbf{j} & \mathbf{k} \\ 0 & 0 & 6 \\ -1 & 3 & 1 \end{vmatrix} = \langle -18, -6, 0 \rangle = -18\mathbf{i} - 6\mathbf{j}$

14. $\mathbf{u} \times \mathbf{v} = \begin{vmatrix} \mathbf{i} & \mathbf{j} & \mathbf{k} \\ \frac{2}{3} & 0 & 0 \\ \frac{1}{3} & 0 & 3 \end{vmatrix} = \langle 0, -2, 0 \rangle = -2\mathbf{j}$

16. $\mathbf{u} \times \mathbf{v} = \begin{vmatrix} \mathbf{i} & \mathbf{j} & \mathbf{k} \\ -1 & 0 & 1 \\ 0 & 1 & -2 \end{vmatrix} = \langle -1, -2, -1 \rangle = -\mathbf{i} - 2\mathbf{j} - \mathbf{k}$

18. $\mathbf{u} \times \mathbf{v} = \begin{vmatrix} \mathbf{i} & \mathbf{j} & \mathbf{k} \\ 1 & 2 & 0 \\ 1 & -3 & 0 \end{vmatrix} = -5\mathbf{k}; \|\mathbf{u} \times \mathbf{v}\| = 5$

Unit vector $= \dfrac{\mathbf{u} \times \mathbf{v}}{\|\mathbf{u} \times \mathbf{v}\|} = \dfrac{1}{5}(-5\mathbf{k}) = -\mathbf{k}$

20. $\mathbf{u} \times \mathbf{v} = \begin{vmatrix} \mathbf{i} & \mathbf{j} & \mathbf{k} \\ 7 & -14 & 5 \\ 14 & 28 & -15 \end{vmatrix} = 70\mathbf{i} + 175\mathbf{j} + 392\mathbf{k}$

$\|\mathbf{u} \times \mathbf{v}\| = \sqrt{70^2 + 175^2 + 392^2} = \sqrt{189189} = 21\sqrt{429}$

Unit vector $= \dfrac{\mathbf{u} \times \mathbf{v}}{\|\mathbf{u} \times \mathbf{v}\|} = \dfrac{1}{21\sqrt{249}}\langle 70, 175, 392\rangle$

$= \dfrac{1}{3\sqrt{429}}\langle 10, 25, 56\rangle$

22. $\mathbf{u} \times \mathbf{v} = \begin{vmatrix} \mathbf{i} & \mathbf{j} & \mathbf{k} \\ 1 & -2 & 2 \\ 2 & -1 & -2 \end{vmatrix} = 6\mathbf{i} + 6\mathbf{j} + 3\mathbf{k}$

$\|\mathbf{u} \times \mathbf{v}\| = \sqrt{36 + 36 + 9} = 9$

Unit vector $= \dfrac{\mathbf{u} \times \mathbf{v}}{\|\mathbf{u} \times \mathbf{v}\|} = \dfrac{1}{9}(6\mathbf{i} + 6\mathbf{j} + 3\mathbf{k}) = \dfrac{2}{3}\mathbf{i} + \dfrac{2}{3}\mathbf{j} + \dfrac{1}{3}\mathbf{k}$

24. $\mathbf{u} \times \mathbf{v} = \begin{vmatrix} \mathbf{i} & \mathbf{j} & \mathbf{k} \\ 1 & 2 & 2 \\ 1 & 0 & 1 \end{vmatrix} = 2\mathbf{i} + \mathbf{j} - 2\mathbf{k}$

Area $= \|\mathbf{u} \times \mathbf{v}\| = \|2\mathbf{i} + \mathbf{j} - 2\mathbf{k}\| = \sqrt{4 + 1 + 4} = 3$

26. $\mathbf{u} \times \mathbf{v} = \begin{vmatrix} \mathbf{i} & \mathbf{j} & \mathbf{k} \\ -2 & 3 & 2 \\ 1 & 2 & 4 \end{vmatrix} = \langle 8, 10, -7\rangle$

Area $= \|\mathbf{u} \times \mathbf{v}\| = \sqrt{8^2 + 10^2 + (-7)^2} = \sqrt{213}$

28. $\mathbf{u} \times \mathbf{v} = \begin{vmatrix} \mathbf{i} & \mathbf{j} & \mathbf{k} \\ 4 & -3 & 2 \\ 5 & 0 & 1 \end{vmatrix} = \langle -3, 6, 15\rangle$

Area $= \|\mathbf{u} \times \mathbf{v}\| = \sqrt{(-3)^2 + 6^2 + 15^2} = \sqrt{270}$

30. $\overrightarrow{AB} = \langle -1 - 3, 8 - 5, 5 - 0\rangle = \langle -4, 3, 5\rangle$ is parallel to

$\overrightarrow{DC} = \langle 1 - 5, 3 - 0, 11 - 6\rangle = \langle -4, 3, 5\rangle$

$\overrightarrow{AD} = \langle 2, -5, 6\rangle$ is parallel to $\overrightarrow{BC} = \langle 2, -5, 6\rangle$

$\overrightarrow{AB} \times \overrightarrow{AD} = \begin{vmatrix} \mathbf{i} & \mathbf{j} & \mathbf{k} \\ -4 & 3 & 5 \\ 2 & -5 & 6 \end{vmatrix} = \langle 43, 34, 14\rangle$

Area $= \|\overrightarrow{AB} \times \overrightarrow{AD}\| = \sqrt{43^2 + 34^2 + 14^2} = \sqrt{3201}$

32. $\mathbf{u} = \langle 2 - 1, 0 - (-4), 2 - 3\rangle = \langle 1, 4, -1\rangle$

$\mathbf{v} = \langle -2 - 1, 2 - (-4), 0 - 3\rangle = \langle -3, 6, -3\rangle$

$\mathbf{u} \times \mathbf{v} = \begin{vmatrix} \mathbf{i} & \mathbf{j} & \mathbf{k} \\ 1 & 4 & -1 \\ -3 & 6 & -3 \end{vmatrix} = \langle -6, 6, 18\rangle$

Area $= \dfrac{1}{2}\|\mathbf{u} \times \mathbf{v}\| = \dfrac{1}{2}\sqrt{(-6)^2 + 6^2 + 18^2} = \dfrac{1}{2}\sqrt{396} = 3\sqrt{11}$

34. $\mathbf{u} = \langle -2-2, -4-4, 0-0 \rangle = \langle -4, -8, 0 \rangle$

$\mathbf{v} = \langle 0-2, 0-4, 4-0 \rangle = \langle -2, -4, 4 \rangle$

$\mathbf{u} \times \mathbf{v} = \begin{vmatrix} \mathbf{i} & \mathbf{j} & \mathbf{k} \\ -4 & -8 & 0 \\ -2 & -4 & 4 \end{vmatrix} = \langle -32, 16, 0 \rangle$

$\text{Area} = \dfrac{1}{2}\|\mathbf{u} \times \mathbf{v}\| = \dfrac{1}{2}\sqrt{(-32)^2 + 16^2} = \dfrac{1}{2}\sqrt{1280} = 8\sqrt{5}$

36. $\mathbf{u} \cdot (\mathbf{v} \times \mathbf{w}) = \begin{vmatrix} 20 & 10 & 10 \\ 1 & 4 & 4 \\ 0 & 2 & 2 \end{vmatrix} = 20(0) - 10(2) + 10(2) = 0$

38. $\mathbf{u} \cdot (\mathbf{v} \times \mathbf{w}) = \begin{vmatrix} 1 & 1 & 3 \\ 0 & 3 & 3 \\ 3 & 0 & 3 \end{vmatrix} = 1(9) - 1(-9) + 3(-9) = -18$

$\text{Volume} = |\mathbf{u} \cdot (\mathbf{v} \times \mathbf{w})| = |-18| = 18$

40. $\mathbf{u} = \langle 2-3, -2-0, 2-0 \rangle = \langle -1, -2, 2 \rangle$

$\mathbf{v} = \langle 4-3, 1-0, 2-0 \rangle = \langle 1, 1, 2 \rangle$

$\mathbf{w} = \langle -1-3, 5-0, 4-0 \rangle = \langle -4, 5, 4 \rangle$

$\mathbf{u} \cdot (\mathbf{v} \times \mathbf{w}) = \begin{vmatrix} -1 & -2 & 2 \\ 1 & 1 & 2 \\ -4 & 5 & 4 \end{vmatrix} = -1(-6) + 2(12) + 2(9) = 48$

$\text{Volume} = 48$

42. $\mathbf{u} \times \mathbf{v} = \begin{vmatrix} \mathbf{i} & \mathbf{j} & \mathbf{k} \\ u_1 & u_2 & u_3 \\ v_1 & v_2 & v_3 \end{vmatrix}$

$= (u_2 v_3 - v_2 u_3)\mathbf{i} - (u_1 v_3 - v_1 u_3)\mathbf{j} + (u_1 v_2 - v_1 u_2)\mathbf{k}$

$\mathbf{u} \cdot (\mathbf{u} \times \mathbf{v}) = u_1(u_2 v_3 - v_2 u_3) - u_2(u_1 v_3 - v_1 u_3) + u_3(u_1 v_2 - v_1 u_2)$

$= 0$

Similarly, $\mathbf{v} \cdot (\mathbf{u} \times \mathbf{v}) = 0$

11.6 Equations of Lines and Planes in Space

2. $x = x_1 + at = 0 + 3t = 3t$

$y = y_1 + bt = 0 - 7t = -7t$

$z = z_1 + ct = 0 - 10t = -10t$

    (a) Parametric equations: $x = 3t, y = -7t, z = -10t$

    (b) Symmetric equations: $\dfrac{x}{3} = \dfrac{y}{-7} = \dfrac{z}{-10}$

4. $x = x_1 + at = 5 + 4t$

$y = y_1 + bt = 0 + 0t$

$z = z_1 + ct = 10 + 3t$

    (a) Parametric equations: $x = 5 + 4t, y = 0, z = 10 + 3t$

    (b) Symmetric equations: $\dfrac{x - 5}{4} = \dfrac{z - 10}{3}$

6. $x = x_1 + at = 10 + 3t$

$y = y_1 + bt = -18 - 3t$

$z = z_1 + ct = 36 + t$

    (a) Parametric equations: $x = 10 + 3t, y = -18 - 3t, z = 36 + t$

    (b) Symmetric equations: $\dfrac{x - 10}{3} = \dfrac{y + 18}{-3} = z - 36$

8. $\mathbf{v} = \langle 8 - 4, 0 - 2, -1 - (-2) \rangle = \langle 4, -2, 1 \rangle$

    (a) $x = 4 + 4t, y = 2 - 2t, z = -2 + t$ (Parametric)

    (b) $\dfrac{x - 4}{4} = \dfrac{y - 2}{-2} = z + 2$ (Symmetric)

10. Use $\mathbf{v} = \langle 3, 2, -1 \rangle$. Then $x = 4 + 3t, y = -3 + 2t, z = -8 - t$.

12. $\mathbf{v} = \langle 2 - 3, 4 - 1, -2 - 0 \rangle = \langle -1, 3, -2 \rangle$

    $x = 3 - t, y = 1 + 3t, z = -2t$, or, $\dfrac{x - 3}{-1} = \dfrac{y - 1}{3} = \dfrac{z}{-2}$

    (a) & (d) satisfy the equations.

14. Equating the two sets of parametric equations

$$\left.\begin{array}{l} 3 - 2t = 3 - s \\ 8 + t = 15 + 4s \\ 12 + 3t = 3 - 3s \end{array}\right\} \Rightarrow \begin{array}{l} -2t + s = 0 \\ t - 4s = 7 \\ 3t + 3s = -9 \end{array}$$

Solving for t and s, you obtain $t = -1$ and $s = -2$. The point of intersection is $(x, y, z) = (5, 7, 9)$.

$$\cos\theta = \frac{\langle -2, 1, 3\rangle \cdot \langle -1, 4, -3\rangle}{\|\langle -2, 1, 3\rangle\| \, \|\langle -1, 4, -3\rangle\|} = \frac{-3}{\sqrt{14}\sqrt{26}} \Rightarrow \theta \approx 99.05°$$

16.

18. $a(x - x_1) + b(y - y_1) + c(z - z_1) = 0$

$$0(x - 2) + 0(y - 3) + 1(z - 5) = 0$$

$$z - 5 = 0$$

20. $0(x - 0) - 3(y - 0) + 5(z - 0) = 0$

$$-3y + 5z = 0$$

22. $\mathbf{n} = \langle 3, 1, -2\rangle \quad 3(x - 1) + 1(y - 3) - 2(z - 1) = 0$

$$3x + y - 2z - 4 = 0$$

24. $\mathbf{u} = \langle 2, 1, 3\rangle \quad \mathbf{u} \times \mathbf{v} = \begin{vmatrix} \mathbf{i} & \mathbf{j} & \mathbf{k} \\ 2 & 1 & 3 \\ -2 & 1 & 3 \end{vmatrix} = \langle 0, -12, 4\rangle$

$\mathbf{v} = \langle -2, 1, 3\rangle$

$$0(x - 2) - 12(y - 1) + 4(z - 3) = 0$$

$$-12y + 4z = 0$$

$$-3y + z = 0$$

26. $\mathbf{u} = \langle 2 - 5, 3 - 1, 10 - (-6)\rangle = \langle -3, 2, 16\rangle$

$\mathbf{v} = \langle 1 - 5, -2 - 1, 4 - (-6)\rangle = \langle -4, -3, 10\rangle$

$$\mathbf{u} \times \mathbf{v} = \begin{vmatrix} \mathbf{i} & \mathbf{j} & \mathbf{k} \\ -3 & 2 & 16 \\ -4 & -3 & 10 \end{vmatrix} = \langle 68, -34, 17\rangle. \text{ Use } \langle 4, -2, 1\rangle = \mathbf{n}$$

$4(x - 5) - 2(y - 1) + 1(z + 6) = 0$

$$4x - 2y + z - 12 = 0$$

28. $\mathbf{n} = \mathbf{k} : 0(x - 2) + 0(y - 5) + 1(z - 3) = 0$

$$z - 3 = 0$$

30. $\langle 1,2,2 \rangle$ and $\langle -4,2,0 \rangle$ are parallel to plane

$$\mathbf{n} = \langle 1,2,2 \rangle \times \langle -4,2,0 \rangle = \begin{vmatrix} i & j & k \\ 1 & 2 & 2 \\ -4 & 2 & 0 \end{vmatrix} = \langle -4,-8,10 \rangle$$

$$-4(x-4) - 8(y-0) + 10(z-0) = 0$$

$$-4x - 8y + 10z + 16 = 0$$

$$-2x - 4y + 5z + 8 = 0$$

32. $\mathbf{n_1} = \langle 1,-3,-1 \rangle, n_z = \langle 2,1,-1 \rangle$

$$\cos\theta = \frac{\mathbf{n_1} \cdot \mathbf{n_z}}{\|\mathbf{n_1}\|\,\|\mathbf{n_z}\|} = 0 \Rightarrow \theta = 90°$$

Solving the two equations simultaneously,

$$x - 3y - z = 3$$

$$2x + y - z = 10$$

Subtracting,

$$-x - 4y = -7$$

$$x = 7 - 4y$$

Let $y = t, x = 7 - 4t, z = x - 3y - 3 = (7 - 4t) - 3t - 3$

$$= 4 - 7t$$

Parametric equations: $x = 7 - 4t, y = t, z = 4 - 7t$

34. $\mathbf{n_1} = \langle 1,-3,1 \rangle, \mathbf{n_2} = \langle 0,2,-1 \rangle$

$$\cos\theta \frac{\mathbf{n_1} \cdot \mathbf{n_2}}{\|\mathbf{n_1}\|\,\|\mathbf{n_2}\|} = \frac{-7}{\sqrt{11}\sqrt{5}} \Rightarrow \theta \approx 160.71°$$

$$x - 3y + z = 8$$

$$2y - 2 = -1$$

Let $y = t$.
Then, $\quad z = 2y + 1 = 2t + 1$

$$x = 8 + 3t - (2t + 1) = 7 + t$$

Parametric equations: $x = 7 + t, y = t, z = 2t + 1$

36.

38.

40.

42.

44.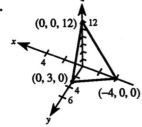

46. $D = \dfrac{|\overrightarrow{PQ} \cdot \mathbf{n}|}{\|\mathbf{n}\|}$ • $P = (2, 0, 0)$ on plane, $Q = (2, -3, 3)$ $\mathbf{n} = \langle 2, -1, 3 \rangle$

$D = \dfrac{|\langle 0, -3, 3 \rangle \cdot \langle 2, -1, 3 \rangle|}{\sqrt{14}} = \dfrac{12}{\sqrt{14}}$

48. (a) $t\mathbf{v}$ line through origin parallel to $\mathbf{v}$

(b) $\mathbf{u} + t\mathbf{v}$ line

(c) $s\mathbf{u} + t\mathbf{v}$ plane containing $\mathbf{u} + \mathbf{v}$.

Review Exercises for Chapter 11

2. $\overrightarrow{AB} = \langle 0 - 4, 5 - 6 \rangle = \langle -4, -1 \rangle$

$\overrightarrow{AC} = \langle -6, -6 \rangle$

$\overrightarrow{AB} + \overrightarrow{AC} = \langle -4 - 6, -1 - 6 \rangle = \langle -10, -7 \rangle$

4. $\mathbf{v} = \langle 15 - 1, 9 - 5 \rangle = \langle 14, 4 \rangle$

6. $\mathbf{v} = \left\langle \dfrac{1}{2} \cos 225°, \dfrac{1}{2} \sin 225° \right\rangle = \left\langle \dfrac{-r2}{4}, \dfrac{-r2}{4} \right\rangle$

8. $\mathbf{v} = \langle 5 \cos 300°, 5 \sin 300° \rangle = \left\langle \dfrac{5}{2}, \dfrac{-5\sqrt{3}}{2} \right\rangle$

10. $3\mathbf{v} = 3(10\mathbf{i} + 3\mathbf{j}) = \langle 30, 9 \rangle$

12. $\dfrac{1}{2}\mathbf{v} = \dfrac{1}{2}(10\mathbf{i} + 3\mathbf{j}) = \left\langle 5, \dfrac{3}{2} \right\rangle$

14. (a) $\mathbf{u} + \mathbf{v} = (-3\mathbf{i} + 5\mathbf{j}) + (10\mathbf{i}) = 7\mathbf{i} + 5\mathbf{j}$

(b) $\mathbf{u} - \mathbf{v} = (-3\mathbf{i} + 5\mathbf{j}) - 10\mathbf{i} = -13\mathbf{i} + 5\mathbf{j}$

(c) $2\mathbf{u} - 3\mathbf{v} = 2(-3\mathbf{i} + 5\mathbf{j}) - 3(10\mathbf{i}) = -36\mathbf{i} + 10\mathbf{j}$

16. $\|\mathbf{v}\| = \sqrt{(-3)^2 + (-3)^2} = \sqrt{18} = 3\sqrt{2}$

$\tan \theta = \dfrac{-3}{-3} = 1 \Rightarrow \theta = 225°$ (Quadrant III)

18. $\mathbf{v} = \langle 8 \cos 90°, 8 \sin 90° \rangle$

   $= \langle 0, 8 \rangle$

20. $\mathbf{v} = 3 \left(\dfrac{1}{\sqrt{3^2 + 4^2}} \right) (3\mathbf{i} + 4\mathbf{j})$

   $= \dfrac{3}{5}(3\mathbf{i} + 4\mathbf{j})$

   $= \dfrac{9}{5}\mathbf{i} + \dfrac{12}{5}\mathbf{j}$

22. $\mathbf{v} = \mathbf{u} + \mathbf{w}$

   $= (2\mathbf{i} - \mathbf{j}) + (\mathbf{i} + 2\mathbf{j})$

   $= 3\mathbf{i} + \mathbf{j}$

24. Midpoint $=$ center $= \left(\dfrac{3+5}{2}, \dfrac{4+8}{2}, \dfrac{2+1}{2} \right) = \left(4, 6, \dfrac{3}{2} \right)$

   Radius $= \sqrt{(4-3)^2 + (6-4)^2 + \left(\dfrac{3}{2} - 2 \right)^2} = \sqrt{1 + 2 + \dfrac{1}{4}} = \dfrac{\sqrt{13}}{2}$

   Sphere: $(x-4)^2 + (y-6)^2 + \left(z - \dfrac{3}{2} \right)^2 = \dfrac{13}{4}$

26. Unit vector $= \dfrac{\overrightarrow{PQ}}{\|\overrightarrow{PQ}\|} = \dfrac{\langle 5, -11, 7 \rangle}{\|\langle 5, -11, 7 \rangle\|} = \dfrac{1}{\sqrt{195}} \langle 5, 11, 7 \rangle$

28. Since $\langle 8, 5, -8 \rangle \cdot \left\langle -2, 4, \dfrac{1}{2} \right\rangle = -16 + 20 - 4 = 0$, the vectors are orthogonal.

30. $\cos \theta = \dfrac{\mathbf{u} \cdot \mathbf{v}}{\|\mathbf{u}\| \, \|\mathbf{v}\|} = \dfrac{-24 - 6 - 216}{\sqrt{369}\sqrt{164}} = \dfrac{-246}{3\sqrt{41} - 2\sqrt{41}} = -1 \Rightarrow \theta = 180°$

32. $\cos \theta = \dfrac{\mathbf{u} \cdot \mathbf{v}}{\|\mathbf{u}\| \, \|\mathbf{v}\|} = \dfrac{12 + 5 - 2}{\sqrt{11}\sqrt{45}} = \dfrac{5}{\sqrt{11}\sqrt{45}} \Rightarrow \theta \approx 47.61°$

34. $\text{proj } \mathbf{u} = \left(\dfrac{\mathbf{u} \cdot \mathbf{v}}{\|\mathbf{v}\|^2} \right) \mathbf{v} = \dfrac{50}{100} \langle 10, 0 \rangle = \langle 5, 0 \rangle$

36. $\text{proj}_{\mathbf{v}} \mathbf{u} = \left(\dfrac{\mathbf{u} \cdot \mathbf{v}}{\|\mathbf{v}\|^2} \right) \mathbf{v} = \dfrac{31}{65} \langle -5, 2, 6 \rangle$

38. $\mathbf{u} \times \mathbf{v} = \begin{vmatrix} \mathbf{i} & \mathbf{j} & \mathbf{k} \\ 20 & 15 & 5 \\ 5 & -3 & 0 \end{vmatrix} = \langle 15, 25, -135 \rangle$

40. $\mathbf{v} \times \mathbf{w} = \begin{vmatrix} \mathbf{i} & \mathbf{j} & \mathbf{k} \\ 2 & 4 & 0 \\ 3 & 4 & 5 \end{vmatrix} = \langle 20, -10, -4 \rangle$

   Unit vector: $\dfrac{\langle 20, -10, -4 \rangle}{\sqrt{516}} = \dfrac{1}{129} \langle 10, -5, -2 \rangle$

42. $\mathbf{u} \times (\mathbf{v} + \mathbf{w}) = \langle 1, -2, -1 \rangle \times \langle 5, 8, 5 \rangle = \begin{vmatrix} \mathbf{i} & \mathbf{j} & \mathbf{k} \\ 1 & -2 & -1 \\ 5 & 8 & 5 \end{vmatrix} = \langle -2, -10, 18 \rangle$

$(\mathbf{u} \times \mathbf{v}) + (\mathbf{u} \times \mathbf{w}) = \begin{vmatrix} \mathbf{i} & \mathbf{j} & \mathbf{k} \\ 1 & -2 & -1 \\ 2 & 4 & 0 \end{vmatrix} + \begin{vmatrix} \mathbf{i} & \mathbf{j} & \mathbf{k} \\ 1 & -2 & -1 \\ 3 & 4 & 5 \end{vmatrix} = \langle 4, -2, 8 \rangle + \langle -6, -8, 10 \rangle = \langle -2, -10, 18 \rangle.$

Hence, $\mathbf{u} \times (\mathbf{v} + \mathbf{w}) = (\mathbf{u} \times \mathbf{v}) + (\mathbf{u} \times \mathbf{w})$

44. The parallelepipe I is determined by the vectors with initial points at $(3, 2, -1)$:

$\mathbf{u} = \langle 1, -3, 7 \rangle$

$\mathbf{v} = \langle -1, 3, 2 \rangle$

$\mathbf{w} = \langle -8, 2, 3 \rangle$

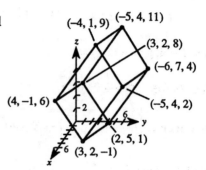

Volume $= \mathbf{u} \cdot (\mathbf{v} \times \mathbf{w})$

$= \begin{vmatrix} 1 & -3 & 7 \\ -1 & 3 & 2 \\ -8 & 2 & 3 \end{vmatrix} = 1(5) + 3(13) + 7(22) = 198.$

46. $\overrightarrow{F} = \langle 0, -200 \rangle$ weight of ingot

$\overrightarrow{F}1 = C \langle \cos 60°, \sin 60° \rangle$

$= C \langle \frac{1}{2}, \frac{r3}{2} \rangle$

$\overrightarrow{F}1 = C \langle \cos 120°, \sin 120° \rangle$

$= C \langle -\frac{1}{2}, \frac{r3}{2} \rangle$

$\overrightarrow{F}1 + \overrightarrow{F}2 + \overrightarrow{F}3 = 0 \Longrightarrow \langle 0, 200 \rangle = C \langle \frac{1}{2}, \frac{\sqrt{3}}{2} \rangle + C \langle \frac{1}{2}, \frac{\sqrt{3}}{2} \rangle$

$200 = 2x \left(\frac{\sqrt{3}}{2} \right) = \sqrt{3}C \Rightarrow C = 200/\sqrt{3}$

$F1 = \frac{200}{\sqrt{3}} \langle \frac{1}{2}, \frac{\sqrt{3}}{2} \rangle = \langle \frac{100}{\sqrt{3}}, 100 \rangle$

$F2 = \frac{200}{\sqrt{3}} \langle -\frac{1}{2}, \frac{\sqrt{3}}{2} \rangle = \langle -\frac{100}{\sqrt{3}}, 100 \rangle$

48. $\mathbf{v} = \langle 4, 3, -6 \rangle \; x = -1 + 4t, \, y = 3 + 3t, \, z = 5 - 6t$

50. $2x - y - 27 = 4$

$x - y - z = 8$

Subtracting, $x - z = -4$ or $x = z - 4$

Let $z = t$

Then $x = t - 4$

$y = x - z - 8 = (t - 4) - t - 8 = -12$

$x = t - 4, y = -12, z = t$

52. $\mathbf{u} = \langle 5, -5, -2 \rangle \mathbf{v} = \langle 3, 5, 2 \rangle$

$\mathbf{n} = \mathbf{u} x \mathbf{v} = \begin{vmatrix} \mathbf{i} & \mathbf{j} & \mathbf{k} \\ 5 & -5 & -2 \\ 3 & 5 & 2 \end{vmatrix} = \langle 0, -16, 40 \rangle$

Plane: $0(x + 1) - 16(y - 3) + 40(z - 4) = 0$

$-2(y - 3) + 5(z - 4) = 0$

$-2y + 5z - 14 = 0$

54. $\mathbf{n} = \mathbf{k}$ normal vector

Plane: $0(x - 5) + 0(y - 3) + 1(z - 2) = 0$

$z - 2 = 0$

56. $D = \dfrac{|\overrightarrow{PQ} \cdot \mathbf{n}|}{\|\mathbf{n}\|}$

$\qquad \mathbf{n} = \langle 1, -10, 3 \rangle$

$\qquad Q = (0, 0, 0)$

$\qquad P = (2, 0, 0)$ in plane

$= \dfrac{|\langle -2, 0, 0 \rangle \cdot \langle 1, -10, 3 \rangle|}{\sqrt{1 + 100 + 9}}$

$= \dfrac{2}{\sqrt{110}} \approx 0.191$

Cumulative Test, Chapters 9—11

1. 8, 12, 16, 20, ...

$a_1 = 8$, $d = 4$, $c = a_1 - d = 8 - 4 = 4$ and $a_{20} = dn + c = 4(20) + 4 = 84$.

Thus, the sum of the first 20 terms is

$$S = \frac{n}{2}(a_1 + a_n)$$
$$= \frac{20}{2}(8 + 84)$$
$$= 920.$$

2. $a_1 = 54$ and $r = -\frac{1}{3}$ $\Rightarrow$ $a_2 = -\frac{1}{3}(54) = -18$

$$a_3 = -\frac{1}{3}(-18) = 6$$
$$a_4 = -\frac{1}{3}(6) = -2$$
$$a_5 = -\frac{1}{3}(-2) = \frac{2}{3}$$

3. $\begin{bmatrix} 1 & 3 & -2 & \vdots & -7 \\ -2 & 1 & -1 & \vdots & -5 \\ 4 & 1 & 1 & \vdots & 3 \end{bmatrix} \Rightarrow \begin{bmatrix} 1 & 3 & -2 & \vdots & -7 \\ 0 & 7 & -5 & \vdots & -19 \\ 0 & -11 & 9 & \vdots & 31 \end{bmatrix} \Rightarrow \begin{bmatrix} 1 & 3 & -2 & \vdots & -7 \\ 0 & -4 & 4 & \vdots & 12 \\ 0 & -11 & 9 & \vdots & 31 \end{bmatrix}$

$\Rightarrow \begin{bmatrix} 1 & 0 & 1 & \vdots & 2 \\ 0 & 1 & -1 & \vdots & -3 \\ 0 & 0 & -2 & \vdots & -2 \end{bmatrix} \Rightarrow \begin{bmatrix} 1 & 0 & 0 & \vdots & 1 \\ 0 & 1 & 0 & \vdots & -2 \\ 0 & 0 & 1 & \vdots & 1 \end{bmatrix}$

$(1, -2, 1)$

4. $3 + 7 + 11 + 15 + \cdots + (4n - 1) = n(2n + 1)$

When $n = 1$, the formula is valid because $3 = 1(2 + 1)$.

Assuming that $3 + 7 + \cdots + (4k - 1) = k(2k + 1)$, we have

$$3 + 7 + \cdots + (4k - 1) + (4(k + 1) - 1) = k(2k + 1) + 4k + 3$$
$$= 2k^2 + 5k + 3$$
$$= (k + 1)(2k + 3)$$
$$= (k + 1)(2(k + 1) + 1).$$

5. $(z - 3)^4 = z^4 + 4(z^3)(-3) + 6(z^2)(-3)^2 + 4(z)(-3)^3 + (-3)^4$

$$= z^4 - 12z^3 + 54z^2 - 108z + 81$$

6. $\displaystyle\sum_{n=1}^{10} 32{,}000(1.05)^{n-1} = 32{,}000 \left[\frac{1 - (1.05)^{10}}{1 - 1.05}\right]$

$$\approx \$402{,}492.56$$

7. $_{10}C_3 = \dfrac{10!}{(10-3)!3!}$

$\quad\quad = 120$

8. (a) $n(S) = 3! = 6$ (b) $n(S) = 2 \cdot 2 \cdot 1 = 4$

$\quad\quad n(E) = 1$ $n(E) = 1$

$\quad\quad P(E) = \frac{1}{6}$ $P(E) = \frac{1}{4}$

9. mean $= 24.17$

median $= 24$

$\mathbf{v} = (21 - 24.17)^2 + (27 - 24.17)^2 + (23 - 24.17)^2 + (25 - 24.17)^2 + (30 - 24.17)^2 + (19 - 24.17)^2$

$\sigma = \sqrt{\mathbf{v}} = 3.67$

10. (a)

       (b)

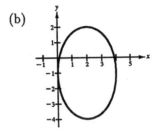

11. Since the foci are at $(0,\ 0)$ and $(0,\ 4)$, the center is $(0,\ 2)$. Since the asymptotes are $y = \pm \frac{1}{2}x + 2$, we have

$$\frac{1}{2} = \frac{a}{b} \text{ or } b = 2a.$$

Hence,

$$b^2 = c^2 - a^2 \Rightarrow (2a)^2 = 2^2 - a^2$$

$$\Rightarrow a^2 = \frac{4}{5} \text{ and } b^2 = \frac{16}{5}.$$

Thus,

$$\frac{(y-2)^2}{4/5} - \frac{x^2}{16/5} = 1.$$

12. $\mathbf{v} = \langle 2 - 6, -3 - 4 \rangle$

use $\langle -4, -7 \rangle$

$x = 2 - 4t, y = -3 - 7t$ (Answer is not unique.)

13. $x = 3 + 3\cos\theta$

$y = 2 + 2\sin\theta$

(a)

(b) $\dfrac{x-3}{3} = \cos\theta, \quad \dfrac{y-2}{2} = \sin\theta$

and

$\cos^2\theta + \sin^2\theta = 1 \Rightarrow \dfrac{(x-3)^2}{9} + \dfrac{(y-2)^2}{4} = 1.$

14. $r = 6 \cos \theta$

$r^2 = 6r \cos \theta$

$x^2 + y^2 = 6x$

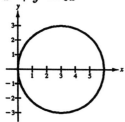

15. $r = 2 \cos 2\theta \cdot \sec \theta$

Vertical asymptote: $x = -2$

16. Focus $(0, \ 0)$, vertex $(1, \ \pi)$. Parabola
$e = 1, \ p = 2$

$$r = \frac{ep}{1 - e \cos \theta} = \frac{2}{1 - \cos \theta}$$

17. $\mathbf{v} = \langle \cos 30°, \sin 30° \rangle$

$$= \langle \frac{\sqrt{3}}{2}, \frac{1}{2} \rangle$$

18. Horizontal component of velocity: $120 \cos 15° \approx 115.91$ ft/sec

19. center $= \sqrt{\frac{4+0}{2}, \frac{4+0}{2}, 8 + 02} = (2, 2, 4)$

radius $= \sqrt{(4-2)^2 + (4-2)^2 + (8-4)^2} = \sqrt{4 + 4 + 16} = \sqrt{24}$

sphere $= (x - 2)^2 + (y - 2)^2 + (z - 4)^2 = 24$

20. $\mathbf{u} \cdot \mathbf{v} = -3(5) + 4(0) + 1(2) = -13$

$$\mathbf{u} \times \mathbf{v} = \begin{vmatrix} \mathbf{i} & \mathbf{j} & \mathbf{k} \\ -3 & 4 & 1 \\ 5 & 0 & 2 \end{vmatrix} = \begin{vmatrix} 4 & 1 \\ 0 & 2 \end{vmatrix} \mathbf{i} - \begin{vmatrix} -3 & 1 \\ 5 & 2 \end{vmatrix} \mathbf{j} + \begin{vmatrix} -3 & 4 \\ 5 & 0 \end{vmatrix} \mathbf{k}$$

$$= (8 - 0)\mathbf{i} - (-6 - 5)\mathbf{j} + (-0 - 20)\mathbf{k}$$

$$= \langle 8, 11, -20 \rangle$$

21. $\mathbf{v} \langle 5, -(-2), 8 - 3, 25 - 0 \rangle = \langle 7, 5, 25 \rangle$

Parametric equations: $x = -2 + 7t, y = 3 + 5t, z = 25t$

22. $\mathbf{u}(-2, 3, 0) \mathbf{u} \times \mathbf{v} = \begin{vmatrix} \mathbf{i} & \mathbf{j} & \mathbf{k} \\ -2 & 3 & 0 \\ 5 & 8 & 25 \end{vmatrix} = \langle 75, 50, -31 \rangle$

$$75(x + 2) + 50(y - 3) - 31(z - 0) = 0$$

$$75x + 150 + 50y - 150 - 31z = 0$$

$$75x + 50y - 31z = 0$$

CHAPTER TWELVE
Limits and an Introduction to Calculus

12.1 Introduction to Limits

2. $\lim\limits_{x \to 4} \left(\dfrac{1}{2}x^2 - 2x + 3\right) = 3$

x	3.9	3.99	3.999	4.0	4.001	4.01	4.1
$f(x)$	2.805	2.980	2.998	3	3.002	3.020	3.205

4. $\lim\limits_{x \to -1} \dfrac{x+1}{x^2 - x - 2} = \dfrac{-1}{3}$

x	-1.1	-1.01	-1.001	-1.0	-0.999	-0.99	-0.9
$f(x)$	-0.3226	-0.3322	-0.3332	?	-0.3334	-0.3344	-0.3448

6. $\lim\limits_{x \to -3} \dfrac{x+3}{x^2 + x - 6} = \dfrac{-1}{5}$

x	-3.1	-3.01	-3.001	-3.0	-2.999	-2.99	-2.9
$f(x)$	-0.1961	-0.1996	-0.19996	?	-0.20004	-0.2004	-0.2041

8. $\lim\limits_{x \to 2} \dfrac{\frac{1}{x+2} - \frac{1}{4}}{x - 2} = \dfrac{-1}{16}$

x	1.9	1.99	1.999	2.0	2.001	2.01	2.1
$f(x)$	-0.0641	-0.0627	-0.0625	?	-0.0625	-0.0623	-0.0610

10. $\lim\limits_{x \to 0} \dfrac{\cos x - 1}{x} = 0$

x	-0.1	-0.01	-0.001	0	0.001	0.01	0.1
$f(x)$	-0.050	-0.005	-0.0005	?	-0.0005	-0.005	-0.050

12. $\lim\limits_{x \to 2} \dfrac{3x^2 - 12}{x - 2} = 12$

14. The limit does not exist because $f(x)$ does not approach a real number as x approaches 1.

16. $\lim\limits_{x \to \pi/3} \sec x = 2$

18. $\lim\limits_{x \to 2} \sqrt[3]{10x + 7} = \sqrt[3]{10(2) + 7} = \sqrt[3]{27} = 3$

20. $\lim\limits_{x \to -5} \dfrac{4}{x + 2} = \dfrac{4}{-5 + 2} = \dfrac{-4}{3}$

22. $\lim\limits_{x \to 8} \dfrac{\sqrt{x + 1}}{x - 4} = \dfrac{\sqrt{8 + 1}}{8 - 4} = \dfrac{3}{4}$

24. $\lim\limits_{x \to e} \ln x = \ln e = 1$

26. $\lim\limits_{x \to 1/2} \dfrac{\tan \pi x}{2} = \dfrac{\tan \frac{\pi}{2}}{2}$. The limit does not exist.

28. $\lim\limits_{x \to 1} \arccos \dfrac{x}{2} = \arccos \dfrac{1}{2} = \dfrac{\pi}{3} \approx 1.0472$.

30. $\lim\limits_{x \to 2} f(x) = 0$. As x approaches 2 from both sides, $f(x)$ approaches 0.

32. (a) $\lim\limits_{x \to c} [-2g(x)] = -2 \lim\limits_{x \to c} g(x) = -2(5) = -10$

(b) $\lim\limits_{x \to c} [f(x) + g(x)] = \lim\limits_{x \to c} f(x) + \lim\limits_{x \to c} g(x) = 4 + 5 = 9$

(c) $\lim\limits_{x \to c} \dfrac{f(x)}{g(x)} = \dfrac{\lim\limits_{x \to i} f(x)}{\lim\limits_{x \to i} g(x)} = \dfrac{4}{5}$

(d) $\lim\limits_{x \to c} \sqrt{f(x)} = \sqrt{\lim\limits_{x \to c} f(x)} = \sqrt{4} = 2$

34. $f(x) = \dfrac{x - 4}{\sqrt{x} - 2}$

$\lim\limits_{x \to 4} f(x) = 4$.

Domain: all $x \geq 0, x \neq 4$.

12.2 Techniques for Evaluating Limits

2. $h(x) = \dfrac{x^2 - 3x}{x}$

$h_2(x) = x - 3$

(a) $\lim\limits_{x \to -2} h(x) = -5$ (b) $\lim\limits_{x \to 0} h(x) = -3$ (c) $\lim\limits_{x \to 3} h(x) = 0$

4. $f(x) = \dfrac{x^2 - 1}{x + 1}$

$f_2(x) = x - 1$

(a) $\lim\limits_{x \to 1} f(x) = 0$ (b) $\lim\limits_{x \to 2} f(x) = 1$ (c) $\lim\limits_{x \to -1} f(x) = -2$

6. $\lim\limits_{x\to 10} (x^2 - 5x - 25) = 10^2 - 5(10) - 25 = 25$

8. $\lim\limits_{x\to -1} e^{x^2-1} = e^{(-1)^2-1} = e^0 = 1$ **10.** $\lim\limits_{x\to \pi/2} \cot t = \cot \dfrac{\pi}{2} = 0$

12. $\lim\limits_{x\to 3} \dfrac{3-x}{x^2-9} = \lim\limits_{x\to 3} \dfrac{3-x}{(x-3)(x+3)} = \lim\limits_{x\to 3} \dfrac{-1}{x+3} = \dfrac{-1}{6}$

14. $\lim\limits_{x\to -2} \dfrac{t^3+8}{t+2} = \lim\limits_{x\to -2} \dfrac{(t+2)(t^2-2t+4)}{(t+2)} = \lim\limits_{t\to -2} (t^2 - 2t + 4) = 12$

16. $\lim\limits_{z\to 0} \dfrac{\sqrt{6-z}-\sqrt{6}}{z} = \lim\limits_{z\to 0} \dfrac{\sqrt{6-z}-\sqrt{6}}{z} \cdot \dfrac{\sqrt{6-z}+\sqrt{6}}{\sqrt{6-z}+\sqrt{6}}$

$$= \lim\limits_{z\to 0} \dfrac{(6-z)-6}{z(\sqrt{6-z}+\sqrt{6})}$$

$$= \lim\limits_{z\to 0} \dfrac{-}{1}\sqrt{6-z}+\sqrt{6} = \dfrac{1}{2\sqrt{6}}$$

18. $\lim\limits_{x\to 2} \dfrac{4-\sqrt{18-x}}{x-2} = \lim\limits_{x\to 2} \dfrac{4-\sqrt{18-x}}{x-2} \cdot \dfrac{4+\sqrt{18-x}}{4+\sqrt{18-x}}$

$$= \lim\limits_{x\to 2} \dfrac{16-(18-x)}{(x-2)(4+\sqrt{18-x})}$$

$$= \lim\limits_{x\to 2} \dfrac{1}{4+\sqrt{18-x}} = \dfrac{1}{8}$$

20. $\lim\limits_{x\to 0} \dfrac{\frac{1}{5-x}-\frac{1}{5}}{x} = \lim\limits_{x\to 0} \dfrac{5-(5-x)}{(5-x)5x} = \lim\limits_{x\to 0} \dfrac{1}{5(5-x)} = \dfrac{1}{25}$

22. $\lim\limits_{x\to \pi/2} \dfrac{1-\sin x}{\cos x} = \lim\limits_{x\to \pi/2} \dfrac{1-\sin x}{\cos x} \cdot \dfrac{1+\sin x}{1+\sin x}$

$$= \lim\limits_{x\to \pi/2} \dfrac{1-\sin^2 x}{\cos x(1+\sin x)} = \lim\limits_{x\to \pi/2} \dfrac{\cos^2 x}{\cos x(1+\sin x)}$$

$$= \lim\limits_{x\to \pi/2} \dfrac{\cos x}{1+\sin x} = 0$$

24. $\lim\limits_{x\to 3+} |x-3| = \lim\limits_{x\to 3-} |x-3| = \lim\limits_{x\to 3} |x-3| = 3$

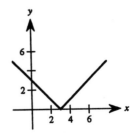

26. $\lim\limits_{x \to 1^-} \dfrac{1}{x^2 - 1}$ does not exist

$\lim\limits_{x \to 1^+} \dfrac{1}{x^2 - 1}$ does not exist

$\lim\limits_{x \to 1} \dfrac{1}{x^2 - 1}$ does not exist

28. $\lim\limits_{x \to 1^-} f(x) = 4 - 1 = 3$

$\lim\limits_{x \to 1^+} f(x) = 3 - 1 = 2$

$\lim\limits_{x \to 1} f(x)$ does not exist

30. $\lim\limits_{x \to 0} \dfrac{\frac{1}{3+x} - \frac{1}{3}}{x} = \lim\limits_{x \to 3} \dfrac{3 - (3 + x)}{3(3 + x)x} = \lim\limits_{x \to 3} \dfrac{-1}{3(3 + x)} = \dfrac{-1}{18}$

32. $\lim\limits_{h \to 0} \dfrac{f(x + h) - f(x)}{h} = \lim\limits_{h \to 0} \dfrac{(3(x + h) - 1) - (3x - 1)}{h} = \lim\limits_{h \to 0} \dfrac{3h}{h} = 3$

34. $\lim\limits_{h \to 0} \dfrac{f(x + h) - f(x)}{h} = \lim\limits_{h \to 0} \dfrac{[4 - 2(x + h) - (x + h)^2] - [4 - 2x - x^2]}{h}$

$= \lim\limits_{h \to 0} \dfrac{4 - 2x - 2h - x^2 - 2xh - h^2 - 4 + 2x + x^2}{h}$

$= \lim\limits_{h \to 0} \dfrac{-2h - 2xh - h^2}{h} = \lim\limits_{h \to 0} (-2 - 2x - h) = -2 - 2x$

36. $\lim\limits_{x \to 0^+} x^2 \ln x = 0$

38. $\lim\limits_{x \to 0} \dfrac{\sin 3x}{x} = 3$

40. $\lim\limits_{x\to 0} \dfrac{1-\cos 2x}{x} = 0$

42. $\lim\limits_{x\to 0} (1+2x)^{1/x} \approx 7.389$

44. (a) can be evaluated by direct substitution

$$\lim_{x\to 0} \frac{x}{\cos x} = \frac{0}{\cos 0} = 0$$

(b) cannot be evaluated by direct substitution

$$\lim_{x\to 0} \frac{1-\cos x}{x} = 0 \text{ (See Section 12.1, Exercise 10)}$$

46. $v = \lim\limits_{t\to 1} \dfrac{s(1)-s(t)}{1-t} = \lim\limits_{t\to 1} \dfrac{(-16+128)-(-16t^2+128)}{1-t}$

$ = \lim\limits_{t\to 1} \dfrac{16(t^2-1)}{1-t} = \lim\limits_{t\to 1} \dfrac{16(t+1)(t-1)}{1-t} = \lim\limits_{t\to 1}(-16(t+1)) = -32\text{ft/sec}$

12.3 The Tangent Line Problem

2. Slope is -1 at (x,y)

4. Slope is -2 at (x,y)

6. Slope ≈ 0

8. Slope ≈ 3

10. The slope is approximately 150 when $t=4$. The rate of change of fish is 150 per month when $t=4$ months.

12. $m_{\text{sec}} = \dfrac{h(-1+h)-h(-1)}{h} = \dfrac{2(-1+h)+5-3}{h} = \dfrac{2h}{h}$

$\phantom{m_{\text{sec}}} m = \lim\limits_{h\to 0} \dfrac{2h}{h} = 2$

14. $m_{\text{sec}} = \dfrac{g(3+h) - g(3)}{h} = \dfrac{(3+h)^2 - 2(3+h) - 3}{h} = \dfrac{h^2 + 4h}{h} = h + 4, h \neq 0$

$m = \lim\limits_{h \to 0} (h+4) = 4$

16. $m_{\text{sec}} = \dfrac{f(3+h) - f(3)}{h} = \dfrac{10(3+h) - 2(3+h)^2 - 12}{h}$

$= \dfrac{-2h - 2h^2}{h} = -2 - 4h, h \neq 0$

$m = \lim\limits_{h \to 0} (-2 - 4h) = -2$

18. $m_{\text{sec}} = \dfrac{g(4+h) - g(4)}{h} = \dfrac{\frac{1}{4+h-2} - \frac{1}{2}}{h} = \dfrac{\frac{1}{2+h} - \frac{1}{2}}{h}$

$= \dfrac{-h}{(2+h)2h} = \dfrac{1}{2(2+h)}, h \neq 0$

$m = \lim\limits_{h \to 0} \left(\dfrac{-1}{2(2+h)} \right) = -\dfrac{1}{4}$

20. $m_{\text{sec}} = \dfrac{h(-1+h) - h(-1)}{h} = \dfrac{\sqrt{-1+h+10} - 3}{h} \cdot \dfrac{\sqrt{h+9} + 3}{\sqrt{h+9} + 3}$

$= \dfrac{(h+9) - 9}{h[\sqrt{h+9} + 3]} = \dfrac{1}{\sqrt{h+9} + 3}, h \neq 0$

$m = \lim\limits_{h \to 0} \dfrac{1}{\sqrt{h+9} + 3} = \dfrac{1}{6}$

22. $m_{\text{sec}} = \dfrac{g(x+h) - g(x)}{h} = \dfrac{(x+h)^3 - x^3}{h} = \dfrac{3x^2 h + 3xh^2 + h^3}{h}$

$= 3x^2 + 3xh + h^2, h \neq 0$

$m = \lim\limits_{h \to 0} (3x^2 + 3xh + h^2) = 3x^2$

At $(1, 1)$, $m = 3(1)^2 = 3$

At $(-2, -8)$, $m = 3(-2)^2 = 12$

24. $m_{\text{sec}} = \dfrac{g(x+h) - g(x)}{h} = \dfrac{\sqrt{x+h-1} - \sqrt{x-1}}{h} \cdot \dfrac{\sqrt{x+h-1} + \sqrt{x-1}}{\sqrt{x+h-1} + \sqrt{x-1}}$

$= \dfrac{(x+h-1) - (x-1)}{h[\sqrt{x+h-1} + \sqrt{x-1}]} = \dfrac{1}{\sqrt{x+h-1} + \sqrt{x-1}}, h \neq 0$

$m = \lim\limits_{h \to 0} \dfrac{1}{\sqrt{x+h-1} + \sqrt{x-1}} = \dfrac{1}{2\sqrt{x-1}}$

At $(5, 2)$, $m = \dfrac{1}{2\sqrt{5-1}} = \dfrac{1}{4}$

At $(10, 3)$, $m = \dfrac{1}{2\sqrt{10-1}} = \dfrac{1}{6}$

26. $f'(x) = \lim\limits_{h \to 0} \dfrac{f(x+h) - f(x)}{h} = \lim\limits_{h \to 0} \dfrac{(-4(x+h)+2) - (-4x+2)}{h}$

$= \lim\limits_{h \to 0} \dfrac{-4h}{h} = 4$

28. $f'(x) = \lim\limits_{h \to 0} \dfrac{f(x+h) - f(x)}{h} = \lim\limits_{h \to 0} \dfrac{((x+h)^2 - 2(x+h) + 3) - (x^2 - 2x + 3)}{h}$

$= \lim\limits_{h \to 0} \dfrac{2xh + h^2 - 2h}{h} = \lim\limits_{h \to 0} (2x + h - 2) = 2x - 2$

30. $h'(s) = \lim\limits_{h \to 0} \dfrac{h(s+h) - h(s)}{h} = \lim\limits_{h \to 0} \dfrac{\frac{1}{\sqrt{s+h+1}} - \frac{1}{s+1}}{h}$

$= \lim\limits_{h \to 0} \dfrac{\sqrt{s+1} - \sqrt{s+h+1}}{h\sqrt{s+h+1}\sqrt{s+1}} \cdot \dfrac{\sqrt{s+1} + \sqrt{s+h+1}}{\sqrt{s+1} + \sqrt{s+h+1}}$

$= \lim\limits_{h \to 0} \dfrac{(s+1) - (s+h+1)}{h\sqrt{s+h+1}\sqrt{s+1}[\sqrt{s+1} + \sqrt{s+h+1}]}$

$= \lim\limits_{h \to 0} \dfrac{-1}{\sqrt{s+h+1}\sqrt{s+1}[\sqrt{s+1} + \sqrt{s+h+1}]}$

$= \dfrac{-1}{(s+1)2\sqrt{s+1}} = \dfrac{-1}{2(s+1)^{3/2}}$

32. $m_{\text{sec}} = \dfrac{f(2+h) - f(2)}{h} = \dfrac{(2+h)^3 - (2+h) - 6}{h}$

$= \dfrac{h^3 + 6h^2 + 11h}{h} = h^2 + 6h + 11, h \neq 0$

$m = \lim\limits_{h \to 0} (h^2 + 6h + 11) = 11$

Tangent line: $y - 6 = 11(x - 2)$

$y = 11x = 16$

34. $m_{\text{sec}} = \dfrac{f(2+h) - f(2)}{h} = \dfrac{\left(2(2+h) + \frac{4}{2+h}\right) - 6}{h}$

$= \dfrac{(4 + 2h)(2+h) + 4 - 6(2+h)}{(2+h)h} = \dfrac{2h + 2h^2}{(2+h)h} = \dfrac{2 + 2h}{2+h}, h \neq 0$

$m = \lim\limits_{h \to 0} \left(\dfrac{2 + 2h}{2+h}\right) = 1$

Tangent line: $y - 6 = 1(x - 2)$

$y = x + 4$

36.

x	-2	-1.5	-1	-0.5	0	0.5	1	1.5	2
f(x)	1	1.225	1.414	1.581	1.732	1.871	2	2.121	2.236
f'(x)	0.5	0.408	0.354	0.316	0.289	0.267	0.25.	0.236	0.224

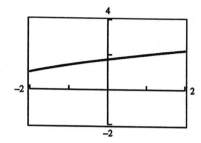

$$f(x) = \sqrt{x+3}$$

$$f'(x) = \frac{1}{2\sqrt{x+3}}$$

38.

x	-2	-1.5	-1	-0.5	0	0.5	1	1.5	2
f(x)	0	-0.7	-1	-1.071	-1	-0.833	-0.6	-0.318	0
f'(x)	-2	-0.92	-0.333	-0.020	0.25	0.407	0.52	0.603	0.667

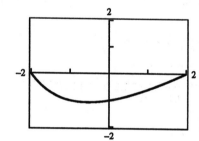

$$f(x) = \frac{x^2 - 4}{x + 4}$$

$$f'(x) = \frac{x^2 + 8x + 4}{(x + 4)^2}$$

40. $f'(x) = \lim\limits_{h \to 0} \dfrac{f(x+h) - f(x)}{h} = \lim\limits_{h \to 0} \dfrac{((x+h)^2 - 4(x+h) + 3) - (x^2 - 4x + 3)}{h}$

$\qquad = \lim\limits_{h \to 0} \dfrac{2xh + h^2 - 4h}{h} = 2x - 4$

$f'(x) = 0 = 2x - 4 \Rightarrow x = 2.$

f has a horizontal tangent at $(2, -1)$.

42. $f'(x) = \lim\limits_{h \to 0} \dfrac{f(x+h) - f(x)}{h} = \lim\limits_{h \to 0} \dfrac{\frac{1}{3}(x+h)^3 - 3(x+h) - \frac{1}{3}(x^3 - 3x)}{h}$

$\qquad = \lim\limits_{h \to 0} \dfrac{x^2 h + xh^2 + \frac{1}{3}x^3 - h}{h} = x^2 - 1$

$f'(x) = 0 = x^2 - 1 \Rightarrow x = \pm 1$

f has horizontal tangents at $\left(-1, \frac{2}{3}\right)$ and $\left(1, -\frac{2}{3}\right)$

44. $f'(x) = \lim\limits_{h \to 0} \dfrac{f(x+h) - f(x)}{h} = \lim\limits_{h \to 0} \dfrac{3(x+h)^4 + 4(x+h)^3 - (3x^4 + 4x^3)}{h}$

$\qquad = \lim\limits_{h \to 0} \dfrac{(12x^3h + 18x^2h^2 + 12xh^3 + 3h^4) + (12x^2h + 12xh^2 + 4h^3)}{h}$

$\qquad = 12x^3 + 12x^2$

$f'(x) = 0 = 12x^3 + 12x^2 = 12x^2(x+1) \Rightarrow x = 0, -1$

f has horizontal tangents at $(0, 0)$ and $(-1, -1)$

46. matches (d) (derivative decreases until origin, then increases.)

48. matches (a) (derivative approaches $-\infty$ when x approaches o.

12.4 Limits at Infinity and Limits of Sequences

2. $\lim\limits_{x \to \infty} \dfrac{4}{2x + 3} = 0$

4. $\lim\limits_{x \to \infty} \dfrac{1 - 5x}{1 + 2x} = -\dfrac{5}{2}$

6. $\lim\limits_{x \to -\infty} \dfrac{x^2 + 1}{x^2} = 1$

8. $\lim\limits_{y \to \infty} \dfrac{4y^4}{y^2 + 1}$ Does not exist.

10. $\lim\limits_{x \to \infty} \dfrac{2x^2 - 6}{(x - 1)^2} = \lim\limits_{x \to \infty} \dfrac{2x^2 - 6}{x^2 - 2x + 1} = 2$

12. $\lim\limits_{x \to \infty} \left[3 + \dfrac{2x^2}{(x + 5)^2} \right] = 3 + 2 = 5$

14. $\lim\limits_{x \to \infty} \left[\dfrac{x}{2x + 1} + \dfrac{3x^2}{(x - 3)^2} \right] = \dfrac{1}{2} + 3 = \dfrac{7}{2}$

16. $\lim\limits_{n \to \infty} \dfrac{3}{n^2} = 0$

18. $\lim\limits_{n \to \infty} \dfrac{1 - n^2}{n - 1}$ Does not exist.

20. matches (a)

22. matches (b)

24.

x	10^0	10^1	10^2	10^3	10^4	10^5	10^6
$f(x)$	-0.162	-0.0167	-0.00167	-1.67×10^{-4}	-1.7×10^{-5}	-1.7×10^{-6}	-2×10^{-7}

$\lim\limits_{x \to \infty} (3x - \sqrt{9x^2 + 1})$

26.

x	10^0	10^1	10^2	10^3	10^4	10^5	10^6
$f(x)$	0.508	0.5008	0.50008	0.5	0.5	0.5	0.5

$$\lim_{x \to \infty} 4(4x - \sqrt{16x^2 - x}) = \frac{1}{2}$$

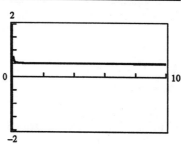

28. $\displaystyle\lim_{n \to \infty} \frac{b + \theta a(n-1)}{1 + \theta(n-1)} = \frac{\theta a}{\theta} = a$

30. $\dfrac{2}{3}, \dfrac{5}{4}, \dfrac{8}{5}, \dfrac{11}{6}, 2 \bullet \displaystyle\lim_{n \to \infty} \dfrac{3n-1}{n+2} = 3$

32. $\dfrac{2}{3}, \dfrac{8}{19}, \dfrac{18}{85}, \dfrac{32}{261}, \dfrac{50}{631} \bullet \displaystyle\lim_{n \to \infty} \dfrac{2n^2}{n^4 + n + 1} = 0$

34. $-3, -1, 1, 3, 5 \bullet \displaystyle\lim_{n \to \infty} (2n - 5)$ does not exist.

36. $\dfrac{1}{6}, \dfrac{1}{20}, \dfrac{1}{42}, \dfrac{1}{72}, \dfrac{1}{110} \bullet \displaystyle\lim_{n \to \infty} \dfrac{(2n-1)!}{(2n+1)!} = \displaystyle\lim_{n \to \infty} \dfrac{1}{(2n+1)(2n)} = 0$

38. $1, -\dfrac{1}{4}, \dfrac{1}{9}, -\dfrac{1}{16}, \dfrac{1}{25} \bullet \displaystyle\lim_{n \to \infty} \dfrac{(-1)^{n+1}}{n^2} = 0$

40. $\dfrac{1}{5}, -\dfrac{1}{2}, \dfrac{5}{7}, -\dfrac{7}{8}, 1 \bullet \displaystyle\lim_{n \to \infty} (-1)^{n+1} \left(\dfrac{2n-1}{n+4} \right)$ does not exist.

42. $\displaystyle\lim_{n \to \infty} An = 12$

n	10^0	10^1	10^2	10^3	10^4	10^5	10^6
An	20	12.8	12.08	12.008	12.0008	12.00008	12.000008

44. $\displaystyle\lim_{n \to \infty} An = 4$

n	10^0	10^1	10^2	10^3	10^4	10^5	10^6
An	16	4.84	4.0804	4.008004	4.00080004	4.00008	4.000008

46. $\displaystyle\lim_{n \to \infty} An = \dfrac{3}{4}$

n	10^0	10^1	10^2	10^3	10^4	10^5	10^6
An	1	0.7975	0.754975	0.75049975	0.75005	0.750005	0.7500005

48. Diverges

50. Converges to 6

12.5 The Area Problem

2. $\displaystyle\sum_{i=1}^{30} i^2 = \frac{n(n+1)(2n+1)}{6} = \frac{30(31)(61)}{6} = 9455$

4. $\displaystyle\sum_{k=1}^{50} (2k+1) = 2\sum_{k=1}^{50} k + \sum_{k=1}^{50} 1 = 2\frac{50(51)}{2} + 50 = 2600$

6. $\displaystyle\sum_{j=1}^{10} (j^3 - 3j^2) = \frac{10^2(11)^2}{4} - 3\left(\frac{10(11)(21)}{6}\right) = 1870$

8. $\displaystyle S(n) = \sum_{i=1}^{n} \frac{i}{n^2} = \frac{1}{n^2}\frac{n(n+1)}{2} = \frac{n+1}{2n}$

n	10^0	10^1	10^2	10^3
$S(n)$	1	0.55	0.505	0.5005

$\displaystyle\lim_{n\to\infty} S(n) = \frac{1}{2}$

10. $\displaystyle S(n) = \sum_{i=1}^{n} \frac{i^3}{n^4} = \frac{1}{n^4}\left[\frac{n^2(n+1)^2}{4}\right] = \frac{n^2+2n+1}{4n^2}$

n	10^0	10^1	10^2	10^3
$S(n)$	1	0.3025	0.255025	0.25050025

$\displaystyle\lim_{n\to\infty} S(n) = \frac{1}{4}$

12. $\displaystyle S(n) = \sum_{i=1}^{n} \frac{24i^2}{n^3} = \frac{24}{n^3}\left[\frac{n(n+1)(2n+1)}{6}\right] = \frac{4}{n^2}(2n^2+3n+1) = \frac{8n^2+12n-4}{n^2}$

n	10^0	10^1	10^2	10^3
$S(n)$	24	9.24	8.1204	8.012004

$\displaystyle\lim_{n\to\infty} S(n) = 8$

14. $\displaystyle S(n) = \sum_{i=1}^{n} \left[3 - 2\left(\frac{i}{n}\right)\right]\frac{1}{n} = \frac{1}{n}\left[3n - \frac{2}{n}\frac{n(n+1)}{2}\right] = 3 - \frac{n+1}{n}$

n	10^0	10^1	10^2	10^3
$S(n)$	1	1.9	1.99	1.999

$\displaystyle\lim_{n\to\infty} S(n) = 2$

16. $\displaystyle S(n) = \sum_{i=1}^{n} \left(\frac{4}{n} + \frac{2i}{n^2}\right)\left(\frac{2i}{n}\right) = \frac{2}{n}\left[\frac{4}{n}\frac{n(n+1)}{2} + \frac{2}{n^2}\frac{n(n+1)(2n+1)}{6}\right]$

$\displaystyle = \frac{2}{n}\left[\frac{4n^2+4n}{2n} + \frac{2(2n^2+2n+1)}{6n}\right] = \frac{16n^2+20n+2}{3n^2}$

n	10^0	10^1	10^2	10^3
$S(n)$	12.667	6.007	5.40007	5.34

$\displaystyle\lim_{n\to\infty} S(n) = \frac{16}{3}$

18. $S(n) = \sum_{i=1}^{n} \left[4 - \left(\frac{3i}{n}\right)^2 \right]\left(\frac{3i}{n^2}\right) = \frac{3}{n^2}\left(4\frac{n(n+1)}{2} - \frac{9}{n^2}\frac{n^2(n+1)^2}{4} \right)$

$= \frac{3}{n^2}\left[2n(n+1) - \frac{9(n+1)^2}{4} \right] = \frac{-3}{4}\left(\frac{n^2 - 10n - 9}{n^2} \right)$

n	10^0	10^1	10^2	10^3
$S(n)$	13.5	−0.0675	−0.6743	−0.7425

$\lim_{n\to\infty} S(n) = \frac{-3}{4}$

20. Width of each rectangle is $\frac{1}{2}$. The height is obtained by evaluating f at the right hand endpoint of each interval.

$A \approx \sum_{i=1}^{6} f\left(\frac{i}{2}\right)\left(\frac{1}{2}\right) = \sum_{i=1}^{6}\left(4 - \frac{i}{2}\right)\frac{1}{2} = \frac{27}{2}\cdot\frac{1}{2} = \frac{27}{4}$

22. The width of each rectangle is 1. The height is obtained by evaluating f at the right hand endpoint of each interval.

$A \approx \sum_{i=1}^{5} f(-2+i)(1) = \sum_{i=1}^{5}[(-2+i)^2 + 1] = 20$

24. The width of each rectangle is $\frac{1}{2}$. The height is obtained by evaluating f at the right hand endpoint of each interval.

$A \approx \sum_{i=1}^{4} f\left(2 + \frac{i}{2}\right)\left(\frac{1}{2}\right) = \sum_{i=1}^{4}\frac{1}{2}\left(2 + \frac{i}{2} - 2\right)^3\left(\frac{1}{2}\right)$

$= \sum_{i=1}^{4}\frac{1}{4}\left(\frac{i}{2}\right)^3 = 3.125$

26. The width of each rectangle is $\frac{3}{n}$. The height is $f\left(\frac{3i}{n}\right) = 2\left(\frac{3i}{n}\right) + 1$.

$A \approx \sum_{i=1}^{n}\left[2\left(\frac{3i}{n}\right) + 1\right]\left(\frac{3}{n}\right)$

(Note: exact area is 12)

n	4	8	20	50
Approximate area	14.25	13.125	12.45	12.18

28. The width of each rectangle is $\frac{3}{n}$. The height is $f\left(\frac{3i}{n}\right) = 9 - \left(\frac{3i}{n}\right)^2$.

$A \approx \sum_{i=1}^{n}\left(9 - \left(\frac{3i}{n}\right)^2\right)\frac{3}{n}$

(Note: exact area is 18)

n	4	8	20	50
Approximate area	14.344	16.242	17.314	17.7282

30. The width of each rectangle is $\dfrac{2-(-1)}{n}=\dfrac{3}{n}$. The height is $f\left(-1+\dfrac{3i}{n}\right)=3-\dfrac{1}{4}\left(-1+\dfrac{3i}{n}\right)^3$.

$$A\approx\sum_{i=1}^{n}\left[3-\frac{1}{4}\left(-1+\frac{3i}{n}\right)^3\right]\frac{3}{n}$$

(Note: exact area is $8\frac{1}{16}\approx 8.0625$)

n	4	8	20	50
Approximate area	7.113	7.614	7.8895	7.994

32. $A\approx\displaystyle\sum_{i=1}^{n}f\left(\frac{3i}{n}\right)\left(\frac{3}{n}\right)$

$=\displaystyle\sum_{i=1}^{n}\left[\frac{1}{3}\left(\frac{3i}{n}\right)+1\right]\frac{3}{n}$

$=\dfrac{3}{n^2}\displaystyle\sum_{i=1}^{n}i+\dfrac{3}{n}\sum_{i=1}^{n}1$

$=\dfrac{3}{n^2}\dfrac{n(n+1)}{2}+\dfrac{3}{n}(n)=3\dfrac{n^2+n}{2n^2}+3$

$A=\displaystyle\lim_{n\to\infty}\left(3\dfrac{n^2n}{2n^2}+3\right)=\dfrac{3}{2}+3=\dfrac{9}{2}$

34. $A\approx\displaystyle\sum_{i=1}^{n}f\left(-2+\frac{4i}{n}\right)\left(\frac{4}{n}\right)$

$=\displaystyle\sum_{i=1}^{n}\left[\left(-2+\frac{4i}{n}\right)+2\right]\frac{4}{n}$

$=\dfrac{16}{n^2}\displaystyle\sum_{i=1}^{n}i=\dfrac{16}{n^2}\dfrac{n(n+1)}{2}$

$A=\displaystyle\lim_{n\to\infty}\left(\dfrac{16}{n^2}\dfrac{n(n+1)}{2}\right)=8$

36. $A\approx\displaystyle\sum_{i=1}^{n}f\left(1+\frac{3i}{n}\right)\left(\frac{3}{n}\right)$

$=\displaystyle\sum_{i=1}^{n}\left[\frac{1}{4}\left(1+\frac{3i}{n}\right)^2+1\right]\frac{3}{n}$

$=\displaystyle\sum_{i=1}^{n}\left[\frac{5}{4}+\frac{3}{2}\frac{i}{n}+\frac{9}{4}\frac{i^2}{n^2}\right]\frac{3}{n}$

$=\dfrac{15}{4n}\displaystyle\sum_{i=1}^{n}1+\dfrac{9}{2n^2}\sum_{i=1}^{n}i+\dfrac{27}{4n^3}\sum_{i=1}^{n}i^2$

$=\dfrac{15}{4n}(n)+\dfrac{9}{2n^2}\dfrac{n(n+1)}{2}+\dfrac{27}{4n^3}\dfrac{n(n+1)(2n+1)}{6}$

$A=\displaystyle\lim_{n\to\infty}\left[\dfrac{15}{4}+\dfrac{9}{4}\dfrac{n(n+1)}{n^2}+\dfrac{9}{8n^3}n(n+1)(2n+1)\right]$

$=\dfrac{15}{4}+\dfrac{9}{4}+\dfrac{9}{4}=\dfrac{33}{4}$

38. $A \approx \sum\limits_{i=1}^{n} f\left(2\dfrac{i}{n}\right)\left(\dfrac{2}{n}\right)$

$= \sum\limits_{i=1}^{n} \left[4\left(\dfrac{2i}{n}\right) - \left(\dfrac{2i}{n}\right)^3\right]\left(\dfrac{2}{n}\right)$

$= \dfrac{16}{n^2}\sum\limits_{n=1}^{n} i - \dfrac{16}{n^4}\sum\limits_{i=1}^{n} i^3$

$= \dfrac{16}{n^2}\dfrac{n(n+1)}{2} - \dfrac{16}{n^4}\dfrac{n^2(n+1)^2}{4}$

$A = \lim\limits_{n\to\infty} \left[8\dfrac{n(n+1)}{n^2} - 4\dfrac{n^2(n+1)^2}{n^4}\right] = 8 - 4 = 4$

40. $y = (-3.0x10^{-6})x^3 + 0.002x^2 - 1.05x + 400$

Note that $y = 0$ when $x = 500$.

Using a graphing utility, or the limit process, you can calculate the area to be 105,208.33 square feet.

Review Exercises for Chapter 12

2. $\lim\limits_{x\to 3} \dfrac{\frac{x}{x+1} - \frac{3}{4}}{(x-3)}$

x	2.9	2.99	2.999	3	3.001	3.01	3.1
$f(x)$	0.0641	0.0627	0.0625	?	0.0625	0.0623	0.0610

4. $\lim\limits_{x\to -2} (5 - 2x - x^2) = 5 - 2(-2) - (-2)^2 = 5$

6. $\lim\limits_{x\to 4} \dfrac{x^3 - 64}{x^2 - 16} = \lim\limits_{x\to 4} \dfrac{(x-4)(x^2+4x+16)}{(x-4)(x+4)} = \lim\limits_{x\to 4} \dfrac{x^2 + 4x + 16}{x+4}$

8. $\lim\limits_{v\to 2} \dfrac{\sqrt{v+7} - 3}{v} = \dfrac{3-3}{2} = 0$

10. $\lim\limits_{x\to 6+} \dfrac{|x-6|}{x-6} = 1$ and $\lim\limits_{x\to 6-} \dfrac{|x-6|}{x-6} = -1$

Hence, $\lim\limits_{x\to 6} \dfrac{|x-6|}{x-6}$ does not exist.

12. $\lim\limits_{h\to 0}\dfrac{f(x+h)-f(x)}{h}=\lim\limits_{h\to 0}\dfrac{(x+h)^3+5-(x^3+5)}{h}$

$=\lim\limits_{h\to 0}\dfrac{x^3+3x^2h+3xh^2+h^3-x^3}{h}$

$=\lim\limits_{h\to 0}(3x^2+3xh+h^2)=3x^2$

14. $m=\lim\limits_{h\to 0}\dfrac{g(x+h)-g(x)}{h}=\lim\limits_{h\to 0}\dfrac{\sqrt{x+h}-\sqrt{x}}{h}\cdot\dfrac{\sqrt{x+h}+\sqrt{x}}{\sqrt{x+h}+\sqrt{x}}$

$=\lim\limits_{h\to 0}\dfrac{(x+h)=x}{h[\sqrt{x+h}+\sqrt{x}]}=\lim\limits_{h\to 0}\dfrac{1}{\sqrt{x+h}+\sqrt{x}}=\dfrac{1}{2\sqrt{x}}$

(a) At $(1, 1)$, $m=\dfrac{1}{2\sqrt{1}}=\dfrac{1}{2}$

(b) At $(4, 2)$, $m=\dfrac{1}{2\sqrt{4}}=\dfrac{1}{4}$

16. $m=\lim\limits_{h\to 0}\dfrac{f(x+h)-f(x)}{h}=\lim\limits_{h\to 0}\dfrac{\frac{1}{4}(x+h)^4-\frac{1}{4}x^4}{h}$

$=\lim\limits_{h\to 0}\dfrac{\frac{1}{4}[x^4+4x^3h+6x^2h^2+4xh^3+h^4-x^4]}{h}$

$=\lim\limits_{h\to 0}\dfrac{1}{4}[4x^3+6x^2h+4xh^2+h^3]=x^3$

(a) At $(-2,4)$, $m=(-2)^3=-8$

(b) At $\left(1,\dfrac{1}{4}\right)$, $m=(1)^3=1$

18. $f'(x)=\lim\limits_{h\to 0}\dfrac{f(x+h)-f(x)}{h}=\lim\limits_{h\to 0}\dfrac{3(x+h)-3x}{h}=3$

20. $f'(x)=\lim\limits_{h\to 0}\dfrac{f(x+h)-f(x)}{h}=\lim\limits_{h\to 0}\dfrac{\left(\frac{1}{2}(x+h)^2+3\right)-\left(\frac{1}{2}x^2+3\right)}{h}$

$=\lim\limits_{h\to 0}\dfrac{\frac{1}{2}\left(x^2+2xh+h^2\right)-\frac{1}{2}x^2}{h}=\lim\limits_{h\to 0}\left(xh+\tfrac{1}{2}h^2\right)=x$

22. $g'(s)=\lim\limits_{h\to 0}\dfrac{g(s+h)-g(s)}{h}=\lim\limits_{h\to 0}\dfrac{\frac{4}{s+h+5}-\frac{4}{s+5}}{h}$

$=\lim\limits_{h\to 0}\dfrac{4s+20-4s-4h-20}{(s+h+5)(s+5)h}=\lim\limits_{h\to 0}\dfrac{-4}{(s+h+5)(s+5)}$

$=\dfrac{-4}{(s+5)^2}$

24. $\lim\limits_{x\to\infty}\dfrac{7x}{14x+2}=\dfrac{7}{14}=\dfrac{1}{2}$

26. $\lim\limits_{x\to\infty}\dfrac{x^2}{2x+3}$ does not exist

28. $\displaystyle\lim_{x\to\infty}(x-2)^{-3} = \lim_{x\to\infty}\frac{1}{(x-2)^3} = 0$

30. $\displaystyle\lim_{n\to\infty}\left(\frac{2}{n}\right)\left[n+\frac{2}{n}\left(\frac{n(n-1)}{2}-n\right)\right]$

$\displaystyle = \lim_{n\to\infty}\left[2+2\frac{(n-1)}{n}-\frac{4}{n}\right] = 2+2 = 4$

32. $\displaystyle A\approx\sum_{i=1}^{6}\left[2\left(3+\frac{i}{2}\right)-6\right]\left(\frac{1}{2}\right)$

$\displaystyle = \sum_{i=1}^{6}(i)\frac{1}{2} = \frac{21}{2}$ approximate area

$\displaystyle A = \lim_{n\to\infty}\sum_{i=1}^{n}\left[2\left(3+\frac{3i}{n}\right)-6\right]\left(\frac{3}{n}\right)$

$\displaystyle = \lim_{n\to\infty}\sum_{i=1}^{n}\frac{18i}{n^2} = \lim_{n\to\infty}\frac{18}{n^2}\sum_{i-1}^{n}i$

$\displaystyle = \lim_{n\to\infty}\frac{18}{n^2}\frac{n(n+1)}{2} = 9$ exact area

34. $\displaystyle A\approx\sum_{i=1}^{4}8\left(\frac{i}{4}-\left(\frac{i}{4}\right)^2\right)\frac{1}{4}$

$\displaystyle = 2\sum_{i=1}^{4}\left(\frac{i}{4}-\frac{i^2}{16}\right) = 5-\frac{15}{4} = \frac{5}{4}$ approximate area

$\displaystyle A = \lim_{n\to\infty}\sum_{i=1}^{n}8\left(\left(\frac{i}{n}\right)-\left(\frac{i}{n}\right)^2\right)\frac{i}{n}$

$\displaystyle = \lim_{n\to\infty}\left[\frac{8}{n^2}\sum_{i=1}^{n}i-\frac{8}{n^3}\sum i^2\right]$

$\displaystyle = \lim_{n\to\infty}\left[\frac{8}{n^2}\frac{n(n+1)}{2}-\frac{8}{n^3}\frac{n(n+1)(2n+1)}{6}\right] = 4-\frac{8}{3} = \frac{4}{3}$ exact area

36. $A \approx \sum_{i=1}^{6} \left[4 - \left[\left(1 + \frac{i}{2} \right) - 2 \right]^2 \right] \left(\frac{1}{2} \right)$

$\qquad = \sum_{i=1}^{6} \left[4 - \left(\frac{i}{2} - 1 \right)^2 \right] \left(\frac{1}{2} \right)$

$\qquad = \sum_{i=1}^{6} \left[4 - \frac{i^2}{4} + i - 1 \right] \left(\frac{1}{2} \right)$

$\qquad = \sum_{i=1}^{6} \left(3 + i - \frac{i^2}{4} \right) \frac{1}{2} = 8.125$

$A = \lim_{n \to \infty} \sum_{i=1}^{n} \left[4 - \left[\left(1 + \frac{3i}{n} \right) - 2 \right]^2 \right] \left(\frac{3}{n} \right)$

$\qquad = \lim_{n \to \infty} \sum_{i=1}^{n} \left[4 - \left(\frac{3i}{n} - 1 \right)^2 \right] \left(\frac{3}{n} \right)$

$\qquad = \lim_{n \to \infty} \frac{3}{n} \sum_{i=1}^{n} \left(3 + \frac{6i}{n} - \frac{9i^2}{n^2} \right)$

$\qquad = \lim_{n \to \infty} \left[\frac{9}{n} \sum_{i=1}^{n} 1 + \frac{18}{n^2} \sum_{i=1}^{n} i - \frac{27}{n^3} \sum_{i=1}^{n} i^2 \right]$

$\qquad = \lim_{n \to \infty} \left[\frac{9}{n}(n) + \frac{18}{n^2} \frac{n(n+1)}{2} - \frac{27}{n^3} \frac{n(n+1)(2n+1)}{6} \right]$

$\qquad = 9 + 9 - 9 = 9 \text{ exact area}$

Answers to Discussion Problems

Section 1.1

Let x be the amount of sales.

Michael's wage: $M = 3100 + 0.07x$

Janet's wage: $J = 3400 + 0.06x$

The graphs of both wage equations are lines. The lines intersect at $(x, y) = (30,000, 5200)$, which indicates that it is possible they receive the same wage (if sales $= 30,000$).

Section 1.2

(a) $\dfrac{7000 - 5000}{10} = 200$ students per year

(b)

Year	1982	1986	1990	1992	1994
Enrollment	5000	5800	6600	7000	7400

(c)

The slope is the *instantaneous* rate of change of enrollment.

Section 1.3

Examples vary.

Section 1.4

Examples vary.

Section 1.5

$g(t)$ seems to be a better model because it is simpler.

Section 1.6

$$(g \circ f)(x) = g(f(x))$$
$$= g(x - 1500)$$
$$= 0.88(x - 1500)$$
$$= 0.88x - 1320 \text{ (rebate first)}$$

$$(f \circ g)(x) = f(g(x))$$
$$= f(0.88x)$$
$$= 0.88x - 1500 \text{ (discount first)}$$

If $x = 15,800$, then $(g \circ f)(15,800) = 12,584$ and $(f \circ g)(x) = 12,404$, which indicates that the price is lower if the discount is taken first.

Section 1.7
1. No

2. Yes

Section 2.1
1. How often you get paid and your monthly salary are red herrings.

2. The amount of time the trip took and the time it took to stop for a snack are red herrings.

Section 2.2
No, this does not mean that -99.1 is a good approximation. In fact, the exact answer is -99.0.

Section 2.3
$(1)(1)(1) = 1$

$$\left(\frac{-1+\sqrt{3}i}{2}\right)\left(\frac{-1+\sqrt{3}i}{2}\right)\left(\frac{-1+\sqrt{3}i}{2}\right) = \left(\frac{-1+\sqrt{3}i}{2}\right)\left(\frac{-1+\sqrt{3}i}{2}\right) = 1$$

$$\left(\frac{-1-\sqrt{3}i}{2}\right)\left(\frac{-1-\sqrt{3}i}{2}\right)\left(\frac{-1-\sqrt{3}i}{2}\right) = \left(\frac{-1+\sqrt{3}i}{2}\right)\left(\frac{-1-\sqrt{3}i}{2}\right) = 1$$

Section 2.4
$$P = R - C$$
$$= xp - (9.5x + 450{,}000)$$
$$= (7{,}307{,}025)(12.95) - [9.5(7{,}307{,}025) - 450{,}000]$$
$$= 25{,}659{,}236.25$$

Section 2.5
Computer program.

Section 2.6
Answers vary.

Section 3.1

$$y = x^2 + bx + c \qquad\qquad y = x^2 + bx + 3$$
$$3 = (0)^2 + b(0) + c \qquad 0 = (1)^2 + b(1) + 3$$
$$3 = c \qquad\qquad\qquad 0 = b + 4$$
$$\qquad\qquad\qquad\qquad -4 = b$$

$$y = x^2 - 4x + 3$$

Section 3.2
(a) Matches the second graph in the first row
(b) Matches the second graph in the second row
(c) Matches the first graph in the first row
(d) Matches the first graph in the second row

Section 3.3

(a) Yes. $x^2 - 2 = 0$
(b) Yes
(c) No; yes, $x^3 + \sqrt{2}x^2 + 2x + 2\sqrt{2}$

Section 3.5

No, the graph is a line with a hole at $x = 0$.

Section 3.6

Answers vary.

Section 4.1

(a) Linear growth, $f(n) = 3n, n = 1, 2, 3, 4, 5, \ldots$
(b) Exponential growth, $f(n) = 3^n, n = 1, 2, 3, 4, 5, \ldots$

Section 4.2

(a) $x = 10^{10}$
(b) $x = 10^{1000}$
(c) $x = 10^{10,000,000}$

Section 4.3

x	y	$\dfrac{\ln x}{\ln y}$	$\ln \dfrac{x}{y}$	$\ln x - \ln y$
1	2	0	-0.693	-0.693
3	4	0.792	-0.288	-0.288
10	5	1.431	0.693	0.693
4	0.5	-2	2.079	2.079

Section 4.4

The graphs of $y = \log_{10} 10^x$, $y = \ln e^x$ and $y = x$ are identical on $(-\infty, \infty)$. The graphs of $y = 10^{\log_{10} x}$, $y = e^{\ln x}$, and $y = x$ are identical on $(0, \infty)$.

Section 4.5

The quadratic model is better.

Section 4.6

Answers vary.

Section 5.1
Computer program.

Section 5.2
The student did not put the calculator in radian mode.

Section 5.3
Answers vary.

Section 5.4
Answers vary.

Section 5.6
Answers vary.

Section 5.7
1. h   **2.** c   **3.** f   **4.** b   **5.** g   **6.** a **7.** e   **8.** d

Section 5.8
Answers vary.

Section 6.1
$1 + \tan^2 u = \sec^2 u$

Section 6.2
1. Right. $\dfrac{1}{2 \csc x} = \dfrac{1}{2} \cdot \dfrac{1}{1/\sin x} = \dfrac{1}{2} \sin x$

2. Right. $(1 - \cos x)(1 + \cos x) = 1 - \cos^2 x = 1 - (1 - \sin^2 x) = \sin^2 x$

3. Right. $\dfrac{\sin^2 x}{1 - \cos x} = \dfrac{(1 - \cos x)(1 + \cos x)}{1 - \cos x} = 1 + \cos x$

4. Right. $\cot^2 x + 2 = (\csc^2 x - 1) + 2 = \csc^2 x + 1$

5. Right. $(\sec x - \tan x)(\sec x + \tan x) = \sec^2 x - \tan^2 x = 1$

The student needs a stronger grasp of trigonometric identities.

Section 6.3
Equations 1 and 2. Eighter $-2 \le \sqrt{b^2 - 4c} + b \le 2$ or $-2 \le \sqrt{b^2 - 4c} - b \le 2$ must be true.

Section 6.4
$$
\begin{aligned}
\sin(u + v) &= \cos\left(\frac{\pi}{2} - (u + v)\right) \\
&= \cos\left(\frac{\pi}{2} - u - v\right) \\
&= \cos\left(\left(\frac{\pi}{2} - 2\right) - v\right) \\
&= \cos\left(\frac{\pi}{2} - u\right)\cos v + \sin\left(\frac{\pi}{2} - u\right)\sin v \\
&= \sin u \cos v + \cos u \sin v
\end{aligned}
$$

Section 6.5

$\cos 3x = \cos(2x + x)$

$\qquad = \cos 2x \cos x - \sin 2x \sin x$

$\qquad = (2\cos^2 x - 1)\cos x - (2\sin x \cos x)\sin x$

$\qquad = 2\cos^3 x - \cos x - 2\sin^2 x \cos x$

$\qquad = 2\cos^3 x - \cos x - 2(1 - \cos^2 x)\cos x$

$\qquad = 2\cos^3 x - \cos x - 2(\cos x - \cos^3 x)$

$\qquad = 2\cos^3 x - \cos x - 2\cos x + 2\cos^3 x$

$\qquad = 4\cos^3 x - 3\cos x$

Section 6.6

Yes. It is probably easier to use the right-triangle definitions of sine, cosine, or tangent to solve the triangle.

Section 6.7

Triangle with given 50° angle: Area $= \frac{1}{2}(2)(4)\sin 50° \approx 3.064$ ft^2

Triangle with sides 2 ft, 3 ft, and 4 ft: Area $= \sqrt{\frac{9}{2}\left(\frac{9}{2} - 2\right)\left(\frac{9}{2} - 3\right)\left(\frac{9}{2} - 4\right)} \approx 2.905$ ft^2

Triangle with height of 2 ft and side 4 ft: Area $= \frac{1}{2}(4)(2) = 4$ ft^2

Section 6.8

$e^{a+bi} = e^a(\cos b + i\sin b)$

Let $a = 0$ and $b = \pi$

$e^{0+\pi i} = e^0(\cos \pi + i\sin \pi)$

$\quad e^{\pi i} = -1$

$e^{\pi i} + 1 = 0$

Section 7.1

(a) Two solutions

(b) One solution

(c) No solution

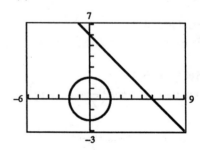

Section 7.2

(a) $k = -8$

(b) $k = 2$ (Solution is not unique.)

(c) No. Both equations will have a slope of 1 for any k, so there will be either an infinite number of solutions or no solution.

Section 7.3
Answers vary.

Section 7.4
Since $\dfrac{x^2+1}{x(x-1)}$ is an improper fraction, begin by dividing to obtain $\dfrac{x^2+1}{x(x-1)} = 1 + \dfrac{x+1}{x(x-1)}$, and decompose $\dfrac{x+1}{x(x-1)}$ into its partial fractions by the usual methods.

Section 7.5

1. $x \geq 0$

$y \geq 0$

$y \leq -\frac{1}{2}x + 2$

$y \leq -x + 3$

2. $x \geq 0$

$y \geq 0$

$y \geq -\frac{1}{3}x + 3$

$y \leq -2x + 8$

Section 7.6
Answers are not unique.

1. $z = -x + y$ **2.** $z = 2x + 3y$ **3.** $z = x - y$ **4.** $z = -x - y$

Section 8.1
Gaussian elimination with back-substitution:

$$\begin{bmatrix} 3 & -2 & 1 & \vdots & -6 \\ -1 & 1 & -2 & \vdots & 1 \\ 2 & 2 & -3 & \vdots & -1 \end{bmatrix} \quad \begin{matrix} R_2 \\ R_1 \end{matrix} \quad \begin{bmatrix} -1 & 1 & -2 & \vdots & 1 \\ 3 & -2 & 1 & \vdots & -6 \\ 2 & 2 & -3 & \vdots & -1 \end{bmatrix}$$

$$-1R_1 \quad \begin{bmatrix} 1 & -1 & 2 & \vdots & -1 \\ 3 & -2 & 1 & \vdots & -6 \\ 2 & 2 & -3 & \vdots & -1 \end{bmatrix}$$

$$\begin{matrix} -3R_1 + R_2 \\ -2R_1 + R_3 \end{matrix} \quad \begin{bmatrix} 1 & -1 & 2 & \vdots & -1 \\ 0 & 1 & -5 & \vdots & -3 \\ 0 & 4 & -7 & \vdots & 1 \end{bmatrix}$$

$$-4R_2 + R_3 \quad \begin{bmatrix} 1 & -1 & 2 & \vdots & -1 \\ 0 & 1 & -5 & \vdots & -3 \\ 0 & 0 & 13 & \vdots & 13 \end{bmatrix}$$

$$R_3 \div 13 \quad \begin{bmatrix} 1 & -1 & 2 & \vdots & -1 \\ 0 & 1 & -5 & \vdots & -3 \\ 0 & 0 & 1 & \vdots & 1 \end{bmatrix}$$

Using back-substitution, we get $x = -1, y = 2,$ and $z = 1$.

Gauss-Jordan elimination:

Use the first four steps of Gaussian elimination, above.

$$R_2 + R_1 \quad \begin{bmatrix} 1 & 0 & -3 & \vdots & -4 \\ 0 & 1 & -5 & \vdots & -3 \\ 0 & 0 & 13 & \vdots & 13 \end{bmatrix}$$
$$4R_2 + R_3$$

$$R_3 \div 13 \quad \begin{bmatrix} 1 & 0 & -3 & \vdots & -4 \\ 0 & 1 & -5 & \vdots & -3 \\ 0 & 0 & 1 & \vdots & 1 \end{bmatrix}$$

$$3R_3 + R_1 \quad \begin{bmatrix} 1 & 0 & 0 & \vdots & -1 \\ 0 & 1 & 0 & \vdots & 2 \\ 0 & 0 & 1 & \vdots & 1 \end{bmatrix}$$
$$5R_3 + R_2$$

Section 8.2

If $C = \begin{bmatrix} c_{11} & 0 & 0 \\ 0 & c_{22} & 0 \\ 0 & 0 & c_{33} \end{bmatrix}$, then $C^2 = \begin{bmatrix} c_{11}{}^2 & 0 & 0 \\ 0 & c_{22}{}^2 & 0 \\ 0 & 0 & c_{33}{}^2 \end{bmatrix}$

$$A^2 = \begin{bmatrix} 1 & 0 & 0 \\ 0 & 4 & 0 \\ 0 & 0 & 0 \end{bmatrix}$$

$$B^2 = \begin{bmatrix} 4 & 0 & 0 & 0 \\ 0 & 1 & 0 & 0 \\ 0 & 0 & 9 & 0 \\ 0 & 0 & 0 & 4 \end{bmatrix}$$

Section 8.3

Invertible

Section 8.4

$12a + 6b + 4c - 4c - 12a - 6b = 0$

Section 9.1

Yes

Section 9.2

No. Yes, the three sequences are arithmetic. The last sequence is not arithmetic.

Section 9.3

The first sequence is geometric, whereas the second is arithmetic.

Section 9.4

$A_n = 180(n - 2)$

Section 9.5

$A_n = 2^n$

$A_{10} = 2^{10} = 1024$

Section 9.6

Answers vary.

Section 9.7
Answers vary.

Section 9.8
Answers vary.
(a) 4, 4, 10   (b) 4, 4, 6, 7, 9   (c) 1, 2, 3, 4, 6, 6, 20

Section 9.9
(a) 4.68   (b) 4.79

Section 10.2
Yes, because a circle is a special case of an ellipse where $a = b = r$.

Section 10.3
The cone is the light from the lamp. The hyperbola is where the light shines directly on the wall.
The cone is the pencil point. The hyperbola is where the shaved portion of the pencil meets the unshaved portion.
The cone is the plane's exhaust. The hyperbola is the horizontal surface of the earth.

Section 10.5
$x = \sin t$ and $y = \cos t$ (other answers possible)

Section 10.6
Answers vary.

Section 10.7
(a) cardoid ($a = b = 3$)   (b) dimpled limaçon $\left(1 < \dfrac{a}{b} = \dfrac{4}{3} < 2\right)$

(c) inner loop $\left(\dfrac{a}{b} = \dfrac{2}{3} < 1\right)$ (d) convex $\left(\dfrac{a}{b} = \dfrac{3}{1} \geq 2\right)$

Section 11.1
(a) Less. $\|\mathbf{u}\| = \|\mathbf{v}\| \approx 575.9$ pounds
(b) Less. $\|\mathbf{u}\| \approx 244.9$ pounds, $\|\mathbf{v}\|$ 273.2

Section 11.2
Since $\|\mathbf{u}\|$ and $\|\mathbf{v}\|$ are nonnegative, the sign of $\mathbf{u} \cdot \mathbf{v} = \|\mathbf{u}\|\,\|\mathbf{v}\| \cos\theta$ will be determined by the sign of $\cos\theta$.
(a) Minus, since the sign of $\cos 115°$ is minus.
(b) Plus, since the sign of $\cos 75°$ is plus.

Section 11.3
The graph of $ax + by + cz = 0$ is a plane.

Section 11.4
$$\begin{aligned}
1.\ \mathbf{u} + \mathbf{v} &= \langle u_1, u_2, u_3 \rangle + \langle v_1, v_2, v_3 \rangle \\
&= \langle u_1 + v_1, u_2 + v_2, u_3 + v_3 \rangle \\
&= \langle v_1 + u_1, v_2 + u_2, v_3 + u_3 \rangle \\
&= \langle v_1, v_2, v_3 \rangle + \langle u_1, u_2, u_3 \rangle \\
&= \mathbf{v} + \mathbf{u}
\end{aligned}$$

The proofs of the other properties are similar.

Section 11.5

The vector **z** determined by $\mathbf{v} \times \mathbf{w}$ is orthogonal to **v** and **w** and thus is perpendicular to the plane determined by **v** and **w**. If $\mathbf{u} \cdot \mathbf{z} = 0$, then **z** is orthogonal to **u**. Thus, **u** must lie in the same place as **v** and **w**.

Section 12.1

(a) Graph of g.

$\lim_{x \to 1} g(x) = 2, \lim_{x \to 2} g(x) = 3.$

$g(1)$ does not exist, but $\lim_{x \to 1} g(x)$ does exist.

(b) Graph of f.

$\lim_{x \to 1} f(x) = 2, \lim_{x \to 2} f(x) = 3$

(c) Graph of h.

$\lim_{x \to 1} h(x) = 2, \lim_{x \to 2} h(x) = 3$

$h(1)$ does not exist, but $\lim_{x \to 1} h(x)$ does exist. $h(2)$ does not exist, but $\lim_{x \to 1} h(x)$ does exist.

Section 12.2

(a) The limit is 0.
(c) The limit does not exist.

(b) The limit is 1.
(d) Use the cancellation technique to find the limit.

Section 12.3

$$f'(t) = \lim_{h \to 0} \frac{f(t+h) - f(t)}{h}$$

$$= \lim_{h \to 0} \frac{\dfrac{3}{t+h} - \dfrac{3}{t}}{h}$$

$$= \lim_{h \to 0} \frac{\dfrac{3t - 3(t+h)}{t(t+h)}}{h}$$

$$= \lim_{h \to 0} \frac{\dfrac{3t - 3t - 3h}{t(t+h)}}{h}$$

$$= \lim_{h \to 0} \left(\frac{-3h}{t(t+h)} \cdot \frac{1}{h} \right)$$

$$= \lim_{h \to 0} \frac{-3}{t^2 + th}$$

$$= -\frac{3}{t^2}$$

$$f'(6) = -\frac{3}{36} = -\frac{1}{12}$$

Section 12.4

(d) is the fastest to converge.
(a) is the slowest to converge.

Section 12.5

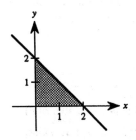

$A = \frac{1}{2}bh$

$\quad = \frac{1}{2}(2)(2)$

$\quad = 2$

Width: $\dfrac{2}{n}$ Height: $f\left(\dfrac{2i}{n}\right) = 2 - \dfrac{2i}{n} = \dfrac{2(n-i)}{n}$

$A \approx \displaystyle\sum_{i=1}^{n} \left(2 - \dfrac{2i}{n}\right)\left(\dfrac{2}{n}\right)$

$\quad = \dfrac{4}{n} \displaystyle\sum_{i=1}^{n} \left(1 - \dfrac{i}{n}\right)$

$\quad = \dfrac{4}{n^2} \displaystyle\sum_{i=1}^{n}(n-i)$

$\quad = \dfrac{4}{n^2} \left(\displaystyle\sum_{i=1}^{n}n - \displaystyle\sum_{i=1}^{n}i\right)$

$\quad = \dfrac{4}{n^2} \left(n^2 - \dfrac{n(n+1)}{2}\right)$

$\quad = \dfrac{4}{n^2} \left(\dfrac{2n^2 - n^2 - n}{2}\right)$

$\quad = \dfrac{4}{n^2} \left(\dfrac{n^2 - n}{2}\right) = \dfrac{2(n-1)}{n} = \dfrac{2n-2}{n}$

$A = \displaystyle\lim_{n\to\infty} \dfrac{2n-2}{2} = 2$

32. (a) $\theta = \dfrac{7\pi}{4}$

$$\theta' = 2\pi - \dfrac{7\pi}{4} = \dfrac{\pi}{4}$$

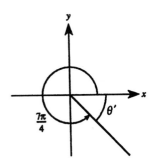

(b) $\theta = \dfrac{8\pi}{9}$

$$\theta' = \pi - \dfrac{8\pi}{9} = \dfrac{\pi}{9}$$

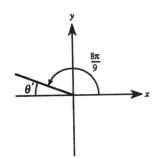

34. (a) $\theta = \dfrac{11\pi}{3}$

$$\theta' = 4\pi - \dfrac{11\pi}{3}$$

$$= \dfrac{12\pi}{3} - \dfrac{11\pi}{3} = \dfrac{\pi}{3}$$

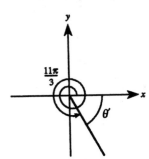

(b) $\qquad \theta = -\dfrac{7\pi}{10}$

$$2\pi - \dfrac{7\pi}{10} = \dfrac{13\pi}{10} \text{ (coterminal angle)}$$

$$\theta' = \dfrac{13\pi}{10} - \pi = \dfrac{13\pi}{10} - \dfrac{10\pi}{10}$$

$$= \dfrac{3\pi}{10}$$

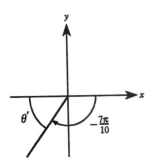

36. (a) $\theta' = 60°$, Quadrant IV

$$\sin 300° = -\sin 60° = -\dfrac{\sqrt{3}}{2}$$

$$\cos 300° = \cos 60° = \dfrac{1}{2}$$

$$\tan 300° = -\tan 60° = -\sqrt{3}$$

(b) $\theta' = 30°$, Quadrant IV

$$\sin 330° = -\sin 30° = -\dfrac{1}{2}$$

$$\cos 330° = \cos 30° = \dfrac{\sqrt{3}}{2}$$

$$\tan 330° = -\tan 30° = -\dfrac{\sqrt{3}}{3}$$

38. (a) $\theta' = 45°$, Quadrant IV

$$\sin(-405°) = -\sin 45° = -\dfrac{\sqrt{2}}{2}$$

$$\cos(-405°) = \cos 45° = \dfrac{\sqrt{2}}{2}$$

$$\tan(-405°) = -\tan 45° = -1$$

(b) $\theta' = 60°$, Quadrant III

$$\sin(-120°) = -\sin 60° = -\dfrac{\sqrt{3}}{2}$$

$$\cos(-120°) = -\cos 60° = -\dfrac{1}{2}$$

$$\tan(-120°) = \tan 60° = \sqrt{3}$$

40. (a) $\sin \dfrac{\pi}{4} = \dfrac{\sqrt{2}}{2}$

$\cos \dfrac{\pi}{4} = \dfrac{\sqrt{2}}{2}$

$\tan \dfrac{\pi}{4} = 1$

(b) $\theta' = \dfrac{\pi}{4}$, Quadrant III

$\sin \dfrac{5\pi}{4} = -\sin \dfrac{\pi}{4} = -\dfrac{\sqrt{2}}{2}$

$\cos \dfrac{5\pi}{4} = -\cos \dfrac{\pi}{4} = -\dfrac{\sqrt{2}}{2}$

$\tan \dfrac{5\pi}{4} = \tan \dfrac{\pi}{4} = 1$

42. (a) $\theta' = \dfrac{\pi}{2}$, Quadrantal Angle

$\sin\left(-\dfrac{\pi}{2}\right) = -\sin \dfrac{\pi}{2} = -1$

$\cos\left(-\dfrac{\pi}{2}\right) = \cos \dfrac{\pi}{2} = 0$

$\tan\left(-\dfrac{\pi}{2}\right) = \tan \dfrac{\pi}{2}$ is undefined

(b) $\sin \dfrac{\pi}{2} = 1$

$\cos \dfrac{\pi}{2} = 0$

$\tan \dfrac{\pi}{2}$ is undefined

44. (a) $\theta' = \dfrac{\pi}{3}$, Quadrant III

$\sin \dfrac{10\pi}{3} = -\sin \dfrac{\pi}{3} = -\dfrac{\sqrt{3}}{2}$

$\cos \dfrac{10\pi}{3} = -\cos \dfrac{\pi}{3} = -\dfrac{1}{2}$

$\tan \dfrac{10\pi}{3} = \tan \dfrac{\pi}{3} = \sqrt{3}$

(b) $\theta' = \dfrac{\pi}{3}$, Quadrant IV

$\sin \dfrac{17\pi}{3} = -\sin \dfrac{\pi}{3} = -\dfrac{\sqrt{3}}{2}$

$\cos \dfrac{17\pi}{3} = \cos \dfrac{\pi}{3} = \dfrac{1}{2}$

$\tan \dfrac{17\pi}{3} = -\tan \dfrac{\pi}{3} = -\sqrt{3}$

46. (a) $\sec 225° \approx -1.4142$

(b) $\sec 135° \approx -1.4142$

48. (a) $\csc 330° = -2.0000$

(b) $\csc 150° = 2.0000$

50. (a) $\cot(1.35) \approx 0.2245$

(b) $\tan(1.35) \approx 4.4552$

52. (a) $\sin(-0.65) \approx -0.6052$

(b) $\sin(5.63) \approx -0.6077$

54. (a) $\cos \theta = \dfrac{\sqrt{2}}{2} \Rightarrow$ reference angle is $45°$ or $\dfrac{\pi}{4}$ and θ is in Quadrant I or IV.

Values in degrees: $45°$, $315°$

Values in radians: $\dfrac{\pi}{4}$, $\dfrac{7\pi}{4}$

(b) $\cos \theta = -\dfrac{\sqrt{2}}{2} \Rightarrow$ reference angle is $45°$ or $\dfrac{\pi}{4}$ and θ is in Quadrant II or III.

Values in degrees: $135°$, $225°$

Values in radians: $\dfrac{3\pi}{4}$, $\dfrac{5\pi}{4}$

56. (a) $\sec\theta = 2 \Rightarrow$ reference angle is 60° or $\dfrac{\pi}{3}$ and θ is in Quadrant I or IV.

Values in degrees: 60°, 300°

Values in radians: $\dfrac{\pi}{3}, \dfrac{5\pi}{3}$

(b) $\sec\theta = -2 \Rightarrow$ reference angle is 60° or $\dfrac{\pi}{3}$ and θ is in Quadrant II or III.

Values in degrees: 120°, 240°

Values in radians: $\dfrac{2\pi}{3}, \dfrac{4\pi}{3}$

58. (a) $\sin\theta = \dfrac{\sqrt{3}}{2} \Rightarrow$ reference angle is 60° or $\dfrac{\pi}{3}$ and θ is in Quadrant I or II.

Values in degrees: 60°, 120°

Values in radians: $\dfrac{\pi}{3}, \dfrac{2\pi}{3}$

(b) $\sin\theta = -\dfrac{\sqrt{3}}{2} \Rightarrow$ reference angle is 60° or $\dfrac{\pi}{3}$ and θ is in Quadrant III or IV.

Values in degrees: 240°, 300°

Values in radians: $\dfrac{4\pi}{3}, \dfrac{5\pi}{3}$

60. (a) $\cos\theta = 0.8746 \Rightarrow \theta' \approx 29.00$

Quadrant I: $\theta = \cos^{-1} 0.8746 \approx 29.00°$

Quadrant IV: $\theta = 360° - \theta' \approx 331.00°$

(b) $\cos\theta = -0.2419 \Rightarrow \theta' \approx 76.00°$

Quadrant II: $\theta = \cos^{-1}(-0.2419) \approx 104.00°$

Quadrant III: $\theta = 180° + \theta' \approx 256.00°$

62. (a) $\sin\theta = 0.0175 \Rightarrow \theta' \approx 0.018$

Quadrant I: $\theta = \sin^{-1} 0.0175 \approx 0.018$

Quadrant II: $\theta = \pi - \theta' \approx 3.124$

(b) $\sin\theta = -0.6691 \Rightarrow \theta' \approx 0.733$

Quadrant III: $\theta = \pi + \theta' \approx 3.875$

Quadrant IV: $\theta = 2\pi - \theta' \approx 5.550$

64. (a) $\cot\theta = 5.671 \Rightarrow \theta' \approx 0.175$

Quadrant I: $\theta = \cot^{-1} 5.671 \approx 0.175$

Quadrant III: $\theta = \pi + \theta' \approx 3.316$

(b) $\cot\theta = -1.280 \Rightarrow \theta' \approx 0.663$

Quadrant II: $\theta = \pi - \theta' \approx 2.478$

Quadrant IV: $\theta = 2\pi - \theta' \approx 5.620$

66. $\tan\theta = \dfrac{3}{2}$

$\sec^2\theta = 1 + \tan^2\theta$

$\sec^2\theta = 1 + \left(\dfrac{3}{2}\right)^2$

$\sec^2\theta = 1 + \dfrac{9}{4} = \dfrac{13}{4}$

$\sec\theta < 0$ in Quadrant III.

$\sec\theta = -\dfrac{\sqrt{13}}{2}$

68. $\sec\theta = -\dfrac{9}{4}$

$1 + \tan^2\theta = \sec^2\theta$

$\tan^2\theta = \sec^2\theta - 1$

$\tan^2\theta = \left(-\dfrac{9}{4}\right)^2 - 1$

$\tan^2\theta = \dfrac{81}{16} - 1$

$\tan^2\theta = \dfrac{65}{16}$

$\tan\theta > 0$ in Quadrant III.

$\tan\theta = \dfrac{\sqrt{65}}{4}$

72. $S = 23.1 + 0.442t + 4.3\sin\left(\dfrac{\pi t}{6}\right)$

(a) $S = 23.1 + 0.442(2) + 4.3\sin\left(\dfrac{2\pi}{6}\right)$

$\approx 27.7, \quad 27{,}700$ units

(b) $S = 23.1 + 0.442(14) + 4.3\sin\left(\dfrac{14\pi}{6}\right)$

$\approx 33.0, \quad 33{,}000$ units

(c) $S = 23.1 + 0.442(9) + 4.3\sin\left(\dfrac{9\pi}{6}\right)$

$\approx 22.8, \quad 22{,}800$ units

(d) $S = 23.1 + 0.442(21) + 4.3\sin\left(\dfrac{21\pi}{6}\right)$

$\approx 28.1, \quad 28{,}100$ units

70. $1 + \tan^2\theta = \sec^2\theta$

$\tan^2 20° - \sec^2 20° = -1$

74. As θ increases from $0°$ to $90°$, x will decrease from 10 centimeters to 0 centimeters, and y will increase from 0 centimeters to 10 centimeters. Therefore, $\sin\theta = y/10$ will increase from 0 to 1 and $\cos\theta = x/10$ will decrease from 1 to 0. Thus, $\tan\theta = y/x$ will increase without bound. When $\theta = 90°$, the tangent will be undefined.